Food Engineering Series

Springer's *Food Engineering Series* is essential to the Food Engineering profession, providing exceptional texts in areas that are necessary for the understanding and development of this constantly evolving discipline. The titles are primarily reference-oriented, targeted to a wide audience including food, mechanical, chemical, and electrical engineers, as well as food scientists and technologists working in the food industry, academia, regulatory industry, or in the design of food manufacturing plants or specialized equipment.

H. Umesh Hebbar • Richa Sharma
Ram Saran Chaurasiya • Shivendu Ranjan
K.S.M.S. Raghavarao
Editors

Engineering Aspects of Food Quality and Safety

Springer

Editors
H. Umesh Hebbar
Food Engineering Department
CSIR-Central Food Technological Research Institute
Mysuru, Karnataka, India

Ram Saran Chaurasiya
String Bio Private Limited
Bengaluru, Karnataka, India

K.S.M.S. Raghavarao
Department of Chemical Engineering
IIT Tirupathi
Tirupati, India

CSIR-Central Food Technological Research Institute
Mysuru, Karnataka, India

Richa Sharma
School of Chemistry
University of Edinburgh
Edinburgh, United Kingdom

Shivendu Ranjan
NanoBio Research Lab
School of Nano Science and Technology
Indian Institute of Technology Kharagpur
Kharagpur, West Bengal, India

ISSN 1571-0297
Food Engineering Series
ISBN 978-3-031-30685-3 ISBN 978-3-031-30683-9 (eBook)
https://doi.org/10.1007/978-3-031-30683-9

This Springer imprint is published by the registered company Springer Nature Switzerland AG
The registered company address is: Gewerbestrasse 11, 6330 Cham, Switzerland

Preface

For the living world, food is synonymous with existence. Food habits of humans vary wildly with the environment, nutritional needs and cultural practices. As a result, the art of food science has historically been much diversified. Focus has been on processing and preservation of agricultural produce and finished food. With understanding of food chemistry, this art has taken the form of well-studied technologies. The goal of food technology remains the same; however, its scope has expanded to meet the rising urbanization and globalization.

Food technology is built on principles of physics, chemistry and biological sciences. Its methodologies, however, need to encompass a wide range of subjects – including engineering, process design, agriculture, economy and consumer demands. A food technologist has to be aware of the edible flora and fauna while also master the knowledge of industrial tools and machinery.

A critical aspect of the world of food engineering is the key to maintain the quality and safety of raw and prepared food. All human nutrition is obtained from food; hence, it is the single most important factor that dictates human health. With varied production and processing methods, there is increased risk of occurrence of unsafe agents, removal of nutritious components and deviation from expected quality. Infections from food are one of the leading causes of sickness and death worldwide. Such threats have worked in tandem to awaken the interests of engineers working towards quality and safety assurance.

The search for comprehensive accounts of principles and methods on food quality engineering revealed a literature gap of more than 5 years. The idea of the book evolved from the need to put together concepts of biochemistry and engineering of food quality and adulterants, methods to ensure food safety in both pre- and post-harvest stages and techniques to detect and quantify food safety threats. The chapters begin with fundamentals and then discuss state-of-the-art engineering technologies in the topics. Several of them compare conventional techniques with upcoming innovations, while charting a path for future engineers. Ample references encourage further reading.

Engineering Aspects of Food Quality and Safety targets postgraduate and doctoral students, academicians, researchers and industrial experts interested in the fields of food chemistry, microbiology, food safety and food quality, risk assessment/management, materials science and nanoscience, composites technology, chemical engineering, commercialization and regulatory policy. We believe this book will help readers to solve fundamental and applied problems and achieve innovation in the field of food technology. We hope that this volume will prove to be an educational resource for faculty and students as well as a blueprint for further research by food engineers and scientists.

Mysuru, Karnataka, India — H. Umesh Hebbar
Edinburgh, United Kingdom — Richa Sharma
Bengaluru, Karnataka, India — Ram Saran Chaurasiya
Kharagpur, West Bengal, India — Shivendu Ranjan
Tirupati, India — K.S.M.S. Raghavarao

Contents

About the Editors

H. Umesh Hebbar is presently working as Chief Scientist and Head, Food Engineering Department, CSIR-Central Food Technological Research Institute, (CFTRI) Mysore, Ministry of Science and Technology, Government of India. He has Chemical Engineering Degree and Ph.D. in Biotechnology with 6 years of industrial and 26 years of research and teaching experience. His research works are in the area of electromagnetic radiation-based processing of foods and liquid-liquid extraction for downstream processing of biomolecules. He was the Chief-Editor of *Indian Food Industry Mag* journal, published by the Association of Food Scientists and Technologist (I), and was serving as member of *Board of Studies* in many of the reputed Institutes.

Richa Sharma is a biotechnologist experienced in teaching and research in the field of analytical techniques for food, environment and healthcare. She completed her Ph.D. from Central Food Technological Research Institute, India, on food bioanalysis. She has been awarded the CSIR-GATE fellowship and the INSA Visiting Scientist Fellowship to conduct research in India. Dr. Sharma secured the Newton-Bhabha International Fellowship (awarded by the Royal Society, UK, and Science and Engineering Research Board, India) to work in the University of Edinburgh, Scotland on biomedical diagnostics and therapy. Her work components include bionanotechnology, nucleic acid, enzyme, peptide, antimicrobial and synthetic dye based labelling and detection techniques for clinical and food samples. She is an scientific communicator and has to her credit 20 research publications, several oral paper presentations, invited talks and public engagement activities.

Ram Saran Chaurasiya has completed his B.Tech., M.Tech. and Ph.D. in Engineering (Biotechnology). He has 3 years of industrial experience and 2 years of teaching and research experience post Ph.D. His major area of research is biochemical engineering in general and downstream processing (DSP) and waste management in particular. His expertise includes development and scale-up of the DSP processes for separation and purification of biomolecules having commercial applications, and its product development. He has delivered several lectures as invited speaker at various reputed institutes in India. He has published several papers in various International journals besides one book chapter. He has presented his research work in various international/national conferences. Apart from that, he has participated in several international conferences, workshops and technical seminars. He has received the prestigious INSA visiting scientist fellowship in 2019, Young Scientist Award in 2019 by the Department of Science and Technology, Madhya Pradesh, and Best Poster Award in 2013 in the 7th International Food Convention by AFST (I), India.

Shivendu Ranjan is currently working as Assistant Professor at the Indian Institute of Technology Kharagpur (IIT Kharagpur) and focusing on Nanomedicine, Nanotoxicology, and Nano-nutraceuticals. He is a Life Fellow of many reputed scientific societies like (i) Indian Chemical Society; (ii) The Linnean Society of London, (iii) Indian Engineering Teachers Association, and (iv) Bose Science Society. He is also an Associate Editor of *Environmental Chemistry Letters* (Springer journal of 13.615 Impact Factor). After graduating from VIT Vellore, Dr. Ranjan has served many organizations like (i) the Indian Institute of Food Processing and Technology (MoFPI, Govt. of India); (ii) E-Spin Nanotech, SIIC, IIT Kanpur, (iii) University of Johannesburg and (iv) Gujarat Biotechnology Research Centre (DST-Govt of Gujarat). He has one international patent and several scientific publications.

K.S.M.S. Raghavarao is a Chemical Engineer specialized in Food Engineering and Biotechnology. He obtained B.Tech. from Andhra University and direct Ph.D. from ICT, Mumbai (formerly UDCT). After PDF at NIST-Colorado and a brief stay at NIT, Warangal, he joined CFTRI in 1990 and continued there till he took over as the Director of CSIR-CFTRI. After superannuation, he has joined recently IIT Tirupati as Professor of Chemical Engineering. Out of 30 significant achievements, 25 were converted into processes/technologies, out of which 15 were transferred to Industry. For example, Virgin Coconut Oil (VCO) technology transferred to about 60 industries. He has guided 22 students/fellow scientists for Ph.D. in Food Engineering/Biotechnology and is guiding 5 at the moment in Chemical Engineering. He has been recipient of prestigious NASI-Reliance and VASVIK awards for applied research. He has received prestigious National Fellowships of Academies (FNAE, FASc, FNAAS). He has over 160 publications (with 'h' index of 54) and citations over 6500, about 28 International patents and 55 Indian patents. He initiated several new research areas at CFTRI and now at IIT Tirupati. He has successfully completed as PI/Co-PI 50 Grant-in-aid projects and 25 industrially sponsored projects.

Chapter 1
Food Quality: Engineering Perspective

Rajat Suhag, Ashutosh Upadhyay, and Anurag Mishra

1 Introduction

Increasing number of consumers today seek high-quality, nutritious products at reasonable costs. This phenomenon is universal across both developing and developed economies. Food quality and safety concerns have grown as cross-border trading of food goods has expanded. As a result, all countries have a shared obligation to assure not only adequate supplies in terms of nutrition, accessibility, and price, but also with the appropriate quality standards. There are emerging microbial risk in food chain on one hand, there are improved and novel food processes made available through the development in process engineering and newer technologies. Hence, continuous efforts are needed to improve the effectiveness of current regulatory systems with more sensitive, robust, efficient, and cost-effective process technologies to assure food safety, quality, and traceability while maintaining other functional, and sensory features in accordance with regulations and customer needs (S. J. Liu et al. 2020). Food engineering is a discipline with a large scope that includes all unit operations of food processing along with measurement of physical properties of foods and interaction of the food component in a food matrix. Engineering aspects pertaining to food quality are also equally

R. Suhag
Department of Food Science and Technology, National Institute of Food Technology Entrepreneurship and Management, Sonipat, Haryana, India

Faculty of Science and Technology, Free University of Bozen-Bolzano, Bolzano, BZ, Italy

A. Upadhyay (✉)
Department of Food Science and Technology, National Institute of Food Technology Entrepreneurship and Management, Sonipat, Haryana, India
e-mail: ashutosh@niftem.ac.in

A. Mishra
Group Head – Quality Assurance and Food Safety, Wilmar Sugars, Belagavi, UP, India

H. U. Hebbar et al. (eds.), *Engineering Aspects of Food Quality and Safety*, Food Engineering Series, https://doi.org/10.1007/978-3-031-30683-9_1

comprehensive. High quality food production at all levels requires clear understanding and engineering perspective of food quality. This chapter is an attempt for highlighting various engineering technologies available today and their relation to food quality.

2 What Is Quality?

Quality can be described as a fundamental instrument for evaluating a natural attribute of a good or service that allows it to be compared to other similar goods or services. The term quality has a wide range of connotations, but it essentially refers to a set of an object's characteristic features that allow it to satisfy stated or implicit needs. Furthermore, a customer's impression of a product or service determines its quality. It is the mindset of a customer who accepts a product or service and recognizes its potential to suit his or her demands.

Quality can be characterized in utilitarian terms as "fitness for use" or, more specifically for foodstuffs, "fitness for consumption," which leads to "customer" or "consumer" satisfaction, as defined by ISO standards. As a result, quality can be defined as the standards necessary to meet the consumer's demands and expectations (Peri 2006). Figure 1.1 shows an analytical model that defines food quality as a combination of customer requirements. Consumer requirements include the following (Peri et al. 2004):

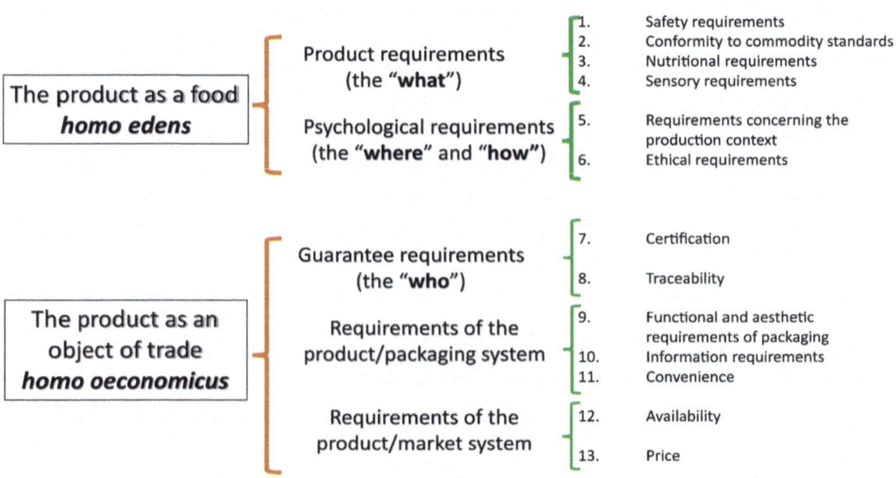

Fig. 1.1 An analytical model of food quality. (Adapted from Peri et al. 2004)

1. ***Safety requirements:*** This might be summed up as the lack of "risk factors." Any violation of safety rules poses a health risk to consumers and is penalized by law.
2. ***Commodity requirements:*** refers to a product's adherence to its specifications. Laws, voluntary norms, and conventional behaviors all play a role. Any failure to comply with these rules shall be regarded as fraudulent and a crime punishable by law.
3. ***Nutritional requirements:*** are obviously vital, as the primary goal of eating is to meet nutritional requirements.
4. ***Sensory requirements:*** because sensory requirements are perceived, they are a crucial way for products and customers to interact. Our sensory experiences take place in a space that is strongly associated with other brain functions and contents, such as memory, culture, values, emotions, and so on, because it is the brain that converts sensations into perceptions.
5. ***Requirements concerning the production context:*** Consumers are influenced by indications about a product's origin or tradition, as well as the usage of organic agriculture. This is mostly a psychological and affective effect that aligns a cuisine with expectations based on memories, culture, and our perceptions of life, nature, and the environment.
6. ***Ethical requirements:*** These are related to the value system that influences consumer behavior. Organic agriculture, environmental protection, biodiversity protection against mass production, animal welfare, and so forth are all ethical standards.
7. ***Guarantee requirements***: Certification and traceability procedures, which are frequently mentioned in contemporary European food regulations, are nothing more than tools that provide consumer assurances. They are based on the certification of behaviors and, in the end, people, as opposed to traditional certification methods focused on product analysis. Trust is earned through a person-to-person relationship rather than a product-to-person relationship.
8. ***The requirements of the product/packaging system:*** make it easier to recognize, sell, and use a product. Aesthetic standards for its presentation, as well as consumer information given via the label, are all part of the package requirements. Whether it comes to transportation, preservation, preparation, or usage of a product, ease of use has become a critical factor (convenience foods). Consumers like things that are easy to handle or use, and marketing gurus have found that their demand for convenience is the most fruitful field.
9. ***Requirements of the product/market system:*** These include having the product available at the appropriate time, in the right place, and in the correct quantity. They also include its price, as the price-to-quality ratio is the final synthesis of a consumer's impressions, which determines preferences and decision.

3 Quality Assessment of Raw Material

Plants and animals are the usual source of foods. Traditional manufacturing lacks operational flexibility to cope up raw material fluctuations inherent in biological resources, leading to product quality variations. Unlike other process parameters, manufacturers do not have direct control over raw materials attributes. The quality is heavily dependent on third-party suppliers as raw materials might significantly differ from one lot to the other. In such situations, mapping raw material qualities can enable dynamic control and monitoring (Rathore 2014). An appropriate response to this challenge can reduce unwanted deviations of important quality attributes of the finished product (Calvo et al. 2018; Rathore et al. 2014). Use of standard analytical or microbial methods is an essential first tool for achieving success in this regard. Adopting a quality-engineering approach, this data can be put to statistical analysis and modelling in order to ensure effective manufacturing management (Caroço et al. 2019). An example in this regard is that of monitoring water quality and keeping it the same before addition to the bread dough and it has been tried by a few brad manufactures. A portable, low-cost equipment capable of performing real-time product testing, evaluations, and quality assurances is eagerly awaited by food industry.

Near infrared (NIR) and Fourier transform infrared (FTIR) spectroscopy are attractive analytical techniques for rapid quality assurance and assessments of raw materials, and ingredients because of their simple, non-invasive, and non-destructive qualities (Fakayode et al. 2020). Moreover, combining multivariate regression analysis with NIR spectroscopy improves the usability and authenticity of the method for assessing the quality of raw ingredients with greater sensitivity and accuracy.

Because of the convenience of onsite analysis, portable NIR sensors have received a lot of interest from food scientists. Pork quality has been assessed using portable NIR sensors (Horcada et al. 2020). The spectra were collected using a portable micro electrical mechanical apparatus to examine live animal skin, carcass tops, fresh meat, and fat tissue samples. This helps classify pork into four quality-based classes which are controlled by the genetics and/or feeding style of the pigs. Researchers have established NIR sensing techniques for quality analysis of intact pineapple fruit (Chia et al. 2020) and dry matter as well as fat contents of cheese (Eskildsen et al. 2019). Wojtkiewicz et al. (2019) created a self-calibrating NIR-based instrument that minimizes the effect of data on the instrument response function. In another study, Caroço et al. (2019) employed a process analytical technology (PAT) tool in conjunction with NIR spectroscopy to highlight the conceptual and performance differences of several quality evaluation methodologies for citrus peel as raw materials for pectin manufacturing.

4 Novel Processing Technologies

4.1 Thermal Process Engineering for Food Quality

Traditionally, there is history of thermal treatment to produce a microbiologically safe product though traditional thermal processing, still in vogue, may compromise on nutritional and organoleptic qualities. Foods that has been agitated retain their sensory and nutritional properties better than those foods that have not been agitated. Rotary thermal processing is becoming commonplace as the need of greater quality processed products grows (Abbatemarco and Ramaswamy 1994; Rattan and Ramaswamy 2014; Singh et al. 2016). The ideal agitation frequencies for rotational and reciprocating agitation have been the subject of numerous studies. Recent studies have indicated that reciprocation frequency of 1.5–2 Hz or rotational speed of 2–2.5 Hz is ideal (Erdogdu et al. 2017; Pratap Singh & Ramaswamy 2016). There are several studies where heat transfer and other quality aspects have been dealt by the researchers. Increasing agitation in liquid foods such as tomatoes (Pratap Singh et al. 2017), banana puree (Batmaz and Sandeep 2015), white sauce (Walden and Emanuel 2010) and others leads to a shorter processing time, improved product quality and nutritive value, and lower energy costs. For such items, a high agitation frequency is preferred. However, when food particles like potatoes, radish, and green beans (Singh et al. 2015; You et al. 2016) are processed, highly intense agitation is shown to be deleterious to textural properties. In such situations, reduced agitation intensity or covering fluid of higher viscosity is recommended for such commodities (liquid with particles). Besides, the impacts on color and nutritional indicators with greater agitation intensity are positive since the processing periods are shorter. Reciprocity has, in most circumstances, proved capable of overcoming the constraints placed by centrifugal forces all through rotational agitation. Despite recent rapid advancements in agitation processes, a major challenge at present is reducing the damage caused by agitation to food (which becomes mushy once prepared). Tattiyakul et al. (2002) and Singh et al. (2016), examined the notion of regulated or intermittent agitation and demonstrated that halting agitation can work better in several situations to curb negative effects on quality. Kinetic models and key control parameters are significantly important for optimization and assessment of thermal processes. Ball's formula has been a classical basis for several of recent and advanced methods (Stoforos 2010).

4.2 Alternative Non-thermal Technologies for Quality

Food safety, good nutrition, appealing palate and convenience are the factors dominant in food production and processing. Thermal processing was well studies and proved its efficacy in maintaining microbial safety of foods. However, overcooking was a dominating quality concern in thermally processed foods.

Minimal or non-thermal approach in processing was explored and projected as a solution to the problem of overcooking and compromise on nutrition due to overheating because instant cooling is always a limiting factor. In present times, global food system has a deep connection with Sustainable Development Goals (SDGs) of United Nations. Non-thermal processing is said to have a vital role for sustainable food preservation. Common non-thermal processing methods are discussed below as a new engineering approach for food quality assurance (Table 1.1).

Table 1.1 Characteristics of novel processing technologies used to ensure food safety

Process	Operating mode	Applicability	Microbial inactivation	Quality	Packaging requirements
Thermal processing	Batch Continuous	Solid foods Liquid foods	Vegetative microbes Spores Protozoa Algae Viruses	Impact heat sensitive food components like flavor, vitamins and nutrients Deactivate enzymes	In-packaging processing Aseptic filling after processing
HPP	Batch Semi-continuous	Solid foods Liquid foods	Vegetative microbes Potentially spores (in combination with thermal processing)	Maintains food quality Scope for developing novel products Varying influence on enzymes inactivation	In-package processing (high barrier packaging with at least one interface flexible enough to transfer pressure)
PEF	Continuous	Liquid foods Semi-liquid foods	Vegetative microbes	Minimal harmful effects Varying influence on enzymes inactivation	Aseptic filling post processing
Irradiation	Batch	Solid foods Liquid foods	Vegetative microbes Spores Parasites	Development of some off-flavor Some loss of vitamins Texture modification	In-packaging processing (radiation transmitting package)
Ultraviolet disinfection	Batch Continuous	Air Water Few liquid surfaces Food contact surfaces	Vegetative microbes Protozoa Algae Viruses	Development of some off-flavors in certain food products	Aseptic filling post processing

4.2.1 High Pressure Processing (HPP)

HPP is an innovative non-thermal technique in which food is exposed to pressures ranging from 100 to 800 MPa and temperatures as low as 20 °C for a few seconds to minutes (Balasubramaniam et al. 2015). In addition to reducing the food microorganisms, causing food deterioration, undue damage to the food is also avoided by application of HPP (Khan et al. 2018). Ultra-high pressures of HPP systems are thought to modify the anatomy of bacteriological cells as well as obstruct enzymatic processes, potentially weakening and killing food-borne pathogens (Bisconsin-Junior et al. 2014). HPP can extend the shelf life as well as preventing negative alterations in nutritional and sensory quality of foods (Bahrami et al. 2020). HPP has also used in liquid food preparations like juices (Morales-de la Peña et al. 2019). This technique has been applied to treat high water activity (>0.8) foods (Jung and Tonello-Samson 2018). Meats, fruits and veggies, dairy products (Verma et al. 2020), beverages, juices, fish and seafood (Bolumar et al. 2015) are some of the primary food groups reported to have been treated via HPP. Multi pulse HPP technology, which involves the application of numerous short high-pressure treatments across a number of cycles, is a specialized use of HPP technology. Application of multi pulsed HPP at 300 MPa for 3 pulses and static HPP pressure of 600 MPa led to a 57% and 31% inactivation of polyphenol oxidase enzyme, respectively in carrot juice (Szczepańska et al. 2020). Clearly, using multi-pulse at low pressures is more cost effective than using standard HPP (Marszałek et al. 2018). In several foods, HPP has been demonstrated to have a significant antibacterial impact. According to Usaga et al. (2021), for juices and beverages using HPP for 3 min at 600 MPa is sufficient to deactivate *E. coli, S. enterica* and *L. monocytogenes.* Also, HPP of ham at 450–600 MPa for 5–10 min and followed by storage at 4–12 °C for 60 days was sufficient to inactivate *L. monocytogenes* (Pérez-Baltar et al. 2020). HPP treatment of 594 MPa for 233 s considerably reduced *B. cereus* and *S. aureus* in human milk, indicating the potential of HPP to replace conventional pasteurization technique (Rocha-Pimienta et al. 2020).

Aside from the inactivation of foodborne pathogens, HPP has several advantages over traditional thermal processing. Refrigerated fresh fruit juice and HPP treated fruit juice showed comparable vitamin C content, indicating minimal effect of HPP processing on nutritional components of food products (Bhattacharjee et al. 2019). HPP has also been shown to increase the bioavailability of bioactives and trace elements, resulting in improved health benefits. Additionally, HPP may aid in the preservation of good fats, the reduction of salt consumption, and the reduction of allergy and toxin production capability in foods (Picart-Palmade et al. 2019).

4.2.2 Pulse Electric Field (PEF)

PEF works, mainly, on two basic mechanisms, electroporation and electrical breakdown, working together to achieve pathogenic annihilation (Bahrami et al. 2020).

Voltage ranging from 10–80 kV/cm is typically supplied for durations of microseconds, resulting in the appropriate electrical field critical for elimination of microorganisms. The presence of charged particles enables electric energy to flow into all parts of food product. PEF can also be used as a preservative for fluid foods like juices (Morales-de la Peña et al. 2019). Electroporation leads to the creation of pores mostly on external and the internal cell layer of a microorganism (Santhirasegaram et al. 2016). The pathogen's quasi membrane is damaged when the procedure is paired with an electrical breakdown. The cytoplasmic contents of the cell, thus, seep out because of the combination of the electroporation and electrical breakdown and the microbial cell dies (Pal 2017; Soltanzadeh et al. 2020). A variety of factors influence the performance of PEF treatment, such as field strength or intensity, time interval, food media conductivity rate, pulse, pH, temperature applied throughout the process, energy applied, pathogen type and polarization (Bhattacharjee et al. 2019; Pal 2017). Foods like apple, apple sauce, orange, tomato, pea soup, carrot juices, salad dressing, eggs (Khouryieh 2021), milk and milk products have all been treated with PEF. PEF has been used in the fruits and vegetable business to improve the physico-chemical, rheological, and antioxidant features of juices (Alirezalu et al. 2020).

PEF has also shown considerable control of microorganisms in various food products (Salehi 2020). Application of PEF at moderate intensity and pulse width of 2.7 kV/cm and 15–1000 s, respectively had a lot of potential for inactivating microorganisms for fruit juice processing (Timmermans et al. 2019). Furthermore, when contrasted to role of SO_2 during wine malolactic fermentation, PEF demonstrated to be significant in microbial stability (González-Arenzana et al. 2019). Pallarés et al. (2021) found that applying HPP and PEF combination, as hurdles, had a substantial effect on reducing grape juice aflatoxins. The combination of hurdles resulted in the reduction of 72% and 84% for aflatoxin B_2 and G_1, respectively. These findings highlight PEF's potential for microbial inactivation, which could be beneficial to both food processors and consumers in assuring product safety.

4.2.3 Pulsed Light (PL)

PL is one of the modern food processing procedures used in food industry. It involves a wide range of short but highly energetic pulses, derived from white light's spectrum, which is identified to contain visible, infrared, and UV light, power of PL can be 1000 times in comparison to the regular, somewhat continuous UV light (Dong et al. 2020; Jiménez-Sánchez et al. 2017). In addition, a wavelength of 200–1000 nm pulsed ultraviolet light (Dong et al. 2020; Vong and Liu 2016) can be used to produce a range of high-power Pulsed light in a relatively short period of time. Moreover, UV light is considered to be the most destructive segment of the PL spectrum for microorganisms (Morales-de la Peña et al. 2019), apart from being cost effective. Structure of food allergens changes under PL, causing proteins to clump together. PL can also be used to inactivate bacteria by changing their structures

(Dong et al. 2020), blocking the cytoplasmic membrane, stopping biocatalysis, and eventually destroying the genetic makeup (Ramos-Villarroel et al. 2012). Liquid foods and foods with simple external conformations can also benefit from PL treatment. Fish, fruits, vegetables, and meat are among the items that PL has been found to be useful (Mahendran et al. 2019).

PL like other non-thermal techniques, can inactivate or eliminate pathogens in foods. Direct and in-package application of PL (1.05 J/cm^2) on Romaine lettuce with a thickness of 0.00254–0.00762 cm resulted in *E. coli* reductions of 2.18 to 2.68 log CFU/g (Mukhopadhyay et al. 2021). Using PL intensity ranging from 8.2–12.5 J/cm^2 to the exterior of lettuce have been shown to be efficient in inactivating the bacterial development of *L. monocytogenes, E. coli, S. aureus and S. enteritidis* (Tao et al. 2019). Furthermore, PL treatments ranging from 0.35 to 3.6 J/cm^2 resulted in reduction of *L. inocua* by 3–4 log on sausages and roughly 1 log in chicken cold cuts and boiling ham (Kramer et al. 2019). It's important to note that PL might be combined with other cutting-edge techniques to inactivate germs on food surfaces. For example, an exceptional 98.2% decrease was achieved in aflatoxins of peanuts by the combined impact of citric acid PL treatment. This could not be achieved by using procedures separately (Abuagela et al. 2019).

4.2.4 Radurization and Ultraviolet (UV) Disinfection

Food irradiation is a novel, energy saving and green preservation method involving exposing the food products to a variety of non-ionizing and ionizing radiations. It is applied to increase a product's shelf life without compromising its nutritional properties. Radurization is the process of applying a substantial dose of ionizing radiation to food in order to improve its shelf life by lowering the quantity of viable spoilage microorganisms. The needed dose ranges from 0.4 to 10 kGy. Irradiation of food is one of many methods that can help to improve the safety of food. Ionizing energy is used to irradiate packed or bulk foods. Unlike traditional heat pasteurization, deactivation of microbes is performed at relatively low temperature, this procedure is also referred to as "cold pasteurization" (Farkas and Mohácsi-Farkas 2011; Pedreschi and Mariotti-Celis 2020).

The impact of irradiation on foods is determined by the radiation doses. Inhibition of potato sprouting, delayed ripening in fruits and vegetables and fruit disinfection (insects and parasites) are all possible with low doses (0.05–0.15 kGy). The inactivation of pathogens like *Campylobacter jejuni*, *Salmonella* spp., *Staphylococcus aureus, E. coli* and *Listeria monocytogenes* moderate radiation doses of 1.0–10 kGy is adequate. Decontaminating food items such as spices and herbs need higher doses of 10–50 kGy. On an industrial basis, doses of 30 to 50 kGy are used to sterilize foods for astronaut and medical meals (Bisht et al. 2021; Ihsanullah and Rashid 2017).

The RADURA symbol has been used since the 1960s as a quality indicator, especially for food exposed to ionized radiation. The logo was used by the Dutch pilot plant to identify their irradiated products and to promote a high-quality produce with a long shelf life. The logo was prominently displayed in supermarkets in the Netherlands where irradiated mushrooms (one of the first goods on the market) were on sale, and customers received a brochure with details about the procedure and the benefits of the treated products (Ehlermann 2009).

UV treatment is another novel non-thermal technique used for improving the quality and safety of foods through decontaminations. UV light is a type of energy that is deemed non-ionizing and has germicidal effects at wavelengths between 200 and 280 nm. UV light has a wavelength range of 100 to 400 nm, with UV-A = 315–400 nm; UV-B = 280–315 nm; UV-C = 200–280 nm; and UV-Vacuum = 100–200 nm being the most common types (Vasuja and Kumar 2018). UV radiation, in principle, works by damaging the pathogen's genetic element, preventing reproduction, replication, and, as a result, spread (Xuan et al. 2017). To inactivate different microorganisms, different types of food products usually require varying levels of UV radiation called as UV-inactivation dosage (mJ/cm^2). For example, bacteria, yeast, fungus, protozoa, and algae require UV-inactivation doses of 1–10, 2–8, 20–200, 100–150, and 300–400 mJ/cm^2, respectively, indicating that algae is perhaps the maximum tolerant in comparison to other pathogenic organisms (Unluturk et al. 2010). As a result, the efficiency of UV radiation is influenced by a variety of parameters, including the UV source and dose, length of exposure, type of the food, orientation of the equipment, and the type of the pathogen (Delorme et al. 2020).

UV-C has been shown to have the ability to prevent infections in several investigations, with UV light wavelengths between 100 and 280 nm being deemed germicidal. The highest log reduction was 3.18 CFU/g in walnuts infected with Salmonella and exposed to UV radiation (8 cm for 45 s). Because the physicochemical qualities of the walnuts were not changed, this showed to be a vital alternative to the less desirable chemical and thermal approaches (Izmirlioglu et al. 2020). Application of UV-C treatment to raw milk reduced the total yeast-mould and aerobic bacteria count by 3 and 2 logs, respectively. Furthermore, UV therapy reduced inoculated *E. coli, Salmonella, S. aureus and L. monocytogenes* by 2–3 logs. Nevertheless, it was stated that UV radiation should be combined with other technologies to obtain far more overall decrease in microbial contamination (Atik and Gumus 2021). UV-C inactivation dosage of 127.2 mJ/cm^2 for a duration of 30 seconds to raw salmon was efficient in lowering the microbial contamination (Pedrós-Garrido et al. 2018). This came about as a result of greater UV-C dosages, which caused disruptions in sensory properties. Nonetheless, the role of UV-C in bacterial stability is demonstrated by these findings, further research is important in order to establish the possible use of other technologies with UV-C as an obstacle in order to ensure that the sensory and physico-chemical aspects of food products being treated are retained during microbial inactivation.

4.3 Hurdle Approach

Due to the limits of other microbial inactivation approaches, hurdle technology, also called as collective treatment, was established, and implemented to improve the sensory and nutritional attributes and safety of products (Khan et al. 2017). It is described as "the planned and smart usage of prevailing and innovative preservation procedures to produce a set of protective variables (hurdles) that no microbes existing should be able to overcome". The principal process of hurdle technology is to destabilize the homeostasis of specific microorganisms by continuously exposing them to physical, chemical, or ecological stress (Oh et al. 2019). Various food products have also been preserved using hurdle technologies developed by researchers (Qiu et al. 2019). Barbosa-Cánovas et al. (2020) found that the hurdle technology influences the development and stability of bacteria in foods primarily through 4 processes: (i) homeostasis; (ii) metabolic exhaustion; (iii) stress reaction; and (iv) multi-target preservation. Homeostasis is described as "the tendency of organisms internal states to be regular and stable" (Tsironi et al. 2020). As a result, the efficient and practical strategy to disrupt microbial homeostasis is to use several barriers. The "auto-sterilization" of foods is aided by the metabolic exhaustion of bacteria (Alzamora et al. 2018). Due to limited energy availability, auto-sterilization happens when microorganisms become metabolically fatigued and use their current energy to attain a homeostatic state. The stress reaction of microorganisms is the third mechanism. Due to any hurdle effect and malnutrition, certain bacteria get more pathogenic or resilient to distress by releasing stress shock proteins (SSP). As a result, many stressors acquired at the same time make activating the genes to create SSP difficult for the bacterium. Microorganisms experience increased metabolic depletion as a result of this mechanism (Rostami et al. 2016). Leistner and Gorris (1995) proposed the notion of multi-target food preservation as a way to prevent bacteria from producing SSP. When stressors or barriers attack distinct sites of microbes, disrupting homeostasis and metabolically exhausting bacteria, a synergistic impact is generated (Leistner 2000). To enhance storage stability and confirm microbiological safety, hurdle technology is commonly used in foods such as meat, fruits and vegetables, dairy goods, seafoods, bakery items, pasta, tinned products, salads, spices and juices. As a result, hurdle technology strives to eradicate unwanted bacteria while preserving sensory and nutritional properties.

5 New Approaches for Quality Measurement of Foods

5.1 Rheological and Textural Quality of Foods

Describing and perfectly, comprehending the rheological properties of food products is critical for a variety of applications in food industry, including uniform characterisation of raw ingredients, novel products, and optimal industrial operations. In recent decades, traditional rheological methodologies, and approaches tailored to specific food products and measurement aims, have gotten a lot of attention, allowing for a better knowledge of the raw ingredients, handling, and basic role in a multifaceted food composition.

Small-amplitude oscillatory shear (SAOS) assessments have been the greatest tool for understanding the dynamic behaviour of food products with time. SAOS calculates material functions such as elastic and viscous modulus (G' and G''), complex viscosity (η^*), and yield stress based on the linear viscoelastic properties of multifaceted liquid foods. Without harming the specimen's structure, SAOS can evaluate the alteration in microstructures and the jamming of biomaterials in-situ. Measurement can monitor and understand the process mechanism and kinetics of probable phase transitions in dietary ingredients (e.g., starch gelatinization and protein denaturation). The SAOS measurement offers an advantage over the DSC analysis (Ahmed et al. 2018). SAOS measurements provide a strong theoretical foundation, allowing the building of rheological equations for evaluating structural and mechanical theories over a broad frequency range. In rotational rheometers for SAOS investigations, different configurations (e.g., parallel plate, cone plate, and concentric cylinder) are used, due to the varying nature of the material (Djalili-Moghaddam et al. 2004; Song et al. 2017). Food engineers strive to create the links between food rheology and processing and then use rheological data to improve process or product. Logical and semi-empirical simulation, and mathematical system flow calculations, utilise rheological observations. Mixing, spinning, extrusion, dispersion, injection moulding, spraying and coating are examples of common food processing flow operations (Fischer and Windhab 2011).

Linking microstructural interpretation and physiological food structure assessments, like rheological qualities, is important for food quality management because it needs knowledge of the food breakdown pathway through chewing and the association of instrumental evaluations to human judgements. The link between the applied force and the deformation imparted to the food is the essence of the mechanical and food rheology method for texture investigations. Texture Profile Analysis (TPA), developed by Szcznesiak and colleagues for determining the mechanical qualities of foods, was one of the most important achievements in food texture studies. Industrial and academic researchers continue to utilize TPA to characterize the texture of solid and semisolid foods. The procedure is a simple double compression test, in which the food item is compressed twice to a restricted level to minimize structural damage. The force displacement curve recorded indicates the material's resistance to deformations and can be utilized to understand several important mechanical texture characteristics, like brittleness, hardness, adhesiveness, springiness, etc. (Chen and Stokes 2012).

However, it is important to remember that rheometer and TPA readings only account for a portion of the texture assessment of food during the early stages of oral processing, when it compresses under the forces produced by the teeth or tongue. Food tends to transform as it is chewed and masticated, with smaller particles, higher moisture absorption, and softness. At a late stage of oral processing, the perception of food texture is likely to be controlled by a mix of fluid flow and surface qualities;

hence, rheology characteristics turn out to be less important, nonetheless lubrication and surface friction become to be more important for textural sensations. As a result, food oral tribology is currently emerging as a new experimental technique for exploring the relationship between food structure, texture, and mouthfeel, alongside food rheology.

Tribology is a branch which deals with investigating the friction and concerning to tearing, wearing and lubrication between 2 interacting surfaces (Prakash 2017). The approach, in which the tongue and upper-palate operate as 2 contacting surfaces that move in tandem, has recently been used for oral food preparation. Food passes through several phases in the mouth while being chewed, mingling with saliva, and lastly transforming into a bolus suitable to be ingested. The entire procedure necessitates a full grasp of physiological principles governing complex oral movements, as well as their relationship to mechanoreceptor stimulation (Selway and Stokes 2013). Investigation of smoothness perception, an important textural element associated with the availability of fat or oil and creamy sensation, is one of the most important applications of food tribology (Ahmed 2018).

5.2 Glass Transition

Amorphous materials go through a glass transition, which is a well-known state shift. Glass transition happens when a liquid-like material cools to the point where it solidifies into a glassy form over a temperature range (vitrification). During the heating process, the glass transition results in greater molecular translation mobility and the accompanying liquid-like appearance of the substance. In the development of dehydration and freezing technologies, the importance of the glass transition to food and medicines processing and stability management has been recognized.

Differential scanning calorimetry (DSC) is possibly the utmost valuable and generally appropriate of all characterisation systems for researching glass transition. DSC equipment is found throughout most biochemical and materials engineering facilities. In glasses, for determining kinetic and thermodynamic properties and phase changes, DSC has become a worldwide standard instrument. DSC is far more than a tool for characterizing materials: it has evolved into a sophisticated and comprehensive methodology for various facets of glass study during the last two decades. The exceptional sensitivity of DSC to minor and major energy variations generated by phase and structural transformations in glasses throughout heating, annealing, cooling, and pressurization has led to its widespread application. During investigations thermal processing rates has been greatly expanded since the introduction of Flash DSC, allowing measurements on ultrashort time scales. With the invention of temperature-modulated DSC, researchers were able to identify intersecting changeover and identify faint and minor transformations thanks to the capacity to segregate the heat capacity and kinematic aspects of the DSC signal (Zheng et al. 2019).

In the wide range of temperature and pressures where the food products are manufactured, preserved, and eaten, individual constituents of food products may undergo phase transitions. The thermal, physical, and material characteristics of food products are affected by the molecular alterations that occur during phase transitions. As a result, understanding phase transitions is critical for effective food product control, distribution, and storage. DSC is mostly utilized to govern first-order changes like protein denaturation when testing proteins in food products. This process entails the transitory or permanent collapse of the initial organized arrangement in favour of a disorderly configuration of peptide chains. The temperature at which at least 50% of the structure of proteins has been destroyed is known as the denaturation temperature (Leon et al. 2003; Leyva-Porras et al. 2020). The biological and physical qualities of proteins are lost when they are denatured, resulting in variations in viscosity, solubility, and coefficient of diffusion. According to Arora et al. (2021), microwave roasting of Black rice significantly affected the glass transition temperature. DSC thermogram showed endothermic curve, indicating melting of the amylopectin crystallites in the starch granules.

5.3 Interpretation of the Food Images and Spectral Signature

The food industry is primarily concerned with providing safe products, which necessitates a continuous obligation to the development and execution of methods and schemes to manage numerous factors in food items. The majority of presently offered analytical methods are time-consuming and deleterious. As a result, it's vital to develop non-invasive, effective, and rapid assessment way to monitor food safety and quality (Du et al. 2020).

The advancements in computer, optical, and image sensor technology, as well as their applicability in agriculture, have resulted in sensing technologies such as imaging systems that can do automatic and speedy quality monitoring on the processing line with minimal human intervention (Gowen et al. 2009). The initial applications of image processing for food and agricultural products were the use of a red-blue-green colour vision system for colours and sizes classification systems and then the detection of defects (Singh et al. 2012). Color imaging can perform quality analysis, but it is inadequate when it comes to detecting tiny ingredients or contaminants. Multispectral photography configurations have been devised to detect properties of interest using a restricted number of narrow wavelengths, frequently below 20 (Yang 2012). The wavelengths of these waves can range from visible to near-infrared. Depending on the imagery technology, the combination of filters employed, and the attributes under study, the number of wavelengths utilized to create a multispectral image cube might vary. For example, if the characteristic of attention is already defined, photographs in certain wavelength ranges are accurately available for the inquiry. Liu et al. (2014) used multi-spectrum imagery for the purposes of determining different quality parameters (firmness, total soluble solids) and maturity indices of strawberry products, with 19 wavelengths spanning from 405 to 970 nm. A spectral signature is an image of wavelength-dependent reflectance curves that is unique to each plant species and circumstance. According to Xie and He (2016), hyperspectral imaging and spectral signature has the capacity to perceive strength and colour of eggshell as well as detect fractured eggs. Improving the acquisition of images and reducing the specularity of samples would enable to use the latest spectral imaging technology in on-line industrial applications to cover various jobs related to food quality assessment processes.

5.4 Food Authentication

Food authentication, which is a procedure of verifying the authenticity of foods (i.e., adherence to their tag data), may assist to combat food fraud. The authenticity of food should be implemented throughout the entire value chain, which include consumers, in order to be effective. Analytical methods like chromatography, mass spectrometry, spectroscopy, immunology, molecular techniques, wet chemistry, sensor-based technologies, and microscopy are used in current food authenticity approaches. These approaches are typically costly, subjective, and inconclusive (Kiani et al. 2016). Research lab processes such as wet chemistry and chromatographic can provide more precise and efficient testing using chemical change and exclusion in the gas or liquid phase. These processes, on the other hand, are often time taking, entail complex sample preparation, and require specialist expertise (Leiva-Valenzuela et al. 2018), making them unsuitable for on-site food quality monitoring.

Computer vision system (CVS) has shown to be highly useful in activities like predicting contents (Cano Marchal et al. 2013; Shafiee et al. 2014), freshness evaluation (Srivastava et al. 2015), and defect identification (Arakeri and Lakshmana 2016; Rong et al. 2017) as an affordable and competent technique for quality valuation. Its goal is to use a soft sensor to substitute the human visual system and to repeatedly get an advanced understanding of food authenticity by acquiring, processing, and analysing digital images. CVS uses simple instruments such as a digital camera and a webcam to record food photos under controlled lighting (Di Rosa et al. 2017). Several image processing techniques are often used to remove background, select area of interest (ROI), and extract picture features in order to improve the usefulness of information regarding food exterior qualities such as color, size, and shape. The association between image data and qualitative/quantitative information about food authenticity is then determined using pattern recognition techniques, allowing prediction for new food samples.

6 Novel Packaging Technologies for Ensuring Food Quality Post Manufacturing

6.1 Modified Atmosphere Packaging (MAP)

MAP has often been used to increase the storage of perishable items (Qiu et al. 2019). Packaging is especially critical during the COVID-19 epidemic to maintain safety and freshness of foods during lengthier transportation interruptions owing to lockdown. Through reduced metabolic, microbiological, and enzymatic activities, MAP technology extends the storage life of food product by changing the gas composition inside it (Rennie and Sunjka 2017). By infusing a specific composite gas, the gas composition inside the packaging can be changed and maintained (Badillo & Segura-Ponce 2020). In other cases, the fresh produce respiration rate inside the MAP is so high that the material's penetrability is inadequate to pass CO_2 and O_2, resulting in an anaerobic setting conducive to anaerobic spoiling (Larsen & Liland 2013). With sufficient numbers and sizes, pierced box was intended for such outcomes (Kartal et al. 2012). Active MAP is a yet another strategy for maintaining sound, safe and healthy products within the package when including an active component contributes to the uptake or discharge of substances for a longer storage stability of food product (Yildirim et al. 2018).

The most significant innovation in MAP is the introduction of intelligent technology, which uses indicators to identify chemical reactions caused by microbial activity. Smart labels, radiofrequency (RF) identification tags, sensors, time-temperature integrator (TTI), security tags, and other indications are among them (Zhang et al. 2015). Polyaniline films, for instance, were employed for monitoring fish freshness (Kuswandi et al. 2012), whilst TTI was used to suggest quality and safety of spiced chicken and ground beef Ellouze & Augustin (2010), fresh produce (Riva et al. 2001), golden drop (Thai dessert) (Nopwinyuwong et al. 2010) and mushroom foods (Bobelyn et al. 2006). The creation of anti-SRS-CoV-2 packaging materials as theorized using polymer-containing nanoparticles (Sportelli et al. 2020), as found to inhibit human Norovirus and hepatitis A virus, would also be an attractive topic for investigation (Randazzo et al. 2018).

6.2 Active Packaging

Active packaging is amongst the most effective options for extending the shelf life of foods by utilizing the capabilities of food packaging. Some of the qualities of Active packaging cannot be obtained through traditional packaging. Active packaging is defined as packaging systems that interact with food in such a manner that they "deliberately contain components that would release or absorb substances into or from the packed food or the environment surrounding the food," according to European regulation (European Commission 2009). The key concern that arose from the active packaging design is the pace of active chemical movement. Because of their low molecular weight, most active chemicals in active packaging have a high potential for quick release, resulting in a loss of active packaging sustained activity during the shelf life of food. Some recent studies have focused on keeping the active agent in the package system for a longer time and extending the length of its release (Yildirim et al. 2018).

Controlled-release active packaging (CRP) is a new type of packaging that may release active compounds at varied speeds, making it ideal for preserving food quality and safety during long periods of storage. Regulated release is a type of slow-release in which the active ingredient is released in a controlled manner over time (Mastromatteo et al. 2010). Active food packaging technology is arising from the development of the CRP as it not only wants to extend the duration of the active supply of compounds, but also to promote the reproductivity and prediction of releases rates (Chen et al. 2019).

To achieve optimum concentrations on food surfaces, regulated release of active agents through food packaging materials is necessary. Some research has been done to help create the CRP for use in food packaging. The CRP is unusual in that it concentrates on the kinetics and method of controlled release, including what to release, when and how to trigger the release, how much to release, and how quickly to release. A very slow pace would lead to inadequate active compounds to retard food deterioration and very quick rate, caused by degradation or interaction with food components, would lead to an excess of active compounds and their losses (Chen et al. 2019).

6.3 Intelligent or "Smart" Packaging

Smart packaging, according to numerous researches, is described as packaging that contains both active and intelligent systems that work together. It can detect changes while stored (temperature or humidity decreases/increases) and is able to halt degradation of quality. The concept of "smart packaging" was developed by combining active packaging components such as antioxidants, carbon dioxide emitters, antibacterial agents, humidity, ethylene, and oxygen scavengers with intelligent devices (Latos-Brozio and Masek 2020; Schaefer and Cheung 2018). Smart packaging comprises gadgets that may heat or chill food inside and display nutritional information on an electronic display in real time (Kalpana et al. 2019).

Research has showed some of the smart packaging used in the canning and beverage sectors. For bottles, cans, or carton containers, a mechanism included in the package that can vary the temperature of the food within was invented. It can reduce the temperature of the product by 18 degrees Celsius in a short period of time (two or three minutes) before eating. The concept behind this packaging is to absorb heat from the liquid within by employing vapours generated by releasing a quantity of pressurized water from a vinyl bag that evaporates instantly (Brody 2002). Warming cups for coffee, tea, soup, and hot chocolate are another example of smart packaging. The exothermic interaction between water and calcium oxide provides the foundation in this scenario. Calcium and water are placed separately in the cup. Customers are instructed to invert the cup and combine the contents, resulting in an exothermic reaction. The material used to make the cup allows it to hold the temperature for about 20 min (de Abreu et al. 2012).

Smart packaging is still in its early stages of research, but it has the potential to prevent foodborne illness while simultaneously assisting in the reduction of environmental issues. Smart packaging is projected to improve food safety and quality in the near future, as research in this subject continues to grow, evolve, and mature.

7 Quality Engineering

The term itself explains that Quality Engineering is "When we club Engineering aspects in Quality Assurance or Control". In modern food manufacturing or processing, Quality engineering is transforming the overall concepts of Quality Assurance and Food Safety.

7.1 Quality Engineering Tools

7.1.1 Cause Effect Diagram

It is widely used tool for identifying the root cause of the identified issue in the process. It's also known as Fish Bone Diagram or Ishikawa Diagram. In this, we identify various probable factors, which may contribute in the identified issues. To build a graphic, write out the question on a white board or a worksheet, then create a box around with a horizontal arrow heading to it. Eventually, a comprehensive schematic resembling a fishbone will be created and to extend an opportunity for the team to discuss possible causes for quality problems.

7.1.2 Pareto Chart

It is popularly known as the 80:20 rule. It identifies that which the 20% issues are causing 80% defects in the product / service. In general, this tool gives an idea that out of whole bunch of issues that all need to be priorities to have maximum Benefits.

7.1.3 Histogram

A histogram is a graphical representation of graph with numerous statistical uses. Histograms show the number of observations that fall inside a given set of variables, allowing for a visual explanation of numeric values. These value ranges are referred to as classes or bins. The usage of a bar depicts the frequency of data that falls into each category. The bigger the bar, the more measured values in that bin occur often.

7.1.4 Check Sheet

The term "check sheet" is frequently used to refer to a "defect concentration diagram," which is used to gather and evaluate data on the prevalence with which a quality issue or defect arises. For example, if one wants to analyse the weight variations in the product, a check sheet may be designed in which data of 20 units are entered in each 15 minutes for each line moreover, each shift. With this data then one can analyse that which machine is having the greatest number of over/ under weight units and at which time.

7.1.5 Control Chart

Control chart is a graphical representation to which is used to study that over the period how the process has changed. In the chart, generally it has average line with upper and lower control limits. These upper and lower control limits are taken from the historical data and when we plot the data with these lines then it shows the variation against average. Upper and Lower Control Limit. This type of chart is used generally to identify the issues which come after a period.

7.1.6 Scatter Diagram

A scatter diagram is a chart that is used to observe the association between data and to display them graphically. The variables' values are displayed by dots. Scatter plots employ Cartesian coordinates to depict the values of variables in a data set since the placement of the dots on the vertical and horizontal axis informs the value of the relevant data point. Scatter diagram is used for demonstration of relationship between two variables, Identification of Correlation relationship and identification of data patterns.

7.1.7 Flow Chart

This tool is used to visualize the sequence of the process, events or workflow. Flow chart gives a clear idea about the relations of the steps and process boundaries. Using flow chart enables user to get into the depth of the process or event. This is one of the most used Quality Engineering Tool.

7.2 *Quality Management Systems (QMS)*

QMS defines how a company can meet the needs of its consumers, customers and other parties involved in its operations. QMS is a framework of multiple aspects affecting the overall quality performance of the product, service etc. QMS is collective tool consisting of Context of Quality Management System (requirements of QMS), Leadership (including decision makers accountable for quality; leaders must show Commitment to the Quality and Focus for Continual Improvement). Important tools of QMS, including engineering aspects, are discussed below:

7.2.1 Advanced Product Quality Planning (APQP)

Quality Plan is a detailed framework of achieving the overall objectives with respect to Quality Assurance / Control. Quality Plan explains the tasks and their flow that needs to be done by the functionaries responsible for maintaining quality of the Product, directly or indirectly. Precisely, it is a structured approach to product and process design in order to satisfy the customer. In a food industry setup, Quality Assurance Team will define a quality plan consisting of activities like testing, inspection, analysis etc. to ensure that material in process or final product is meeting the agreed specifications. It may pertain to a section of subsection of the plant like 'production quality plan' that will focus on zero defect production. Similarly maintenance team may have a quality plan leading to a production free from defects that could be possible due to poor maintenance. This becomes more important in the present era of automation. Maintenance team can add on the activities like Preventive maintenance, Breakdown maintenance etc.

7.2.2 Failure Mode and Effects Analysis (FMEA)

Globally regulatory agencies are bring a food recall plan as a mandatory requirement for food manufacturers to implement stated quality and safety specification related to food products. FMEA is a method which allows businesses to predict errors designing phase. It locates potential flaws in a product design and production process. FEMA should always be used when creating a new product, process, or service; when planning a change or upgrade to a current process; and so on.

7.2.3 Statistical Process Control (SPC)

Engineers at shop floor needs to employ statistical methods while monitoring process and quality operations to conform to quality standards, preventing rejection and wastage. The military used the SPC method extensively during World War II. Later, the notion became popular across Japanese enterprises. SPC is now a commonly utilized quality tool in a variety of sectors. In SPC data is obtained from shop floor in real-time to plot on a Control Chart to ensure the variation within the control limits.

8 Food Quality and Traceability

To boost consumer trust and buy willingness, the food chain must become more sustainable. Identifying and managing sources of contamination requires tracking and verifying data from the entire food supply chain, which contributes to agri-food chain sustainability management (Olsen and Borit 2018). Conventional internet of things tracing systems are used to monitor and maintain detailed info at all stages of the manufacturing process, transportation, and consumption employing systems such as Wireless Sensor Networks, Radio Frequency Identification, Near Field Communication, and others. It also has the ability to give valuable data for food quality and authenticity management. Based on the centralized servers/customers paradigm, however, stakeholders and consumers must rely on a single information point to save, transmit and share traceability data Khan & Salah (2018). As a result, most consumers find it difficult to obtain complete transaction details and track product sources (Imeri and Khadraoui 2018). Customers and food supply chain partners must be well-informed about the product life cycle in order to ensure that food is safe, sustainable, and of excellent quality. Current food traceability systems, on the other hand, are ineffective in establishing confidence among players in the chain (Zhao et al. 2019). Improving food traceability requires an innocent and operative agri-food information management system.

To address food safety and quality concerns, increase traceability, openness, protection, resilience, and honesty (Banerjee et al. 2018; Feng et al. 2019). As a result, privacy protection and data leakage issues are critical in agro – based tracking, which has become an urgent issue for growers, suppliers, cold storage management, authorities, and buyers (Caro et al. 2018). No one entity in the supply chain may modify current statistics, making blockchain technology a potential tool for building trust mechanisms to solve transparency and security challenges. Blockchain is a decentralized and distributed system that consists of a network of time-stamped blocks linked by a cryptographic algorithm. It was widely accepted as a remedy to the inherent difficulties of trustworthiness in the area of data transparency and the prohibition of information handling (Galvez et al. 2018).

9 Conclusion

Food consumption is linked to a variety of health risks, which originate at various points throughout the food supply chain. To effectively and innovatively handle these potential risks, a multidisciplinary strategy encompassing microbiology, chemistry, and engineering is required. Food safety engineering is an excellent example of this multidisciplinary approach, which aims to provide complete solutions to both chronic and emergent food safety issues. Researchers in this new field of study should be familiar with food's microbiological and chemical concerns, as well as capable of combining science and engineering to reduce or eliminate these dangers. The greatest solution to the numerous food safety concerns is intervention techniques founded on the collaboration of microbiologists, chemists, and engineers.

References

Abbatemarco C, Ramaswamy HS (1994) End-over-end thermal processing of canned vegetables: effect on texture and color. Food Res Int 27:327–334. https://doi.org/10.1016/0963-9969(94)90188-0

Abuagela MO, Iqdiam BM, Mostafa H, Marshall SM, Yagiz Y, Marshall MR, Gu L, Sarnoski P (2019) Combined effects of citric acid and pulsed light treatments to degrade B-aflatoxins in peanut. Food Bioprod Process 117:396–403. https://doi.org/10.1016/j.fbp.2019.08.011

Ahmed J (2018) Advances in rheological measurements of food products. Curr Opin Food Sci 23: 127–132. https://doi.org/10.1016/j.cofs.2018.10.007

Ahmed J, Al-Ruwaih N, Mulla M, Rahman MH (2018) Effect of high pressure treatment on functional, rheological and structural properties of kidney bean protein isolate. LWT-Food Sci Technol 91:191–197. https://doi.org/10.1016/j.lwt.2018.01.054

Alirezalu K, Munekata PES, Parniakov O, Barba FJ, Witt J, Toepfl S, Wiktor A, Lorenzo JM (2020) Pulsed electric field and mild heating for milk processing: a review on recent advances. J Sci Food Agric 100:16–24. https://doi.org/10.1002/jsfa.9942

Alzamora SM, López-Malo A, Guerrero SN, Tapia MS (2018) The hurdle concept in fruit processing. In: Food Engineering Series. https://doi.org/10.1007/978-1-4939-3311-2_5

Arakeri MP, Lakshmana (2016) Computer vision based fruit grading system for quality evaluation of tomato in agriculture industry. Procedia Comput Sci 79:426–433. https://doi.org/10.1016/j.procs.2016.03.055

Arora S, Virdi IK, Sharanagat VS, Kheto A, Dhua S, Suhag R, Kumar R, Kumar Y, Patel A (2021) Roasting of black rice (Oryza Sativa L.): change in physico-functional, thermo-pasting, antioxidant and anthocyanin content. J Food Meas Charact, 15(3), 2240–2250. https://doi.org/10.1007/s11694-021-00828-7

Atik A, Gumus T (2021) The effect of different doses of UV-C treatment on microbiological quality of bovine milk. LWT 136:110322. https://doi.org/10.1016/j.lwt.2020.110322

Badillo, GM, & Segura-Ponce LA (2020) Classic and Reaction-Diffusion Models Used in Modified Atmosphere Packaging (MAP) of Fruit and Vegetables. In Food Engineering Reviews.https://doi.org/10.1007/s12393-020-09214-3

Bahrami A, Moaddabdoost Baboli Z, Schimmel K, Jafari SM, Williams L (2020) Efficiency of novel processing technologies for the control of listeria monocytogenes in food products. Trends Food Sci Technol 96:61–78. https://doi.org/10.1016/j.tifs.2019.12.009

Balasubramaniam VMB, Martínez-Monteagudo SI, Gupta R (2015) Principles and application of high pressure-based technologies in the food industry. Annu Rev Food Sci Technol 6:435–462. https://doi.org/10.1146/annurev-food-022814-015539

Banerjee M, Lee J, Choo KKR (2018) A blockchain future for internet of things security: a position paper. Digit Commun Networks 4:149–160. https://doi.org/10.1016/j.dcan.2017.10.006

Barbosa-Cánovas GV, Yildiz S, Oner ME, Candoğan K (2020) Selected novel food processing technologies used as hurdles. In: Food Engineering Series. https://doi.org/10.1007/978-3-030-42660-6_24

Batmaz E, Sandeep KP (2015) Integration of ResonantAcoustic® mixing into thermal processing of foods: a comparison study against other in-container sterilization technologies. J Food Eng 165: 124–132. https://doi.org/10.1016/j.jfoodeng.2015.06.013

Bhattacharjee C, Saxena VK, Dutta S (2019) Novel thermal and non-thermal processing of watermelon juice. Trends Food Sci Technol 93:234–243. https://doi.org/10.1016/j.tifs.2019.09.015

Bisconsin-Junior A, Rosenthal A, Monteiro M (2014) Optimisation of high hydrostatic pressure processing of Pêra Rio Orange juice. Food Bioprocess Technol 7:1670–1677. https://doi.org/10.1007/s11947-013-1176-7

Bisht B, Bhatnagar P, Gururani P, Kumar V, Tomar MS, Sinhmar R, Rathi N, Kumar S (2021) Food irradiation: effect of ionizing and non-ionizing radiations on preservation of fruits and vegetables– a review. Trends Food Sci Technol 114:372–385. https://doi.org/10.1016/j.tifs.2021.06.002

Bobelyn E, Hertog MLATM, Nicolaï BM (2006) Applicability of an enzymatic time temperature integrator as a quality indicator for mushrooms in the distribution chain. Postharvest Biol Technol 42:104–114. https://doi.org/10.1016/j.postharvbio.2006.05.011

Bolumar T, Georget E, Mathys A (2015) High pressure processing (HPP) of foods and its combination with electron beam processing. In: Electron beam pasteurization and complementary food processing technologies. https://doi.org/10.1533/9781782421085.2.127

Brody AL (2002) Packages that heat and cool themselves-packaging. Food Technol-Chicago, 56(7), 80–83

Calvo NL, Maggio RM, Kaufman TS (2018) Characterization of pharmaceutically relevant materials at the solid state employing chemometrics methods. J Pharm Biomed Anal 147:538–564. https://doi.org/10.1016/j.jpba.2017.06.017

Cano Marchal P, Martínez Gila D, Gámez García J, Gómez Ortega J (2013) Expert system based on computer vision to estimate the content of impurities in olive oil samples. J Food Eng 119:220–228. https://doi.org/10.1016/j.jfoodeng.2013.05.032

Caro MP, Ali MS, Vecchio M, Giaffreda R (2018) Blockchain-based traceability in Agri-Food supply chain management: A practical implementation. In: 2018 IoT vertical and topical summit on agriculture – Tuscany, IOT Tuscany 2018. https://doi.org/10.1109/IOT-TUSCANY.2018.8373021

Caroço RF, Bevilacqua M, Armagan I, Santacoloma PA, Abildskov J, Skov T, Huusom JK (2019) Raw material quality assessment approaches comparison in pectin production. Biotechnol Prog 35:e2762. https://doi.org/10.1002/btpr.2762

Chen J, Stokes JR (2012) Rheology and tribology: two distinctive regimes of food texture sensation. Trends Food Sci Technol 25:4–12. https://doi.org/10.1016/j.tifs.2011.11.006

Chen X, Chen M, Xu C, Yam KL (2019) Critical review of controlled release packaging to improve food safety and quality. Crit Rev Food Sci Nutr 59:2386–2399. https://doi.org/10.1080/10408398.2018.1453778

Chia KS, Jam MNH, Gan Z, Ismail N (2020) Pre-dispersive near-infrared light sensing in non-destructively classifying the brix of intact pineapples. J Food Sci Technol 57:4533–4540. https://doi.org/10.1007/s13197-020-04492-5

Commission E (2009) Commission Regulation (EC) No 450/2009 of 29 May 2009 on active and intelligent materials and articles intended to come into contact with food. Off J Eur Union 135:3–11

de Abreu DAP, Cruz JM, Losada PP (2012) Active and intelligent packaging for the food industry. Food Rev Intl 28:146–187. https://doi.org/10.1080/87559129.2011.595022

Delorme MM, Guimarães JT, Coutinho NM, Balthazar CF, Rocha RS, Silva R, Margalho LP, Pimentel TC, Silva MC, Freitas MQ, Granato D, Sant'Ana AS, Duart MCKH, Cruz AG (2020) Ultraviolet radiation: an interesting technology to preserve quality and safety of milk and dairy foods. Trends Food Sci Technol 102:146–154. https://doi.org/10.1016/j.tifs.2020.06.001

Di Rosa AR, Leone F, Cheli F, Chiofalo V (2017) Fusion of electronic nose, electronic tongue and computer vision for animal source food authentication and quality assessment – a review. J Food Eng 210:62–75. https://doi.org/10.1016/j.jfoodeng.2017.04.024

Djalili-Moghaddam M, Ebrahimzadeh R, Toll S (2004) Study of geometry effects in torsional rheometry of fibre suspensions. Rheol Acta 44:29–37. https://doi.org/10.1007/s00397-004-0363-x

Dong X, Wang J, Raghavan V (2020) Critical reviews and recent advances of novel non-thermal processing techniques on the modification of food allergens. Crit Rev Food Sci Nutr 61:196–210. https://doi.org/10.1080/10408398.2020.1722942

Du Z, Zeng X, Li X, Ding X, Cao J, Jiang W (2020) Recent advances in imaging techniques for bruise detection in fruits and vegetables. Trends Food Sci Technol 99:133–141. https://doi.org/10.1016/j.tifs.2020.02.024

Ehlermann DAE (2009) The RADURA-terminology and food irradiation. Food Control 20:526–528. https://doi.org/10.1016/j.foodcont.2008.07.023

Ellouze M, Augustin JC (2010) Applicability of biological time temperature integrators as quality and safety indicators for meat products. Int J Food Microbiol 138:119–129. https://doi.org/10.1016/j.ijfoodmicro.2009.12.012

Erdogdu F, Tutar M, Sarghini F, Skipnes D (2017) Effects of viscosity and agitation rate on temperature and flow field in cans during reciprocal agitation. J Food Eng 213:76–88. https://doi.org/10.1016/j.jfoodeng.2017.05.030

Eskildsen CE, Sanden KW, Wubshet SG, Andersen PV, Øyaas J, Wold JP (2019) Estimating dry matter and fat content in blocks of Swiss cheese during production using on-line near infrared spectroscopy. J Near Infrared Spectrosc 27:293–301. https://doi.org/10.1177/0967033519855436

Fakayode SO, Baker GA, Bwambok DK, Bhawawet N, Elzey B, Siraj N, Macchi S, Pollard DA, Perez RL, Duncan AV, Warner IM (2020) Molecular (Raman, NIR, and FTIR) spectroscopy and multivariate analysis in consumable products analysis1. Appl Spectrosc Rev 55:647–723. https://doi.org/10.1080/05704928.2019.1631176

Farkas J, Mohácsi-Farkas C (2011) History and future of food irradiation. Trends Food Sci Technol 22:121–126. https://doi.org/10.1016/j.tifs.2010.04.002

Feng H, Chen J, Zhou W, Rungsardthong V, Zhang X (2019) Modeling and evaluation on WSN-enabled and knowledge-based HACCP quality control for frozen shellfish cold chain. Food Control 98:348–358. https://doi.org/10.1016/j.foodcont.2018.11.050

Fischer P, Windhab EJ (2011) Rheology of food materials. Curr Opin Colloid Interface Sci 16:36–40. https://doi.org/10.1016/j.cocis.2010.07.003

Galvez JF, Mejuto JC, Simal-Gandara J (2018) Future challenges on the use of blockchain for food traceability analysis. TrAC, Trends Anal Chem 107:222–232. https://doi.org/10.1016/j.trac.2018.08.011

González-Arenzana L, Portu J, López N, Santamaría P, Gutiérrez AR, López R, López-Alfaro I (2019) Pulsed electric field treatment after malolactic fermentation of Tempranillo Rioja wines: influence on microbial, physicochemical and sensorial quality. Innov Food Sci Emerg Technol 51:57–63. https://doi.org/10.1016/j.ifset.2018.05.019

Gowen AA, Taghizadeh M, O'Donnell CP (2009) Identification of mushrooms subjected to freeze damage using hyperspectral imaging. J Food Eng 93:7–12. https://doi.org/10.1016/j.jfoodeng.2008.12.021

Horcada A, Valera M, Juárez M, Fernández-Cabanás VM (2020) Authentication of Iberian pork official quality categories using a portable near infrared spectroscopy (NIRS) instrument. Food Chem 318:126471. https://doi.org/10.1016/j.foodchem.2020.126471

Ihsanullah I, Rashid A (2017) Current activities in food irradiation as a sanitary and phytosanitary treatment in the Asia and the Pacific region and a comparison with advanced countries. Food Control 72:345–359. https://doi.org/10.1016/j.foodcont.2016.03.011

Imeri A, Khadraoui D (2018) The Security and Traceability of Shared Information in the Process of Transportation of Dangerous Goods. In: 2018 9th IFIP International Conference on New Technologies, Mobility and Security, NTMS 2018 – Proceedings. https://doi.org/10.1109/NTMS.2018.8328751

Izmirlioglu G, Ouyang B, Demirci A (2020) Utilization of pulsed UV light for inactivation of salmonella Enteritidis on shelled walnuts. LWT 134:110023. https://doi.org/10.1016/j.lwt.2020.110023

Jiménez-Sánchez C, Lozano-Sánchez J, Segura-Carretero A, Fernández-Gutiérrez A (2017) Alternatives to conventional thermal treatments in fruit-juice processing. Part 1: techniques and applications. Crit Rev Food Sci Nutr 57:501–523. https://doi.org/10.1080/10408398.2013.867828

Jung S, Tonello-Samson C (2018) Chapter 7: high hydrostatic pressure food processing: potential and limitations. In: RSC Green Chemistry. https://doi.org/10.1039/9781782626596-00251

Kalpana S, Priyadarshini SR, Maria Leena M, Moses JA, Anandharamakrishnan C (2019) Intelligent packaging: trends and applications in food systems. Trends Food Sci Technol 93:145–157. https://doi.org/10.1016/j.tifs.2019.09.008

Kartal S, Aday MS, Caner C (2012) Use of microperforated films and oxygen scavengers to maintain storage stability of fresh strawberries. Postharvest Biol Technol 71:32–40. https://doi.org/10.1016/j.postharvbio.2012.04.009

Khan I, Tango CN, Miskeen S, Lee BH, Oh DH (2017) Hurdle technology: a novel approach for enhanced food quality and safety – a review. Food Control 73:1426–1444. https://doi.org/10.1016/j.foodcont.2016.11.010

Khan MA, Salah K (2018) IoT security: review, blockchain solutions, and open challenges. Futur Gener Comput Syst 82:395–411. https://doi.org/10.1016/j.future.2017.11.022

Khan MK, Ahmad K, Hassan S, Imran M, Ahmad N, Xu C (2018) Effect of novel technologies on polyphenols during food processing. Innovative Food Sci Emerg Technol 45:361–381. https://doi.org/10.1016/j.ifset.2017.12.006

Khouryieh HA (2021) Novel and emerging technologies used by the U.S. food processing industry. Innovative Food Sci Emerg Technol. https://doi.org/10.1016/j.ifset.2020.102559

Kiani S, Minaei S, Ghasemi-Varnamkhasti M (2016) Fusion of artificial senses as a robust approach to food quality assessment. J Food Eng 171:230–239. https://doi.org/10.1016/j.jfoodeng.2015.10.007

Kramer B, Wunderlich J, Muranyi P (2019) Inactivation of listeria innocua on packaged meat products by pulsed light. Food Packag Shelf Life 21:100353. https://doi.org/10.1016/j.fpsl.2019.100353

Kuswandi B, Jayus, Restyana A, Abdullah A, Heng LY, Ahmad M (2012) A novel colorimetric food package label for fish spoilage based on polyaniline film. Food Control 25:184–189. https://doi.org/10.1016/j.foodcont.2011.10.008

Larsen H, Liland KH (2013) Determination of O_2 and CO_2 transmission rate of whole packages and single perforations in micro-perforated packages for fruit and vegetables. J Food Eng 119:271–276. https://doi.org/10.1016/j.jfoodeng.2013.05.035

Latos-Brozio M, Masek A (2020) The application of natural food colorants as indicator substances in intelligent biodegradable packaging materials. Food Chem Toxicol 135:110975. https://doi.org/10.1016/j.fct.2019.110975

Leistner L (2000) Basic aspects of food preservation by hurdle technology. Int J Food Microbiol 55:181–186. https://doi.org/10.1016/S0168-1605(00)00161-6

Leistner L, Gorris LGM (1995) Food preservation by hurdle technology. Trends Food Sci Technol 6:41–46. https://doi.org/10.1016/S0924-2244(00)88941-4

Leiva-Valenzuela GA, Mariotti M, Mondragón G, Pedreschi F (2018) Statistical pattern recognition classification with computer vision images for assessing the furan content of fried dough pieces. Food Chem 239:718–725. https://doi.org/10.1016/j.foodchem.2017.06.095

Leon A, Rosell CM, De Barber CB (2003) A differential scanning calorimetry study of wheat proteins. Eur Food Res Technol 217:13–16. https://doi.org/10.1007/s00217-003-0699-y

Leyva-Porras C, Cruz-Alcantar P, Espinosa-Solís V, Martínez-Guerra E, Piñón-Balderrama CI, Martínez IC, Saavedra-Leos MZ (2020) Application of differential scanning calorimetry (DSC) and modulated differential scanning calorimetry (MDSC) in food and drug industries. Polymers 12(1):5. https://doi.org/10.3390/polym12010005

Liu C, Liu W, Lu X, Ma F, Chen W, Yang J, Zheng L (2014) Application of multispectral imaging to determine quality attributes and ripeness stage in strawberry fruit. PLoS One 9. https://doi.org/10.1371/journal.pone.0087818

Liu SJ, Wu YN, Chan L (2020) Application of Metabonomics approach in food safety research-a review. Food Rev Intl 36:547–558. https://doi.org/10.1080/87559129.2019.1655571

Mahendran R, Ramanan KR, Barba FJ, Lorenzo JM, López-Fernández O, Munekata PES, Roohinejad S, Sant'Ana AS, Tiwari BK (2019) Recent advances in the application of pulsed light processing for improving food safety and increasing shelf life. Trends Food Sci Technol 88:67–79. https://doi.org/10.1016/j.tifs.2019.03.010

Marszałek K, Woźniak Ł, Barba FJ, Skąpska S, Lorenzo JM, Zambon A, Spilimbergo S (2018) Enzymatic, physicochemical, nutritional and phytochemical profile changes of apple (Golden delicious L.) juice under supercritical carbon dioxide and long-term cold storage. Food Chem. https://doi.org/10.1016/j.foodchem.2018.06.109

Mastromatteo M, Mastromatteo M, Conte A, Del Nobile MA (2010) Advances in controlled release devices for food packaging applications. Trends Food Sci Technol 21:591–598. https://doi.org/10.1016/j.tifs.2010.07.010

Morales-de la Peña M, Welti-Chanes J, Martín-Belloso O (2019) Novel technologies to improve food safety and quality. Curr Opin Food Sci 30:1–7. https://doi.org/10.1016/j.cofs.2018.10.009

Mukhopadhyay S, Sokorai K, Ukuku DO, Jin T, Fan X, Olanya OM, Leng J, Juneja V (2021) Effects of direct and in-package pulsed light treatment on inactivation of E. coli O157:H7 and reduction of microbial loads in Romaine lettuce. LWT. https://doi.org/10.1016/j.lwt.2020.110710

Nopwinyuwong A, Trevanich S, Suppakul P (2010) Development of a novel colorimetric indicator label for monitoring freshness of intermediate-moisture dessert spoilage. Talanta 81:1126–1132. https://doi.org/10.1016/j.talanta.2010.02.008

Oh DH, Khan I, Tango CN (2019) Hurdle enhancement of electrolyzed water with other techniques. In: Electrolyzed water in food: fundamentals and applications. https://doi.org/10.1007/978-981-13-3807-6_10

Olsen P, Borit M (2018) The components of a food traceability system. Trends Food Sci Technol 77:143–149. https://doi.org/10.1016/j.tifs.2018.05.004

Pal M (2017) Pulsed electric field processing: an emerging technology for food preservation. J Exp Food Chem. https://doi.org/10.4172/2472-0542.1000126

Pallarés N, Berrada H, Tolosa J, Ferrer E (2021) Effect of high hydrostatic pressure (HPP) and pulsed electric field (PEF) technologies on reduction of aflatoxins in fruit juices. LWT 142:111000. https://doi.org/10.1016/j.lwt.2021.111000

Pedreschi F, Mariotti-Celis MS (2020) Irradiation kills microbes: can it do anything harmful to the food? In: Genetically modified and irradiated food: controversial issues: facts versus perceptions. https://doi.org/10.1016/B978-0-12-817240-7.00014-0

Pedrós-Garrido S, Condón-Abanto S, Clemente I, Beltrán JA, Lyng JG, Bolton D, Brunton N, Whyte P (2018) Efficacy of ultraviolet light (UV-C) and pulsed light (PL) for the microbiological decontamination of raw salmon (Salmo salar) and food contact surface materials. Innov Food Sci Emerg Technol 50:124–131. https://doi.org/10.1016/j.ifset.2018.10.001

Pérez-Baltar A, Serrano A, Montiel R, Medina M (2020) Listeria monocytogenes inactivation in deboned dry-cured hams by high pressure processing. Meat Sci 160:107960. https://doi.org/10.1016/j.meatsci.2019.107960

Peri C (2006) The universe of food quality. Food Qual Prefer 17(1–2):3–8. https://doi.org/10.1016/j.foodqual.2005.03.002

Peri C, Lavelli V, Marjani A (2004) Qualità nelle aziende e nelle filiere agroalimentari. Gestione e certificazione dei sistemi per la qualità, per la rintracciabilità e per l'igiene. HOEPLI EDITORE

Picart-Palmade L, Cunault C, Chevalier-Lucia D, Belleville MP, Marchesseau S (2019) Potentialities and limits of some non-thermal technologies to improve sustainability of food processing. Front Nutr 5. https://doi.org/10.3389/fnut.2018.00130

Prakash S (2017) From rheology to tribology: applications of tribology in studying food oral processing and texture perception. In: Advances in food rheology and its applications. https://doi.org/10.1016/B978-0-08-100431-9.00004-8

Pratap Singh A, Ramaswamy HS (2016) Simultaneous optimization of heat transfer and reciprocation intensity for thermal processing of liquid particulate mixtures undergoing reciprocating agitation. Innov Food Sci Emerg Technol 33:405–415. https://doi.org/10.1016/j.ifset.2015.10.011

Pratap Singh A, Singh A, Ramaswamy HS (2017) Effect of reciprocating agitation thermal processing (RA-TP) on quality of canned tomato (Solanum lycopersicum) puree. J Sci Food Agric 97:2411–2418. https://doi.org/10.1002/jsfa.8054

Qiu L, Zhang M, Tang J, Adhikari B, Cao P (2019) Innovative technologies for producing and preserving intermediate moisture foods: a review. Food Res Int 116:90–102. https://doi.org/10.1016/j.foodres.2018.12.055

Ramos-Villarroel AY, Aron-Maftei N, Martín-Belloso O, Soliva-Fortuny R (2012) Influence of spectral distribution on bacterial inactivation and quality changes of fresh-cut watermelon treated with intense light pulses. Postharvest Biol Technol 69:32–39. https://doi.org/10.1016/j.postharvbio.2012.03.002

Randazzo W, Fabra MJ, Falcó I, López-Rubio A, Sánchez G (2018) Polymers and biopolymers with antiviral activity: potential applications for improving food safety. Compr Rev Food Sci Food Saf 17(3):754–768. https://doi.org/10.1111/1541-4337.12349

Rathore AS (2014) QbD/PAT for bioprocessing: moving from theory to implementation. Curr Opin Chem Eng 6:1–8. https://doi.org/10.1016/j.coche.2014.05.006

Rathore AS, Mittal S, Pathak M, Arora A (2014) Guidance for performing multivariate data analysis of bioprocessing data: pitfalls and recommendations. Biotechnol Prog 30:967–973. https://doi.org/10.1002/btpr.1922

Rattan NS, Ramaswamy HS (2014) Comparison of free/bi-axial, fixed axial, end-over-end and static thermal processing effects on process lethality and quality changes in canned potatoes. LWT Food Sci Technol 58:150–157. https://doi.org/10.1016/j.lwt.2014.02.056

Rennie TJ, Sunjka PS (2017) Modified atmosphere for storage, transportation, and packaging. In: Novel postharvest treatments of fresh produce. https://doi.org/10.1201/9781315370149

Riva M, Piergiovanni L, Schiraldi A (2001) Performances of time-temperature indicators in the study of temperature exposure of packaged fresh foods. Packag Technol Sci 14:1–9. https://doi.org/10.1002/pts.521

Rocha-Pimienta J, Martillanes S, Ramírez R, Garcia-Parra J, Delgado-Adamez J (2020) Bacillus cereus spores and Staphylococcus aureus sub. Aureus vegetative cells inactivation in human milk by high-pressure processing. Food Control. https://doi.org/10.1016/j.foodcont.2020.107212

Rong D, Rao X, Ying Y (2017) Computer vision detection of surface defect on oranges by means of a sliding comparison window local segmentation algorithm. Comput Electron Agric 137:59–68. https://doi.org/10.1016/j.compag.2017.02.027

Rostami Z, Ahmad MA, Khan MU, Mishra AP, Rashidzadeh S, Shariati MA (2016) Food preservation by hurdle technology: a review of different hurdle and interaction with focus on foodstuffs. J Pure Appl Microbiol 10:2633–2639. https://doi.org/10.22207/JPAM.10.4.20

Salehi F (2020) Physico-chemical properties of fruit and vegetable juices as affected by pulsed electric field: a review. Int J Food Prop 23:1036–1050. https://doi.org/10.1080/10942912.2020.1775250

Santhirasegaram V, Razali Z, Somasundram C (2016) Safety improvement of fruit juices by novel thermal and nonthermal processing. In: Food hygiene and toxicology in ready-to-eat foods. https://doi.org/10.1016/B978-0-12-801916-0.00012-1

Schaefer D, Cheung WM (2018) Smart packaging: opportunities and challenges. Procedia CIRP. https://doi.org/10.1016/j.procir.2018.03.240

Selway N, Stokes JR (2013) Insights into the dynamics of oral lubrication and mouthfeel using soft tribology: differentiating semi-fluid foods with similar rheology. Food Res Int 54:423–431. https://doi.org/10.1016/j.foodres.2013.07.044

Shafiee S, Minaei S, Moghaddam-Charkari N, Barzegar M (2014) Honey characterization using computer vision system and artificial neural networks. Food Chem 159:143–150. https://doi.org/10.1016/j.foodchem.2014.02.136

Singh A, Pratap Singh A, Ramaswamy HS (2016) A controlled agitation process for improving quality of canned green beans during agitation thermal processing. J Food Sci 81:E1399–E1411. https://doi.org/10.1111/1750-3841.13308

Singh A, Singh AP, Ramaswamy HS (2015) Effect of processing conditions on quality of green beans subjected to reciprocating agitation thermal processing. Food Res Int 78:424–432. https://doi.org/10.1016/j.foodres.2015.08.040

Singh CB, Jayas DS, Paliwal J, White NDG (2012) Fungal damage detection in wheat using short-wave near-infrared hyperspectral and digital colour imaging. Int J Food Prop 15:11–24. https://doi.org/10.1080/10942911003687223

Singh PA, Singh A, Ramaswamy HS (2017) Effect of reciprocating agitation thermal processing (RA-TP) on quality of canned tomato (Solanum lycopersicum) puree. J Sci Food Agric. https://doi.org/10.1002/jsfa.8054

Stoforos NG (2010) Thermal process calculations through ball's original formula method: a critical presentation of the method and simplification of its use through regression equations. Food Engineering Reviews 2(1):1–16. https://doi.org/10.1007/s12393-010-9014-4

Soltanzadeh M, Peighambardoust SH, Gullon P, Hesari J, Gullón B, Alirezalu K, Lorenzo J (2020) Quality aspects and safety of pulsed electric field (PEF) processing on dairy products: a comprehensive review. Food Rev Int 1:1–22. https://doi.org/10.1080/87559129.2020.1849273

Song HY, Salehiyan R, Li X, Lee SH, Hyun K (2017) A comparative study of the effects of cone-plate and parallel-plate geometries on rheological properties under oscillatory shear flow. Korea Aust Rheol J 29:281–294. https://doi.org/10.1007/s13367-017-0028-9

Sportelli MC, Izzi M, Kukushkina EA, Hossain SI, Picca RA, Ditaranto N, Cioff N (2020) Can nanotechnology and materials science help the fight against sars-cov-2? Nano 10. https://doi.org/10.3390/nano10040802

Srivastava S, Vaddadi S, Sadistap S (2015) Quality assessment of commercial bread samples based on breadcrumb features and freshness analysis using an ultrasonic machine vision (UVS) system. J Food Meas Charact 9:525–540. https://doi.org/10.1007/s11694-015-9261-4

Szczepańska J, Barba FJ, Skąpska S, Marszałek K (2020) High pressure processing of carrot juice: effect of static and multi-pulsed pressure on the polyphenolic profile, oxidoreductases activity and colour. Food Chem 307:125549. https://doi.org/10.1016/j.foodchem.2019.125549

Tao T, Ding C, Han N, Cui Y, Liu X, Zhang C (2019) Evaluation of pulsed light for inactivation of foodborne pathogens on fresh-cut lettuce: effects on quality attributes during storage. Food Packag Shelf Life 21:100358. https://doi.org/10.1016/j.fpsl.2019.100358

Tattiyakul J, Rao MA, Datta AK (2002) Heat Transfer to Three Canned Fluids of Different Thermo-Rheological Behaviour Under Intermittent Agitation. Food and Bioproducts Processing, 80(1), 20–27. https://doi.org/10.1205/096030802753479070

Timmermans RAH, Mastwijk HC, Berendsen LBJM, Nederhoff AL, Matser AM, Van Boekel MAJS, Nierop Groot MN (2019) Moderate intensity pulsed electric fields (PEF) as alternative mild preservation technology for fruit juice. Int J Food Microbiol 298:63–73. https://doi.org/10.1016/j.ijfoodmicro.2019.02.015

Tsironi T, Houhoula D, Taoukis P (2020) Hurdle technology for fish preservation. Aquac Fish 5: 65–71. https://doi.org/10.1016/j.aaf.2020.02.001

Unluturk S, Atilgan MR, Baysal AH, Unluturk MS (2010) Modeling inactivation kinetics of liquid egg white exposed to UV-C irradiation. Int J Food Microbiol 142:341–347. https://doi.org/10.1016/j.ijfoodmicro.2010.07.013

Usaga J, Acosta Ó, Churey JJ, Padilla-Zakour OI, Worobo RW (2021) Evaluation of high pressure processing (HPP) inactivation of Escherichia coli O157:H7, salmonella enterica, and listeria

monocytogenes in acid and acidified juices and beverages. Int J Food Microbiol 339:109034. https://doi.org/10.1016/j.ijfoodmicro.2020.109034
Vasuja S, Kumar VS (2018) Ultra violet irradiation and its applications in food processing industries: a review. Int J Trend Res Dev, 5(1), 343–346
Verma DK, Thakur M, Kumar J, Srivastav PP, Al-Hilphy ARS, Patel AR, Suleria HAR (2020) High pressure processing (HPP): fundamental concepts, emerging scope, and food application. In: Emerging thermal and nonthermal Technologies in food processing. Apple Academic Press, pp 225–257. https://doi.org/10.1201/9780429297335-9
Vong WC, Liu SQ (2016) Biovalorisation of okara (soybean residue) for food and nutrition. Trends Food Sci Technol 52:139–147. https://doi.org/10.1016/j.tifs.2016.04.011
Walden R, Emanuel J (2010) Developments in in-container retort technology: the Zinetec Shaka® process. In: Case studies in novel food processing technologies. https://doi.org/10.1533/9780857090713.4.389
Wojtkiewicz S, Gerega A, Zanoletti M, Sudakou A, Contini D, Liebert A, Durduran T, Dehghani H (2019) Self-calibrating time-resolved near infrared spectroscopy. Biomed Opt Express 10:2657. https://doi.org/10.1364/boe.10.002657
Xie C, He Y (2016) External characteristic determination of eggs and cracked eggs identification using spectral signature. Sci Rep 6. https://doi.org/10.1038/srep21130
Xuan XT, Ding T, Li J, Ahn JH, Zhao Y, Chen SG, Ye XQ, Liu DH (2017) Estimation of growth parameters of listeria monocytogenes after sublethal heat and slightly acidic electrolyzed water (SAEW) treatment. Food Control 71:17–25. https://doi.org/10.1016/j.foodcont.2016.06.018
Yang C (2012) A high-resolution airborne four-camera imaging system for agricultural remote sensing. Comput Electron Agric 88:13–24. https://doi.org/10.1016/j.compag.2012.07.003
Yildirim S, Röcker B, Pettersen MK, Nilsen-Nygaard J, Ayhan Z, Rutkaite R, Radusin T, Suminska P, Marcos B, Coma V (2018) Active packaging applications for food. Compr Rev Food Sci Food Saf 17:165–199. https://doi.org/10.1111/1541-4337.12322
You J, Singh AP, Singh A, Ramaswamy HS (2016) Quality retention enhancement in canned potato and radish using reciprocating agitation thermal processing. Int J Food Eng 12:491–500. https://doi.org/10.1515/ijfe-2015-0276
Zhang M, Meng X, Bhandari B, Fang Z, Chen H (2015) Recent application of modified atmosphere packaging (MAP) in fresh and fresh-cut foods. Food Rev Intl 31:172–193. https://doi.org/10.1080/87559129.2014.981826
Zhao G, Liu S, Lopez C, Lu H, Elgueta S, Chen H, Boshkoska BM (2019) Blockchain technology in Agri-food value chain management: a synthesis of applications, challenges and future research directions. Comput Ind 109:83–99. https://doi.org/10.1016/j.compind.2019.04.002
Zheng Q, Zhang Y, Montazerian M, Gulbiten O, Mauro JC, Zanotto ED, Yue Y (2019) Understanding glass through differential scanning calorimetry. Chem Rev 119:7848–7939. https://doi.org/10.1021/acs.chemrev.8b00510

Chapter 2
Chemical Adulterants in Food: Recent Challenges

Prasanna Vasu and Asha Martin

1 Introduction

Food is very much essential for the sustenance of life. It gives adequate nourishment to human body and maintains good health. In earlier days, the main emphasis was on adequate nutrition, i.e., people should get enough food for their growth and maintenance, which culminated in several revolutions like Green Revolution, and White Revolution. Now the concept has changed to safety and optimal nutrition, which indicate that along with adequacy, food also should be safe and wholesome. Thus, the food which is developed for taste and hunger has now transitioned to health, nutrition and performance specific with a value-addition of being 'Functional Foods' (Katiyar et al. 2014). Unsafe food constitutes a major safety concern and growing public health risks causing acute and life-long chronic diseases (Spink and Moyer 2011).

Food quality and safety are of prime importance to both consumers and food manufacturers. Food quality that influence product value to the consumers, focuses on the deterioration of food or its unintentional spoilage, mainly attributed to deviation in specific product characteristics from the reference standards. Additionally, food safety also focuses on the unintentional contamination of food during production, processing, storage, or transportation, by known or unknown chemicals, or microorganisms (Spink and Moyer 2011). According to World Health Organization (WHO) and Food and Agricultural Organization (FAO), food safety is defined

P. Vasu
Department of Food Safety and Analytical Quality Control Laboratory, CSIR – Central Food Technological Research Institute, Mysore, Karnataka, India

A. Martin (✉)
Department of Biochemistry, CSIR – Central Food Technological Research Institute, Mysore, Karnataka, India
e-mail: ashamartin@cftri.res.in

H. U. Hebbar et al. (eds.), *Engineering Aspects of Food Quality and Safety*, Food Engineering Series, https://doi.org/10.1007/978-3-031-30683-9_2

as 'absence or acceptable safe levels of contaminants, adulterants, naturally occurring toxins, or any other substance that may make food injurious to health on an acute or chronic basis'. Food safety is basically an assurance that food will not cause any negative or adverse effect on health, when prepared, or consumed as per its intended use. For avoiding food safety issues, following the good manufacturing (GMP) and good hygiene practices (GHP) are deemed necessary.

Today, humanity is facing many challenges worldwide primarily driven by the rapidly growing world population, which has crossed 7.9 billion (UN Report 2021). Consequently, the rising global demand for food runs parallelly with increased competition for diminishing resources such as land, water and energy. Globalization and climate change contributed immensely for the spread of pathogens generating increased uncertainty in the food supply (FAO 2015). The spread of human influenza viruses that spillover from poultry and swine is a good example (Hosseini et al. 2010). Climate change has the potential to severely impact food production and animal welfare by altering the distribution and severity of pests and diseases in crops and livestock (Skovgaard 2007). Studies have indicated a strong association between climate change i.e., increasing air and water temperatures resulting in altered and extended summer season and non-cholera *Vibrio* species (spp.) infections (Smith and Fazil 2019). The spread of health risks in a globalized world is characterized by increasing transmission opportunities for pathogens that impact agricultural and livestock production (Giangaspero et al. 2019). The foremost sustainable development goal is to ensure agri-food security, which can be achieved by producing a sufficient and safe food supply. However, food security may become hampered by alarmingly increasing food safety incidents, most of which are mainly due to fraudulent practice of adulteration.

Adulteration is an ancient cheat, where food contains prohibited substances in crude form, either partly or wholly added or substituted in it. Since ages, several notable adulteration incidents made a huge impact in societies and food industries and have culminated in serious health risks to consumers. Scientists and researchers have found references to food adulteration in the laws of Moses and the early literature of China, Greece and Rome (Forte 1966). Several ancient texts in India, especially in Charak Samhita, there are references pertaining to quality of food for maintaining a good health. In 375 BC, the great economist Chanakya in his 'Arthashastra' referred to adulteration of food and punishments given to traders indulging in fraudulent activities (Sattigeri and Appaiah 2010). Food adulteration has been the main focus of the very first food laws of Roman times. Some food adulteration cases which involved olive oil, wine, spices, and tea dates back thousands of years. The economic adulteration of food that increased considerably in the 1800s and early 1900s, resulted from the rampant reporting of milk diluted with water, maple syrup diluted with cane sugar or glucose, coffee diluted with chicory and other roasted vegetable products, and spices mixed with ground wheat and corn. It was during this period of public resentment which made US government to pass the first prophylactic legislation preventing the debasement of foods (Forte 1966).

Adulterated food is dangerous as it may be toxic affecting health and it could deprive the essential nutrients required for proper growth and development. The most common foods which are subjected to illegitimate adulteration include honey, milk and dairy products, oil, fish, meat, wine and alcoholic beverages, grain-based foods, fruit juices, organic foods, spices, coffee, tea, and many others (Zambonin 2021). As it is not possible to cover all the chemical adulterants, selected examples of important food adulterants are discussed in general and the issues associated with four of the most highly adulterated commodities in the world namely milk, spice, olive oil and honey are described in detail in this chapter.

2 Food Fraud, Adulteration and the Regulatory Framework

Food on sale for human consumption should be safe, wholesome, unadulterated, uncontaminated, properly labelled and fit for consumption. As per the United States Pharmacopeia Convention (USP) 'food fraud' refers to the fraudulent addition of non-authentic substances or removal or replacement of authentic substances without the purchaser's knowledge for economic gain of the seller. The European Commission defines food frauds as "intentional actions by businesses or individuals for the purpose of deceiving purchasers and gaining undue advantage there from, in violation of the rules referred to in Article 1(2) of Regulation (EU) 2017/625 (the agri-food chain legislation)". Food fraud is a collective term and includes economically motivated food adulteration. Food adulteration is defined as a process by which the quality or the nature of food is lessened through the addition of a foreign or an inferior substance or by the removal of valuable substances from food articles. Most countries have strict and comprehensive legislations governing food adulteration. In USA, adulterated food is generally defined as unsafe, impure or unwholesome food. Codex Alimentarius, the Codex Food Standards, is the 'food book,' having a set of internationally recognised voluntary standards, codes of practice and guidelines which protect consumers from unsafe food and fraudulent practices. Established in 1962, Codex Alimentarius is a science based international organization, and its standards and guidelines are recognized globally for their key role in protecting the consumers and facilitating international trade. Harmonization of food standards is important for getting the countries of the world to agree on food codes (Spink 2014). People everywhere have the basic right of access to good quality, nutritious and safe food. Codex Alimentarius supports to reach this aim by combining consumer protection with food production and trade at global, national, regional and local levels. WTO recognized Codex Alimentarius as an international reference point for the resolution of disputes pertaining to food safety and consumer protection. It provides reassurance that foods when produced in accord with codes of hygienic practice and complying with its standards and guideline are safe and nutritious and provide adequate health protection. The adoption of the Food Safety and Standards Act in 2006 (FSSA) and the establishment of the Food Safety and Standards Authority of India (FSSAI), has enabled India to rise to the challenges concerning food safety and adulteration. According to the FSSA, 2006 and FSS

Regulations 2011, adulterant is defined as 'any material which is or could be employed for making the food unsafe or sub-standard or mis-branded or containing extraneous matter.' Violation of these provisions may lead to regulatory action against any concerned food business operator under the act, rules and regulations.

3 Types of Food Adulteration

One of the most common frauds according to the European Commission, is food adulteration. The different ways of food adulteration are "replacing a nutrient, an ingredient, a food or part of a food with another one with lower value" (substitution), "mixing an ingredient with high value with an ingredient with a lower value" (dilution), "adding unknown and undeclared compounds to food products in order to enhance their quality attributes" (unapproved enhancement), and "hiding the low quality of food ingredients or products" (concealment) (Spink and Moyer 2011; Valletta et al. 2021). Some examples of the types of adulteration routinely encountered are extensively reviewed by Bansal et al. (2017), and are listed below;

1. **Replacement**: Partial or complete replacing of a valuable food ingredient or authentic constituent with cheaper substitute. For examples: inclusion of melamine to milk to inflate protein contents; addition of water and citric acid to lemon juice to increase titratable acidity; and over treating frozen fish with ice to adds extra water and weight. Substituting sheep's or goat's milk with less expensive cow's milk; substituting common wheat for durum wheat, substituting synthetically produced vanillin for botanically derived natural vanillin, and labeling of Greek or Turkish olive oil as Italian olive oil. Black pepper is adulterated with papaya seeds (Fig. 2.1a, b) and black berries due to their similar size. Argemone seeds are used as adulterant in mustard due to their remarkable resemblance (Fig. 2.1c, d) to add bulk and weight. Cinnamon (*Cinnamomum verum*) is adulterated with an inferior quality cassia cinnamon (*Cinnamomum aromaticum*) (Fig. 2.1g, h), which has health implications due to the presence of high levels of coumarins.
2. **Addition**: Masking inferior quality ingredient by adding small amounts of a non-authentic substance. For example, adding red color additive (such as Sudan Red dyes) for enhancing the poor color quality of paprika, malachite green dyes to old vegetables, and metanil yellow colour to turmeric powder. Coal tar synthetic colours like metanil yellow are added to tea powder to mask the substandard tea leaves (Fig. 2.1e, f).
3. **Removal**: Removing or intentionally omitting a valuable constituent without consumer's knowledge. For example, removing non-polar constituents like lipids and flavour compounds from paprika, to produce "defatted" paprika, which lacks valuable flavouring compounds, and removal of valuable and expensive oleoresin from chili powder.

Fig. 2.1 Some of the highly adulterated foods along with their major adulterants found. Pepper (**a**) is adulterated with papaya seeds (**b**); Mustard (**c**) is adulterated with Argemone seeds (**d**); Tea powder (**e**) and tea powder adulterated with metanil yellow (**f**); Cinnamon (**g**) is adulterated with Cassia bark (**h**)

4 Food Adulteration: Causes, Health Hazards and Mitigation Strategies

4.1 Causes

Adulterants can enter the food supply chain from various sources. Adulterants can be classified into two main categories namely intentional and unintentional or accidental or incidental adulterants. Adulteration may be unintentional or accidental or incidental in which the presence of unwanted substances is caused mainly due to ignorance, negligence, or lack of proper facilities and hygiene during processing of food. On the other hand, the intentional adulteration is a willful act on the part of fraudulent adulterator whose intention is to increase profit margin by inclusion or addition of inferior substances having similar properties with those foods in which these adulterants are added, and hence is difficult to detect. In fact, any commodity that commands a premium price in the market and is either expensive or has some high volume sales, is a target for adulteration. Adulteration of milk and milk products, edible oils and fats, and spices is most common due to the high total sales value of these food items. Food articles, which are in a powder, minced, or paste form, are more susceptible to adulteration, since it is more difficult for the naked eye to detect adulteration in these foodstuffs. Adulteration of foods sold loose by the retailer is also more common as compared to the packaged food items. However, in some cases, packaged food may also be adulterated, as it is easy to tamper the package. The main causes of adulteration are; (1) availability of too many similar products, (2) poor buying practices, (3) bargaining mentality of consumer, (4) easy access to adulterants, (5) profit motive of traders, (6) food insecurity, (7) increased urbanization, (8) high population demands, (9) illiteracy of general public, (10) lack of effective food laws and enforcement, and (11) lack of suitable test methods (England et al. 2020; Chachan et al. 2021; Bansal et al. 2017).

4.2 Health Hazards of Food Adulterants

An exhaustive discussion of this vital topic is beyond the scope of this chapter but it is important to know the health implications. There are three important adverse effects humans might have following consumption of adulterated food viz., (a) it may be toxic and affect health (b) it may deprive nutrients required for proper health, or (c) it may cause intoxication or allergy in sensitized individuals (Bansal et al. 2017). Some specific effects of metallic and other food contaminants and adulterants on human health are as follows;

(1) Lead chromate contaminated turmeric powder and spices can cause anaemia, paralysis, brain damage and abortions. Lead contaminated water and processed food can cause lead poisoning complications such as foot drop, constipation, anaemia, insomnia, and mental retardation.

(2) Cobalt contaminated water and liquors can cause cardiac damage.
(3) Copper, tin and zinc contaminated food can cause colic, diarrhoea and vomiting in humans.
(4) Mercury contaminated grains or fish can cause brain damage, paralysis and death.
(5) Cadmium contaminated in fruit juices, soft drinks, water and shell-fish can cause itai-itai (ouch-ouch) disease, increased salivation, acute gastritis, liver and kidney damage, and prostate cancer.
(6) Mineral oil adulterated edible oil and fats can cause several types of cancers.
(7) Food contaminated with permitted food colour beyond the safe limit and non-permitted colour such as metanil yellow, may cause hyperactivity, allergies, liver damage, infertility, anaemia, cancer and birth defects.
(8) Ice creams are adulterated with pepperonil, ethyl acetate, butraldehyde, nitrate, washing powder, which are dangerous to human health, causing terrible diseases affecting lungs, kidneys and heart.
(9) Chick pea flour is prone to adulteration with kesari dhal which contains a neurotoxin β-N-oxalyl L-α, β-diaminopropanoic acid (β -ODAP). Prolonged consumption of kesari dhal leads to neurolathyrism, a neurodegenerative disease (Thippeswamy et al. 2007).
(10) Consumption of mustard oil adulterated with argemone oil leads to epidemic dropsy characterized by oedema of the legs, erythema, and gastrointestinal disturbances which are attributed to the presence of sanguinarine, a toxic alkaloid in argemone.

4.3 Mitigation Strategies

Recently, food industries have started to conduct mapping of raw materials vulnerabilities by routine analysis to identify and prioritize major risks and to develop food adulteration mitigation strategies. For example, in case of carcinogenic 3-MCPD (3-monochloropropane-1,2- diol) and its fatty acid ester derivatives, found in food stuffs like bread, infant formula, refined vegetable oils, etc., the food industries have taken action to implement control of its contents and derivatives, regardless of regulatory requirements. Raw materials with the low 3-MCPD esters are being supplied by the oil and fat industries (Jala et al. 2015).

Some key mitigation strategies for addressing food fraud and adulteration are as follows;

- Proper surveillance and enforcement of the implemented food laws.
- Periodical examination and monitoring of the adulteration activities with hazard records.
- Conducting periodical training programmes for concerned personnel of food safety
- Conducting consumer awareness programmes by organizing exhibitions, seminars, or training programmes.

- Strict actions and punishment to those who are involved in food adulteration and fraud.

5 Methods of Detection of Food Adulterants and Emerging Challenges

In addition to economic problems, adulteration leads to serious health complications to the consumers. Since the methods of foods adulteration have become more advanced, very efficient and improved detection techniques for harmful and toxic adulterants are required. Thus, many analytical methods have been used for detecting food adulteration, which include titrimetry, chromatography, spectroscopy, and recent advanced techniques like stable isotope ratio analysis, metabolomics, proteomics, enzymatic methods, and DNA-based techniques (Schieber 2018).

The increasing number of emerging food adulterants has raised concerns about food safety which culminated in tremendous improvements in novel analytical methodologies. Sensitive, fast and cost-effective detection techniques must be developed and utilized to prevent food fraud and to improve food safety. Various chemical, biochemical and molecular-based analytical techniques are summarized in Table 2.1. These sensitive methods, which are based mainly on chromatography, spectroscopy, and stable isotope ratio mass spectrometry are designed for detection of adulterants in foods. However, its applicability in industries is hampered due to high capital cost, and need of expertise and training in handling such instrument (Gonzalez et al. 2003). Many analytical methods for detecting food adulteration require elaborate, difficult, or time-consuming sample preparation steps prior to analysis involving high-end technologies (Banerjee et al. 2017). The various steps involved are; extraction, cleanup, chromatographic separation and selective detection.

HPLC (High-performance liquid chromatography) is the most widely used technique, which can separate various chemical constituents from mixtures. Some examples where adulterant has been detected by HPLC include hazelnut oil in olive oil (Blanch et al. 1998), flavones glycosides in citrus juices (Mouly et al. 1998), phenolic pigments in black tea liquors (McDowell et al. 1995), proline isomers and amino acids in wines and so on (Calabrese et al. 1995). We have previously reported a HPLC method for the detection of Khesari dhal (grass pea) in pulses and processed foods (Thippeswamy et al. 2007; Mishra et al. 2009). This method is based on the detection of β-ODAP, a neurotoxin found in khesari dhal. Representative HPLC chromatogram of detection of khesari dhal adulteration in chickpea is depicted in Fig. 2.2.

Gas chromatography (GC) is mainly used to separate volatile organic compounds. GC coupled with mass spectroscopy (MS) and Fourier transform infrared spectroscopy (FTIR) has been extensively utilized for quantification of adulterants.

Table 2.1 Analytical techniques routinely used for detection of food adulteration

Analytical techniques	Adulterants	Type of adulteration	Category adulteration
Chromatography-based			
TLC	Sudan red dyes in spices	Intentional	Concealment
HPLC	Kesari dhal (β-ODAP) in dhals,	Intentional	Substitution
	Argemone oil (Sanguinarine) in mustard oil	Intentional	Substitution
	Synthetic dyes in sweets, spices	Intentional	Concealment
LC-MS/MS	Meat adulteration	Intentional	Substitution
GC	Lard in virgin coconut oil	Intentional	Substitution
GC-MS	Hazelnut oil in extra olive oil	Intentional	Substitution
Spectroscopy-based			
UV –visible	Synthetic colors in foods	Intentional	Concealment
	γ-oryzanol in mustard oil	Incidental/ Intentional	Substitution/ Incidental
FT-IR	Roasted coffee with roasted barley	Intentional	Substitution
NMR	Melamine in milk	Intentional	Substitution
AAS	Heavy metals in foods	Incidental	Incidental
ICP-AES	Mercury, lead, cadmium, arsenic in foods	Incidental	Incidental
Stable isotope EA-IRMS & LC-IRMS	C3 and C4 sugars adulteration in honey	Intentional	Substitution
Electrophoresis-based			
Capillary Electrophoresis	Ordinary rice in basmati rice	Incidental/ Intentional	Substitution
Immunological -based			
ELISA	Bovine milk in goat's milk	Intentional	Substitution
DNA-based			
PCR, RT-PCR	Meat adulteration	Intentional	Substitution
	Basmati rice adulteration	Intentional	Substitution
	GM crops and food derived therefrom	Incidental/ Intentional	Incidental/ Substitution
	Spice adulteration	Intentional	Substitution

Additionally, GC is used for authentication and identification, i.e., discriminate among different varieties of the same product. It is employed to differentiate wines from same or different geographical regions. Volatile compounds in wine such as 1-propanol, 2-methyl-1-propanol, 2-propen-1-ol, and 3-methyl-1-butanol were quantified by GC or GC-MS for pattern classification (Nogueira and Nascimento 1999). GC-MS is used to detect adulteration of ground-roasted coffee with roasted barley (Oliveira et al. 2009; Matute et al. 2007).

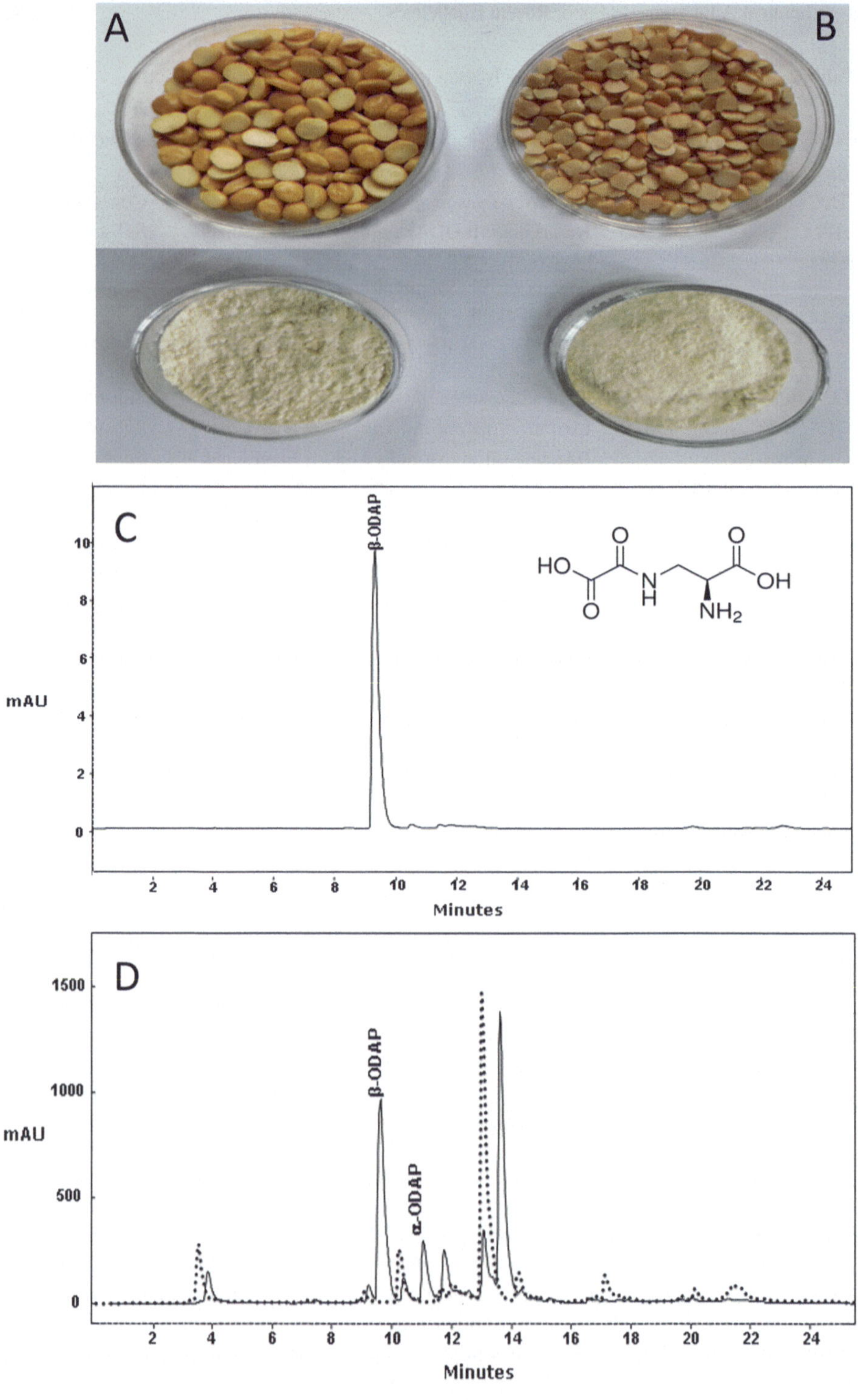

Fig. 2.2 Detection of kesari dhal (Grass pea) adulteration in chick pea flour by reverse phase HPLC. Panel A: Chickpea and grass pea unprocessed intact seeds (upper panel) can be

A wide range of spectroscopic techniques are in vogue for detection of food adulteration. UV spectroscopy has been used to determine added synthetic colors in food stuffs. Spectrophotometer can be utilized to determine γ- oryzanol content (%) in oils using a wavelength of 315 nm (Codex Alimentarious 2010). Among the non-destructive spectroscopic techniques, near infrared spectroscopy (NIR) rapidly detects adulterants in raw material, though it will not identify the adulterants. Fourier transform infrared spectroscopy (FTIR) has shown excellent potential for detecting adulterants in milk (Ozen and Mauer 2002). Another non-destructive method, nuclear magnetic resonance (NMR) not only detects food adulterants but also provide structural identification and confirmation (Bansal et al. 2017). NMR is based on the detection of proton nuclei, and has been used for the authentication of honey (Hatzakis 2019). Atomic Absorption Spectrometry can be utilized as method to analyze metals such as lead in foods except oils, fats and extremely fatty products. Similarly, Inductive Coupled Plasma–Mass Spectrometry (ICP-MS) can be exhaustively used for analysis of metals like lead in all the food articles (FAO/WHO 2014).

Electrophoresis has been widely utilized in the detection of food adulteration, especially to detect and quantify added whey proteins in milk and dairy beverages (De-Souza et al. 2000). Capillary electrophoresis has the capability to detect various adulterants and to analyze authenticity. Adulteration of cow milk in goat milk products, the origin of species of milk, and adulteration in rice and basmati rice are few examples (Cartoni et al. 1999; Vemireddy et al. 2007). Enzyme-linked immunosorbent assay (ELISA) are immunoassays extensively used for detection of food adulterants, owing to its high sensitivity and speed. ELISA kits have been developed for the quantification of bovine milk adulteration in goat's milk (Xue et al. 2010). DNA-based methods especially the polymerase chain reaction (PCR) have the potential to match the chemical techniques and thus laboratories are now taking advantage of the rapid development in DNA techniques. However, only a few DNA based methods have been proved robust enough to be employed for adulteration detection.

6 Adulteration Associated with Specific Food Commodities

Milk, spices, olive oil and honey are the four important food commodities which are frequently adulterated. The high nutritional value, gap between demand and supply, the perishable nature of these commodities, and more importantly dearth of

Fig. 2.2 (continued) distinguished on the basis of morphology, however, when finely milled (lower panel) they are indistinguishable. Panel B**:** Representative chromatogram for β-ODAP (inset) standard at 340 nm with a retention time of 9.2 min. Panel C: Representative chromatogram of kesari flour (solid line) and chick pea flour (dashed line) extracts at 340 nm analytical methods are grounds for the increased food fraud of these commodities across the world. Issues associated with adulteration of each of these commodities are discussed separately in the ensuing section.

6.1 Milk and Dairy Products

The most commonly consumed food in almost every part of the world is undoubtedly milk with a global production of 883 metric tonnes in 2019 (FAO 2021). Milk safety issues arise when some unscrupulous traders adulterate it with the objective of increasing profits thereby modifying its chemical composition and bringing down the nutritional value of milk (Hansen and Ferrão 2018; Nascimento et al. 2017). Urea and melamine are nitrogen rich compounds added extraneously to milk to give a false impression of high protein content. Limits ranging from 0.5 ppm to 2.5 ppm in different milk and food products have been prescribed for melamine, under the requirements of the Food safety and Standards Act 2006 and regulations 2011. The Codex Alimentarius Commission has recommended MRLs for melamine as 1.0 mg/kg for infant formula and 2.5 mg/kg for other foods including animal feed (Jooste et al. 2014). Prolonged consumption of milk adulterated with melamine causes kidney failure. The most infamous cases of food adulteration occurred in 2008 in China, which is mainly due to melamine-adulterated infant formula resulted in over 300,000 infants being affected with 52,000 hospitalizations and 6 reported deaths involving kidney stones and liver failure. The standard Kjeldahl method applied for determining protein measures the total nitrogen content. Therefore, melamine, which is a rich source of nitrogen, is used ostensibly to fabricate results (Moy and Han 2014).

Consumption of milk adulterated with urea also causes with gastro-intestinal disorders and renal failure. As urea is naturally found in milk, FSSAI allows upto 700 ppm in milk to account for the endogenous presence of urea in milk (Khan et al. 2015; Food safety and Standards Act 2006 and regulations 2011). Many times to prevent the rejection of poor-quality milk, traders add neutralizers such as sodium bicarbonate, sodium carbonate, sodium hydroxide and calcium hydroxide. The addition of such neutralisers aids in concealing the pH and acidity values. However, such addition of neutralizers can be injurious to the consumer causing disruptions of key hormone signalling pathways that regulates development and reproduction (Aiello et al. 2019). Further, to increase the product shelf life preservatives like formaldehyde, hydrogen peroxide, hypochlorite, dichromate, and salicylic acid have been added to milk affecting consumer health adversely (Nagraik et al. 2021). Adding non-milk fat like vegetable oils and fats to dairy products and milk is an ancient but illegal practice in several countries (Molkentin 2007).

6.2 Spices

Spices are commodities that are traded globally and are an integral part of several cuisines (Anibal et al. 2016). In spices, the fraudulent practices involves "bulking up" the spices with starch or other plant substances or the addition of non-permissible synthetic colours. Synthetic colours are intentionally added to spices so as to enhance their natural appearance and for masking their low quality. To make their products more desirable to consumers, food industry often resorts to synthetic food colours as they are cheaper. However, these synthetic dyes can be potentially genotoxic, mutagenic and carcinogenic, posing a grave risk to human health. Owing to the adverse health implications of synthetic dyes they are banned worldwide and only few are permitted by regulatory agencies. The permissible synthetic dyes include allura red, brilliant blue, carmosine, erythrosine, Fast green, indigo carmine, ponceau 4R, sunset yellow, and tartrazine, which are used to colour a multitude of food products. Turmeric is an important spice used in Asian cuisine extensively (Kar et al. 2018). Owing to its superior medicinal properties, it has been used in traditional medicine since time immemorial. Spices like turmeric have an antiseptic effect on the body, where quality of these ingredients is very important. As per clause 2.9.18 of the FSS Act 2006, Regulations 2011, turmeric must be devoid of lead chromate, extraneously added starch and colouring matter. However, it is often adulterated with dyes such as lead chromate, tartrazine and metanil yellow (Bandara et al. 2020). The food industry uses chilli powder as a natural colouring and flavouring agent. As per clause 2.9.3.2 of the FSS Act 2006, Regulations 2011, chilli powder must be free from extraneous colouring matter. Yet, over the last two decades, chilli powder was found to be falsely manipulated by addition of non-permissible colouring matter especially in marketplaces where spices are often sold loose and non-branded (Lohumi et al. 2017; Abbasi et al. 2021). Adulteration of chilli powder with Sudan dye was first brought into limelight in 2003 where Sudan 1 dye was found to be present in chilli powder that was incorporated into many food products such as curry, sauces etc., resulting in the food recall of these products in England. Sudan dyes are classified are Class III carcinogen and are banned in articles of food by most regulators. Saffron, one of the costliest spices in the world, is extremely susceptible to adulteration. As per clause 2.9.17 of the FSS Act 2006, Regulations 2011, saffron must be free from any added colouring matter. However, saffron powder is often adulterated with plant materials and synthetic dyes to decrease saffron concentration as well as for colour retention (Dai et al. 2020).

6.3 Olive Oil

Olive oil, the main edible vegetable oil of the Mediterranean countries such as Italy, Greece, and Spain, is much valued for its flavour and aroma (Di Giovacchino 2000). Distinctive sensory and health beneficial properties have resulted in olive oil commanding a high market price globally. Cultivar, geographical origin and quality are important factors that decides price of olive oil and consumer liking. The limited production capacity of olive oil coupled with high production costs adds to the market price. Olive oil, is the generic term used for the oil that is obtained exclusively from the fruit of the olive tree (*Olea europaea* L.). The various categories of olive oil globally available include, virgin olive oil, extra virgin olive oil (EVOO) and Olive-pomace oil. Amongst these, EVOO which is produced by employing either a cold press procedure or a centrifugation process without using thermal or chemical treatments is deemed as the premium and best quality oil (Aparicio et al. 2013; Tiryaki and Ayvaz 2017). An increased awareness of the health benefits of olive oil has led to an increase in consumption of olive oil in India (https://www.prnewswire.com/news-releases/indias-olive-oil-industry-forecast-to-2025-301004630.html). The immense popularity of olive oil amongst consumers has made it a highly susceptible to adulteration and a preferred commodity for fraudsters for financial gain. Olive oil adulteration is indeed a growing concern worldwide and poses a serious threat to the economy of the nation as well as health of consumers. The commonest way to adulterate olive oil is by blending it with less expensive vegetable oils such as edible sunflower oil, hazel nut oil or lower-grade olive oils. EVOO is frequently adulterated by lower-grade olive oils making it very difficult to detect the adulteration. In addition to the menace of adulteration, authentication of the geographical origin of olive oil is also an issue that needs to be addressed. Declaration of the geographic origin of olive oil on the label is important especially for EVOO as consumers perceive it as an added assurance of quality and authenticity (Conte et al. 2020). Moreover, olive oil originating from highly valued geographical regions are sold at significantly higher prices than similar merchandises of unheard origin (Camin et al. 2017). It is very important to certify the authenticity and justify the high cost of olive oil. To take action against fraudsters, robust analytical methods that unequivocally establish adulteration or authenticity of olive oil are needed. These methods facilitate the regulators to take decision to withdraw fraudulent products from the market. Legislations have been introduced worldwide to regulate the quality of olive oil, to safeguard consumer health and to prevent unfair competition in the industry. These include standards specified by the Codex Alimentarius, the International Olive Council and the European Union (Conte et al. 2020; (EEC) 2568/91). As per clause 2.2.8 of the FSS Act 2006, Regulations 2011, olive oils shall be free from rancidity, suspended or other foreign matter, separated water, added colouring or flavouring substances or mineral oil. Olive oil is the only food commodity requiring sensory analysis in the European Union to validate the labelling claims for EVOO and virgin olive oil categories (Mariotti 2014). A recent review confirmed that most common infringements includes selling virgin olive oil as extra virgin, and blends of olive oil with other edible vegetable oils such as rapeseed, palm, corn and sunflower being sold as olive oil (Casadei et al. 2021). Rigorous methodology is needed to authenticate the olive oil marketing categories assigned by the producers for enabling fair trade as well as ensuring consumer safety.

6.4 Honey

Honey is a high-value, globally consumed, complex food product composed mainly of fructose and glucose, the remaining being maltose, sucrose, complex carbohydrates, proteins, pigments and phenolics (Simsek et al. 2012; Dong et al. 2018). Traditionally it has been used as an integral ingredient in medicines from time immemorial. Several studies have associated honey with many medicinal benefits, such as antioxidant and anti-inflammatory properties, antimicrobial, antihypertensive, hypoglycemic, hepatoprotective and gastro-protective properties (Soares et al. 2017). According to Codex Alimentarius, honey is defined as "the natural sweet substance produced by honey bees from the nectar of plants or from secretions of living parts of plants or excretions of plant-sucking insects on the living parts of plants, which the bees collect, transform by combining with specific substances of their own, deposit, dehydrate, store and leave in the honey comb to ripen and mature." Honey composition depends on various factors like the geographical origin, botanical origin, the amount of nectar flow, the handling and packing procedure, the beekeepers' manipulations, the climatic conditions, the time of storage and storage conditions (Thrasyvoulou et al. 2018). For human consumption, honey must be of high quality, honey fetches a high market value primarily based on its origin (Tsagkaris et al. 2021). It is considered a premium-priced food product of natural origin. This makes it more prone to intentional or economically motivated adulteration. It is amongst the three highly economically adulterated food commodities in the globe, the first two being olive oil and milk (Spink 2014). Honey being the most economically motivated adulterated food products, has infiltrated news headlines now-a-days, mainly due to the numerous fake honey scandals. This is because, honey is the much sought after natural food product from honey bees, which share the global food market valued at 8.4 billion USD, in 2018. To cater to the high demand, many fraudulent practices are employed resulting in honey adulteration. Substituting pure honey ingredients with cheaper ingredients such as cane sugar, beet sugar, corn, rice, or wheat syrup, though not injurious to health, had huge impact on economy growth and damage consumer confidence (Dong et al. 2018; Kawashima et al. 2019; Tsagkaris et al. 2021).

In a bid to curb adulteration and assess quality FSSAI had come up with honey standards. As per the FSS regulation 2.8.3.1, honey should comply with over 17 parameters which include specific gravity (not less than 1.35), moisture content (not more than 20%), total reducing sugars (not less than 65%), sucrose (not more than 5%), fructose to glucose ratio (within 0.95–1.50), total ash (not more than 0.5%), insoluble impurities, acidity (not more than 0.2% formic acid), free acidity (50 milli-equivalents acid/kg), Hydroxymethyl furfural (HMF, not more than 80 mg/kg), proline (not more than 180 mg/kg), diastase activity (not less than 3 Schade units/g), electrical conductivity (not more than 0.8 mS/cm), water insoluble matters (not more than 0.1%), C_4 sugar adulteration (not more than 7.0%), pollen count and plant elements (not less than 5000/g), specific and trace marker for rice syrup (SMR and TMR should be negative), and $\Delta\delta^{13}C$ Max (± 2.1‰), $\Delta\delta^{13}C$ Fru − Glu (±

1.0‰), and $\Delta\delta^{13}C$ Protein – Honey (≥ -1.0‰) and foreign oligosaccharides (not more than 0.7% peak area). Based on laboratory experience and literature survey, normally honey samples contains 30–40% fructose, 22–40% glucose, 61–83% reducing sugar and 0.25–7.7% sucrose (Simsek et al. 2012). As far as adulteration is concerned, adulteration with sugar syrups is the commonest form of honey adulteration in India. Testing for adulteration can be performed using various methods, including sensory analysis, amino acid profiling, sugar analysis, and others. To assess honey quality, standard methods, including spectrophotometric, refractometric, titrimetric, chromatographic, mass spectrometric, conductometric, and melissopalynological (pollen count) methods are often used. LC-IRMS is specifically used to analyze foreign oligosaccharides, which can be used to detect syrups made from crops like corn, sugar cane, and sugar beet. AFGP (2-Acetylfuran-3-Glucopyranoside), also called as 3-O-α-D-Glucosyl Isomaltol, is the novel specific marker for rice syrup (SMR) adulteration in honey samples, as this molecule is present in rice syrup, and is absent in natural, genuine honey samples (Xue et al. 2013).

The challenges in the utilization of analytical techniques for the detection of adulteration in honey and to trace the geographical origin and authentication of global honey for eliminating the prospect of fake honey infiltrating the supply chains is a difficult task. However, due to recent innovative technological advancements, scientists and analysts can now determine the authenticity of honey by conducting rapid screening tests to detect adulterated honey. Analytical techniques such as thin layer chromatography (TLC), high-performance liquid chromatography (HPLC), liquid chromatography mass spectrometry (LC-MS), near infrared transflectance spectroscopy (NITS), Fourier transform infrared spectroscopy (FTIR), gas chromatography (GC), high performance anion exchange chromatography coupled with pulsed amperometric detector (HPAEC-PAD), Isotope ratio mass spectrometry (IRMS) are useful in providing pertinent information about honey quality, adulteration and authentication. Current analytical approaches carry significant specificity and sensitivity limitations, which can be overcome by mass spectrometry-based techniques. IRMS has emerged and evolved as a promising alternate method to overcome these shortcomings and can be applied efficiently to authenticate honey.

EA-IRMS detects the $\delta^{13}C$ value, the ratio between the two stable isotopes $^{13}C/^{12}C$, which varies in organisms due to the 'kinetic fractionation' of the metabolic processes; for example photosynthesis in C_3 and C_4 plants. EA-IRMS is extensively used to detect C_4 sugar ($\delta^{13}C = -19$ to -12‰) adulteration of honey, which mainly comprised of C_3 sugars ($\delta^{13}C = -32$ to -21‰). Honey is adjudged adulterated if its $\delta^{13}C$ value is heavier than -23.5‰, and C_4 sugar content is >7% (i.e., below the internationally accepted threshold of -1‰, Kawashima et al. 2019). AOAC official method 998.12 (2016), the stable carbon isotope ratio analysis (SCIRA) method for C_4 sugars in honey, provided substantial improvement in the sensitivity, which was mainly achieved by using the extracted honey protein as an internal standard. The SCIRA value ($\delta^{13}C$) for honey protein provides a standard to which the $\delta^{13}C$ of honey is compared, and the difference in $\delta^{13}C$ value gives the measure of the C_4 sugar adulteration. Thus, the $\Delta\delta^{13}C$ Protein – Honey, i.e., the difference $\delta^{13}C$

protein – $\delta^{13}C$ honey should be ≥ −1.0‰ (the range = – 0.9 to 1.5‰), which corresponds to less than 7% corn or cane sugar (C_4 sugars) adulteration (Cabañero et al. 2006). However, EA-IRMS is incapable of detecting honey adulteration with cane sugar up to 5%, cane syrup up to 10%, HFCS up to 10%, or beet sugar up to 20% (Cabañero et al. 2006). Furthermore, EA-IRMS cannot be used to detect honey adulteration by C_3 sugars (example rice syrup, wheat syrup, sugar beet syrup), since there will not be any difference in their $\delta^{13}C$ value when compare to honey. Pioneering work entailing EA/LC-IRMS method by Elflein and Raezke (2008), have made important strides towards the detection of honey that has been adulterated with sugar syrups obtained from different C_3 and C_4 plant sources. Representative LC-IRMS chromatograms of standard sugar mix, pure honey, brown rice syrup, and honey adulterated with 50% rice syrup are shown in Fig. 2.3. The EA/LC-IRMS method and the purity criteria defined exemplifies a significant advancement compared to prevailing methods, in detecting adulteration in honey and its authentication. Recently, Dong et al. (2018) demonstrated the determination of honey authentication with non-extractable protein by utilizing EA- and LC-IRMS. In their work, the $\delta^{13}C$ value of total honey and the major sugars such as glucose, fructose and sucrose are used to identify honey without extracting proteins. The LC-IRMS method can detect C_3 sugar adulteration in honey with a sensitivity of 10%. Additionally, LC-IRMS detects the presence of foreign oligosaccharides, and the contents should not be exceeding 0.7% of relative peak area.

A new analytical technique utilizing high resolution Nuclear Magnetic Resonance (NMR) was developed to test different kinds of honey adulteration with significant improvement in the detection capabilities (Tsagkaris et al. 2021). NMR test scans through various and specific honey markers for the detection of adulteration at the molecular level. For the targeted compounds, specific honey quality markers were identified and quantified (glucose, fructose, sucrose, 5-HMF) whereas, for the untargeted approach, the whole spectrum was processed (Spiteri et al. 2015). Recently, Fiamegos et al. (2020) showed that trace elements determined by energy-dispersive X-ray fluorescence (ED-XRF), and multivariate analysis can be utilized to classify honey according to the botanical variety and geographical origin.

7 Future Perspectives

Despite a great deal of research, food adulteration is a global phenomenon affecting practically all food commodities. Adulteration has been in vogue since pre-historic times underlying the unceasing struggle between the scientific community and unscrupulous traders. Novel adulterants are developed to dodge the state-of-the-art detection methods as exemplified by the melamine adulteration which brought to fore the analytical challenges in detection as well as prevention of fraudulent activities before their occurrence in order to avoid adverse health effects. There is a great demand to update analytical techniques for enhanced food safety and quality control. A detailed study to understand the chemistry and biochemistry of harmful

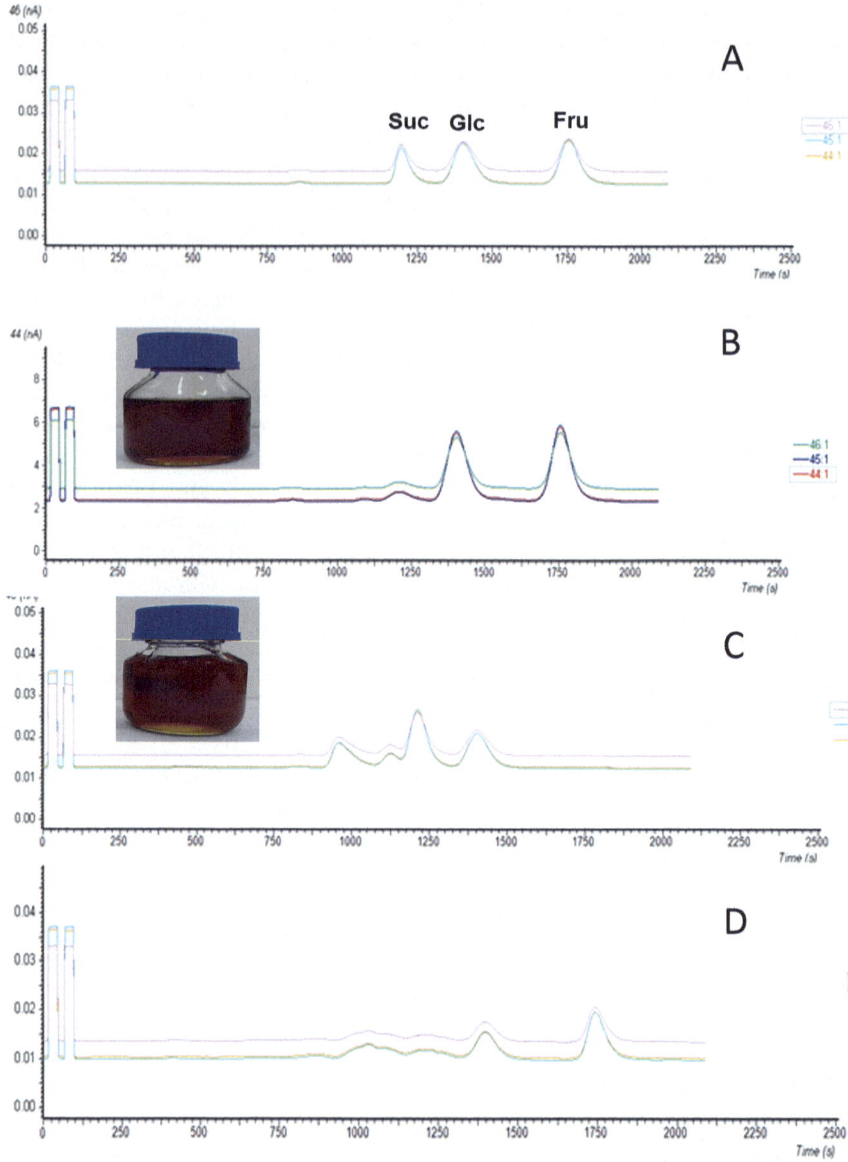

Fig. 2.3 Typical LC-IRMS chromatograms of standard sugar mix (**a**), pure honey (**b**), brown rice syrup (**c**), and honey adulterated with 50% rice syrup (**d**). Pure honey (inset **b**) and brown rice syrup (inset **c**) are difficult to distinguish, while LC-IRMS differentiates easily

chemical adulterants can help in development of strategies to efficiently monitor, reduce, mitigate and detoxify these substances. In addition to targeted ones, untargeted analytical approaches are desirable for rapid detection and identification of unknown adulterants. Concerted efforts need to be made for the production of safe and nutritious food by reducing food adulteration incidents in the entire food supply chain. Greater consumer awareness and education is needed to tackle the menace of adulteration. It is incumbent upon manufacturers, regulators and consumers to ensure a safe and nutritious supply of food. Intensive research efforts are obligatory to safeguard consumers from adverse health effects and law abiding food business operators from monetary loss. Periodic surveillance schemes can alone help identify adulterants and incidents of food adulteration. International compliance with legislation must invariably be the focus for a safe and adulteration free supply of food.

References

Abbasi A, Ansari II, Shakir M (2021) Highly selective and sensitive benzimidazole based bifunctional sensor for targeting inedible azo dyes in red chilli, red food color, turmeric powder, and Cu (Ii) in coconut water. J Fluoresc 31:1353–1361. https://doi.org/10.1007/s10895-021-02766-5

Aiello A, Pizzolongo F, Manzo N et al (2019) A new method to distinguish the Milk adulteration with neutralizers by detection of lactic acid. Food Anal Methods 12:2555–2561. https://doi.org/10.1007/s12161-019-01594-5

Anibal D, Rodríguez CV, Albertengo S, L. et al (2016) UV-visible spectroscopy and multivariate classification as a screening tool for determining the adulteration of sauces. Food Anal Methods 9:3117–3124. https://doi.org/10.1007/s12161-016-0485-7

AOAC official method 998.12 (2016) C4 plant sugars in honey. AOAC Official Methods of Analysis (20th edn)

Aparicio R, Conte LS, Fiebig HJ (2013) Olive oil authentication. In: Aparicio R, Harwood J (eds) Handbook of olive oil: analysis and properties, 2nd edn. Springer, New York, pp 590–653. https://doi.org/10.1007/978-1-4614-7777-8_16

Bandara WGC, Prabbhath GWK, Sahan DW, Dissanayake CB, Herath VR, Godaliyadda GMRI, Ekanayake MPB, Demini D, Madhujith T (2020) Validation of multispectral imaging for the detection of selected adulterants in turmeric samples. J Food Eng 266:109700, ISSN 0260-8774. https://doi.org/10.1016/j.jfoodeng.2019.109700

Banerjee D, Chowdhary S, Chakraborty S, Bhattacharyya R (2017) Recent advances in detection of food adulteration. In: Gupta RK, Dudeja P, Minhas AS (eds) Food safety in the 21st century: public health perspective. Academic Press, pp 145–160

Bansal S, Singh A, Mangal M, Mangal AK, Kumar S (2017) Food adulteration: sources, health risks, and detection methods. Crit Rev Food Sci Nutr 57(6):1174–1189. https://doi.org/10.1080/10408398.2014.967834

Blanch GP, Mar Caja M, Ruiz del Castillo ML, Herraiz M (1998) Comparison of different methods for the evaluation of the authenticity of olive oil and hazelnut oil. J Agric Food Chem 46:3153–3157

Cabañero AI, Recio JL, Ruperez M (2006) Liquid chromatography coupled to isotope ratio mass spectrometry: a new perspective on honey adulteration detection. J Agric Food Chem 54:9719–9727

Calabrese M, Stancher B, Riccobon P (1995) High-performance liquid chromatography determination of proline isomers in Italian wines. J Agric Food Chem 69:361–366

Camin F, Markus LB, Carsten F, Simon DK, Riedl J, Rossmann A (2017) Stable isotope techniques for verifying the declared geographical origin of food in legal cases. Trends Food Sci Technol 61:176–187, ISSN 0924-2244. https://doi.org/10.1016/j.tifs.2016.12.007

Cartoni G, Coccioli F, Jasionowska R, Masci M (1999) Determination of cows' milk in goats' milk and cheese by capillary electrophoresis of the whey protein fractions. J Chromatogr A 846(1–2): 135–141

Casadei E, Valli E, Panni F, Donarski J, Gubern JF, Lucci P, Conte L, Lacoste F, Maquet A, Brereton P, Bendini A, Toschi TG (2021) Emerging trends in olive oil fraud and possible countermeasures. Food Control 124:107902, ISSN 0956-7135. https://doi.org/10.1016/j.foodcont.2021.107902

Chachan S, Kishore A, Kumari K, Sharma A (2021) Trends of food adulteration in developing countries and its remedies. In: Food chemistry: the role of additives, preservatives and adulteration, Chapter 6, pp. 165–187

Codex Alimentariaus (2010) Codex standard for named vegetable oils. CODEX-STAN 210 (Amended 2003, 2005) 8:1–13

Conte L, Bendini A, Valli E, Lucci P, Moret S, Maquet A, Lacoste F, Brereton P, García-González DG, Moreda W, Toschi TG (2020) Olive oil quality and authenticity: a review of current EU legislation, standards, relevant methods of analyses, their drawbacks and recommendations for the future. Trends Food Sci Technol 105:483–493. ISSN 0924–2244. https://doi.org/10.1016/j.tifs.2019.02.025

Dai H, Gao Q, He L (2020) Rapid determination of saffron grade and adulteration by thin-layer chromatography coupled with Raman spectroscopy. Food Anal Methods 13:2128–2137. https://doi.org/10.1007/s12161-020-01828-x

De-Souza EMT, Arruda SF, Brandao PO, Siqueira EMA (2000) Electrophoretic analysis to detect and quantify additional whey in milk and dairy beverages. Cienc Tecnol Aliment 20(3):314–317

Di Giovacchino L (2000) Technological aspects. In: Harwood J, Aparicio R (eds) Handbook of olive oil. Springer, Boston, MA. https://doi.org/10.1007/978-1-4757-5371-4_2

Dong H, Xiao K, Xian Y, Wu Y (2018) Authenticity determination of honeys with non-extractable proteins by means of elemental analyzer (EA) and liquid chromatography (LC) coupled to isotope ratio mass spectroscopy (IRMS). Food Chem 240:717–724

Elflein L, Raezke KP (2008) Improved detection of honey adulteration by measuring differences between 13C/12C stable carbon isotope ratios of protein and sugar compounds with a combination of elemental analyzer – isotope ratio mass spectrometry and liquid chromatography – isotope ratio mass spectrometry (δ13C-EA/ LC-IRMS). Apidologie 39:574–587. https://doi.org/10.1051/apido:2008042

England P, Tang W, Kostrzewa M et al (2020) Discrimination of bovine milk from non-dairy milk by lipids fingerprinting using routine matrix-assisted laser desorption ionization mass spectrometry. Sci Rep 10:5160. https://doi.org/10.1038/s41598-020-62113-9

FAO (2015) Climate change and food security: risks and responses. https://www.fao.org/3/i5188e/I5188E.pdf

FAO (2021) MILK_FACTS_EN-2021_I9966 (fao.org). Accessed 1 July 2021

Fiamegos Y, Dumitrascu C, Ghidotti M, Guntinas MBD (2020) Use of energy-dispersive X-ray fluorescence combined with chemometric modelling to classify honey according to botanical variety and geographical origin. Anal Bioanal Chem 412(463–472):670–678. https://doi.org/10.1016/B978-0-08-100596-5.21788-5

Forte WE (1966) The Food and Drug Administration and the economic adulteration of foods. Indiana Law J 41(3):Article 2

FSS Regulation (2011) https://archive.fssai.gov.in/home/fss-legislation/fss-regulations.html

Giangaspero M, Decaro N, Turno P, Apicella Q, Gargano P, Buonavoglia C (2019) Pathogen spread and globalization: the case of Pestivirus heterogeneity in southern Italy. Res Vet Sci 125: 100–112. https://doi.org/10.1016/j.rvsc.2019.05.017

Gonzalez M, Gallego M, Valcarcel M (2003) Determination of natural and synthetic colorants in prescreened dairy samples using liquid chromatography-diode array detection. Anal Chem 75: 685–693

Hansen L, Ferrão MF (2018) Identification of possible Milk adulteration using physicochemical data and multivariate analysis. Food Anal Methods 11:1994–2003. https://doi.org/10.1007/s12161-018-1181-6

Hatzakis E (2019) Nuclear magnetic resonance (NMR) spectroscopy in food science: a comprehensive review. Compr Rev Food Sci Food Saf 18:189–220

Hosseini P, Sokolow SH, Vandegrif KJ, Kilpatrick AM, Daszak P (2010) Predictive power of air travel and socio-economic data for early pandemic spread. PLoS One 5:e12763

Jala RCR, Zhang X, Huang H, Gao B, Yu L, Xu X (2015) 3-MCPD fatty acid esters. Chemistry, safety, and technological approaches for their reductions. In: Yu L, Wang S, Sun B-G (eds) Food safety chemistry. Toxicant occurrence, analysis and mitigation. CRC Press, pp 113–131

Jooste PJ, Anelich L, Motarjemi Y (2014) Milk and dairy products. Safety of food and beverages. In: Motarjemi Y, Moy GG, Todd ECD (eds) Encyclopedia of food safety. Academic Press\Elsevier, San Diego\London, pp 285–296

Kar S, Tudu B, Bag AK et al (2018) Application of near-infrared spectroscopy for the detection of metanil yellow in turmeric powder. Food Anal Methods 11:1291–1302. https://doi.org/10.1007/s12161-017-1106-9

Katiyar V, Bhanumati S, Rana MK (2014) Food adulteration: the demonic onslaught on health and wellness. In: Rana MK (ed) Herbaceous plants as natural protective food. Scientific Publisher, Jodhpur, pp 498–517

Kawashima H, Suto M, Suto N (2019) Stable carbon isotope ratios for organic acids in commercial honey samples. Food Chem 289:49–55. https://doi.org/10.1016/j.foodchem.2019.03.053

Khan KM, Krishna H, Majumder SK et al (2015) Detection of urea adulteration in milk using near-infrared Raman spectroscopy. Food Anal Methods 8:93–102. https://doi.org/10.1007/s12161-014-9873-z

Lohumi S, Joshi R, Kandpal LM, Lee H, Kim MS, Cho H, Mo C, Seo Y, Rahman A, Cho B (2017) Quantitative analysis of Sudan dye adulteration in paprika powder using FTIR spectroscopy. Food Addit Contam 34(5):678–686. https://doi.org/10.1080/19440049.2017.1290828

Mariotti M (2014) Virgin olive oil: definition and standards. In: Peri C (ed) The extra-virgin olive oil handbook. Wiley, Chichester, pp 11–20

Matute AIR, Soria AC, Castro IM, Sanz ML (2007) A new methodology based on GC-MS to detect honey adulteration with commercial syrups. J Agric Food Chem 55(18):7264–7269

McDowell I, Taylor S, Gay C (1995) The phenolic pigment composition of black tea liquors part I: predicting quality. J Agric Food Chem 69:467–474

Mishra R, Martin A, Gowda LR (2009) Effect of traditional Indian household thermal processing operations on the 3-N-oxalyl L-2-3-diaminopropanoic acid (β-ODAP) in grass pea (*Lathyrus sativus*) as determined by reverse phase high performance liquid chromatography. J Food Compos Anal 22:704–708

Molkentin J (2007) Detection of foreign fat in milk fat from different continents bytriacylglycerol analysis. Eur J Lipid Sci and Technol 109:505–510

Mouly P, Gaydou EM, Auffray A (1998) Simultaneous separation of flavone glycosides and polymethoxylated flavones in citrus juices using liquid chromatography. J Chromatogr A 800: 171–179

Moy GG, Han FF (2014) History of foodborne disease in Asia – examples from China, India, and Japan. History of food safety and related sciences. In: Motarjemi Y, Moy GG, Todd ECD (eds) Encyclopedia of food safety. Academic Press\Elsevier, San Diego\London, pp 22–27

Nagraik R, Sharma A, Kumar D, Chawla P, Avvaru PK (2021) Milk adulterant detection: conventional and biosensor based approaches: a review. Sens Bio-Sens Res 33:100433, ISSN 2214.1804. https://doi.org/10.1016/j.sbsr.2021.100433

Nascimento CF, Santos PM, Rodrigues EP, Rocha FRP (2017) Recent advances on determination of milk adulterants. Food Chem 221:1232–1244, ISSN 0308-8146. https://doi.org/10.1016/j.foodchem.2016.11.034

Nogueira JMF, Nascimento AMD (1999) Analytical characterization of Madeira wine. J Agric Food Chem 47:566–575

Oliveira RCS, Oliveira LS, Franca AS, Augusti R (2009) Evaluation of the potential of SPME-GC-MS and chemometrics to detect adulteration of ground roasted coffee with roasted barley. J Food Compos Anal 22(3):257–261

Ozen BF, Mauer LJ (2002) Detection of hazelnut oil adulteration using FTIR spectroscopy. J Agric Food Chem 50:3898–3901

Sattigeri, Appaiah (2010) Development of Food Legislation Around the World. In: Christine E. Boisrobert, Aleksandra Stjepanovic, Sangsuk Oh, Huub L.M. Lelieveld (eds) Ensuring Global Food Safety. Academic Press, pp5–69. https://doi.org/10.1016/B978-0-12-374845-4.00002-3

Schieber A (2018) Introduction to food authentication. In: Modern techniques for food authentication, 2nd edn. Academic Press, pp 1–21

Simsek A, Bilsel M, Goren AC (2012) 13C/12C pattern of honey from Turkey and determination of adulteration in commercially available honey samples using EA-IRMS. Food Chem 130(4): 1115–1121

Skovgaard N (2007) New trends in emerging pathogens. Int J Food Microbiol 120(3):217–224. https://doi.org/10.1016/j.ijfoodmicro.2007.07.046

Smith BA, Fazil A (2019) How will climate change impact microbial foodborne disease in Canada? Can Commun Dis Rep 45(4):108–113. https://doi.org/10.14745/ccdr.v45i04a05

Soares S, Amaral JS, Oliveira BMPP, Mafra I (2017) A comprehensive review on the Main honey authentication issues: production and origin. Compr Rev Food Sci Food Saf 16:1072–1100

Spink J (2014) Risks of food adulteration. Safety of food and beverages. In: Encyclopedia of food safety. Academic Press, pp 413–416

Spink J, Moyer DC (2011) Defining the public health threat of food fraud. J Food Sci 75(9):57–63

Spiteri M, Jamin E, Thomas F, Rebours A, Lees M, Rogers KM, Rutledge DN (2015) Fast and global authenticity screening of honey using 1H-NMR profiling. Food Chem 189:60–66

Thippeswamy R, Martin A, Gowda LR (2007) A reverse phase high performance liquid chromatography method for analyzing of neurotoxin β -N-oxalyl L-α,β-diaminopropanoic acid in legume seeds. Food Chem 101:1290–1295

Thrasyvoulou A, Tananaki C, Goras G, Karazafiris E, Dimou M, Liolios V, Kanelis D, Gounari S (2018) Legislation of honey criteria and standards. J Apic Res 57(1):88–96. https://doi.org/10.1080/00218839.2017.1411181

Tiryaki YG, Ayvaz H (2017) Quantification of soybean oil adulteration in extra virgin olive oil using portable raman spectroscopy. J Food Meas Charact 11:523–529. https://doi.org/10.1007/s11694-016-9419-8

Tsagkaris AS, Koulis GA, Danezis GP, Martakos I, Dasenaki M, Georgiou CA, Thomaidis NS (2021) Honey authenticity: analytical techniques, state of the art and challenges. RSC Adv 11: 11273–11294

UN Report (2021) World Population Day – the United Nations https://www.un.org "observances" world-population-day. Accessed on 1st July 2022

Valletta M, Ragucci S, Landi N, Di Maro A, Vincenzo Pedone P, Russo R, Chambery A (2021) Mass spectrometry-based protein and peptide profiling for food frauds, traceability and authenticity assessment. Food Chem 365:130456. https://doi.org/10.1016/j.foodchem.2021.130456

Vemireddy LR, Archak S, Nagaraju J (2007) Capillary electrophoresis is essential for microsatellite marker based detection and quantification of adulteration of Basmati rice (Oryzasativa). J Agric Food Chem 55:8112–8117

WHO/FAO (2014) Procedural manual joint FAO/WHO food standards programme, codex alimentarius commission (22nd edn, pp 1–212)

Xue H, Hu W, Son H, Han Y, Yang Z (2010) Indirect ELISA for detection and quantification of bovine milk in goat milk. J Food Sci Tech 31(24):370–373

Xue XF, Wang Q, Li Y, Wu LM, Chen LZ, Zhao J, Liu FM (2013) 2-Acetylfuran-3-glucopyranoside as a novel marker for the detection of honey adulterated with rice syrup. J Agric Food Chem 61:7488–7493. https://doi.org/10.1021/jf401912u

Zambonin C (2021) Maldi-Tof mass spectrometry applications for food fraud detection. Appl Sci 2021(11):3374. https://doi.org/10.3390/app11083374

Chapter 3
Microbial Adulterants in Food: Challenges to Overcome

Bhaskar Das, Bhaskar Kalita, Risha Hazarika, and Sanjukta Patra

1 Introduction

Microbiological contamination of food refers to the inadvertent presence of pathogenic organisms such as bacteria, fungus, yeasts, protozoa, and viruses in food because of their exposure during manufacturing and processing, rendering it unfit for human consumption (Bintsis 2017). Microbial contamination happens most frequently during the voyage from the farm to the processing facility, during processing, storage, transportation, distribution, and before consumption. Microbial poisons are also potential dietary toxins; nevertheless, microbes and their products can also be used to combat pathogenic organisms. Food spoilage is a complex process, and even with contemporary preservation techniques, microbial spoilage causes enormous volumes of food to be wasted. (Alum et al. 2016; Amit et al. 2017). The widespread contamination of food by pathogenic microorganisms, their capacity to grow and survive in low-oxygen environments, and the low microbial dose required for food poisoning epidemics point to a significant public health issue. Ionizing radiation and heating are two well-known and widely utilized microbial control strategies. On the other hand, novel nonthermal and modified thermal technologies have become effective methods for killing pathogenic microbes in food.

Authors Bhaskar Das, Bhaskar Kalita, Risha Hazarika, and Sanjukta Patra have equally contributed to this chapter.

B. Das · S. Patra (✉)
Department of Biosciences and Bioengineering, Indian Institute of Technology Guwahati, Guwahati, Assam, India
e-mail: sanjukta@iitg.ac.in

B. Kalita · R. Hazarika
School of Agro and Rural Technology, Indian Institute of Technology Guwahati, Guwahati, Assam, India

H. U. Hebbar et al. (eds.), *Engineering Aspects of Food Quality and Safety*, Food Engineering Series, https://doi.org/10.1007/978-3-031-30683-9_3

The food processing industry's main challenge is to keep up with innovative microbial decontamination technologies that will allow it to provide safe, high-quality food items with a long shelf life. In fact, during the next 50 years, the development or re-emergence of the novel or unanticipated microbial contaminants will impact the food sector (Tauxe et al. 2010). With the globalization of the food trade, the risk of foodborne illness has increased, leading to import bans, which have hampered the economies of many developing countries worldwide (Bhat 2004). The World Health Organization (WHO), a component of primary health care, has advocated food safety. Keeping this in mind, the present chapter endorses the causes, routes, and mechanisms of microbial food spoilage and foodborne infections, novel technologies for controlling foodborne pathogens, the need to monitor food contamination, and the effect on the economy due to microbial food spoilage.

2 Microbes and Food

2.1 Microbe Assisting Food Production

Microorganisms associated with foods can be categorized as "spoilage," "pathogenic," or "useful." Nature uses microorganisms to carry out fermentation processes. Yeasts, molds, and bacteria have been exploited for thousands of years to make bread, beer, wine, vinegar, yogurt, cheese, fermented fish, meat, and vegetables. Fermentation is one of the oldest methods of food transformation and preservation. This biological process increases food's nutritional and organoleptic characteristics while preserving it (relating to the senses, taste, sight, smell, and touch). A well-run fermentation will favor beneficial organisms over undesired flora to minimize spoilage and increase taste and texture.

Functional microorganisms during fermentation of food transform its chemical constituents resulting in enhancement of nutrients bio-availability and sensory quality of food, imparts bio-preservative effects and improve food safety, degrading toxic/anti-nutritive factors, producing antioxidant and antimicrobial compounds, stimulating probiotic functions along with fortification of health-promoting bioactive compounds (Tamang et al. 2016).

In the food processing industry, microorganisms play a significant role. They are utilized in developing a wide range of food products and are also responsible for food spoiling, resulting in poisoning and illness. Although the microflora of raw materials is typically varied, food preparation frequently imposes a distinct and very particular microbial flora. The natural flora of low-acid canned goods that have been severely heat processed but not sterilized is made up of thermophilic spore-forming bacteria, which are the raw materials' most heat-resistant microbial components. The significant flora of shelf-stable canned cured meats is made up of mesophilic aerobic and anaerobic spore-forming bacteria, which are resistant to the heat treatment used to preserve these items. Small populations of spore-forming bacteria, yeasts, and lactic acid bacteria compromise the natural flora of mayonnaise and salad dressing.

Aerobic spore-forming bacteria predominate in dry spices and in several dry vegetable products. Molds and yeasts are the most common microorganisms found in dried fruits. Yeasts make up the typical flora in carbonated beverages. The nature of the raw materials, production conditions, packaging, and storage of the shelf-stable product are all reflected in the surviving and predominating microflora in each of the examples above. The microflora that survives processing in perishable products may be diverse, but the percentage that develops during storage and causes deterioration is usually rather particular. As a result of contamination from the animal and the processing environment, a heterogeneous flora can be found on raw red meats, poultry, and fish. However, spoiling is caused mainly by a highly specialized group of microbes, particularly *Pseudomonas* and closely related aerobic, psychrotrophic Gram-negative bacteria, during refrigerated storage of such products. Changes in perishable food processing must consider the impact these changes may have on the spoilage flora and the product's usual rotting pattern.

2.2 Food Spoilage

Food spoilage caused by microorganisms is known as microbial spoilage. It's also the leading cause of food poisoning. Microorganisms are ubiquitous and have the potential to spoil food and induce foodborne illness. Microbial contamination of food products occurs most commonly on the journey from the farm to the processing facility, during processing, storage, transportation, and distribution, and prior to consumption. Food spoiling is a complicated process, and even with modern preservation procedures, large amounts of food are lost owing to microbial spoilage. The microbiota that develops during storage and in deteriorating foods can be anticipated, despite the heterogeneity in raw ingredients and processing environments.

Both intrinsic and extrinsic causes can influence microbial spoilage in foodstuffs. The inherent qualities of foods affect the type and rate of microbial decomposition and their projected shelf life or perishability. The essential intrinsic qualities of food spoilage include endogenous enzymes, substrates, light sensitivity, and oxygen. The pH, water activity, nutrient content, and oxidation-reduction potential are all intrinsic factors in food deterioration. Food deterioration is caused by extrinsic variables such as relative humidity, temperature, and the presence and activities of other microorganisms (Amit et al. 2017).

Changes in appearance (color, pockets of gas/swelling), texture (soft and mushy), color, aroma, flavor, or slime development are all signs of microbial deterioration of food. It covers Gram-positive, Gram-negative aerobic bacteria, yeasts, molds, and fungal pathogens, among other spoilage species. Food spoilage has an enormous economic impact, and microbial food spoilage plays a significant role in food waste and loss.

2.2.1 Microbial Spoilage of Raw Foodstuff

Microbiological contamination can be found everywhere in the biosphere, in plants, animals, soil, and water. Many bacteria, such as pseudomonads, lactic, micrococci, and coliforms, grow readily on agricultural and horticultural plants. For example, Raw/fresh produce and highly perishable products like fluid milk account for a disproportionately high percentage of annual losses compared to other commodity groupings. Liquid milk, although pasteurized, is an excellent growing medium for spore formers that have survived the heat process. Produce is frequently harvested before it has fully ripened and continues to breathe throughout transportation. Still, it is not until senescence, when the native protections have been compromised, that produce becomes most susceptible to deterioration.

2.2.2 Microbial Spoilage in Processed Foodstuff

All of the preventative measures and sanitation tactics still apply to processed products, but there is new potential for spoiling mitigation in processing and packaging. Economically, further spoilage complications in food processing have arisen due to shifting consumer preferences. Demands include minimizing antimicrobial compounds and heat processing and changing compositions to incorporate more contamination components.

Groups of Foods Spoilers

Ubiquitous microorganisms found in soil, water, and air, as well as specific sources of contamination such as spoiled raw materials, food waste, biofilm on equipment surfaces, and personal hygiene from food workers or consumers, depending on the ecological microbial niche, cause microbial spoilage. Microorganisms involved in food spoilage can be divided into three major categories: molds, yeasts, and bacteria. Table 3.1. presents the operational conditions of different microorganisms that affect foods.

Yeasts are generally single-celled organisms adapted for life in specialized, usually liquid, environments and do not produce toxic secondary metabolites. They often colonize foods with high sugar or salt content and contribute to maple syrup, pickles, and sauerkraut spoilage. Fruits and juices with a low pH are other targets, and some yeasts grow on the surfaces of meat and cheese.

Molds are essential for recycling dead plant and animal remains in nature but also attack various foods and other materials helpful to humans. They are well adapted for growth on and through solid substrates, generally produce airborne spores, and require oxygen for their metabolic processes. Most molds grow at a pH range of 3 to 8; some can grow at deficient water activity levels (0.7–0.8) on dried foods.

Table 3.1 List of food spoilage microbes

Group of microorganism	Common microbes	Types of foods	Activities/ mechanism	References
Lactic acid bacteria	*Lactobacillus, Weisella, Leuconostoc, Lactococcus, Pediococcus, Streptococcus, Enterococcus.*	Dairy products sourdough beverages (fruit juices, beer)	Acetic acid, gas blowing, post acidification	Sakamoto and Konings (2003)
Acetic acid bacteria	*Acetobacter, Gluconobacter, Acidomonas, Gluconacetobacter, Asaia, Kozakia, Swamina – thania, Saccharibacter, Neoasaia, Granulibacter, Tanticharoenia and Ameyamaea.*	Wine	Strictly aerobic	Bartowsky and Henschke (2008)
Filamentous fungi, Moulds	Zygomycetes	Rot in stored fruits and vegetables	Formation of spores	Rawat (2015)
	Penicillium spp.	Citrus, pear, and apple fruits, refrigerated and processed foods such as jams and margarine	Produce potent mycotoxins (patulin, ochratoxin, citreoviridin, penitrem)	
	Aspergillus spp.	Grains, dried beans, peanuts, tree nuts, and some spices.	Produce mycotoxins: aflatoxins, ochratoxin, territrems, cyclopiazonic acid	
	Fusarium spp.	Harvested grains	Mycotoxins	
Yeast	Zygosaccharomyces	Honey, dried fruit, jams and soy sauce	Producing off-odors and flavors and carbon dioxide	Rawat (2015)
	Saccharomyces spp.	Alcoholic beverages	Producing gassiness, turbidity, and off-flavors associated with hydrogen sulfide and acetic acid.	
	Candida	Fruits, some vegetables, and dairy products		
	Dekkera/ Brettanomyces	Fermented foods, including alcoholic beverages and some dairy products.	Produce volatile phenolic compounds responsible for off-flavors	

Bacteria: Lactic acid bacteria are a group of Gram-positive bacteria, including species of *Lactobacillus*, *Pediococcus*, *Leuconostoc,* and *Oenococcus*; under low oxygen, low temperature, and acidic conditions, these bacteria become the predominant spoilage organisms on a variety of foods. Undesirable changes caused by LAB include greening of meat and gas formation in cheeses (blowing), pickles (bloater damage), and canned or packaged meat and vegetables. Off-flavors described as mousy, cheesy, malty, acidic, buttery, or liver-like may be detected in wine, meats, milk, or juices spoiled by these bacteria.

3 Foodborne Diseases

Hippocrates (460 B.C.) highlighted the correlation between human food consumption and diseases. The ingestion of food contaminated with microbial pathogens and toxins results in foodborne illness (Bintsis 2017). Microbiological contamination of food refers to the unintentional presence of pathogenic microorganisms such as bacteria, fungi, yeasts, protozoa, and viruses owing to its exposure during food production and processing, making it unfit for human consumption. The appearance or re-emergence of the novel or unexpected microbial contaminants will affect the food industry within the next 50 years (Tauxe et al. 2010). According to The Centre for Disease Control and Prevention (CDC), an estimated 76 million people get affected by foodborne diseases, with yearly medical expenses up to 5–6 billion dollars (Kirch 2008). Sudershan et al. (2010) reported that a foodborne illness outbreak affecting 60 people in the Indian city of Hyderabad led to an economic burden of U.S. $ 2070. European Union (E.U.) in 2015 reported 4362 foodborne outbreaks in 26 member states which accounted for 45,874 cases of illness, 3892 hospitalizations, and 17 deaths (EFSA 2015). With the globalization of food trade, the probability of the spread of foodborne illness leads to the ban of imported food, which undoubtedly hampers the economy of many developing nations worldwide (Bhat 2004).

3.1 Microbial Contamination in the Food Industry: Routes and Mechanism

The first significant route for food contamination in the food-processing sector is due to the failure of industry personnel to follow the standard hygiene rules, which include hand-washing and gloves during food manufacture or packaging (Green and Selman 2005). Secondly, using water contaminated with pathogens results in microbial contamination of food (Ashbolt 2004). Along with the above factors, the microbial contamination of instruments/equipment routinely used for food processing is a significant route for microbial contamination of processed food, stressing the need for food industries to follow proper protocols for sterilization of the equipment (Alum et al. 2016). Bacterial, fungal, and viral pathogens frequently contaminate processed food in the industry due to improper handling. Foods like red meats, poultry, and seafood products are more prone to microbial contamination than fruits or vegetables. As per EFSA (European Food Safety Authority), the toxin from bacterial pathogens are ranked second among the microbial pathogens responsible for food- and waterborne outbreaks, followed by viruses responsible for 19.5% and 9.2% of the total episodes in 2015. On the other hand, parasites and other causative agents resulted in less than 3% of the outbreaks, while for 34% of attacks, the causative agent could not be determined. Studies have shown that most foodborne diseases were related to *Norovirus*, nontyphoidal *Salmonella* spp., *Clostridium perfringens,* and *Campylobacter* spp. Foodborne microbial illnesses are categorized into:

(a) **Foodborne infections:** With the advent of next-generation sequencing technology, it has been possible to carry out sequencing of the foodborne pathogens' whole genome, which provides insights into their pathogenicity. Foodborne pathogens such as *Salmonella* spp., *Campylobacter* spp., *E. coli* O157:H7, etc., are ingested via food, followed by colonization of intestinal epithelial cell lining, which spreads further to the lower intestinal tract and liver coupled to release of toxin (Fig. 3.1).
(b) **Foodborne intoxications:** Foodborne intoxications involve ingesting toxins into the human body instead of bacterial or fungal cells. e.g., fungal toxins like fumonisins, aflatoxin B, ochratoxin A, bacterial toxins such as botulinum neurotoxin, cholera toxins Shiga toxin, and Staphylococcus enterotoxin.
(c) **Foodborne toxico-infections:** The toxic foodborne infections involve dormant cells (spores) of *C. perfringens*, *Bacillus anthracis*, *B. cereus*, before, etc., ingested, which, after death, release toxins (Fig. 3.2).

4 Antibiotic Resistance Pathogens (ARGs): An Emerging Challenge for the Food Industry

The application of antibiotics to livestock has resulted in the emergence of antibiotic-resistant foodborne pathogens and livestock bacteria. Foodborne pathogens and other opportunistic pathogens such as *Salmonella*, *E. coli*, and *Campylobacter* exhibiting multidrug resistance have been found in fresh produce and food-producing animals. The pathogens have been resistant to various antibiotics such as azithromycin, tetracycline, nalidixic acid, amikacin, ciprofloxacin, trimethoprim-sulfamethoxazole, and cephalosporin. Antibiotic resistance in foodborne microbes has resulted from the inappropriate use of antibiotics to promote growth and fight infections. One of the commonly used growth promoters in Belgium, India, China, and Brazil is colistin, and as a consequence, colistin-resistance genes have been

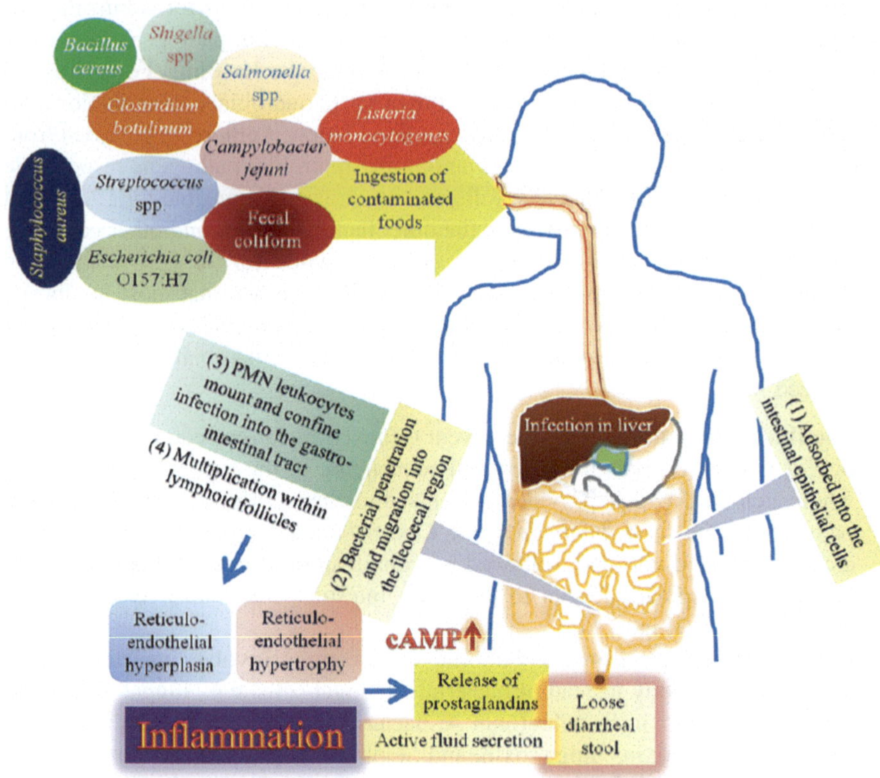

Fig. 3.1 Schematic representation of the mode of infections by foodborne pathogens. The ingestion of pathogenic microbes at a dose equal to/above the infectious amount adheres to the intestine's epithelial cells, followed by dissemination to the liver and lower intestinal tract. This results in the weakening of the immune system or diarrheal syndromes. (Reproduced from Noor 2019)

predominantly found in the environment. Mcr-1 and mcr-2 are in pork carcasses, chicken meat, and mutton in Belgium, Brazil, and India, respectively. The contaminated surfaces during food processing serve as a source of antibiotic resistance gene transfer. The bacteria on contaminated surfaces exhibit antibiotic resistance by taking up antibiotic resistance genes indicating how resistance is spread in the inter-connected environment. In addition, food processing techniques routinely applied to control foodborne pathogens create microbial stress resulting in their inactivation. However, the processing techniques can also serve as a route for transferring antibiotic resistance genes in pathogens with prolonged exposure to such stress (Thakali and MacRae 2021). Low and middle-income countries apply animal waste as agricultural fertilizers or as fish feed without necessary treatment, resulting in contamination of food products such as vegetables, fish, shellfish, and

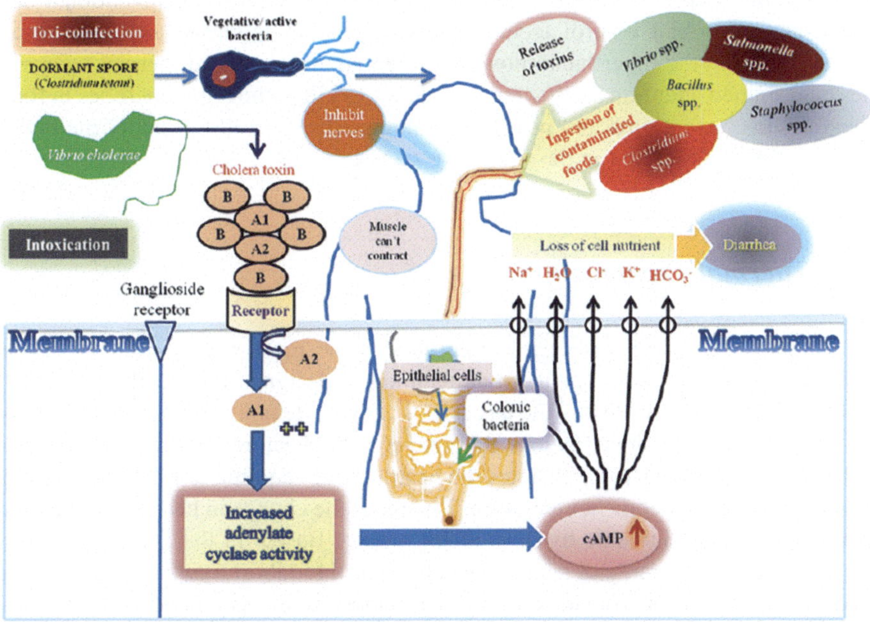

Fig. 3.2 The mode of foodborne microbial intoxication and toxico-infection in host cells. Food contaminated with pathogens such as *Clostridium* spp., *Bacillus* spp., *Vibrio cholera,* and *Staphylococcus* spp. Results in foodborne intoxication and toxico-infection using toxins apart from foodborne infections are described in Fig. 3.1. This diagram shows the route of transmission and mode of toxin action in cholera (intoxication) and botulism (toxico-infection) caused by *Vibrio cholerae* and *Clostridium botulinum*, respectively. (Reproduced from Noor 2019)

water with antibiotic-resistant bacteria. This entry of antibiotic-resistant genes into the food chain ultimately enhances the risk to public health. Tao et al. (2022) reported the prevalence of foodborne pathogens resistant to antibiotics in different food types such as aquatic products, meat, milk, and dairy products. Multidrug-resistant (MDR) pathogens comprise $\geq 36\%$ of foodstuffs, with the highest rate of 52% found in meat. The pathogens resistant to β-lactams at the rate of $\geq 57\%$ were most common among all the food products. Among aquatic food products, fluoroquinolones and sulphonamides-resistant microbes were around 13% while isolates resistant to β-lactams were over six times higher.

5 Strategies to Prevent Microbial Contamination of Food

The ubiquitous contamination of food by pathogenic microbes, their ability to grow and survive under refrigeration/ reduced oxygen conditions, and the low microbial dose required for food poisoning outbreaks indicate a potential risk to public health. The matter has been further complicated by documentation of food poisoning by microbes previously unknown to cause foodborne disease. Conventional and established microbial control techniques like ionizing radiation and heating are widely used. On the other hand, novel nonthermal and modified thermal technologies have gained attention as efficient techniques to kill pathogenic microbes in food. The various conventional and novel technologies for controlling foodborne pathogens have been described in the following sections.

5.1 Thermal Treatment

Heat application to kill pathogenic microbes is the most widely used foodborne pathogen control strategy. The thermal treatment is designed for specific lethality of foodborne pathogens to ensure the target food's shelf life and microbiological safety. D- and Z-values are two factors that govern the heat resistance of foodborne pathogenic bacteria. D-value denotes the heating time required to kill 90% of viable cells or spores at a specific temperature. Z-value is the change in heating temperature required to change D-value by 90% (1 log cycle). Based on the D- and Z-values, thermal processing strategies are designed to control foodborne pathogens in specific food products. The sterilization of food products is being carried out using the following methods:

(a) **Direct heating:** This is carried out either by steam injection or steam infusion.
(b) **Indirect heating:** The food product is subjected to indirect heating by using steam or hot water as the medium of heat in the plate, tubular, or scraped surface heat exchanger.
(c) **Ohmic heating:** This novel sterilization technique utilizes electric current to generate heat in the food products to be sterilized. This technology works on the principle that it is possible to generate heat in the material by passing electric current, using the inherent electrical resistance of the material. Ohmic heating results in rapid and uniform heating of the material. Figure 3.3 shows the essential part of ohmic heating equipment, which comprises a power supply, heating chamber, electrodes, thermocouple, current sensor, and data acquisition system. Ohmic heating results in heating the food rapidly and uniformly by Joule's law of heating. This makes it possible to heat the food material volumetrically and efficiently. Ohmic heating inactivates undesirable enzymes and pathogenic microbes in food in a comparatively shorter time than conventional heating. Thermal effects of ohmic heating results in enzyme inactivation. Apart from this, nonthermal effects such as chemical changes and cell membrane electroporation result in microbial killings during ohmic heating. High microbial pathogens could be treated at lower pH by ohmic heating. However, fat contents compromise the activity of ohmic heating, resulting in low microbial inactivation efficiency. Thus, owing to the advantages of ohmic heating, such as quick and volumetric heating, high energy efficiency, rapid microbial/enzyme inactivation, etc., ohmic heating is an alternative to conventional thermal strategies (Makroo et al. 2020).

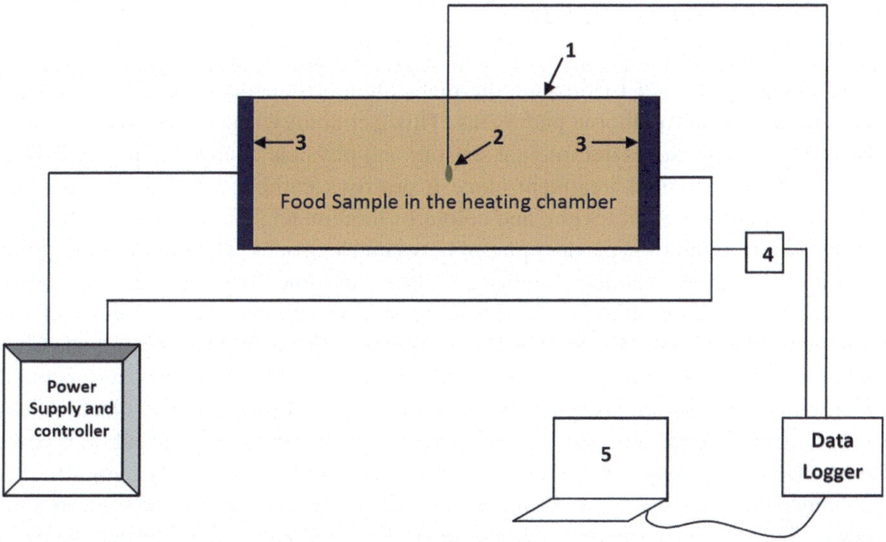

Fig. 3.3 Schematic representation of ohmic heating setup; (1) O.H. chamber (2) Thermocouple (3) Electrode (4) Current sensor (5) Personal computer. (Reproduced from Makroo et al. 2020)

(d) **Microwave heating:** For food sterilization, microwave heating encompasses several advantages over conventional thermal treatment as heating speed, better process control with rapid start and shut down times, and better taste, texture, and nutritional content of the treated food product. Microwave heating of food products occurs by the interaction of microwaves with the ionic or dipolar content of the food. Food's dipolar content comprises water, proteins, and carbohydrates, which are volumetrically distributed in food. Thus, the microwave could cause volumetric heating, which is governed by the material's dielectric properties and the microwave's frequency. The food product dielectric property is responsible for the number of incident microwaves reflected, transmitted, or absorbed by the material (Juneja et al. 2007). Osaili et al. (2021) evaluated the effect of microwave heating on treating pathogenic microbes in tahini, a traditional food associated with multiple foodborne-related outbreaks due to *Salmonella*, *L. monocytogenes,* and *E. coli* contamination. They reported microwave heating as a potential method for lowering the risk of Salmonella spp., E. coli O157:H7, and L. monocytogenes in tahini with no fadverse effect on quality. The treatment showed no effect on acid, peroxide, p-anisidine, or color values of tahini up to 90 °C. Hashemi et al. (2019) reported that microwave treatment showed significant decrease in pathogens as *Escherichia coli, Salmonella Typhimurium, S. Enteritidis and Staphylococcus aureus* and content of vitamin C, β-carotene, phenolic compounds whereas the pH of samples did not show significant changes in cantaloupe juice.

5.2 *Pulsed Electric Fields*

Pulsed electric fields (PEF) are a nonthermal technique dependent on short electrical pulses to inactivate foodborne pathogens. This technology is preferred over thermal processing, which causes detrimental sensory and physical changes in foods. PEF is considered over thermal treatment since it destroys pathogens while keeping the nutritional quality, flavor, texture and colour of food intact. PEF technology delivers pulsing power to the food product placed between electrodes. The high voltage pulse generator, treatment chamber having a system for fluid handling, and necessary devices required to monitor and control the system are important sections of the equipment (Fig. 3.4). The food to be treated is placed in a treatment chamber containing two electrodes to which high voltage electrical pulses are applied which conducts high-intensity electrical pulse to a food product. The electric field to which food is exposed causes the breakdown of membranes in foodborne pathogens. This technology is preferred for pasteurizing juices, milk, yogurt, etc. (Mohamed and Eissa 2012). Bulut et al. (2020) developed a pilot-scale pulsed electric fields system to treat sesame seeds by analyzing their physicochemical properties and *Aspergillus parasiticus* inactivation. The pulsed electric field energy in the range of 0.97 to 17.28 J resulted in maximum reductions of peroxide value (67.4%) and acidity number (85.7%) with no change in color. Applying maximum pulsed electric field energy led to a 60% reduction of *A. parasiticus* counts. The study proved that a pulsed electric field could be used to treat sesame seeds by preserving physicochemical properties and inactivating *A. parasiticus*. Mendes-

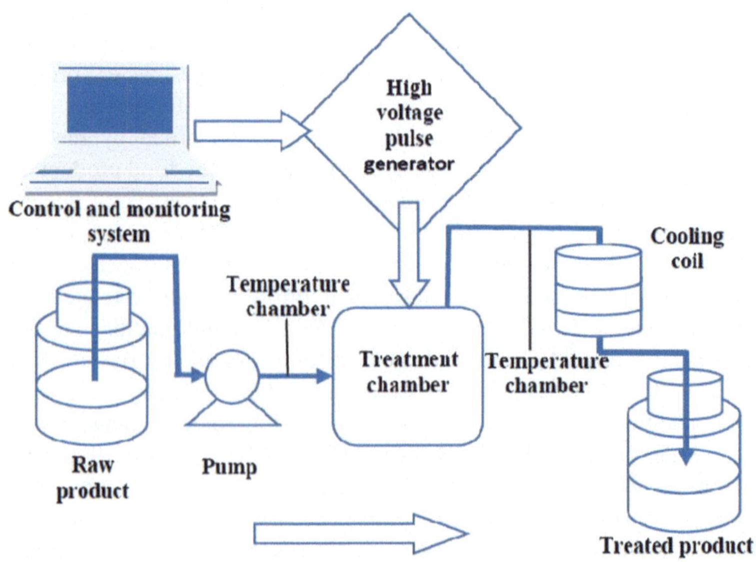

Fig. 3.4 Schematic diagram of a pulsed electric field food processing system. (Reproduced from Mohamed and Eissa 2012)

Oliveira et al. (2021) processed juice inoculated with *Escherichia coli* O157:H7 and *Salmonella Typhimurium* in a continuous PEF process system. They reported an over 5 log reduction of pathogens with no regrowth during storage at 4, 10, and 22 ° C. The native microbes population in treated juices after 24 week-storage was found to be same or lower compared to control samples. PEF showed no significant negative effect against pH, color, and vitamin C retention.

5.3 Electrolyzed Water

Electrolyzed water has been used in Japan for several years due to its antimicrobial properties. The application of electrolyzed water to destroy foodborne pathogens is preferred since it has no hazardous effects on the human body. Electrolyzed water has strong bactericidal effects on foodborne pathogenic bacteria. It is produced by passing the diluted salt solution through an electrolytic cell containing an anode and cathode separated by a membrane. The electrodes are exposed to direct current voltages. Negative ions like chloride and hydroxide move in salt solution towards the anode to release electrons and produce oxygen gas, chlorine gas, hypochlorite ion, hypochlorous acid, and hydrochloric acid. On the other hand, positively charged ions like hydrogen and sodium migrate towards the cathode and accept electrons to produce hydrogen gas and sodium hydroxide. This results in the production of electrolyzed water having properties of low pH (2.3–2.7) and high oxidation-reduction potential (ORP, >1000 mV) at the anode. Liang et al. (2019) analyzed the application of slightly acidic electrolyzed water (SAEW) for microbial reduction in buckwheat. This treatment resulted in molds, yeasts, and total bacteria count reduction in buckwheat seeds and harvested sprouts. SAEW treatment showed strong efficiency for removing *E. coli* and *L. monocytogenes* inoculated on harvested shoots. Medina-Gudiño et al. (2020) analyzed the activity of Neutral Electrolyzed Water (NEW) was tested in vitro and on artificially contaminated eggs against *Salmonella enterica* subsp. *enterica* or *Escherichia coli*. Neutral electrolyzed solution do not damage egg cuticle and causes pore formation in *Salmonella enterica* and *E. coli* surfaces after 30 s treatment.

5.4 High-Pressure Processing Technology

High-Pressure Processing (HPP) is a nonthermal method for processing food where food is exposed to high pressure in the range of 100–800 MPa. It is also considered a cold processing technique since the temperature is in the ambient field. This technique kills foodborne pathogens but the food possesses better texture and color than heat-treated foods. This technique does not result in a change of molecules conferring flavor and nutritional content to the food. This technique can treat liquid and solid food materials except for foods with air pockets. For HPP to be applicable in the control of food pathogens, the pressure should exceed 400 MPa. The microbial resistance to the treatment depends on several factors, such as pressure, temperature, duration of therapy, and the microbe type. The resistance of gram-positive bacteria to HPP is higher than Gram-negative bacteria due to teichoic acid (a bacterial polysaccharide). Spores are highly resistant to vegetative cells owing to the presence of dipicolinic acid. It has been found that heat-resistant microbes are generally resistant to pressure, while cells in the exponential phase are more sensitive to stress than cells in the stationary phase. The control of viral pathogens in food is related to capsid protein denaturation required for attachment to the host cell. The cell membrane damage at HPP is related to compression of the cell membrane on applying pressure with lipid bi-layer expansion following the release of tension. This leads to loss of integrity of cell membrane, making microbes unable to reproduce. The damaged cells cannot control water and ions transport across membranes resulting in cell death (Naik et al. 2013). Stratakos et al. (2019) attempted to study the effect of high-pressure processing on microbiological safety/shelf life and raw milk quality compared to that of conventional heat pasteurization and untreated milk. It was observed that high-pressure processing could achieve 5 log reductions for pathogenic *E. coli*, *Salmonella,* and *L. monocytogenes,* respectively. High-pressure processing prolongs the shelf life of raw milk by decreasing the levels of Total mesophilic aerobic bacteria, Enterobacteriaceae, lactic acid bacteria, and *Pseudomonas* spp. Levels as compared to those in pasteurized and raw milk. The milk processed using high-pressure exhibited that the treated milk did not affect the quality attributes characterizing raw milk, such as color and mouth sensation due to particle size. Considering the increasing demand for raw milk, high pressure processing can be an alternative for microbiologically safe milk while retaining fresh-like characteristics.

5.5 *Ultrasonication*

Ultrasonication is a rapid and non-destructive technology based on mechanical waves applied for the microbial safety of food products. Ultrasound in a liquid medium results in acoustic cavitation, i.e., phenomena of the bubbles' production, growth, and collapse. The propagation of ultrasound waves causes bubbles to oscillate and collapse, leading to thermal, mechanical (collapse pressure, turbulence, and shear stress), and chemical effects (production of free radicals). The results cause significantly high temperatures (5000 K) and pressures (1000 atm). The microbial control by ultrasonication accounts for the production of acoustic cavitations, leading to an increase in membrane permeability, thinning of the cell membrane, confined heating, and production of hydroxyl radicals (Fig. 3.5). FDA recommends ultrasonication for a 5-log reduction of the microbial population. Ultrasonic 100 W power is considered optimum for microbial control, demonstrating its efficiency against Escherichia coli, Listeria monocytogenes, and other pathogens. The effectiveness of this treatment depends on several microbial properties such as size, hydrophobicity, Gram status, and phase of growth. Microbes with soft and thick capsules are resistant to ultrasonication (Majid et al. 2015). Zhang

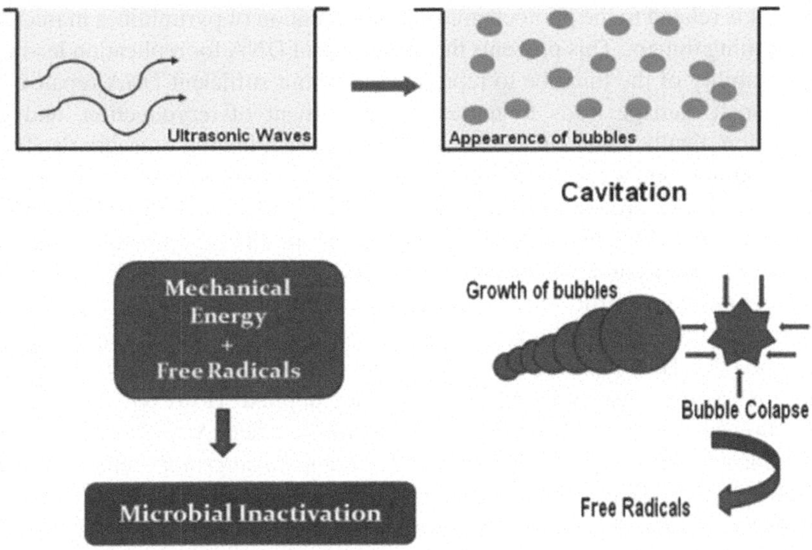

Fig. 3.5 Cavitation and inactivation of microbes by ultrasonication. (Reproduced from de Sao Jose et al. 2014)

et al. (2020) showed the antimicrobial activity of natural compounds in combination with low/high-frequency ultrasound against *E. coli* K12 or *L. innocua*. The synergistic bacterial inactivation was achieved with 0.5–2.0 log reduction on the simultaneous application of U.S. 1 MHz / 20 kHz and citral, carvacrol, or geraniol compared to the sum of individual treatments. Bacterial inactivation was based on ultrasound frequency, bacteria species along with the presence of antimicrobial compounds. The physical impact and dispersion of citral droplets determine the effectiveness of low-frequency ultrasound processes. On the other hand, high frequency causes inactivation by oxidative stress production inside the bacterial cells. The combined treatment showed better bacterial treatment on surfaces of blueberry and washed water used for washing. Tavsanli et al. (2021) showed that ultrasound treatment could serve as an alternative to pasteurization of raw goat milk with 99% inactivation of food pathogens *Brucella melitensis* type 3, *Salmonella Typhimirium*, *Escherichia coli*, *Listeria monocytogenes* and methicilin resistant *Staphylococcus aureus* in goat milk.

5.6 Pulsed-Light System

The pulsed-light system is a nonthermal technology for microbial control in food using short-duration and intense pulses on the food surface or its packaging. The efficiency of microbial decontamination is based on light intensity and the number of pulses the food is exposed to. The high U.V. content in pulsed light is crucial for microbial decontamination. Nucleic acids are the cell target of pulsed light.

U.V. light is related to the photochemical transformation of pyrimidines in microbial DNA forming dimers. This prevents the unzipping of DNA for replication leading to the incapability of the microbe to reproduce. Without sufficient DNA repair mechanisms, such damage leads to mutation, impairment of reproduction, and gene transcription, finally leading to cell death. The decontamination procedure's efficacy depends on the food's composition. This technology is unsuitable for treating food with high oil and protein content since a part of radiation will be absorbed by proteins/oils. Pulsed-light sterilization of packaged products is possible if the packaging is U.V. transparent (Elmnasser et al. 2007). Tao et al. (2019) analyzed the pulsed-light system's efficiency for controlling foodborne pathogens inoculated on the lettuce surface and the treatment impact on lettuce quality through 8 days of storage at 4 °C. They reported that pulsed-light treatment could effectively decrease pathogens' load on leaves based on the fluence applied. However, the pathogens tested showed varied susceptibility to the treatment where *Staphylococcus aureus* was the most sensitive, *Escherichia coli* and *Salmonella enteritidis* being moderately susceptible, while *Listeria monocytogenes* were comparatively resistant. After the refrigeration period, the pulsed-light treated lettuce had low bacterial/yeast/mold count while retaining the food quality by minimizing weight loss and color and preserving the loss of total soluble solids, chlorophyll, and ascorbic acid. Chen et al. (2019) attempted to determine the effect of an intense pulsed light on inactivation of *Cronobacter sakazakii* and *Enterococcus faecium* in powdered food including non-fat dry milk, wheat flour and egg white. They reported up to 5 $\log_{10}$ CFU/g and 2.7 $\log_{10}$ CFU/g reductions for *C. sakazakii* and *E. faecuium*, respectively with no undesirable agglomeration.

6 The Scope and Need for Monitoring Food Contamination

Unsafe food is a global concern to human health and economics, with an estimated 600 million cases of foodborne illness every year. As a result, ensuring food safety is a top priority for public health and a critical step toward achieving food security. Food safety and quality control systems are essential for protecting people's health and well-being, encouraging economic development, and improving livelihoods by facilitating access to domestic, regional, and international markets. Protecting the health of a country's population is one of a government's most important responsibilities, and it's directly tied to the achievement of several Sustainable Development Goals (SDGs). Food safety regulations or monitoring, on the other hand, is essential for ensuring fair practices in the food trade and fostering economic opportunities for all stakeholders involved in the food chain. Controlling foodborne dangers across the entire food chain has become increasingly important in an age of fast-evolving food technologies and ever-increasing global food trade. Food control/monitoring systems must be up to date with the newest advances, function based on risk analysis, and be harmonized with international standards and best practices to meet the complex developing challenges of the twenty-first century.

Due to the water, energy, and material consumption required for the production, processing, storage, and transportation of food that is not being used effectively, this loss of food due to microbial contamination is fundamentally unsustainable (Pleissner 2018). Foodborne illness can be contracted by ingesting microorganisms that cause food poisoning or by eating toxins produced by toxigenic pathogens in food (Bintsis 2017). Livestock and human mobility, land application of raw manure, polluted irrigation water, immature compost application, contaminated soil, and runoff from compost and manure stockpiles on the farm are all sources of pathogen contamination of fresh produce at the farm level (Maurice Bilung et al. 2018; Ssemanda et al. 2018). Environmental samples (soil, feces, water), poorly sanitized food contact surfaces (conveyor belts, knives, slices, etc.) and poorly sanitized non-food contact surfaces (walls, drains, floors, etc.), unhygienic plant design, unregulated traffic patterns, non-sanitized worker's hands, transport trailers, and crates are some of the contamination sources (Perez-Arnedo and Gonzalez-Fandos 2019).

Monitoring food contamination provides information and evidence on the types of contaminants in food. It gives an insight into the increasing trends or drifts in food contamination—this aid in initiating proactive precautionary measures before any severe or threatening health hazard occurs and becomes widespread. Moreover, monitoring programs also help evaluate the feasibility and success of any activity or initiative to minimize contamination. However, monitoring itself is insufficient to solve the problem of contamination. The information obtained must be followed up, like identification of the source of contamination, its control, and elimination. Simultaneously, any food, if found to be contaminated, must immediately be banned by taking appropriate measures. The need/requirement of monitoring food contamination is crucial to ensure public safety and health and proper management of food and agricultural resources to stop or prevent any economic or financial loss. It provides a timely investigation of the changes in the level of contaminants and corroborates that their amount does not exceed any standards. The lack of a proper food contamination monitoring system, therefore, not only poses the risk of severe health complications but also affects the finance and economy. The scope of any monitoring program depends on the availability of resources, its significance concerning health and economy, and any technical limitations like lack of proper analytical tools. Environmental monitoring systems (EMP) are a hands-on method developed to ensure and check food safety by monitoring sanitation and hygiene processes. It is an indicator for preliminary and final product testing since the entire production process must be thoroughly scrutinized to warrant product quality. Environmental sampling is a rational means to check the various contamination sources, maintain cleanliness, and highlight any issues that might require counteractive action.

Commercial technology for the efficient and low-cost detection of microbial contamination in food, industrial wastewater, and clinical samples is in high demand. Microorganisms' optical, electrochemical, metabolic, and physical capabilities have been used in various detection approaches. The necessity for a technology that can generate a quick, reliable, accurate, critical analysis for clinical, industrial, and environmental applications has resulted in significant progress in developing biosensors for microbial detection in recent years. Several instruments have been commercialized as a result of this comprehensive investigation.

Rapid detection and monitoring technologies will be classified as either non-biochemical or bioelectrochemical. Non-biochemical approaches can be split into two categories: standard (dry weight, viable count, and turbidity), specialized (Dry weight, viable count, and turbidity), and sophisticated (Microcalorimetry, Epifluorescence filter technique, Fluorescent-antibody technique, Radiometry, Bio-luminescence, Coulter counter, Electronic particle analysis, Micro-ELISA, Electron microscopy, spectroscopy, etc.). For the assessment of microbial biomass, a variety of electrochemical detection/monitoring technologies (impedimetry and conductivity, fuel cell technology, cyclic and square wave voltammetry) have been presented. Like the preceding strategies, their goal is to lower the time it takes to detect microorganisms, eventually developing a quick, accurate, economic, and repeatable biomass probe (Hobson et al. 1996).

Microbial food contamination is a crucial area of food safety where hazard analysis and critical control points (HACCP) will significantly impact preventing bacteria from being transmitted to humans through food. When foodborne bacteria cause illnesses, one of the main concerns is that the germs may become invasive and necessitate medical intervention, such as using antibiotics. In the context of food microbiological contamination, antibiotic-resistant microorganisms are a subpopulation of organisms that, when present, can be carried inside the food product and offer a significant obstacle to illness treatment and remediation in humans. The hazard of disease and illness caused by food contamination with microbes is far greater than the threat of resistance transfer from animals to people, based on the number of recorded cases. More explicitly, the relationship between the actual ailment caused by an antibiotic-resistant organism and the incidence of a genetic component of resistance being passed to microbes and causing disease must be tracked. Furthermore, because of the potential introduction of antibiotic residues and microbiological infections from countries with less rigorous agricultural production practices and quality-control systems, imported foods may need to be strictly regulated (The Use of Drugs in Food Animals: Benefits and Risks 1999. National Academies Press (U.S.)).

To ensure human health and safety, it is necessary to systematically detect, assess, and regulate the harmful consequences and related probability resulting from microbiological pathogen-contaminated foods. Food safety, on the other hand, is difficult to pin down and quantify. Factors like microbiological safety, chemical safety, and personal and environmental hygiene are crucial to ensure the overall well-being of the food. It necessitates consistent and reasonable efforts by competent authorities, corporate operators, scientists, consumers, and monitoring body/organizations representatives like the Food Safety and Standard Authority of India, 2011. Globally, proposed measures include (i) the implementation of standards and guidelines such as ISO 22000, the Hazard Analysis Critical Control Point (HACCP) scheme, the Good Manufacturing or Management Practices (GMP) scheme, and (ii) the scientific basis for risk management related to food consumption, such as the Food Safety Barometer developed by the Belgian Food Safety Scientific Committee (Saad et al. 2013).

7 Effect of Microbes on the Economy – Effect on the Import and Export Sector

Food spoilage poses a severe challenge to food security, i.e., our ability to provide a sufficient food supply to the world's growing population. In the opinion of a former World Health Organization official, "*This large increasing world population needs food, and we have a moral obligation to utilize all our skills and technologies to increase food production and limit food spoilage.*"

In recent years, an increase in the international food trade, extensive production typically involving multiple sites, and a complex supply chain have contributed to microbial food deterioration. Food hygiene is critical for people's general well-being and daily lives, as well as for economic development, social stability, and the reputation of the government and country (Hussain and Dawson 2013).

Food processing sectors have a disproportionately large share of the global economy. The processed food business is constantly expanding due to technical advancements, rising demand, and changing customer preferences. As a result of this improvement, both developed and developing countries are adopting innovative food processing and delivery technologies. (Amit et al. 2017). The food processing industry loses millions of dollars yearly owing to microbiological contamination and substantial reductions in products that do not fulfill consumer expectations. Food is vulnerable to deterioration and pathogenic microbial activity. Some microorganisms generally attach to solid surfaces with sufficient nutritional content for nourishment and growth in natural habitats. Reduced crop yield and quality, as well as considerable economic losses owing to buyer rejection and mycotoxin contamination of grains, are all consequences of microbial infection of food crops (e.g., poisonous fungal secondary metabolites).

Spoilage is a business concern in these countries, affecting producers' and manufacturers' profits and losses. Spoilage continues to be a severe worry in less developed countries. It is impossible to calculate the actual economic cost of food deterioration. Approximately 30% of manufactured food products are damaged, with microbial food spoilage being the most common cause (Gram et al. 2002).

Food waste is inefficient and expensive, and it has the potential to harm the economy and diminish customer confidence. According to the E.U. 2020 Resource Efficiency Flagship, which presents a strategic framework for more sustainable and efficient use of natural resources, each person wastes approximately 179 kg of food annually. This amounts to around 89 million tonnes per year in total. In Europe, food waste is anticipated to increase to almost 126 million tonnes.

An increase in the globalized food trade in recent years, extensive production often involving many sites, and a complex supply chain all contribute toward an increased number of microbiological food. The presence of foodborne pathogens in a country's food supply impact not only the health of the local population but also poses a risk of pathogen transmission to travelers and customers in other nations where food is exported. Awareness of these issues has prompted international initiatives to harmonize food safety standards, which has complicated international food commerce.

More than 200 ailments, ranging from diarrhea to cancer, are caused by contaminated food carrying pathogenic bacteria, viruses, parasites, or inorganic chemicals. Every year, an estimated 600 million people worldwide become infected after eating tainted food, with 420,000 deaths resulting in the loss of 33 million disability-adjusted life years. With 125,000 deaths annually, children under five account for 40% of the foodborne disease burden. Diarrheal infections are the most frequent illnesses caused by contaminated food, affecting 550 million people each year and resulting in 230,000 fatalities. Different parties involved in the national or international food trade may be defrauded, such as manufacturers, co-packers, distributors, and others along the distribution chain. Because it impacts individuals of all ages, races, genders, and income levels worldwide, food safety in the food market is one of the most critical areas of concern in public health. Food marketing on a local and international scale continues to impact public food safety and health substantially. Food supply systems now span many national borders, putting health concerns on a global scale (Gizaw 2019).

Food loss is significant in the efforts to combat hunger, raise income and improve food security in the world's poorest countries. Food loss due to spoilage or contaminated food hurts the food industry and customers, resulting in financial losses and higher hospitalization costs. Infectious diseases associated with food intake are one of the impediments to economic growth, and they have an impact on a country's productivity as well as medical costs. Apart from the sudden drop in market transactions that commonly precede such occurrences, which can take years to stabilize, the result of such incidences is significant economic consequences as a result of taking products off shelves, reporting consumers, and the cost of lawsuit reparations. Food processing, packaging, and formulation procedures are frequently designed to suppress or control microbial development to prevent spoilage. The most restricted techniques are pH reduction, preservatives, water activity limiting, oxygen tension management, thermal processing, and hermetic packing. These procedures are used in tandem to inactivate or prevent the growth of possible spoiling microorganisms. On the other hand, the lack of competition from different background microbiota enhances the propagation of certain microorganisms that can resist these controls.

8 Conclusion

Food rotting caused by bacteria or microbiological food contamination threatens the global system. Furthermore, the emergence of antibiotic-resistant pathogens has aggravated food safety risks. To keep a circular food supply chain viable, extremely cautious skills and expansion are required to reduce the level and recurrence of food contamination, as well as scientific studies into the eventual demise of contaminants during treatment, techniques for simplified, economic, and reliable monitoring, and policy options to safeguard the framework. Therefore, from the conventional thermal and chemical strategies for food decontamination, the food processing industry implements novel and innovative techniques like pulsed electric fields, nonthermal plasma, electrolyzed water, etc., which will ensure the safety and improved shelf-life of the food. In addition, various food contamination monitoring technologies have also emerged, which detect specific physical and chemical characteristics of the microorganisms leading to their rapid identification and development of control mechanisms. These technologies will make the food industry aware of the various food-related diseases, identify and combat them, and strengthen their manufacturing and packaging processes to deliver the products safely.

Acknowledgments The authors acknowledge the Department of Biotechnology, New Delhi, for research funding of the Indo-EU collaborative project "Strategic Planning for Water Resources and Implementation of Novel Biotechnical Treatment solutions and Good Practices (SPRING)" (Sanction No. BT/IN/EU-WR/60/S.P./2018).

References

Alum EA, Urom SMOC, Ben CMA (2016) Microbiological contamination of food: the mechanisms, impacts and prevention. Int J Sci Technol Res 5:65–78

Amit SK, Uddin MM, Rahman R, Islam SR, Khan MS (2017) A review on mechanisms and commercial aspects of food preservation and processing. Agric Food Secur 6:1–22. https://doi.org/10.1186/s40066-017-0130-8

Ashbolt NJ (2004) Microbial contamination of drinking water and disease outcomes in developing regions. Toxicology 198:229–238

Bartowsky EJ, Henschke PA (2008) Acetic acid bacteria spoilage of bottled red wine — a review. Int J Food Microbiol 125:60–70. https://doi.org/10.1016/j.ijfoodmicro.2007.10.016

Bhat RV (2004) Safety and quality of processed foods. Indian Food Ind 23:58–61

Bintsis T (2017) Foodborne pathogens. AIMS Microbiol 3:529–563. https://doi.org/10.3934/microbiol.2017.3.529

Bulut N, Atmaca B, Akdemir Evrendilek G, Uzuner S (2020) Potential of pulsed electric field to control *Aspergillus parasiti*cus, aflatoxin and mutagenicity levels: sesame seed quality. J Food Saf 40(6):e12855

Chen D, Cheng Y, Peng P, Liu J, Wang Y, Ma Y et al (2019) Effects of intense pulsed light on *Cronobacter sakazakii* and *Salmonella* surrogate *Enterococcus faecium* inoculated in different powdered foods. Food Chem 296:23–28

de São José JFB, de Andrade NJ, Ramos AM, Vanetti MCD, Stringheta PC, Chaves JBP (2014) Decontamination by ultrasound application in fresh fruits and vegetables. Food Control 45:36–50

ECDC E. EFSA (European Food Safety Authority) and ECDC (European Centre for Disease Prevention and Control) (2015) The European Union summary report on trends and sources of zoonoses, zoonotic agents and foodborne outbreaks in. 2014;13:4036

Elmnasser N, Guillou S, Leroi F, Orange N, Bakhrouf A, Federighi M (2007) Pulsed-light system as a novel food decontamination technology: a review. Can J Microbiol 53:813–821

Gizaw Z (2019) Public health risks related to food safety issues in the food market: a systematic literature review. Environ Health Prev Med 24:1–21. https://doi.org/10.1186/s12199-019-0825-5

Gram L, Ravn L, Rasch M, Bruhn JB, Christensen AB, Givskov ME (2002) Food spoilage—interactions between food spoilage bacteria. Int J Food Microbiol 78:79–97. https://doi.org/10.1016/S0168-1605(02)00233-7

Green L, Selman C (2005) Factors impacting food workers' and managers' safe food preparation practices: a qualitative study. Food Protec Trend 25:981–990

Hashemi SMB, Gholamhosseinpour A, Niakousari M (2019) Application of microwave and ohmic heating for pasteurization of cantaloupe juice: microbial inactivation and chemical properties. J Sci Food Agric 99:4276–4286

Hobson NS, Tothill I, Turner AP (1996) Microbial detection. Biosens Bioelectron 11:455–477. https://doi.org/10.1016/0956-5663(96)86783-2

Hussain MA, Dawson CO (2013) Economic impact of food safety outbreaks on food businesses. Foods 2:585–589. https://doi.org/10.3390/foods2040585

Juneja VK, Sheen S, Tewari G (2007) Intervention technologies for food safety and preservation. In: Wilson CL (ed) Microbial food contamination. CRC Press, pp 347–393

Kirch W (2008) Microbiological hazards, food safety. In: Encyclopedia of public health. Springer, p 457

Liang D, Wang Q, Zhao D, Han X, Hao J (2019) Systematic application of slightly acidic electrolyzed water (SAEW) for natural microbial reduction of buckwheat sprouts. LWT – Food Sci Technol 108:14–20

Majid I, Nayik GA, Nanda V (2015) Ultrasonication and food technology: a review. Cogent Food Agric 1:1071022. https://doi.org/10.1080/23311932.2015.1071022

Makroo HA, Rastogi NK, Srivastava B (2020) Ohmic heating assisted inactivation of enzymes and microorganisms in foods: a review. Trends Food Sci Technol 97:451–465

Maurice Bilung L, Sin Chai L, Tahar AS, Ted CK, Apun K (2018) Prevalence, genetic heterogeneity, and antibiotic resistance profile of *Listeria* spp. and *Listeria monocytogenes* at farm level: a highlight of ERIC-and BOX-PCR to reveal genetic diversity. Biomed Res Int 2018:3067494. https://doi.org/10.1155/2018/3067494

Medina-Gudiño J, Rivera-Garcia A, Santos-Ferro L, Ramirez-Orejel JC, Agredano-Moreno LT, Jimenez-Garcia LF et al (2020) Analysis of Neutral Electrolyzed Water anti-bacterial activity on contaminated eggshells with *Salmonella enterica* or *Escherichia coli*. Int J Food Microbiol 320: 108538

Mendes-Oliveira G, Jin TZ, Campanella OH (2021) Microbial safety and shelf-life of pulsed electric field processed nutritious juices and their potential for commercial production. J Food Process Preserv 46:e16249

Mohamed ME, Eissa AH (2012) Pulsed electric fields for food processing technology. Structure and function of food engineering. https://www.intechopen.com/chapters/38363

Naik L, Sharma R, Rajput YS, Manju G (2013) Application of high-pressure processing technology for dairy food preservation-future perspective: a review. J Anim Prod Adv 3(8):232–241. https://doi.org/10.5455/japa.20120512104313

Noor R (2019) Insight to foodborne diseases: Proposed models for infections and intoxications. Biomed Biotechnol Res J 3(3):135

Osaili TM, Al-Nabulsi AA, Al Sheikh YM, Alaboudi AR, Olaimat AN, Al-Holy M et al (2021) Inactivation of Salmonella spp., Escherichia coli O157: H7 and Listeria monocytogenes in Tahini by Microwave Heating. Food 10(12):2972

Perez-Arnedo I, Gonzalez-Fandos E (2019) Prevalence of *Campylobacter* spp. in poultry in three Spanish farms, a slaughterhouse and a further processing plant. Foods 8:111. https://doi.org/10.3390/foods8030111

Pleissner D (2018) Recycling and reuse of food waste. Curr. Opin. Green sustain. Chem 1(13): 39–43

Rawat S (2015) Food spoilage: microorganisms and their prevention. Asian J Plant Sci 5:47–56

Saad M, See TP, Abdullah MF, Nor NM (2013) Use of rapid microbial kits for regular monitoring of food-contact surfaces towards hygiene practices. Procedia Soc Behav Sci 105:273–283. https://doi.org/10.1016/j.sbspro.2013.11.029

Sakamoto K, Konings WN (2003) Beer spoilage bacteria and hop resistance. Int J Food Microbiol 89:105–124. https://doi.org/10.1016/S0168-1605(03)00153-3

Ssemanda JN, Reij MW, van Middendorp G, Bouw E, van der Plaats R, Franz E, Muvunyi CM, Bagabe MC, Zwietering MH, Joosten H (2018) Foodborne pathogens and their risk exposure

factors associated with farm vegetables in Rwanda. Food Control 89:86–96. https://doi.org/10.1016/j.foodcont.2017.12.034
Stratakos AC, Inguglia ES, Linton M, Tollerton J, Murphy L, Corcionivoschi N et al (2019) Effect of high pressure processing on the safety, shelf life and quality of raw milk. Innov Food Sci Emerg Technol 52:325–333
Sudershan RV, Naveen Kumar R, Kashinath L, Bhaskar V, Polasa K (2010) Economic impact of a food borne disease outbreak in Hyderabad-a case study. Indian J Nutr Diet 47:246–251
Tamang JP, Shin DH, Jung SJ, Chae SW (2016) Functional properties of microorganisms in fermented foods. Front Microbiol 7:578
Tao T, Ding C, Han N, Cui Y, Liu X, Zhang C (2019) Evaluation of pulsed light for inactivation of foodborne pathogens on fresh-cut lettuce: effects on quality attributes during storage. Food Pack Shelf Life 21:100358
Tao Q, Wu Q, Zhang Z, Liu J, Tian C, Huang Z, Malakar PK, Pan Y, Zhao Y (2022) Meta-analysis for the global prevalence of foodborne pathogens exhibiting antibiotic resistance and biofilm formation. Front Microbiol 13:906490.
Tauxe RV, Doyle MP, Kuchenmüller T, Schlundt J, Stein CE (2010) Evolving public health approaches to the global challenge of foodborne infections. Int J Food Microbiol 139:16–28. https://doi.org/10.1016/j.ijfoodmicro.2009.10.014
Tavsanli H, Aydin M, Ede ZA, Cibik RECEP (2021) Influence of ultrasound application on the microbiota of raw goat milk and some food pathogens including *Brucella melitensis*. Food Sci Technol Int 28:634–640
Thakali A, MacRae JD (2021) A review of chemical and microbial contamination in food: what are the threats to a circular food system? Environ Res 194:110635
Zhang H, Wang S, Goon K, Gilbert A, Huu CN, Walsh M et al (2020) Inactivation of foodborne pathogens based on synergistic effects of ultrasound and natural compounds during fresh produce washing. Ultrason Sonochem 64:104983

Chapter 4
Enhancing Nutritional Quality of Crops Through Genetic Engineering

Debee Prasad Sahoo and Chetan Panda

1 Introduction

Malnutrition or hidden hunger is one of the most global health challenges of twenty-first century and affects two billion people in developing countries (Boliko 2019). Insufficient intake of micronutrients in the diet leads to weakened immunity and poor health mostly in women, infants, and children. Since our daily diet is mainly based on cereal-based foods, there is necessity to improve yield as well as nutritional content of food crops. A strategy called 'biofortification' has been employed to improve nutritional content in food crops by using agronomic practices, plant breeding and modern biotechnology or genetic engineering (Saltzman et al. 2017).

Agronomic bio-fortification provides temporary micronutrient improvement through fertilizers. The approach is highly recommendable, if the primary motto is to increase micronutrient that can be directly absorbed by the plants such as Zn, but less so for the micronutrients that are synthesized within the plants intrinsically and cannot be absorbed directly (Lyons and Cakmak 2012). Biofortification of crops through conventional breeding approach has worked for long, but it is only limited to sexually compatible plants and also challenging due to lack of genetic variation for micronutrient traits within crop species. Genetic engineering is a promising approach for plant nutritional improvements, as the gene pool for genetic modification is virtually unlimited and an entirely new metabolic pathway can be introduced into a plant to achieve targeted micronutrient (Hirschi 2009; Uncu et al. 2013).

D. P. Sahoo (✉)
School of Biotechnology, KIIT Deemed to be University, Bhubaneswar, Odisha, India
e-mail: debee.sahoo@kiitbiotech.ac.in

C. Panda
Department of Agricultural Biotechnology, College of Agriculture, Odisha University of Agriculture and Technology, Bhubaneswar, Odisha, India

H. U. Hebbar et al. (eds.), *Engineering Aspects of Food Quality and Safety*, Food Engineering Series, https://doi.org/10.1007/978-3-031-30683-9_4

In this chapter, we discuss applications of genetic engineering towards biofortification of crop plants, advancements of nutritionally enriched crops with various micronutrients in last several decades and the future of engineered crops with more precise biotechnological tools for better consumer acceptance.

2 Genetic Engineering Micronutrients and Phytonutrients in Crops

Micronutrients and phytonutrients are essential for human health, but the major food crops we consume in our daily diet are often deficient in these compounds (Mattoo et al. 2010). So, it has been a major challenge for researchers to biofortify the food crops for enhanced nutritional qualities by using several strategies. Genetic engineering strategy has relied on the use of a single or multiple key enzyme genes from plants or microbes and subsequent transfer of foreign genes to crop plants by plant transformation. This strategy enables introduction of novel genes directly into the plant's genome to boost essential nutrients. The organic molecules like vitamins and fatty acids are generally synthesized by the plants. Therefore, to biofortify crop plants with these compounds, the metabolic pathway of the compounds is targeted with an aim to increase their synthesis or to decrease the anti-nutritional factors or to extend the existing metabolic pathway for novel products (Capell and Christou 2004). Furthermore, alternate metabolic pathways from bacteria and other organisms can also be incorporated into crop plants for metabolic engineering (Newell-McGloughlin 2008). By contrast, for mineral biofortification, which are generally obtained by the plant from the environment, mineral transporter genes are used to improve the absorption and utilization efficiency of minerals in crop plants (Zhu et al. 2007). Several transgenic approaches such as overexpression, gene stacking, gene silencing by RNA interference (RNAi) have been successfully used to enhance the nutrient content of food crops which are discussed in following sections.

3 Vitamin Biofortification in Crops

Vitamins in diets are very much essential for proper human growth and development. Human body is unable to synthesize these essential micronutrients and must obtain them from food crops (Garcia-Casal et al. 2017). Unfortunately, major food crops are poor source of these micronutrients (Strobbe et al. 2018). So, the primary goal of genetic engineering is enhancement of vitamin content in major food crops. To date, biofortification of vitamins in crops via genetic engineering have mainly focused on vitamins A, C and E and folate (vitamin B9) enrichment.

3.1 Vitamin A

Vitamin A deficiency (VAD) in the diets is the main cause of blindness and of a compromised immune system (Oliver 2014). β-carotene (provitamin A) is the most common precursor of vitamin A (Grune et al. 2010) and efforts have been made to introduce the correct metabolic steps of β-carotene biosynthesis pathway (Fig. 4.1) in staple crops to facilitate β-carotene synthesis. In last two decades since the generation of Golden Rice, several β-carotene enriched crop plants have been developed through genetic engineering approaches which are summarized in Table 4.1.

Rice being a major staple food lacks provitamin A in its endosperm. Ye et al. (2000) introduced three different genes of β-carotene biosynthetic pathway: phytoene synthase (*PSY*) and lycopene β-cyclase (*LCYB*) from daffodil and phytoene desaturase from bacteria into rice which resulted in 1.6 μg of carotenoids per g of endosperm, with 50% of this as β-carotene. By replacing daffodil *PSY* with maize *PSY1* together with bacterial *CrtI* enhanced carotenoids level upto 37 μg/g of

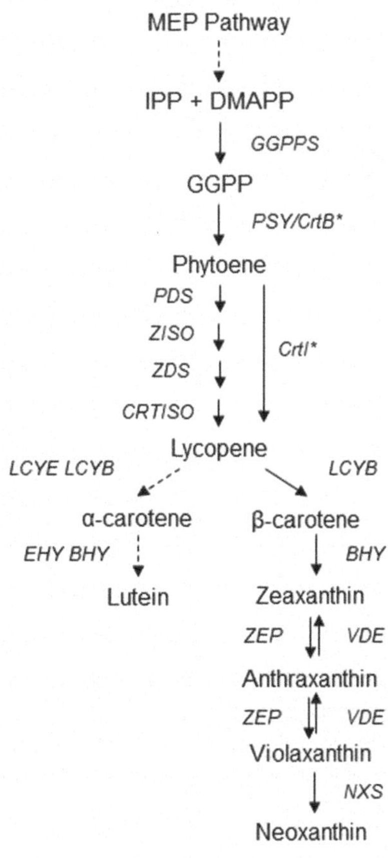

Fig. 4.1 Schematic carotenoid biosynthesis pathway in plants. Bacterial counterparts of the plant enzymes are indicated with stars. Abbreviations: MEP, methylerythritol phosphate; IPP, isopentenyl diphosphate; DMAPP, dimethylallyl diphosphate; GGPPS, geranylgeranyl diphosphate synthase; PSY, phytoene synthase, CrtB, bacterial phytoene synthase; PDS, phytoene desaturase; ZISO, ζ-carotene isomerase; ZDS, ζ-carotene desaturase; CRTISO, carotenoid isomerase; CrtI, bacterial carotene desaturase, LCYB, lycopene β-cyclase; LCYE, lycopene ε-cyclase; BHY, β-carotene hydroxylase; EHY, ε-carotene hydroxylase; ZEP, zeaxanthin epoxidase, VDE, violaxanthin de-epoxidase; NXS, neoxanthin synthase. (Redrawn from Zhu et al. (2020))

Table 4.1 Vitamin biofortification in crop plants

Target compound	Target crop	Target genes	References
β-carotene	Rice	*PSY, Crt1, LCYB*	Ye et al. (2000)
		PSY1, Crt1	Paine et al. (2005)
	Potato	*CrtB*	Ducreux et al. (2005)
		Silencing LCYE	Diretto et al. (2006)
	Corn	*CrtB, Crtl*	Aluru et al. (2008)
		PSY1, Crt1	Naqvi et al. (2009)
	Wheat	*y1, Crt1*	Cong et al. (2009)
	Cassava	*CrtB, DXS*	Sayre et al. (2011)
	Banana	*PSY*	Paul et al. (2017)
Folate	Rice	*GTPCHI, ADCS*	Storozhenko et al. (2007)
		GTPCHI, ADCS, FPGS, FBP	Blancquaert et al. (2015)
		HPPK/DHPS	Gillies et al. (2008)
	Corn	*folE (GTPCHI)*	Naqvi et al. (2009)
	Tomato	*GTPCHI*	de la Garza et al. (2004)
	Potato	*GTPCHI, ADCS*	Blancquaert et al. (2013)
		GTPCHI, ADCS, FPGS, FBP	De Lepeleire et al. (2018)
	Mexican bean	*GTPCHI*	Ramírez Rivera et al. (2016)
Ascorbate	Corn	*DHAR*	Chen et al. (2003)
		DHAR	Naqvi et al. (2009)
	Potato	*GalUR*	Hemavathi et al. (2009)
		GLOase	Hemavathi et al. (2010)
		DHAR	Qin et al. (2011)
	Tomato	*DHAR*	Haroldsen et al. (2011)
		Silencing AO	Zhang et al. (2011a)
		Silencing APX	Zhang et al. (2011b)
Vitamin E	Corn	*HGGT*	Cahoon et al. (2003)
		HPPD, MPBQ MT	Naqvi et al. (2011)
	Rice	*HPPD*	Farre et al. (2012)
		γ-TMT	Zhang et al. (2013)
	Soybean	*MPBQ MT*	Van Eenennaam et al. (2003)
		γ-TMT	Tavva et al. (2007)

which 80% was β-carotene (Paine et al. 2005). Transgenic potato expressing bacterial phytoene synthase (*CrtB*) resulted in higher accumulation of total carotenoids and β-carotene (Ducreux et al. 2005). In another study silencing of lycopene ε-cyclase (*LCYE*) gene resulted in higher accumulation of β-carotene in potato tubers (Diretto et al. 2006). Transgenic maize overexpressing bacterial *CrtB* and *CrtI* accumulated higher levels of carotenoids and β-carotene in kernels (Aluru et al.

2008). Overexpressing two transgenes, maize *PSY1* and bacterial *CrtI* in maize, Naqvi et al. (2009) demonstrated elevated level of carotenoids and β-carotene in transgenic maize endosperm. Cong et al. (2009) engineered wheat with maize phytoene synthase (maize *y1*) and bacterial *CrtI* which resulted in higher accumulation of carotenoids in wheat endosperm. Transgenic cassava overexpressing bacterial *CrtB* achieved 10- to 20- fold carotenoid content whereas co-expressing *CrtB* with 1-deoxyxylulose-5-phosphate synthase (DXS) increased carotenoid level up to 30-fold (Sayre et al. 2011). A *cis*-genic *PSY*-overexpression resulted in 5.5 mg/100 g DW β-carotene content in banana fruits (Paul et al. 2017).

3.2 Vitamin B9 (Folate)

Vitamin B9 or folate is very much required in the diet to prevent birth defects, cardiovascular disease, cancer, and anemia like ailments (Hossain et al. 2004). Human body can't synthesize folates and must be supplied in diets (Basset et al. 2005). Many successful biofortified crops have been developed in the last few decades (Table 4.1) using folate metabolic engineering approach (Fig. 4.2).

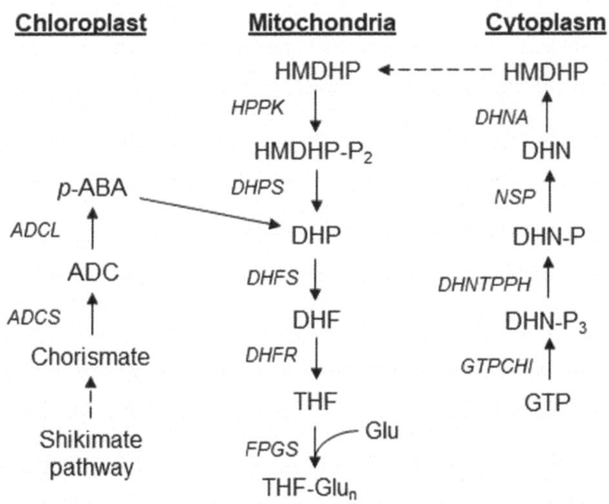

Fig. 4.2 Folate biosynthesis pathway in plants. Abbreviations: ADC, aminodeoxychorismate; ADCS, ADC synthase; ADCL, ADC lyase; *p*-ABA, *para*-aminobenzoate; DHN-P_3, dihydroneopterin triphosphate; DHN-P, dihydroneopterin monophosphate; DHN, dihydroneopterin; HMDHP, 6-hydroxymethyldihydropterin; GTPCHI, GTP cyclohydrolase I; DHNTPPH, dihydroneopterin triphosphate pyrophophohydrolase; NSP, non-specific phosphatase; DHNA, DHN aldolase; HMDHP-P2, HMDHP pyrophosphate; DHP, dihydropteroate; DHF, dihydrofolate; THF, tetrahydrofolate; Glu, glutamate; HPPK, HMDHP pyrophosphokinase; DHPS, DHP synthase; DHFS, DHF synthetase; DHFR, DHF reductase; FPGS, folylpolyglutamate synthetase. (Redrawn from Uncu et al. (2013) and Strobbe and Van Der Straeten (2018))

Introducing a single gene GTP cyclohydrolase I (*GTPCHI*), folate levels were increased in rice (Storozhenko et al. 2007), tomato (de la Garza et al. 2004), maize (Naqvi et al. 2009), potato (Blancquaert et al. 2013), and Mexican bean (Ramírez Rivera et al. 2016). By overexpressing the bifunctional enzyme, 6-hydroxymethyl-7,8 dihydropterin pyrophosphokinase/7,8-dihydropteroate synthase (*HPPK/DHPS*), Gillies et al. (2008) improved folate level in transgenic rice. However, the single gene approach didn't result in much folate content in the transgenic lines. Using multigenic approach, several attempts have been made to increase the folate content. Overexpressing *GTPCHI* along with aminodeoxychorismate synthase (*ADCS*) resulted enhanced folate level in rice (Storozhenko et al. 2007), and potato (Blancquaert et al. 2013). Transgenic rice and potato overexpressing four genes (*GTPCHI*, *ADCS*, *FPGS*, and *FBP*) simultaneously demonstrated not only enhanced folate level but also enhanced folate stability upon storage (Blancquaert et al. 2015; De Lepeleire et al. 2018).

3.3 Vitamin C

Vitamin C (ascorbate) acts as an antioxidant in human body and plays a major role in several enzymatic processes as a cofactor. Plants are the major source of vitamin C as human body can't synthesize this vitamin. Several reports have demonstrated enhanced level of ascorbate by overexpressing or downregulating genes of ascorbate biosynthesis pathway in crop plants (Fig. 4.3, Table 4.1).

Ascorbate recycling in which dehydroascorbate reductase (DHAR) catalyzes the oxidized ascorbic acid back to non-oxidized form has been a major target for vitamin C biofortification of crop plants (Chen et al. 2003). Enhanced level of vitamin C accumulated in maize by overexpressing *DHAR* gene from wheat and rice, respectively (Chen et al. 2003; Naqvi et al. 2009). Overexpression of strawberry D-galacturonic acid reductase in potato doubled the ascorbate content compared to wild type plants (Hemavathi et al. 2009). Transgenic potato expressing a rat L-gulono-γ-lactone oxidase gene accumulated 40% higher ascorbate content (Hemavathi et al. 2010). Overexpression of cytosolic targeted *DHAR* enhanced ascorbate content both in potato tubers and leaves whereas chloroplastic targeted *DHAR* only increased leaf ascorbate content (Qin et al. 2011). Transgenic tomato expressing cytosolic targeted *DHAR* produced enhanced level of ascorbic acid in tomato fruits (Haroldsen et al. 2011). Down-regulating ascorbate oxidase (*AO*) and ascorbate peroxidase (*APX*) using RNAi, Zhang et al. (2011a, b) demonstrated improved ascorbate content in tomato fruit.

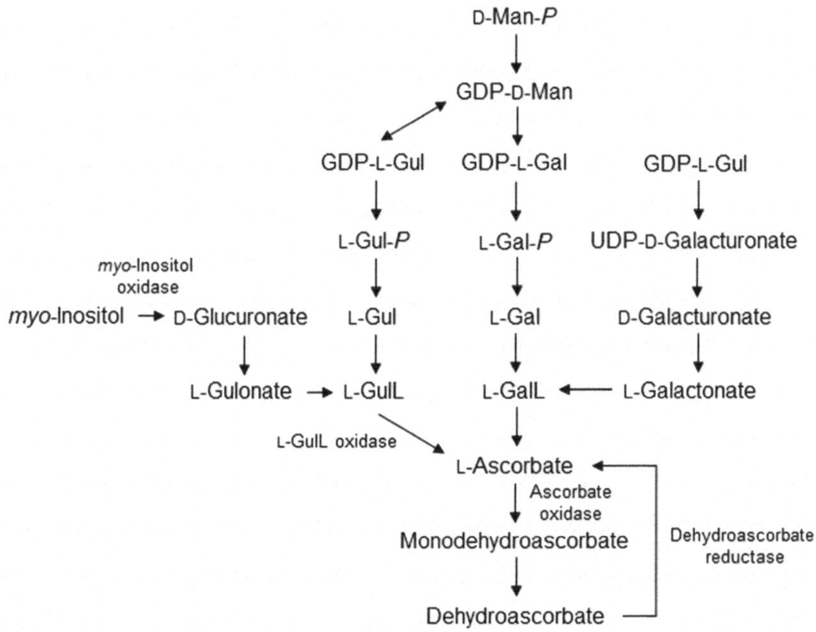

Fig. 4.3 Ascorbate biosynthesis in plants. Abbreviations: Gal, galactose; GalL, galactonolactone; GDP, guanosine diphosphate; Gul, gulose; GulL, gulonolactone; Man, mannose; P, phosphate; UDP, uridine diphosphate. (Redrawn from Zhu et al. (2007))

3.4 Vitamin E

Vitamin E is a group of compounds consisting of four tocopherols and four tocotrienols out of which α-tocopherol has the highest activity (Rocheford et al. 2002). Vitamin E prevents cardiovascular disease and cancer in human body (Bramley et al. 2000). Crop plants are successfully engineered with enhanced level of vitamin E by overexpressing genes involved in tocopherol biosynthesis pathway (Fig. 4.4, Table 4.1).

Transgenic corn overexpressing homogentisic acid geranylgeranyl transferase (*HGGT*) resulted in enhanced level of vitamin E (Cahoon et al. 2003). Genetic engineering of soybean with 2-methyl-6-phytyl benzoquinol methyltransferase (*MPBQ MT*) and γ-tocopherol methyltransferase (*γ-TMT*) increased seed vitamin E level up to five-fold (Van Eenennaam et al. 2003). Overexpression of tocopherol methyl transferase (*γ-TMT*) gene from *Perilla frutescens* resulted in 4.8-fold higher vitamin E activity in transgenic soybean seeds (Tavva et al. 2007). Overexpressing Arabidopsis hydroxyphenylpyruvate dioxygenase (*HPPD*) gene in rice Farre et al. (2012) demonstrated increased vitamin E activity in transgenic rice grains. Transgenic maize lines overexpressing Arabidopsis *HPPD* and *MPBQ methyl transferase*

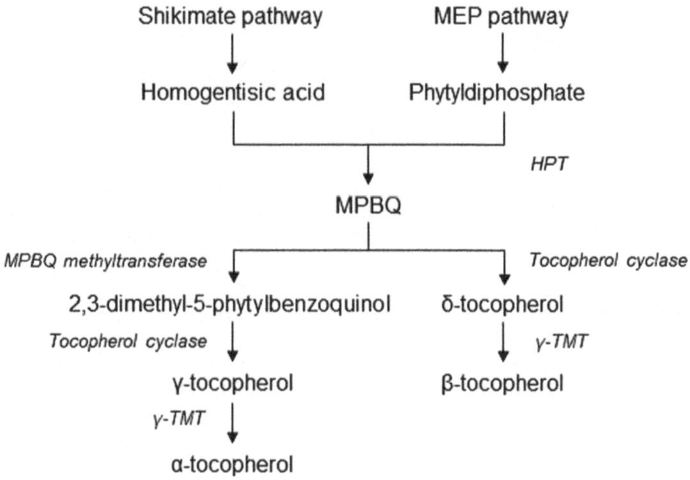

Fig. 4.4 Tocopherol biosynthesis pathway in plants. Abbreviations: HPT, homogentisic acid prenyltransferase; MEP, methylerythritol phosphate; MPBQ, 2-methyl-6-phytylbenzoquinol; γ-TMT, γ-tocopherol methyltransferase. (Redrawn from Zhu et al. (2007))

accumulated three times more γ-tocopherol content compared to wild type with enhanced vitamin E activity (Naqvi et al. 2011). Constitutive overexpression of Arabidopsis *γ-TMT* gene resulted in increased vitamin E level in transgenic rice (Zhang et al. 2013).

4 Mineral Biofortification in Crops

Various minerals take part in maintaining health and proper growth of human body (Welch 2002). Deficiency of essential minerals like iron and zinc could cause immune disorders as well as interfere in normal growth and development of human body (Zimmermann and Hurrell 2002). Plant genes involved in mineral uptake, transport and storage are the major targets for mineral biofortification in crop plants (Table 4.2).

In early work, Goto et al. (1999) introduced soybean ferritin, a major iron storage protein, in rice and observed a three-fold increase of iron content in transgenic rice. When a human lactoferrin gene was overexpressed in rice, a two-fold increase of iron content accumulated in rice grain (Nandi et al. 2002). Soybean ferritin overexpression in rice exhibited simultaneous increase in iron and zinc content throwing light on a common regulating mechanism of mineral transport (Vasconcelos et al. 2003). Transgenic wheat overexpressing ferritin resulted in 50%–85% increase in iron content (Borg et al. 2012). Another strategy was

Table 4.2 Mineral biofortification in crop plants

Target compound	Target crop	Target genes	References
Iron	Rice	*Ferritin*	Goto et al. (1999)
		Synthetic human lactoferrin	Nandi et al. (2002)
		Ferritin	Vasconcelos et al. (2003)
		NAS3	Lee et al. (2009)
		NAS, ferritin, phytase	Wirth et al. (2009)
		Ferritin, NAS1, YSL2	Masuda et al. (2012)
		NAS1	Nozoye (2018)
	Wheat	*Ferritin*	Borg et al. (2012)
Zinc	Rice	*Ferritin*	Vasconcelos et al. (2003)
		NAS3	Lee et al. (2009)
		Ferritin, NAS1, YSL2	Masuda et al. (2012)
		NAS1	Nozoye (2018)

employed by Lee et al. (2009) by overexpressing nicotinamine synthase *(NAS)* gene to enhance iron content in rice. Transgenic rice grains overexpressing *OsNAS3* showed elevated levels of both iron and zinc. In a multigenic approach, transgenic rice overexpressing Arabidopsis *NAS* gene, common bean ferritin and *Aspergillus fumigatus* phytase, demonstrated six-fold increase in iron content (Wirth et al. 2009). In a similar approach by overexpressing soybean ferritin, barley *NAS1*, and rice nicotinamine-metal transporter (*OsYSL2*) genes Masuda et al. (2012) demonstrated elevated level of both iron and zinc content in transgenic rice seeds. Genetic engineering rice with barley *NAS1* resulted in enhanced levels of both iron and zinc (Nozoye 2018).

5 Phytonutrient Biofortification in Crops

Concentrations of phytonutrients have significantly decreased in modern diets due to selective use of food crops. This has resulted in rise of several chronic diseases in human beings and crops fortified with phytonutrients have the potential to reduce the incidence of chronic diseases to maintain a healthy life (Martin and Li 2017). Phytonutrients include polyphenolic compounds like flavonoids and polyunsaturated fatty acids (PUFAs) and have beneficial roles in disease prevention and promoting human health. Several transgenic approaches have been used to develop foods with enriched phytonutrients by exploiting metabolic pathways.

5.1 Flavonoid and Anthocyanin

Flavonoids are a group of polyphenolic compounds (viz., chalcones, flavanones, anthocyanins, flavanols) which have potent antioxidant capacity (Bovy et al. 2002). By using the key genes of flavonoid biosynthesis pathway, some crops have been biofortified with flavonoids and anthocyanins (Fig. 4.5, Table 4.3).

In early work, transgenic tomato expressing petunia chalcone isomerase (*CHI*) resulted in enhanced level of flavonol (Muir et al. 2001). By overexpressing petunia *CHI* along with gerbera flavone synthase gene (*FLS*), Schijlen et al. (2006) enhanced levels of flavones and flavonols. Transgenic canola expressing Arabidopsis gene, production of anthocyanin pigment 1 (*PAP1*), produced high level of anthocyanins (Li et al. 2010). Potato tubers overexpressing anthocyanidin 3-o-glucosyltransferase (*3GT*) accumulated up to three-fold improved anthocyanin content (Wei et al. 2012). In a multigenic approach, by combining maize regulatory genes (*Lc* and *Pl*) and six structural genes of anthocyanin biosynthesis pathway from coleus in a single multigene stacking system, Zhu et al. (2017) generated purple endosperm rice with high anthocyanin content. Purple maize with high anthocyanin content was developed by co-expressing four maize anthocyanin biosynthetic genes (Liu et al. 2018).

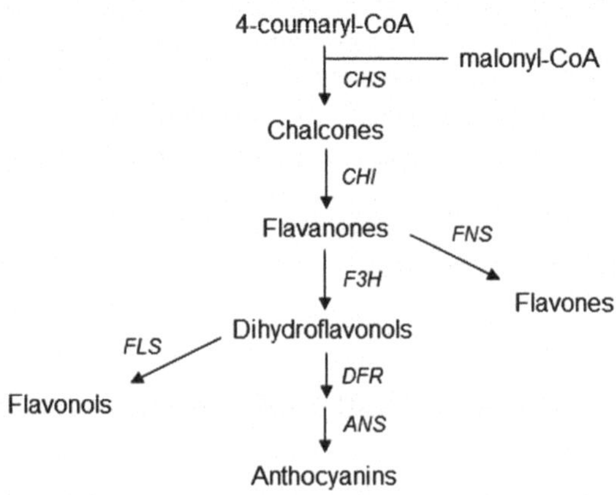

Fig. 4.5 The flavonoid biosynthesis pathway in plants. Abbreviations: CHS, chalcone synthase; CHI, chalcone isomerase; F3H, flavanone 3-hydroxylase; DFR, dihydroflavonol 4-reductase; ANS, anthocyanidin synthase; FLS, flavonol synthase; FNS, flavone synthase. Redrawn from Uncu et al. 2013

Table 4.3 Flavonoid and anthocyanin biofortification in crop plants

Target compound	Target crop	Target genes	References
Flavonoids	Tomato	*CHI*	Muir et al. (2001)
		CHI, FLS	Schijlen et al. (2006)
Anthocyanin	Canola	*PAP1*	Li et al. (2010)
	Potato	*3GT*	Wei et al. (2012)
	Rice	*Lc, Pl, CHS, CHI, F3H, F3'H, DFR, ANS*	Zhu et al. (2017)
	Maize	*C1, R2, ANS, GST*	Liu et al. (2018)

5.2 *Polyunsaturated Fatty Acids (PUFAs)*

Edible oils rich in unsaturated fatty acids are better for human health and cardiovascular system (Kinney et al. 2002). The omega-3 (n-3) long-chain polyunsaturated fatty acids (LC-PUFAs), normally found in marine fish oil are of great importance due to their health benefits (Singh et al. 2005). Plants are unable to synthesize LC-PUFAs, such as eicosapentaenoic acid (EPA, C20:5) and docosahexaenoic acid (DHA, C22:6), and engineering LC-PUFA pathway genes (Fig. 4.6) have shown promising results in oilseed crops.

By overexpressing Δ6 desaturase, Δ6 elongase and Δ5 desaturase genes from *Medicago alpina,* EPA content of 19.6% was achieved in soybean somatic embryos (Kinney et al. 2004). In the same study using two additional genes: Arabidopsis *FAD3* and *M. alpina* Δ17 desaturase higher EPA level was demonstrated (Kinney et al. 2004). Transgenic *Brassica carinata* overexpressing 18-carbon ω3 desaturase (CpDesX) from *Claviceps purpurea* and a 20-carbon ω3 desaturase (Pir-ω3) from *Pythium irregulare* produced increased EPA level in mustard seeds (Cheng et al. 2010). Transgenic camelina expressing genes for EPA and DHA synthesis from marine microbes produced considerable amount of omega-3 LC-PUFAs similar to those found in fish oil (Ruiz-Lopez et al. 2014; Usher et al. 2017).

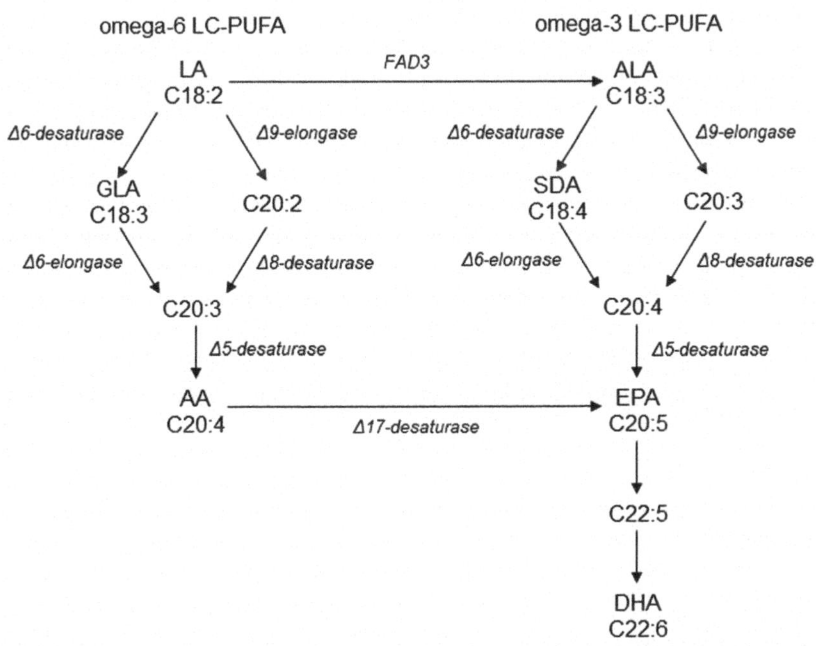

Fig. 4.6 The biosynthesis of long chain polyunsaturated fatty acids. Abbreviations: LA, linoleic acid; GLA, γ-linolenic acid; AA, arachidonic acid; ALA, α-linolenic acid; SDA, stearidonic acid; EPA, eicosapentaenoic acid; DHA, docosahexaenoic acid; *FAD3,* fatty acid desaturase 3. (Redrawn from Uncu et al. (2013))

6 Conclusion and Future Prospects

Genetic engineering has been demonstrated as a powerful tool for enhancing micronutrient and phytonutrient contents in food crops. Nevertheless, commercialization of nutritionally enriched transgenic varieties is facing huge challenges by both regulatory and public acceptance issues and the approval process itself can span decades (Napier et al. 2019). Recent approval of provitamin A-rich "Golden Rice" in the Philippines after a decade long regulatory process marks a breakthrough in the fight against vitamin A deficiency (De Steur et al. 2022); but public opposition to transgenic crops has forced the regulators to adopt overly precautionary policies.

Advances in modern biotechnological tools, such as genome editing, have unlocked new avenues of crop biofortification. Manipulation of innate metabolism genes by genome editing such as the CRISPR/Cas system have shown promising results by altering the nutritional content. For example, CRISPR/Cas technology have been successfully used to generate yellow and purple tomatoes by targeting *phytoene synthase 1* (*PSY1*), and *Anthocyanin 2* (*ANT2*) respectively (Filler Hayut et al. 2017; Čermák et al. 2015). In lettuce, a threefold increase of vitamin C was achieved by removing regulatory DNA regions (Zhang et al. 2018). Moreover, by editing *FAD2* gene increased oleic acid content was observed in *Brassica napus* and peanut (Huang et al. 2020). The examples of genome edited crops with enhanced nutritional quality are growing rapidly.

In addition to that, the regulatory landscape for transgene-free genome-edited crops is changing all around the world. In a landmark regulatory exemption, Government of India, in 2022, has allowed certain genome-edited plants not to go through a stringent biosafety approval mechanism and it would circumvent the procedure applied to genetically engineered crops which involve a 'foreign gene'. This has heralded a new era for scientists in India to work on crop biofortification using genome editing. Moreover, there is an urgent need for dialogue between scientists, policymakers, and the public to encompass the potential of these new technologies and their applications in crop improvements (Carroll and Krainer 2021). Further research and understanding the underlying mechanisms of different metabolic processes, combined with the use of advanced synthetic biology approach and more precise genome editing tools, will enhance future biofortification strategies.

References

Aluru M, Xu Y, Guo R, Wang Z, Li S, White W, Wang K, Rodermel S (2008) Generation of transgenic maize with enhanced provitamin A content. J Exp Bot 59(13):3551–3562. https://doi.org/10.1093/jxb/ern212

Basset GJ, Quinlivan EP, Gregory JF, Hanson AD (2005) Folate synthesis and metabolism in plants and prospects for biofortification. Crop Sci 45(2):449–453

Blancquaert D, Storozhenko S, Van Daele J, Stove C, Visser RGF, Lambert W, Van Der Straeten D (2013) Enhancing pterin and para-aminobenzoate content is not sufficient to successfully biofortify potato tubers and Arabidopsis thaliana plants with folate. J Exp Bot 64(12): 3899–3909. https://doi.org/10.1093/jxb/ert224

Blancquaert D, Van Daele J, Strobbe S, Kiekens F, Storozhenko S, De Steur H et al (2015) Improving folate (vitamin B9) stability in biofortified rice through metabolic engineering. Nat Biotechnol 33(10):1076–1078

Boliko MC (2019) FAO and the situation of food security and nutrition in the world. J Nutr Sci Vitaminol 65(Supplement):S4–S8

Borg S, Brinch-Pedersen H, Tauris B, Madsen LH, Darbani B, Noeparvar S, Holm PB (2012) Wheat ferritins: improving the iron content of the wheat grain. J Cereal Sci 56(2):204–213

Bovy A, De Vos R, Kemper M, Schijlen E, Pertejo MA, Muir S, Collins G, Robinson S, Verhoeyen M, Hughes S, Santos-Buelga C, Van Tunen A (2002) High-flavonol tomatoes resulting from the heterologous expression of the maize transcription factor genes LC and C1. Plant Cell 14:2509–2526. https://doi.org/10.1105/tpc.004218

Bramley PM, Elmadfa I, Kafatos A, Kelly FJ, Manios Y, Roxborough H et al (2000) Vitamin E: a critical review. J Sci Food Agric 80(7):913–938

Cahoon EB, Hall SE, Ripp KG, Ganzke TS, Hitz WD, Coughlan SJ (2003) Metabolic redesign of vitamin E biosynthesis in plants for tocotrienol production and increased antioxidant content. Nat Biotechnol 21(9):1082–1087. https://doi.org/10.1038/nbt853

Capell T, Christou P (2004) Progress in plant metabolic engineering. Curr Opin Biotechnol 15(2): 148–154

Carroll D, Krainer KMC (2021) Attitude and application: judge a crop on its potential and not breeding technology. Mol Plant 14(9):1405–1407

Čermák T, Baltes NJ, Čegan R, Zhang Y, Voytas DF (2015) High-frequency, precise modification of the tomato genome. Genome Biol 16(1):1–15

Chen Z, Young TE, Ling J, Chang SC, Gallie DR (2003) Increasing vitamin C content of plants through enhanced ascorbate recycling. Proc Natl Acad Sci U S A 100(6):3525–3530. https://doi.org/10.1073/pnas.0635176100

Cheng B, Wu G, Vrinten P, Falk K, Bauer J, Qiu X (2010) Towards the production of high levels of eicosapentaenoic acid in transgenic plants: the effects of different host species, genes and promoters. Transgenic Res 19(2):221–229

Cong L, Wang C, Chen L, Liu H, Yang G, He G (2009) Expression of phytoene synthase1 and carotene desaturase crtI genes result in an increase in the total carotenoids content in transgenic elite wheat (Triticum aestivum L.). J Agric Food Chem 57(18):8652–8660

de la Garza RD, Quinlivan EP, Klaus SM, Basset GJ, Gregory JF, Hanson AD (2004) Folate biofortification in tomatoes by engineering the pteridine branch of folate synthesis. Proc Natl Acad Sci 101(38):13720–13725

De Lepeleire J, Strobbe S, Verstraete J, Blancquaert D, Ambach L, Visser RG et al (2018) Folate biofortification of potato by tuber-specific expression of four folate biosynthesis genes. Mol Plant 11(1):175–188

De Steur H, Stein AJ, Demont M (2022) From Golden Rice to Golden diets: how to turn its recent approval into practice. Glob Food Sec 32:100596

Diretto G, Tavazza R, Welsch R, Pizzichini D, Mourgues F, Papacchioli V, Beyer P, Giuliano G (2006) Metabolic engineering of potato tuber carotenoids through tuber-specific silencing of lycopene epsilon cyclase. BMC Plant Biol 6. https://doi.org/10.1186/1471-2229-6-13

Ducreux LJM, Morris WL, Hedley PE, Shepherd T, Davies HV, Millam S, Taylor MA (2005) Metabolic engineering of high carotenoid potato tubers containing enhanced levels of β-carotene and lutein. J Exp Bot 56(409):81–89. https://doi.org/10.1093/jxb/eri016

Farre G, Sudhakar D, Naqvi S, Sandmann G, Christou P, Capell T, Zhu C (2012) Transgenic rice grains expressing a heterologous ρ-hydroxyphenylpyruvate dioxygenase shift tocopherol synthesis from the γ to the α isoform without increasing absolute tocopherol levels. Transgenic Res 21(5):1093–1097

Filler Hayut S, Melamed Bessudo C, Levy AA (2017) Targeted recombination between homologous chromosomes for precise breeding in tomato. Nat Commun 8(1):1–9

Garcia-Casal MN, Peña-Rosas JP, Giyose B, Consultation Working Groups (2017) Staple crops biofortified with increased vitamins and minerals: considerations for a public health strategy. Ann N Y Acad Sci 1390(1):3–13

Gillies SA, McIntosh SR, Henry R (2008) A transgenic cereal crop with enhanced folate: rice expressing wheat HPPK/DHPS. Sydney University Press

Goto F, Yoshihara T, Shigemoto N, Toki S, Takaiwa F (1999) Iron fortification of rice seed by the soybean ferritin gene. Nat Biotechnol 17(3):282–286. https://doi.org/10.1038/7029

Grune T, Lietz G, Palou A, Ross AC, Stahl W, Tang G, Thurnham D, Yin SA, Biesalski HK (2010) β-carotene is an important vitamin A source for humans. J Nutr 140(12):2268S–2285S. https://doi.org/10.3945/jn.109.119024

Haroldsen VM, Chi-Ham CL, Kulkarni S, Lorence A, Bennett AB (2011) Constitutively expressed DHAR and MDHAR influence fruit, but not foliar ascorbate levels in tomato. Plant Physiol Biochem 49(10):1244–1249

Hemavathi, Upadhyaya CP, Young KE, Akula N, soon Kim H, Heung JJ, Oh OM, Aswath CR, Chun SC, Kim DH, Park SW (2009) Over-expression of strawberry d-galacturonic acid reductase in potato leads to accumulation of vitamin C with enhanced abiotic stress tolerance. Plant Sci 177(6):659–667. https://doi.org/10.1016/j.plantsci.2009.08.004

Hemavathi, Upadhyaya CP, Akula N, Young KE, Chun SC, Kim DH, Park SW (2010) Enhanced ascorbic acid accumulation in transgenic potato confers tolerance to various abiotic stresses. Biotechnol Lett 32(2):321–330. https://doi.org/10.1007/s10529-009-0140-0

Hirschi KD (2009) Nutrient biofortification of food crops. Annu Rev Nutr 29:401–421

Hossain T, Rosenberg I, Selhub J, Kishore G, Beachy R, Schubert K, Danforth D (2004) Enhancement of folates in plants through metabolic engineering. www.pnas.orgcgi, https://doi.org/10.1073pnas.0401342101

Huang H, Cui T, Zhang L, Yang Q, Yang Y, Xie K et al (2020) Modifications of fatty acid profile through targeted mutation at BnaFAD2 gene with CRISPR/Cas9-mediated gene editing in Brassica napus. Theor Appl Genet 133(8):2401–2411

Kinney AJ, Cahoon EB, Hitz WD (2002) Manipulating desaturase activities in transgenic crop plants. Biochem Soc Trans 30(6):1099–1103

Kinney A, Cahoon E, Damude H, Hitz W, Liu ZB, KolarC (2004) Production of very long chain polyunsaturated fatty acids in oilseed plants. U.S. Patent Application No. 10/776,311

Lee S, Jeon US, Lee SJ, Kim YK, Persson DP, Husted S, Schjørring JK, Kakei Y, Masuda H, Nishizawa NK, An G (2009) Iron fortification of rice seeds through activation of the nicotianamine synthase gene. Proc Natl Acad Sci U S A 106(51):22014–22019. https://doi.org/10.1073/pnas.0910950106

Li X, Gao MJ, Pan HY, Cui DJ, Gruber MY (2010) Purple canola: Arabidopsis PAP1 increases antioxidants and phenolics in Brassica napus leaves. J Agric Food Chem 58(3):1639–1645

Liu X, Li S, Yang W, Mu B, Jiao Y, Zhou X, Zhang C, Fan Y, Chen R (2018) Synthesis of seed-specific bidirectional promoters for metabolic engineering of anthocyanin-rich maize. Plant Cell Physiol 59(10):1942–1955. https://doi.org/10.1093/pcp/pcy110

Lyons G, Cakmak I (2012) Agronomic biofortification of food crops with micronutrients. Fertilizing crops to improve human health: a scientific review 1:97–122

Martin C, Li J (2017) Medicine is not health care, food is health care: plant metabolic engineering, diet and human health. New Phytol 216(3):699–719. https://doi.org/10.1111/nph.14730

Masuda H, Ishimaru Y, Aung MS, Kobayashi T, Kakei Y, Takahashi M, Higuchi K, Nakanishi H, Nishizawa NK (2012) Iron biofortification in rice by the introduction of multiple genes involved in iron nutrition. Sci Rep 2(Ii):1–7. https://doi.org/10.1038/srep00543

Mattoo AK, Shukla V, Fatima T, Handa AK, Yachha SK (2010) Genetic engineering to enhance crop-based phytonutrients (nutraceuticals) to alleviate diet-related diseases. In: Bio-farms for nutraceuticals. Springer, pp 122–143

Muir SR, Collins GJ, Robinson S, Hughes S, Bovy A, Ric De Vos CHV et al (2001) Overexpression of petunia chalcone isomerase in tomato results in fruit containing increased levels of flavonols. Nat Biotechnol 19(5):470–474

Nandi S, Suzuki YA, Huang J, Yalda D, Pham P, Wu L, Bartley G, Huang N, Lönnerdal B (2002) Expression of human lactoferrin in transgenic rice grains for the application in infant formula. Plant Sci 163(4):713–722. https://doi.org/10.1016/S0168-9452(02)00165-6

Napier JA, Haslam RP, Tsalavouta M, Sayanova O (2019) The challenges of delivering genetically modified crops with nutritional enhancement traits. Nature Plants 5(6):563–567

Naqvi S, Zhu C, Farre G, Ramessar K, Bassie L, Breitenbach J, Conesa DP, Ros G, Sandmann G, Capell T, Christou P (2009) Transgenic multivitamin corn through biofortification of endosperm with three vitamins representing three distinct metabolic pathways. Proc Natl Acad Sci U S A 106(19):7762–7767. https://doi.org/10.1073/pnas.0901412106

Naqvi S, Farre G, Zhu C, Sandmann G, Capell T, Christou P (2011) Simultaneous expression of Arabidopsis ρ-hydroxyphenylpyruvate dioxygenase and MPBQ methyltransferase in transgenic corn kernels triples the tocopherol content. Transgenic Res 20(1):177–181. https://doi.org/10.1007/s11248-010-9393-6

Newell-McGloughlin M (2008) Nutritionally improved agricultural crops. Plant Physiol 147(3): 939–953

Nozoye T (2018) The nicotianamine synthase gene is a useful candidate for improving the nutritional qualities and Fe-deficiency tolerance of various crops. Front Plant Sci 9:340

Oliver MJ (2014) Science of medicine why we need GMO crops in agriculture science of medicine-national review. December. http://www.goldenrice

Paine JA, Shipton CA, Chaggar S, Howells RM, Kennedy MJ, Vernon G, Wright SY, Hinchliffe E, Adams JL, Silverstone AL, Drake R (2005) Improving the nutritional value of Golden Rice through increased pro-vitamin a content. Nat Biotechnol 23(4):482–487. https://doi.org/10.1038/nbt1082

Paul JY, Khanna H, Kleidon J, Hoang P, Geijskes J, Daniells J et al (2017) Golden bananas in the field: elevated fruit pro-vitamin A from the expression of a single banana transgene. Plant Biotechnol J 15(4):520–532

Qin A, Shi Q, Yu X (2011) Ascorbic acid contents in transgenic potato plants overexpressing two dehydroascorbate reductase genes. Mol Biol Rep 38(3):1557–1566

Ramírez Rivera NG, García-Salinas C, Aragão FJL, Díaz de la Garza RI (2016) Metabolic engineering of folate and its precursors in Mexican common bean (Phaseolus vulgaris L.). Plant Biotechnol J 14(10):2021–2032. https://doi.org/10.1111/pbi.12561

Rocheford TR, Wong JC, Egesel CO, Lambert RJ (2002) Enhancement of vitamin E levels in corn. J Am Coll Nutr 21(sup3):191S–198S

Ruiz-Lopez N, Haslam RP, Napier JA, Sayanova O (2014) Successful high-level accumulation of fish oil omega-3 long-chain polyunsaturated fatty acids in a transgenic oilseed crop. Plant J 77(2):198–208

Saltzman A, Birol E, Oparinde A, Andersson MS, Asare-Marfo D, Diressie MT et al (2017) Availability, production, and consumption of crops biofortified by plant breeding: current evidence and future potential. Ann N Y Acad Sci 1390(1):104–114

Sayre R, Beeching JR, Cahoon EB, Egesi C, Fauquet C, Fellman J et al (2011) The BioCassava plus program: biofortification of cassava for sub-Saharan Africa. Annu Rev Plant Biol 62:251–272

Schijlen E, Ric de Vos CH, Jonker H, Van Den Broeck H, Molthoff J, Van Tunen A et al (2006) Pathway engineering for healthy phytochemicals leading to the production of novel flavonoids in tomato fruit. Plant Biotechnol J 4(4):433–444

Singh SP, Zhou XR, Liu Q, Stymne S, Green AG (2005) Metabolic engineering of new fatty acids in plants. Curr Opin Plant Biol 8(2):197–203. https://doi.org/10.1016/j.pbi.2005.01.012

Storozhenko S, De Brouwer V, Volckaert M, Navarrete O, Blancquaert D, Zhang GF, Lambert W, Van Der Straeten D (2007) Folate fortification of rice by metabolic engineering. Nat Biotechnol 25(11):1277–1279. https://doi.org/10.1038/nbt1351

Strobbe S, Van Der Straeten D (2018) Toward eradication of B-vitamin deficiencies: considerations for crop biofortification. Front Plant Sci 9:443

Strobbe S, De Lepeleire J, Van Der Straeten D (2018) From in planta function to vitamin-rich food crops: the ACE of biofortification. Front Plant Sci 9:1862

Tavva VS, Kim YH, Kagan IA, Dinkins RD, Kim KH, Collins GB (2007) Increased α-tocopherol content in soybean seed overexpressing the Perilla frutescens γ-tocopherol methyltransferase gene. Plant Cell Rep 26(1):61–70

Uncu AO, Doganlar S, Frary A (2013) Biotechnology for enhanced nutritional quality in plants. Crit Rev Plant Sci 32(5):321–343. https://doi.org/10.1080/07352689.2013.781453

Usher S, Han L, Haslam RP, Michaelson LV, Sturtevant D, Aziz M et al (2017) Tailoring seed oil composition in the real world: optimising omega-3 long chain polyunsaturated fatty acid accumulation in transgenic Camelina sativa. Sci Rep 7(1):1–12

Van Eenennaam AL, Lincoln K, Durrett TP, Valentin HE, Shewmaker CK, Thorne GM et al (2003) Engineering vitamin E content: from Arabidopsis mutant to soy oil. Plant Cell 15(12): 3007–3019

Vasconcelos M, Datta K, Oliva N, Khalekuzzaman M, Torrizo L, Krishnan S, Oliveira M, Goto F, Datta SK (2003) Enhanced iron and zinc accumulation in transgenic rice with the ferritin gene. Plant Sci 164(3):371–378. https://doi.org/10.1016/S0168-9452(02)00421-1

Wei Q, Wang QY, Feng ZH, Wang B, Zhang YF, Yang Q (2012) Increased accumulation of anthocyanins in transgenic potato tubers by overexpressing the 3GT gene. Plant Biotechnol Rep 6(1):69–75

Welch RM (2002) The impact of mineral nutrients in food crops on global human health. Plant Soil 247:83–90

Wirth J, Poletti S, Aeschlimann B, Yakandawala N, Drosse B, Osorio S, Tohge T, Fernie AR, Günther D, Gruissem W, Sautter C (2009) Rice endosperm iron biofortification by targeted and synergistic action of nicotianamine synthase and ferritin. Plant Biotechnol J 7(7):631–644. https://doi.org/10.1111/j.1467-7652.2009.00430.x

Ye X, Al-Babili S, Kloti A, Zhang J, Lucca P, Beyer P, Potrykus I (2000) Engineering the provitamin A (β-carotene) biosynthetic pathway into (carotenoid-free) rice endosperm. Science 287(5451):303–305

Zhang Y, Li H, Shu W, Zhang C, Zhang W, Ye Z (2011a) Suppressed expression of ascorbate oxidase gene promotes ascorbic acid accumulation in tomato fruit. Plant Mol Biol Report 29(3): 638–645

Zhang YY, Li HX, Shu WB, Zhang CJ, Ye ZB (2011b) RNA interference of a mitochondrial APX gene improves vitamin C accumulation in tomato fruit. Sci Hortic 129(2):220–226

Zhang GY, Liu RR, Xu G, Zhang P, Li Y, Tang KX, Liang GH, Liu QQ (2013) Increased α-tocotrienol content in seeds of transgenic rice overexpressing Arabidopsis γ-tocopherol methyltransferase. Transgenic Res 22(1):89–99. https://doi.org/10.1007/s11248-012-9630-2

Zhang H, Si X, Ji X, Fan R, Liu J, Chen K et al (2018) Genome editing of upstream open reading frames enables translational control in plants. Nat Biotechnol 36(9):894–898

Zhu C, Naqvi S, Gomez-Galera S, Pelacho AM, Capell T, Christou P (2007) Transgenic strategies for the nutritional enhancement of plants. Trends Plant Sci 12(12):548–555. https://doi.org/10.1016/j.tplants.2007.09.007

Zhu Q, Yu S, Zeng D, Liu H, Wang H, Yang Z, Xie X, Shen R, Tan J, Li H, Zhao X, Zhang Q, Chen Y, Guo J, Chen L, Liu YG (2017) Development of "purple endosperm rice" by engineering anthocyanin biosynthesis in the endosperm with a high-efficiency transgene stacking system. Mol Plant 10(7):918–929. https://doi.org/10.1016/j.molp.2017.05.008

Zhu Q, Wang B, Tan J, Liu T, Li L, Liu YG (2020) Plant synthetic metabolic engineering for enhancing crop nutritional quality. Plant Commun 1(1):100017

Zimmermann MB, Hurrell RF (2002) Improving iron, zinc and vitamin A nutrition through plant biotechnology. Curr Opin Biotechnol 13(2):142–145. https://doi.org/10.1016/S0958-1669(02)00304-X

Chapter 5
Mechanization in Pre-harvest Technology to Improve Quality and Safety

Aparna Kumari, Shubhendra Singh, Aparnna V. P., Pallawi Joshi, Anil Kumar Chauhan, Meenakshi Singh, and S. Hemalatha

1 Introduction

Growing demand for industrialization, production, housing and infrastructure causes agricultural land to be transformed into non-agricultural uses. The area for cultivation is restricted. The reach is limited. As per Agriculture Census (2015–16), number of operational holdings in the country is estimated at 14.64 Crore. Over the years, the average keeping size has reduced for all operating groups (small and marginal, medium and large). In promoting agricultural development, the supply of labour is essential to agriculture. In the present chapter, the importance of mechanization in farming is explained taking an example of India, whose major share of GDP is sourced by agriculture. Agricultural workers would therefore play a significant role in agricultural development in the region (Khandetod 2018).

According to World population prospects (2021), India is the world's second most populated nation, with an estimated 1.37 billion in 2019 population and an annual growth rate of 1.3%. Approximately two-thirds of the population still resides in rural areas, where approximately 50% still depend on agriculture for their livelihood. Optimum input-use efficiency and sustainable productivity increase are

A. Kumari · S. Singh · Aparnna V. P. · P. Joshi · A. K. Chauhan (✉)
Department of Dairy Science and Food Technology, Institute of Agricultural Science, Banaras Hindu University, Varanasi, India
e-mail: shubhendra.singh10@bhu.ac.in; pallawi28@bhu.ac.in; achauhan@bhu.ac.in

M. Singh
Technology Management Directorate (Socio-economic Ministry Interface), Council for Scientific and Industrial Research, New Delhi, India
e-mail: meenakshi@csir.res.in

S. Hemalatha
Department of Pharmaceutics, Indian Institute of Technology, Banaras Hindu University, Varanasi, India
e-mail: shemalatha.phe@iitbhu.ac.in

H. U. Hebbar et al. (eds.), *Engineering Aspects of Food Quality and Safety*, Food Engineering Series, https://doi.org/10.1007/978-3-031-30683-9_5

key challenges faced by the Indian agriculture sector (Dwivedy 2011). The degree of mechanization and investment in new technologies play a critical role in resolving these challenges. The demand for sustainable mechanization and services will continue to increase naturally in the face of the increasing population's demand for food, feed and biological industrial raw materials (Khandetod 2018). Studies reveal that adopting appropriate mechanization in agricultural operations can increase cropping intensity by 5–20% and farm productivity and food production by 10–15% (Tiwari et al. 2019). In developing countries like India where it needs to not only find ways to promote precision farming through mechanization and automation but also ensure effective adoption of technology. The Government of India has executed various schemes, policies supporting greater mechanization of Indian agriculture and the Sub Mission on Agricultural Mechanization (SMAM) is a significant initiative by Government of India (GoI) in this direction. The objective of these initiatives is to enhance agriculture productivity and double farmers' income by 2022–23.

1.1 The Role of Mechanization in Agricultural Development

For sustainable productivity growth in agriculture, it is essential to implement mechanization (machinery, drones, robots, AGVs, etc.) in the agriculture sector. Mechanization saves time and work, decreases seed expense, reduces losses from post-harvest and raises crop production and farm incomes in the long run. According to Sims and Kienzle (2006), the goals of mechanization include:

1. Enhanced agriculture productivity per unit area because of improvement in timeliness of farm operations.
2. Increasing the yield and cropping intensity by utilizing properly the natural resources such as land and water.
3. Completion of tasks such as sowing of seeds or transplanting of seedlings, providing irrigation, harvesting, threshing, etc. that are difficult to perform without mechanical aids.
4. Improvement of the product quality and work.
5. Increasing agriculture export quality of agricultural product in pre-harvest operation.
6. Reduction of work in farming activities, thereby making the farm work more attractive
7. To improve the quality of farm machinery.
8. Strengthen Research and Development.
9. Create linkage with manufacturers for speedy commercialization of farm machinery.
10. To mechanize farm of all categories.

Based on the source of power, technological levels of mechanization have been broadly classified as draught animal technology, hand-tool technology and mechanical power technology (Gifford 1992). Farm and yield of crops are directly

Table 5.1 Current status of mechanization and contribution of agriculture to GDP in India and other developed countries

Country	Mechanization level (%)	Contribution to GDP (%)
India	40	15.9
U.K.	95	1
Europe	95	1.4
Russia	85	3.1
Brazil	75	4.3
China	50	7.5

Source: World Bank Open data 2019, Feder Unacoma, PwC analysis

influenced by the correct usage of equipment and machinery. The Government of India (GoI) in 2018 has set an ambitious target of increasing farm power availability from 2.02 kW/ha (2016–17) to 4.0 kW/ha by the end of 2030.

Mechanization technologies (over 10 ha) were first adopted by large farmers and then medium-sized farmers (with 4 to 10 ha farm size). In particular, after the Green Revolution of the 1960s, Indian farmers gradually reacted to farm mechanization technology. The use of agricultural machinery depends on various tractive and stationary operations on its agricultural power sources. In India, farm mechanization is still in its initial stages, with 40–45% of the mechanization level, which is very low compared to any developed country, where the mechanization level has reached more than 90% as shown in Table 5.1. Even though the mechanization growth rate is very slow, India's overall food grain production grew from over 50 million tonnes in 1950–51 to 283 million tonnes in 2018–19 (Tiwari et al. 2019). Governments across the world are trying to focus on increasing food productivity. The demand for agricultural machine and equipment with advanced technology is expected to rise, as machine or equipment with advanced technology will act as a promoter to increase agricultural yield. In the recent past year, the adoption rates of farm equipment have increased, as shown by the sales of tractors and the rise in farm power availability (FPA). In 2009, tractors' domestic sales were 3 lakh units, which increased by 7.8 lakh units in 2019 (Industry reports, Tractor Manufacturers Association 2019, PwC analysis). Average farm power availability in India has also risen from 1.1 kW/ha in 1995–96 to 2.02 kW/ha in 2017–18 (NABAD report 2018). Presently, India is one of the largest manufacturers of equipment such as harvesters, tractors and tillers due to adoption of pre-harvest mechanization. The farm equipment market in India was estimated to be worth USD 3 billion in FY19 and is expected to reach USD 18 billion by FY25 – growing at a CAGR of 6% between FY19 and FY25 (Singh 2006).

2 Problems in Pre-harvest Issues

The significance and complexity of the pre-harvest phases of food production have grown over time in preharvest food safety studies and activities. Research and policy have grown as awareness about pathogenesis, virulence and transmission of pathogens and pollutants in nutrition has increased. Increased efforts have been made to develop quick, sensible and precise methods to detect or screen foodborne pathogens at pre-harvest using new methods and technologies. Furthermore, epidemiological research gained momentum through sponsored research programs (Torrence 2003) that allowed longitudinal cohort, broad case control and ecological studies to be conducted. The findings of these studies are still published with essential data on pre-harvest food safety and on future action, control or prevention strategies. Food safety and inspection services (FSIS) from the USDA have moved towards more preharvest regulations (*Escherichia coli* and *Salmonella*) (Berghaus et al. 2013; LeJeune et al. 2001). The European Union has developed international standards for pre-harvest efficiency, in particular for poultry and Salmonella.

The World Health Organization and the Food and Agriculture Organization of UN and Codex (WHO/FAO) have held expert consultations, studies and regular programs on pre-harvest action on foodborne specific pathogens in food animals. Recommendations on difference of monitoring, study and cooperation across various phases of food production and regulatory fragmentation and use of Hazard Analysis and Critical Control Point (HACCP) and Good Agriculture Practices were made available in some studies. The reports also offered recommendations for research to improve our knowledge of pre-harvest food pathogens and to bridge our understanding of pathogenesis and transmission to manufacture and retail, as well as to establish effective response, control and prevention strategies. The government must identify a scheme owner who will be in charge of planning and implementing the GAP programme. As a result of the formation of multi-stakeholder committees, diverse interests relating to the Scheme are represented by these committees. According to the FSSAI, being a food regulator, it monitors and regulates the contaminants and maximum residue limits for various pesticides in farm produce once the food enters in the food value chain. FSSAI (Food Safety and Standards Authority of India) in coordination with State Food Safety Commissioners need to organize a training and awareness programme for primary farmers to address the issue of pesticide contamination and residues in post-harvesting, which is a major concern in the country.

The key steps remain: the ability to identify a pre-harvest "criteria" or success goal and to show what measures and controls or preventive strategies are healthy. These reports provide us with ambitious and optimistic guidance for the future, both from a research approach and strategic approach to "traditional" areas, such as agricultural produce, horticultural produce, meat and poultry.

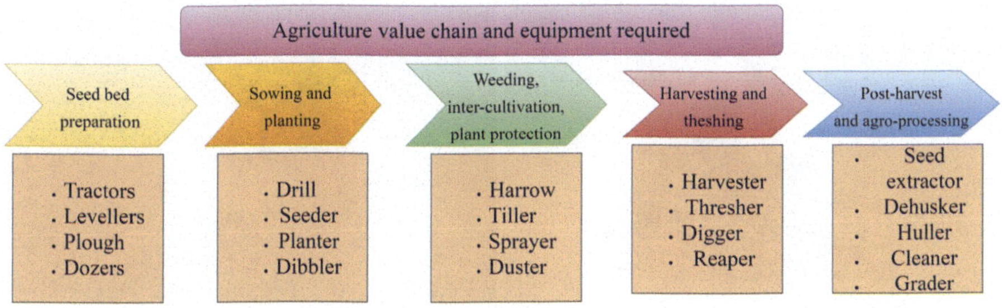

Fig. 5.1 Agricultural value chain and types of equipment required

3 Pre-harvest Mechanization and Types of Equipment Required

For successful agricultural practices, farmers perform various steps in the field; these include seedbed preparation, sowing and planting, weeding, inter cultivation, plant protection, harvesting and threshing and performing all these steps various mechanized equipment are required as shown in Fig. 5.1. Predominant preharvest machinery is tractors, sprayers, threshers, combine harvesters, rotavators, power tillers and multi-crop planters, either manufactured domestically or imported from the manufacturing country. Percentage of operation mechanized in soil working and seedbed preparation is (40%), seeding and planting (29%), plant protection (34%), irrigation (37%) and harvesting and threshing (60–70%) in wheat and rice field (Ahmed et al. 2020). Tractors have the largest share of India's farm machinery market, contributing 80% of the total farm machinery sold in the country, while others such as threshers, rotavators and power tiller have 3.7%, 2.5% and 1.4% market share, respectively (Sekhar and Bhatt 2014). Some of preharvest machinery that are currently used in agricultural value chain is discussed below:

3.1 Seedbed Preparation Equipment

The main problems with seedbed preparation are requirement of more time, money, small or scattered holdings for operating big machinery and less efficiency. Seedbed preparation is more difficult in the dry conditions of the land during harvest time. However, the use of mechanized seedbed preparation equipment solves the problems, creates an appropriate soil structure for a seedbed and provides optimum environmental conditions for plant growth. It allows good retention of rainfall, rapid infiltration, adequate air capacity and exchange within the soil, and minimizes soil erosion (Kishore et al. 2017). This mechanized equipment are tractor-drawn roller, rotavator, cultivator and check basin former (Fig. 5.2a–b), which are superior to the traditional method.

Fig. 5.2 Mechanized seedbed preparation equipment (**a**) Tractor mounted pulverizing roller attachment to tiller (**b**) Tractor mounted rotary tiller. (Source-AICRP – FIM)

Fig. 5.3 Power tiller with rotary attachment

3.2 Power Tiller

Planting and bed preparation are critical operations and consume more than 60% of the total energy requirement in mechanized sector. Therefore, its appropriate management can play a critical role in reducing energy consumption. Considering the adverse effects of extensive tractor passes on fields during bed preparation, such as erosion and soil compaction, research interest in selecting appropriate implements and increased farm efficiency is growing every day. Power tiller is mainly designed for tilling of seedbed in small farms and hill farming and also for carrying out spraying operations in horticultural crops and food crops. The concept of the power tiller (walking type tractor as shown in Fig. 5.3) was introduced in the year 1920 and Japan was the first country to use it on a large scale. During the year 1950–1965, the production of power tillers increased rapidly. It was introduced in India in 1963 (Hendriadi and Salokhe 2002). Traditional power tiller is operated by directing the two handles of tiller by hands. Power tiller is also called a single axle walking type tractor, though a seat is provided in some designs. Nowadays some advanced power tiller has an optional riding facility. They have been specially developed and designed for use on the small or medium size field where four-wheel tractors are not readily available. However, a power tiller is mainly used for seedbed preparation in low land paddy fields. Power tiller is also used as a power source for other agricultural operations such as sowing, seedbed preparation and fertilizer application. Tillers are also valuable for intercultural operations in wide-spaced row crops (more than 1.0 m row to row spacing) and harvesting of cereal crops under upland conditions, including transportation of farm products and power source for stationary farm operations (Cherian et al. 2016). Given the right set of implements and attachments, the power tiller can perform most field operations in intensive cultivation. The lightweight of the power tiller is a favourable factor for working in wet and dryland conditions. External attachments can be made on the tiller depending upon the nature of the work. By far, tillers can be considered as one of the most useful, adaptable and multipurpose machines ever developed for the medium-to-large scale farmers of developing countries.

3.3 Mini (Low Horse Power) Tractors

The Indian model is focused on broad public funding in terms of grants to purchase four-wheel tractors (4WTs) and other machines – marketing regulations (such as minimum price guarantee) and large-scale infrastructure investments (Singh 2006; Hazell 2009). This level of public support is currently unlikely to occur in most of sub-Saharan Africa (SSA) due to fiscal limitations. Besides, the model led to great inequity, favoured medium-sized to large-scale farmers and left many small-scale farmers unavailable (Biggs et al. 2011). These differences are the result of significant regional disparities, the highest mechanisation of Punjab & Haryana and the lowest Eastern (e.g., Bihar, Orissa) states (Singh 2006). Now-a-days mini tractors (Fig. 5.4) are used as multipurpose and inexpensive sources of power coupled with the promotion of energy saving technologies such as conservation agriculture (CA), whilst ensuring the profitability for farmers, service providers and other private sector actors in the supply chain. It especially used in secondary tillage for interculture, making trenches, channel farming etc., where use of regular sized tractors is not possible. These types of mini tractors are ideal for horticultural crops due to its small size, which also allows it to manoeuvre between trees in orchids. Some of the mini tractor can also be used for plant protection aspects in plantation crops (Cherian et al. 2016).

Fig. 5.4 Mini tractor. (Adopted from Cherian et al. 2016)

3.4 Mechanized Seedling Machines and Seedling Transplanters

Manual seed sowing is laborious and time-consuming. There are lots of disadvantages in conventional seed sowing methods like no control over the depth of seed placement, wastage of seed, variation in spacing, no uniformity in the distribution of seed placement, resulting in improper germination of seeds leading to loss in overall productivity. The objective of mechanization with seed planter is to place seeds at desired depth with constant seed spacing, cover the seed with suitable soil and plant the seed in the ridge, furrow, flat arrow and flatbed method, as multi-cropping with an added advantage of reduced human effort. Some of the mechanized seedling equipment's are discussed below.

3.4.1 Rotary Dibbler

It is a manually operated push-type device for dibbling of bold and medium-sized seeds such as soybean, maize, Bengal gram, pigeon pea and sorghum in a prepared seedbed. It consists of penetrating jaws with a rotating dibbling head, seed hopper with cell-type wooden roller, covering-cum-transport wheel and a handle (Fig. 5.5a). The seed to seed distance is based upon the polygon plate's size to which jaws are attached. Its cost of operation is Rs. 460/ha and implement costs Rs. 2300/–. It covers 0.6–1.0 ha/day and labour requirement is 27 man-h/ha (AICRP – FIM, 2017).

3.4.2 Power Tiller Mounted Air Assisted Seed Drill

It has been developed for drilling tiny seeds at the desired seed rate (Fig. 5.5b). The unit consists of an air drive and blower, feeding device and seed hopper, seed tubes, seed distributor head, furrow opener ground wheel and furrow closer. Blower is used to blow air through a vertical distributor tube to the distributor head. Seed is metered

Fig. 5.5 Mechanized seedling application equipment (**a**) Rotary dibbler (**b**) Power tiller mounted air assisted seed drill (**c**) Tractor Operated Small Seed Planter (**d**) Self-propelled rice transplanter. (Source-AICRP – FIM)

by the ground wheel into the air stream in controlled manner. The stream of seed is distributed to the 9 furrow openers by the distributor head. This ensures uniform drilling of fine seeds. The percentage of germination varies widely for small seeds such as sesame, ragi, maize. The rows spacing can be adjusted from 600 mm for two rows to 300 mm for four rows. The adequate field capacity of the machine is 0.25 hectare/h. The cost of operation with this machine is Rs. 750/ha as against Rs. 1200/ha for the conventional method (AICRP – FIM 2017).

3.4.3 Tractor Operated Small Seed Planter

Planting small seeds like onion, PAU, Ludhiana has designed and developed a tractor (26.11 kW) operated six-row planter. It consists of a seed hopper for each row, inclined plate type metering mechanism, three-point hitch system and shovel type furrow openers (Fig. 5.5c). The seed hopper capacity is about 1.5 kg and the metering plate of 130 mm diameter is made of plastic. Lugged ground wheel provided the power to the metering mechanism. The row-to-row spacing of the machine is 150 mm, whereas plant-to-plant spacing can be changed either by changing the sprockets or changing the plate with different notches. The cost and effective field capacity of the machine are Rs. 5090/ha and 0.16 ha/h, respectively, at 2.0 km/h speed of onion seeder. It reduces about 50% of operation cost and 81% of labor requirement compared to the conventional method of onion sowing.

3.4.4 Rice Transplanters

Rice transplanting is a tedious and very time-consuming job requiring about 25&300 man-hour/ha which is roughly 25% of the total labour requirement of the crop. A rice transplanter is an indispensable machine for rice farming. Figure 5.5d show the self-propelled rice transplanter. It is suitable for transplanting paddy seedlings in puddled soils. It is a single wheel driven, riding type machine and fitted with diesel engine. It transplants seedlings from mat type nursery in eight rows in a single pass. A propeller shaft from the gearbox provides power to the transplanting mechanism mounted over the float. The float facilitates the transplanter to slide over the puddled surface. The tray containing mat type nursery for 8 rows is moved sideways by a scroll shaft mechanism, which converts rotary motion received from the engine through belt-pulley, gear and universal joint shaft into linear motion of a rod connected to the seedling tray having provision to reverse the direction of movement of tray after it reaches the extreme position at one end. Fixed fork with knock out lever type planting fingers (cranking type) are moved by a four bar linkage to give the designed locus to the tip of the planting finger. The cost of the equipment is approximately Rs. 2.25 lakh. It can transplant 1.2–1.5 ha/day with the help of 5 persons by working at a speed of 1.1–1.5 km/h. The cost of operation with the transplanter is Rs. 3000/ha as compared to Rs. 5000/ha by traditional method. It saves about 65% labour and 40% cost of operation as compared to manual transplanting. Four row self-propelled walking type and six/eight row four wheels driven riding type self-propelled transplanter are also commercially available. The use of rice transplanters minimizes the direct human contact with the crop thereby reducing the chances of disease and pest infestation which can transmit from biological sources.

3.5 Mechanized Fertilizer Application Equipment

In the conventional method, fertilizer is mainly spread on the soil surface by hand that increases farmers effort and labour costs and decreases fertilization uniformity, resulting in lower profits and total income for the farmer (Si-yuan et al. 2014; De-feng et al. 2015). Using a precision hand operated spreader, one can assure an even application of fertilizer and it is simple to use. As part of the International Maize and Wheat Improvement Center's (CIMMYT) Cereal Systems Initiative for South Asia, a programme designed to help farmers in Nepal adapt to climate change, CIMMYT has endorsed this technology. Manually application of fertilizer harms health and causes skin diseases. Increasing the degree of mechanization in fertilization will solve these problems. So the agriculture scientists and researchers have developed some mechanized fertilizer application equipment. The fertilizer application by mechanical means is shown in Fig. 5.6a–c.

Fig. 5.6 Mechanized fertilizer application equipment (**a**) Tractor-operated fertilizer dibbler (**b**) Fertilizer band placement cum earthing up the machine (**c**) GPS based variable rate granular fertilizer applicator. (Source-AICRP – FIM)

3.5.1 Tractor-Operated Fertilizer Dibbler

The tractor-operated fertilizer dibbler (Fig. 5.6a) which is most popular and widely used. It is a semi-mounted implement attached to the tractor hitch system and used to spread the fertilizer in the band form. The principal components are fertilizer metering device, revolving spade, fertilizer placement funnel, pressing device and soil cover. The lever can control the fertilizer application rate, by opening and closing the aperture. The cost of the machine is Rs. 45,000/– and field capacity is 0.2 ha/h. The operation cost is Rs.1550/ha and results in a saving of 60% compared to the conventional method. This method also reduces the direct contact of human with the crop thereby reducing the chances of pest and disease occurrence.

3.5.2 Fertilizer Band Placement cum Earthing Up the Machine

The third one is a tractor operated (26 kW and above) fertilizer band placement cum earthing up the machine (Fig. 5.6b). It has been developed and designed at GBPUAT, Pantnagar, India. The machine is suitable for simultaneous distribution of fertilizer, earthing up and cutting of weeds in crops such as sugarcane, maize, potato, etc., having more than 0.50 m row to row spacing. Its urea fertilizer application rate ranges from 60–250 kg/ha. It is suitable for topdressing of fertilizer at 50–100 mm from the plant. The field capacity and efficiency of the machine are 0.56 ha/h and 82.4%, respectively. The estimated cost of the machine is Rs. 50,000. There is a considerable saving in time, effort and fertilizer over the traditional method.

3.5.3 GPS-Based Variable Rate Granular Fertilizers (NPK) Applicator

The fifth one is a GPS-based variable rate granular fertilizers (NPK) applicator. It has been developed at CIAE Bhopal and IIT Kharagpur to ensure the ideal application of fertilizers as basal dose. It consists of a differential, micro-processor, global positioning system (DGPS), micro-controller, power supply, DC motor actuator and fluted roller fitted metering mechanism and threaded screw arrangement (Fig. 5.6c). The fertilizer application rate is changed according to the prescribed application rate at the identified grid with a variation of coefficient of 11.7–15.0%. The fertilizer application accuracy ranges from 89.3% to 98.1% at various discharge rates.

3.6 Mechanized Sprayer Application Equipment

Many types of sprayers like a hydraulic sprayer, power sprayer, knapsack sprayer, hand compression sprayer, bucket sprayer, foot-operated sprayer, rocker sprayer and dusters like knapsack type, plunger type, power-operated and rotary type dusters are available in different shapes and sizes for plant protection work (Kishore et al. 2017). Many advanced types of sprayer are also available, shown in Fig. 5.7a–d.

Fig. 5.7 Sprayer application equipment (**a**) Multi orchard sprayer (**b**) Tractor mounted boom sprayer (**c**) Tractor operated aero blast sprayer (**d**) Ultrasonic sensor based spraying system. (Source-AICRP – FIM)

3.6.1 Multi Orchard Sprayer

It is appropriate for spraying chemicals in orchards like citrus, pomegranate, grapes, etc. It consists of a horizontal triplex piston (HTP) pump, chemical tank with hydraulic agitation system, cut-off device, transport wheels with trailed type main chassis and boom equipped with turbo nozzles (Fig. 5.7a). Turbo nozzles produce droplets of 100–150 micron in size and create 883–1766 kPa pressure. The angle of booms can be adjusted depending upon row spacing and the plant size. The booms sprays are mounted behind the operator. The boom covers 50% of the tree canopy on either side of the sprayer. It can cover 0.40–0.70 ha/h at a forward speed of 1.20–1.50 km/h and costs Rs. 60,000/−.

3.6.2 Tractor Mounted Boom Sprayer

The sprayer essentially consists of a tank made of plastic or fiber glass, suction pipe with strainer, pump assembly, pressure regulators and gauges, spray boom fitted with nozzles, air chamber and delivery pipe (Fig. 5.7b). The sprayer is attached to three-point linkages of the tractor and utilizes the PTO power of tractor to operate pump of the sprayer. It uses a high discharge and high-pressure pump as the number of nozzles may go up to 20 depending upon the type of crop and structure of the sprayer.

3.6.3 Tractor Operated Aero Blast Sprayer

The machine consists of a 400-liter capacity tank, fan, pump, filling unit, control valve, spraying nozzles and spout adjustable handle (Fig. 5.7c). Chemical solutions are release with the help of nozzles into the stream of the air blast. The air blast sprays pesticide solution in the form of microscopic particles throughout its swath on one side of the tractor. The main blast covers the swath's major portion through the main spout and auxiliary nozzles cover the tractor's swath area. The orientation of the air outlet can change its width of coverage and direction. The machine works at a speed of 1.5 km/h and covers about 1.7 ha/h. Depending upon different valve settings, the sprayer's application range can be varied from 100–400 l/ha. The effective width of the sprayer is about 13.0 m. The unit price of an aero blast sprayer is 1.00 lakh. The cost of operation of this machine is Rs. 500/ha as against Rs. 700/ha by the conventional method.

3.6.4 Ultrasonic Sensor Based Spraying System

IIT Kharagpur developed a sensor-based tractor-mounted automatic spraying system to detect plant canopy and spraying of liquid pesticide over the spotted plant canopy (Fig. 5.7d). The developed sprayer consists of solenoid valves, ultrasonic sensors, micro-controller board, spray pump, nozzle, one-way valves, relief valve, pressure gauge and 12 V batteries. The number of plants covered and the field capacity are 1370 plants/h and 0.88 ha/h, respectively. The application rate is found minimum (200 l/ha) for hollow cone nozzle with sensor and maximum (500 l/ha) for turbo nozzles without sensor. The percentage saving of liquid with the sprayer is 45–50% and 25–30% with hollow cone and turbo nozzles, respectively.

3.7 Importance of Pre-harvest Machinery in Improving Quality and Safety of Foods

Food quality and safety is a global public health concern. This is an important agricultural priority to achieve safe and quality food. Post-harvest losses in fruit and vegetables can be either qualitative or quantitative (Arah et al. 2015). As per Aayog (2020), more than 30% of the horticultural produce produced in India each year is wasted and costs the country a total of one lakh crore rupees. Nowadays, horticulture researches is also increasingly moving from quantity to quality of product but the improvement rate to enhance the quality of fruit and vegetable varieties is very slow hence there is high amount of postharvest losses such as nutrient status, financial income to producers and consumer acceptance (Oko-Ibom and Asiegbu 2007). Consumer quality acceptance depends on extrinsic and intrinsic features, which vary from the consumer's expectations and demands. The extrinsic factor of the produce contains external qualities such as colour, shape and size, as well as free from blemishes. Internal factors comprise texture, acidity, sweetness, flavour, aroma, nutritional value and free from harmful components. These factors cannot be improved after harvesting; it can only be maintained. Postharvest managements are not only the preeminent way of maintaining produce quality because such treatments are costly, increases the chances of damage through handling and encourage cultivator to give less attention to quality (Mirdehghan and Rahimi 2016). So, pre-harvest application is considered a good alternative of coping with the mentioned problem. Preharvest food safety must be primarily focused on protecting human health and being considered in the context of a farm-to-table food safety approach. There is various pre-harvest food safety programmes are organized to fulfil public health demands because traditional food safety systems have failed to effectively address the emerging and new foodborne pathogens, gain market access and strengthen consumer confidence.

Introduction of pre- harvest mechanization systems have shown to improve input-use efficiency, lowering labor costs and reduction in time required. Laser-assisted land leveling equipment's help achieve high accuracy with only a +/− 1 cm variation in height, which typically results in a 40–60% water savings, a 10–15% increase in yields, a reduction in agrochemical use, and even a slight improvement in grain quality (Khandetod 2018). This significantly reduces the cost of rice production and, due to improved water management, can also help reduce greenhouse gas emissions. The use of equipment such as a drum seeder, or a more complex power tiller or four-wheel tractor, can reduce the problems caused by manual transplanting, such as uneven crop establishment and hampered mechanical weed control. The use of machinery such as a 12 V-powered spreader for granular fertilizer and improved sprayers aids in mitigating the negative effects of fertilizers by facilitating the precise application of fertilizers to crops.

4 Pre-harvest Factors That Affect Quality and Safety of Food

Pre-harvest factors are critical to ensure the production of quality agricultural commodities. Understanding the effects of pre-harvest factors on the quality and physiology of agricultural commodities is very important because it could help in the implementation of the best practices to increase the overall quality of horticulture produce. Pre-harvest factors such as genetics (variety), cultural practices, and environmental and physiological factors affect the external and internal quality of vegetables and fruit (Hewett 2006). These factors profoundly influence postharvest performance and the final quality of vegetables and fruit (Fig. 5.8). Scientists, market personnel, extension specialists and farmers will have to work together to provide knowledge of best practices and enable tools for cultivators to ensure customized pre-harvest conditions so that they can produce high-quality crops that satisfy and reward conscious consumers. Hence, it is essential to know which preharvest factors most affect harvest quality that simultaneously increases the rate of postharvest deterioration and, subsequently, the consumers' decision to purchase the product in the marketplace.

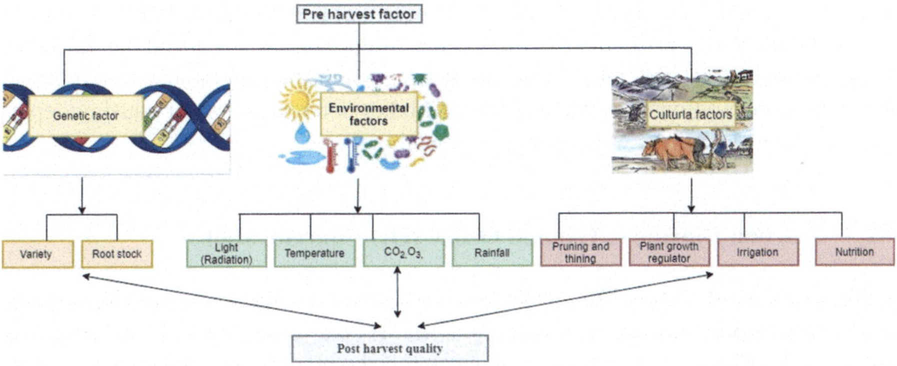

Fig. 5.8 Pre-harvest factors on the quality and safety of food

4.1 Genetic Factors

The species and cultivars are the main factors that decide the potential quality of produce concerning parameters such as color, weight, shape, size, dry matter, taste as well as biochemical composition, production levels and the ripening rates of horticulture crops) (Tyagi et al. 2017). To control the quality parameters genetically, plant breeders have taken a keen interest in selecting an appropriate cultivar that can give higher quality and higher yield with greater market acceptability (Scalzo and Mezzetti 2010).

4.2 Environmental Factors

4.2.1 Light (Radiation)

Light is a major factor in the rate of photosynthesis and the quality of light is responsible for the morphology and growth pattern of the plant (Madakadze and Kwaramba 2004). Plants commonly respond to 350–780 nm wavelengths of light and different wavelengths can affect some physiological processes in the plant. Many researchers have studied the effect of light on biochemical compounds and the nutritional composition of vegetables (Lee and Kader 2000; Woolf et al. 1999; Madakadze and Kwaramba 2004) and have concluded that optimal levels of light are required for nutritional composition and radical scavenging activity of crops.

4.2.2 Temperature (°C)

Crops generally have specific temperature requirements which affect nutrient uptake, and changes in optimum temperature can cause physiological, morphological and biochemical changes resulting in lower yields (Nonnecke 1989). The rates of biochemical reactions increase at higher temperatures, catalyzed by different enzymes and affect the mineral accumulation. Higher temperatures can increase the evapotranspiration rate which creates a higher demand for water causing depletion of water in soil and water stress in plants.

4.2.3 Carbon Dioxide (CO_2) and Ozone (O_3) Concentrations

Concentrations of $CO_2$and O_3 affect crop production in many aspects like production, organoleptic and nutritional composition (Felzer et al. 2007; Lloyd and Farquhar 2008). Through the stomates, ozone enters the plant tissues and causes cellular injury, particularly in the palisade cells (Mauzerall and Wang 2001). The damage changes in membrane permeability result in reduced growth, visible injury and, ultimately, reduced yield (Krupa and Manning 1988). Adding CO_2 to the greenhouse environment is generally recommended when farmers employ supplemental lighting during the darker months of the year.

4.2.4 Rainfall and Water Availability

Water has a significant role in biological activities, transports food nutrients to the plant, and water stress conditions can affect the growth and processes like photosynthesis (Madakadze and Kwaramba 2004). Crops have different requirements for water at different stages of growth. Distribution and the total amount of rainfall received annually are essential for agriculture growth. Generally, drip and sprinkler irrigation systems are used for water distribution in crops. Adaptation of automation where a computer-controlled system is used to link the irrigation automation can be useful so that several inputs can be examined at the same time.

4.2.5 Harvesting Time and Season

Harvesting season greatly influenced the quality of products, and vegetables and fruits collected in the off-season, provide the grower more remunerative price. Harvesting at an inappropriate maturity leads to ripening failure, lower organoleptic properties, and excessive softening. During rainy seasons the rate of microbial contamination is high. Thus, crops should be harvested when the temperature is mild because the high-temperature rate of respiration is high.

4.3 Soil

Crops grow in fertile soil, which has a high water-holding capability, is well-drained, with optimum acidity and alkalinity and contains essential nutrients (P, N, Ca, K, S, Mg, Fe, Mn, Cl, B, Zn, Mo and Cu) that are necessary for growth. The ideal soil for most fruits and vegetables is the medium clay loams supplied with organic matter or other nutrients. Most fruits and vegetables generally accept a pH range of between 5.5–7.5 while peppers and tomatoes can tolerate slightly more acid soils. Sandy soils are nematodes host that hampers the growth of production and is very difficult to control. So, mechanization in pre-harvest techniques is crucial for the normal growth of the plant.

5 Recent Advanced Practices Used to Reduce the Effect of Pre-harvest Factor

5.1 Photoselective Nettings

The Photoselective netting shows an innovative and advances agro-technical practices, which improves the conventional net covering to a more advanced level. It would be more cost effective to improve phytochemical contents, yield and postharvest quality through the photoselective colored shade nettings by change the temperature (microclimate) and light quality. They are based on the combination of various light dispersive, reflective elements and chromatic additives into the netting materials during development. Photoselective pearl nets improved the color, sensory properties and marketable yield as well as retained higher vitamin C and radical scavenging activity of tomatoes and sweet peppers (Sivakumar and Jifon 2018). Overall, photoselective shade appear to be a cost-effective practices for changing crop microclimate properties to improve yield, eating quality as well as bioactive or functional properties that are connected with human health and well-being.

5.2 Deficit Irrigation

Deficit irrigation (DI) methods are popular and suitable in the drought-sensitive area where unstable climate situations or supply of water is very low (Fereres and Soriano 2007). A deficit irrigation strategy gives opportunities for saving water without compromising fruit and vegetable production. Many research reported that it saves water from 43–65%, with a higher quality of the product but a small reduction in the yield of produce (Mirás-Avalos et al. 2016). Generally, the vegetable and fruit yield reduced in deficit irrigation system by their weight and size, but quality parameters, such as ascorbic acid, anthocyanin and sugars contents in fruit increased by water restrictions (Roccuzzo et al. 2014). Fruit and vegetables such peach, grape, brinjal, spinach and orange shows improved nutritional composition and antioxidant activity by adoption of DI (Mirás-Avalos et al. 2016; Permanhani et al. 2016; Roccuzzo et al. 2014). Thus deficit irrigation strategies in recent pre-harvest practices used to improve quality and safety of agriculture produce those grow in water stress condition.

5.3 Mulching

In areas where saline water used for irrigation, salts move up during the drying phase, which creates an adverse environment for the growth of plant. To minimize upward salt movement, modifying soil temperature, rate of evaporation and releasing nutrients in the soil profile as well as improving aeration, mulching (surface cover) with plastics or biological material could be an effective option (Abd El-Mageed et al. 2016; Sharma et al. 2005). Crop residues act as an adequate soil cover that creates a barrier against water vapor evaporation losses, increases infiltration and slow surface runoff. The reduced water evaporation losses result in a reduc-tion in a salt buildup in crop root zone better and soil moisture regimes (Pang et al. 2010), which are desirable steps pre-harvesting crop where saline environments.

5.4 *Bacterial Endophytes as Biocontrol Agents*

Agricultural diseases cause significant losses in the production of vegetables and fruits, during their cultivation, handling, transportation and storage. Synthetic fungicides are used to control agriculture diseases but not suitable alternative to control diseases due to their high environmental and economic costs and leave their residue in soil, atmosphere and water in addition to inducing resistance in phytopathogenic strains (Sharma et al. 2009). Thus, the development of environmentally friendly and efficient technologies that reduce or eliminate synthetic fungicides' effect in agriculture is highly desirable (Santoyo et al. 2019). Bacterial endophytes colonize and inhabit internal plant tissues without causing any apparent damage. Within the horticulture plant, these bacteria exert multiple beneficiary functions such as the production of metabolites or phytohormones' action, including direct stimulation of plant growth. Bacterial endophytes also protect the plant through biocontrol pathogens by inducing plant innate immune system. Finally, we conclude that the use of bacterial endophytic as biological agents during both pre and post-harvest stages protecting their agricultural produce and crop plants with pathogens.

5.5 *Pre-harvest Screening by Using NIRS Technology*

Recent years, consumers have become attentive of the presence of nitrates (NO3–) in foods, because nitrates cause a severe threat to human health, due to the change of nitrate (NO3–) to nitrite (NO2–), which may produce met-haemoglobin due to the oxidation of ferrous in haemoglobin (Elia et al. 1998). Methemoglobin reduced capability to deliver oxygen to tissues leads to severe toxic effects and may even prove lethal when the percentage of methemoglobin accounts for over 70% of total haemoglobin (Santamaria 2006). All this has prompted greater attention to vegetable and fruits quality and safety concerns. The nitrate concentration in vegetables (spinach) when harvested is the key to check the final quality of the harvested produce. NIRS (Near infrared spectroscopy) is non-destructive technology for the study of chemical composition of vegetables at field level. Perez-Marin et al. (2019) had done a pre-harvest screening on-vine of spinach quality and safety using NIRS technology. This technology changes the conventional analytical methods. In NIRS technology a single spectrum allows to characterize of different chemical composition with-in seconds without sample preparation, thus allowing real-time decision making. Thus, the NIR spectroscopy can be used as a screening tool to analyze spinach production for enhancing the quality and safety parameters.

6 Conclusion

Mechanization in agriculture has become increasingly significant and has played a crucial role in agricultural production worldwide. Machines for pre-harvest factors like soil preparation and seeding or vegetative propagation and irrigating or fertilizing and pruning play a significant role in determining the harvest characteristics and post-harvest losses of crops. If agricultural machinery is used effectively, it is possible to conserve and properly utilize natural resources, as well as to reduce the overall cost of production. Population growth has led to an increase in the demand for agricultural products. Therefore, it is anticipated that the global market for agricultural equipment and pre-harvest supplies will continue to expand. Growth in the pre-harvest market has been steady around the world as a result of the growing population, increasing demand for food, increasing farm mechanization levels, and farmers' acceptance of farm mechanization techniques and implements. Agriculture equipment has proven to be beneficial in terms of increased yield and quality of crop production. Using farm equipment has resulted in a rise in farmers' adoption of the technology. It is now possible to produce environmentally friendly agricultural equipment at a lower cost and with a higher degree of customization based on location and crop type. A steady increase in agricultural equipment sales is seen as an excellent business opportunity because of these positive attributes. The existing systems of pre-harvest machinery can be scaled up for large-scale production. Along with this, awareness about pre-harvest mechanized practices and implementation of multi-crop planters in agricultural production can help farmers shift to mechanized pre-harvest systems. The gap between industrial pre- harvest machine producers and farmers needs to be shortened for the ease of its adaptation to agricultural farms. Prototype demonstrations, fabrications, and commercialization of agricultural mechanization technologies are critical to ensuring their availability, accessibility, and affordability for smallholder farmers.

References

Aayog N (2020) Annual Report 2020–21 of Niti Aayog

Abd El-Mageed TA, Semida WM, Abd El-Wahed MH (2016) Effect of mulching on plant water status, soil salinity and yield of squash under summer-fall deficit irrigation in salt affected soil. Agric Water Manag 173:1–12

Ahmed S, Bagal YS, Mahajan R, Sharma LK (2020) Impact of Farm Mechanization on Crop Productivity in Sub-Tropical Areas of Jammu and Kashmir. Int J Curr Microbiol App Sci 9(5): 1168–1173

Arah IK, Amaglo H, Kumah EK, Ofori H (2015) Preharvest and postharvest factors affecting the quality and shelf life of harvested tomatoes: a mini review. Int J Agron 2015:478041. https://doi.org/10.1155/2015/478041

Berghaus RD, Thayer SG, Law BF, Mild RM, Hofacre CL, Singer RS (2013) Enumeration of Salmonella and Campylobacter spp. in environmental farm samples and processing plant carcass rinses from commercial broiler chicken flocks. Appl Environ Microbiol 79(13): 4106–4114

Biggs S, Justice S, Lewis D (2011) Patterns of rural mechanisation, energy and employment in South Asia: reopening the debate in economic and political weekly, vol XLVI No 9, Mumbai, India

Cherian A, Punnan J, Varghese A (2016) A review on power tiller attachments. In: Kerala technological congress (KETCON-2016 technology for sustainability)

De-feng Z, Yu-ping Z, Hui-zhe C, Jing X, Yi-kai Z (2015) Innovation and practice of high-yield rice cultivation technology in China. Sci Agric Sin 48:3404–3414

Dwivedy N (2011) Challenges faced by the Agriculture Sector in Developing Countries with special reference to India. Int Rural Stud 18:2–7

Elia A, Santamaria P, Serio F (1998) Nitrogen nutrition, yield and quality of spinach. J Sci Food Agric 76(3):341–346

Felzer BS, Cronin T, Reilly JM, Melillo JM, Wang X (2007) Impacts of ozone on trees and crops. Compt Rendus Geosci 339:784–798

Fereres E, Soriano MA (2007) Deficit irrigation for reducing agricultural water use. J Exp Bot 58(2):147–159

Gifford RC (1992) Agricultural engineering in development: concepts and principles, vol 1. Food & Agriculture Org

Hazell PB (2009) The Asian green revolution, vol 911. International Food Policy Research Institute

Hendriadi A, Salokhe VM (2002) Improvement of a power tiller cage wheel for use in swampy peat soils. J Terramechanics 39(2):55–70

Hewett EW (2006) An overview of preharvest factors influencing postharvest quality of horticultural products. Int J Postharvest Technol Innov 1(1):4–15

Khandetod YP (2018) Mechanization in horticulture crops: present status and future scope. J Adv Agric Technol 3:92–103

Kishore KK, Vamsi Krishna P, Srikanth D (2017) Automatic feeding system for aquaculture. In 2017 Third International Conference on Sensing, Signal Processing and Security (ICSSS), pp. 426–429. IEEE, 2017

Krupa SV, Manning WJ (1988) Atmospheric ozone: formation and effects on vegetation. Environ Pollut 50(1–2):101–137

Lee SK, Kader AA (2000) Preharvest and postharvest factors influencing vitamin C content of horticultural crops. Post Harvest Biol Tec 20(3):207–220

LeJeune JT, Besser TE, Hancock DD (2001) Cattle water troughs as reservoirs of Escherichia coli O157. Appl Environ Microbiol 67(7):3053–3057

Lloyd J, Farquhar GD (2008) Effects of rising temperatures and [CO2] on the physiology of tropical forest trees. Philos Trans R Soc B Biol Sci 363(1498):1811–1817

Madakadze RM, Kwaramba J (2004) Effect of preharvest factors on the quality of vegetables produced in the tropics. In: Production practices and quality assessment of food crops volume 1. Springer, Dordrecht, pp 1–36

Mauzerall DL, Wang X (2001) Protecting agricultural crops from the effects of tropospheric ozone exposure: reconciling science and standard setting in the United States, Europe, and Asia. Annu Rev Environ Resour 26(1):237–268

Mirás-Avalos JM, Pérez-Sarmiento F, Alcobendas R, Alarcón JJ, Mounzer O, Nicolás E (2016) Using midday stem water potential for scheduling deficit irrigation in mid–late maturing peach trees under Mediterranean conditions. Irrig Sci 34(2):161–173

Mirdehghan SH, Rahimi S (2016) Pre-harvest application of polyamines enhances antioxidants and table grape (Vitis vinifera L.) quality during postharvest period. Food Chem 196:1040–1047

Nonnecke IL (1989) Vegetable production. An AVI Book. Publisher van Nostrand Reinhold, pp 614–623

Oko-Ibom GO, Asiegbu JE (2007) Aspects of tomato fruit quality as influenced by cultivar and scheme of fertilizer appication. Agro-Science 6(1):71–81

Pang HC, Li YY, Yang JS, Liang YS (2010) Effect of brackish water irrigation and straw mulching on soil salinity and crop yields under monsoonal climatic conditions. Agric Water Manag 97(12):1971–1977

Perez-Marin D, Torres I, Entrenas JA, Vega M, Sánchez MT (2019) Pre-harvest screening on-vine of spinach quality and safety using NIRS technology. Spectrochim Acta A Mol Biomol Spectrosc 207:242–250

Permanhani M, Costa JM, Conceição MAF, De Souza RT, Vasconcellos MAS, Chaves MM (2016) Deficit irrigation in table grape: eco-physiological basis and potential use to save water and improve quality. Theor Exp Plant Physiol 28(1):85–108

Roccuzzo G, Stagno F, Ferlito F, Intrigliolo F, Cirelli GL, Consoli S (2014, August) Deficit irrigation for enhancing' Tarocco'orange fruit quality. In: XXIX International horticultural congress on horticulture: sustaining lives, livelihoods and landscapes (IHC2014): 1112, pp 179–186

Santamaria P (2006) Nitrate in vegetables: toxicity, content, intake and EC regulation. J Sci Food Agr 86(1):10–17

Santoyo G, Sánchez-Yáñez JM, de los Santos-Villalobos S (2019) Methods for detecting biocontrol and plant growth-promoting traits in Rhizobacteria. In: Methods in rhizosphere biology research. Springer, Singapore, pp 133–149

Scalzo J, Mezzetti B (2010) Biotechnology and breeding for enhancing the nutritional value of berry fruit. In: Biotechnology in functional foods and nutraceuticals, pp 61–80

Sekhar CSC, Bhatt Y (2014) Effect of farm mechanization on agricultural growth and comparative economics of labour and machinery in India. Institute of Economic Growth, New Delhi, India

Sharma AR, Singh R, Dhyani SK (2005) Conservation tillage and mulching for optimizing productivity in maize-wheat cropping system in the outer western Himalaya region–a review. Indian J Soil Conserv 33(1):35–41

Sharma RR, Singh D, Singh R (2009) Biological control of postharvest diseases of fruits and vegetables by microbial antagonists: A review. Biol Control 50(3):205–221

Sims BG, Kienzle J (2006) Farm power and mechanization for small farms in Sub-Saharan Africa, Agricultural and food engineering technical report, vol 3. FAO, Rome

Singh G (2006) Agricultural machinery industry in India (Manufacturing, marketing and mechanization promotion). In: Status of Farm Mechanization in India, pp 154–174

Sivakumar D, Jifon J (2018) Influence of photoselective shade nettings on postharvest quality of vegetables. In: *Preharvest* Modulation of Postharvest Fruit and Vegetable Quality. Academic, pp 121–138

Si-yuan C, Guang-qiao C, Wen-yi Z, Xiao-xing Z (2014) Suitable mechanical transplanting spacing promoting rice growth and increasing rice yield. Trans Chin Soc Agric Eng 30:37–43

Tiwari PS, Singh KK, Sahni RK, Kumar V (2019) Farm mechanization–trends and policy for its promotion in India. Indian J Agric Sci 89(10):1555–1562

Torrence ME (2003) US Federal activities, initiatives, and research in food safety. In: Microbial food safety in animal agriculture: current topics, pp 3–9

Tyagi S, Sahay S, Imran M, Rashmi K, Mahesh SS (2017) Pre-harvest factors influencing the postharvest quality of fruits: a review. Curr J Appl Sci Technol 23:1–12

Woolf AB, Ferguson IB, Requejo-Tapia LC, Boyd L, Laing WA, White A (1999) Impact of sun exposure on harvest quality of 'Hass' avocado fruit. Revista Chapingo Serie Horticultura 5:353–358

Chapter 6
Non-thermal Processing of Foods: Recent Advances

M. L. Bhavya and H. Umesh Hebbar

1 Introduction

Food processing is an important and vital element for the extension of shelf-life of food. The food processing includes traditional methods like drying, pickling, cooking, fermentation and modern/advanced methods like retort processing, high pressure processing (HPP), ohmic heating and many more. The preliminary and traditional purpose of food processing was to keep food safe for long time, while the present-day technologies not only aim at safety and shelf-life of food but also focus on the quality of the product. The limitations of conventional food processing techniques include, loss of nutrients; change in colour, texture, flavour; formation of off-flavours and toxic compounds; oxidation, and low consumer's acceptance (Lee et al. 2016b; Proctor 2018; Hernández-Hernández et al. 2019). In this age of urbanization, consumers around the world are more conscious of the importance of nutrition and the safety of processed foods. In the present competitive world, developing sustainable, low energy, environmental friendly processing technology is the main challenge for food researchers. In this view, non-thermal food processing techniques like HPP, pulsed electric field, cold plasma (CP), ozone, ultrasound (US), light-based processing etc., are gaining prominence as they have exhibited potential to meet the growing demands of health-conscious consumers (Zhang et al. 2019). Focus on the emerging non-thermal technologies has been increased and research based on these technologies have been widely developed in first world countries like Europe and U.S.A, nevertheless, even developing countries have also begun to emphasize on these technologies (Hernández-Hernández et al. 2019). The recent

M. L. Bhavya · H. U. Hebbar (✉)
Food Engineering Department, CSIR – Central Food Technological Research Institute, Mysuru, Karnataka, India
e-mail: hebbar@cftri.res.in

H. U. Hebbar et al. (eds.), *Engineering Aspects of Food Quality and Safety*, Food Engineering Series, https://doi.org/10.1007/978-3-031-30683-9_6

advances in these areas have been focused in the present chapter pertaining to the application of novel technologies/techniques for microbial load reduction, retaining product quality, physicochemical parameters, and also sensorial attributes.

2 High Pressure Processing

High pressure processing (HPP) is one of the non-thermal food processing techniques where food product is subjected to pressure at or above 100 mega Pascal (MPa). During HPP, the product is pre-packed in a packaging material which can withstand the high pressure (vacuum packed or flexible packages-plastic bottles) and placed in a sealed pressurized chamber filled with potable water (Fig. 6.1). The hydrostatic pressure is applied (usually from 400 to 600 MPa) for a set period and transmitted to the food instantly and equally throughout the package. The application of HPP in food processing was first reported in 1899 by Bert H. Hite for improving the stability of milk (Hite 1899). However, this technology reached food market in Japan, only in 1990 (Yuste et al. 2001). HPP has been employed for inactivating microbes and enzymes while retaining physicochemical attributes of food (freshness, colour, aroma, texture, bioactive compounds) (Abera 2019). Number of studies reported so far have indicated that instantaneous and uniform transmission of pressure (isostatic transmission) during HPP resulted in desired change, without deforming food matrix even under high pressures (100–1000 MPa/−20 °C to 60 °C). HPP can be performed even at lower temperatures which is said to facilitate quality retention due to preservation of primary structure of macromolecules by HPP, as covalent bonds are kept intact (Knorr 1993). Also, HPP was reported to be an energy efficient process since it doesn't require extra energy or there will be no loss of energy after desired pressure is reached.

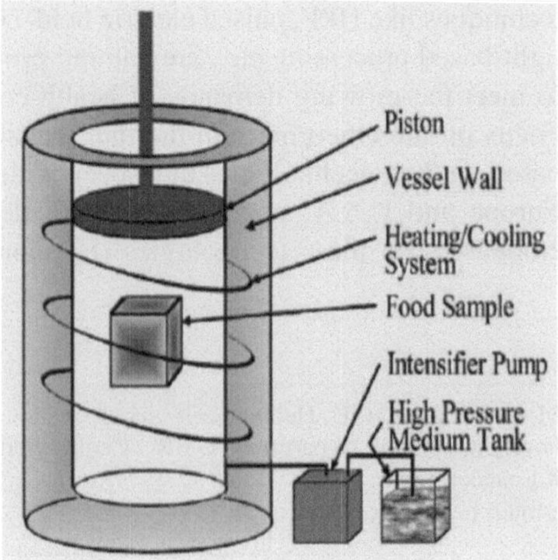

Fig. 6.1 Schematic representation of high pressure processing equipment. (Khaliq et al. 2021)

2.1 Microbial Load Reduction

HPP has been used for sterilization or decontamination of foods such as vegetables, fruits, milk, meat, poultry, fish, beverages and many more (Georget et al. 2015; Hurtado et al. 2015; Stratakos et al. 2019; Balamurugan et al. 2020). HPP causes microbial inactivation due to (i) disruption of cell wall and membrane, (ii) modification/denaturation of biological polymers like enzymes, proteins (iii) chemical reactions and enzyme catalyzed conversion processes (Oey et al. 2008). The degree of inactivation is affected by the processing conditions, such as magnitude of pressure, exposure period, processing temperature (higher are more effective), alternating pressure cycles (better than continuous pressurization) and composition of medium (Yuste et al. 2001). The effect of HPP treatment was influenced by the type and nature of microorganism. Gram-negative bacteria and rod shaped ones were more sensitive to HPP compared to Gram-positive and cocci shaped bacteria (Hoover et al. 1989). Spores and prokaryotic microorganisms were found to be more resistant to HPP than eukaryotic and vegetative cells (Hoover et al. 1989). Also, cells in stationary or death phase were more resistant to HPP than that in logarithmic phase.

Recent works on the effect of HPP on microbial inactivation of foods is summarized in Table 6.1. The HPP is a potential technology which is used to extend the shelf life of different range of food products such as milk, fruit and vegetable smoothies, soy-smoothies, and coconut water by 7, 26, 45, and 120 days (stored at 4 °C), respectively (Andrés et al. 2016a; Fernandez et al. 2019; Stratakos et al. 2019; Raghubeer et al. 2020). Rode and Hovda (2016) reported that the effect of HPP treatment induced changes both in chemical and microbiological attributes in different fish species (salmon, cod and mackerel) with extension of shelf-life. Mackerel fillets processed by HPP (500 MPa for 2 min or 5 min) showed significant microbial inactivation without affecting the nutritional attributes and lipid oxidation of the pelagic fish, while had some changes in texture (increase in hardness, chewiness and springiness) and colour (increased lightness and decreased redness) (de Alba et al. 2019). HPP has been successfully employed for inactivation of microbial population in the food making it an alternative technology for conventional thermal processes. A few products developed using HPP technology are already available in the market.

2.2 Enzyme Inactivation

Enzymes accelerates several processes (oxidation, browning and ripening) resulting in the loss of flavour, colour and texture. In this regard, several studies focus on the efficacy of HPP treatment to inactivate enzymes in solid (vegetables), semisolid (smoothie) and liquid matrices (juice) have been reported. Reports have shown that the matrix (cube, puree, and extract) greatly influences the degree of enzyme inactivation (Tribst et al. 2016). Paciulli et al. (2016) mentioned that HPP partially

Table 6.1 Microbial inactivation in foods using high pressure processing technology

Treatment conditions	Micro-organism	Matrix/ medium	Log reduction	Reference
350 and 450 MPa at 10 °C for 5 min, 600 MPa at 10 °C for 3 min	Total viable counts Psychrophilic bacteria Total coliforms Moulds and yeasts	Fruit smoothies	~2 to 4 log CFU/g	Hurtado et al. (2015)
550 MPa for 6 min	Total plate count	Carrot juice	4.30 log CFU/mL	Zhang et al. (2016)
200, 300 and 400 MPa for 30 min, 27 s, and 12 s	*S. cerevisiae* ascospores	Beer	2.5 log CFU/mL	Milani et al. (2016)
500 MPa for 5 min at 20 ± 1 °C	Total aerobic count Yeast and mold	Cucumber juice	2 and 1 log CFU/mL	Liu et al. (2016)
450 and 600 MPa for milk-smoothies; 550 and 650 MPa for soymilk smoothies for 3 min at 20 °C	Aerobic mesophilic bacteria	Milk-and soy-smoothies	>3 log CFU/mL	Andrés et al. (2016a)
600 MPa for 5 min	Total viable bacteria Yeast and moulds count	Elephant apple juice	>3 and 2 log CFU/mL	Nayak et al. (2017)
600 MPa for 5 min at 6 °C	*L. monocytogenes*	Cheese	~4 log CFU/g	Evert-Arriagada et al. (2018)
400 and 600 MPafor 5 min at 15 °C	Aerobic bacterial count	Aronia berry purée	2.31 and 2.75 log CFU/g, respectively	Yuan et al. (2018)
500 MPa at 20 °C for 2 min	*L. innocua* *E. coli*	Carrot juice	4 and 5 log CFU/mL	Pokhrel et al. (2019)
500 MPa at 10 °C for 5 min	Total viable count	Mackerel fillets	~2.5 log CFU/g	de Alba et al. (2019)
600 MPa for 3 and 5 min	*E. coli* *Salmonella* *L. monocytogenes*	Milk	~5.1 to 6.8 log CFU/mL	Stratakos et al. (2019)
593 MPa for 3 min at 4 ± 1 °C	*E. coli* O157:H7 *Salmonella* *L. monocytogenes*	Coconut water	~4 to 5 log CFU/mL	Raghubeer et al. (2020)
600 MPa for 3 min	*E. coli* O157:H7	Fermented sausage	4.8 log CFU/g	Balamurugan et al. (2020)
400 MPa for 3 min at pH 4.3 and 2.9°brix	*E. coli* O157:H7 *Salmonella* *L. Monocytogenes*	Açaí juice	>6 log CFU/mL	Gouvea et al. (2020)
300 MPa for 5 min	Mesophilic and psychrophilic bacteria	Mackerel mince	1.69 log CFU/g and 2.23 log CFU/g, respectively	Cropotova et al. (2020)

(continued)

Table 6.1 (continued)

Treatment conditions	Micro-organism	Matrix/ medium	Log reduction	Reference
500 MPa for 15 min at 34 °C	Viable total mold and yeast, *Lactococcus lactis* subsp. cremoris, and *lactobacillus paracasei*	Fermented turnip juice (shalgam)	3.09, 2.51, and 2.68 log CFU/mL, respectively	Ates et al. (2021)
350 MPa for 4 to 12 min	*Salmonella* spp.	Ground chicken meat	1.46 to 3.54 log CFU/g	Chai and Sheen (2021)

inactivated the undesirable enzymes, viz., polyphenol oxidase (PPO) and peroxidase (POD) by 10–25% in beetroot. Lower enzyme inactivation during HPP compared to the thermal treatment (blanching) can be attributed to pressurization induced enzyme release, extractability during cell membrane breakdown and release of more active enzymatic isoforms. Rinaldi et al. (2020) have described higher residual pectin methylesterase (PME) activity (~82%) in HPP treated pineapple, indicating the resistance of PME to HPP.

HPP treatment (600 MPa for 5 or 30 min at 25 °C) inactivated PPO and POD by 55 to 98% in extract and puree of tubers like cocoyam and Peruvian carrot, without affecting the colour (Tribst et al. 2016). While, HPP treatment did not show any significant reduction in PPO, POD and PME activities in fruit smoothies, but thermal processing inactivated all the enzymes by ~90% (Hurtado et al. 2015). Yi et al. (2017) also indicated that PPO and POD were inactivated by less than 50% in HPP treated apple juice. Szczepańska et al. (2020) observed that POD (57% inactivation at 300 MPa × 3 pulses) was more resistant to HPP than PPO (31% inactivation at 600 MPa) in carrot juice. HPP treatment was found to reduce the activity of PME, POD and PPO by 85, 31–45 and 10%, respectively in mixed fruit and vegetable smoothie (Fernandez et al. 2018, 2019). These studies have shown that though HPP inactivates enzymes, the degree of inactivation depended on nature of enzyme.

2.3 *Physical Attributes*

In consumers' perspective, physical attributes such as appearance, colour, and texture are essential. Effect of HPP on turbidity and viscosity of liquid/semisolid foods, particle size, pH and color have been studied. It was found that turbidity and viscosity of HPP treated fruit smoothies were lower than the thermally treated samples indicating the higher tendency for clarification of pressurized smoothies (Hurtado et al. 2015). HPP treatment significantly reduced the viscosity of aronia berry purée and slightly affected total soluble solids (TSS) content and particle size distribution of the samples (Yuan et al. 2018). During storage, titratable acidity, pH and TSS of HPP treated elephant apple juice, cranberry juice and gooseberry juice did not show any significant change (Nayak et al. 2017; Gomes et al. 2017; Raj et al. 2019). The viscosity of the elephant apple juice was found to increase after HPP treatment due to the activity of PME (Nayak et al. 2017). HPP treatment was found to have no significant effect on the soluble sugars, minerals and organic acids of smoothies compared to untreated samples, suggesting the application of HPP as an alternative to the traditional thermal methods for processing (Andrés et al. 2016a).

Increase in pH values were observed for HPP treated beef samples at higher pressure levels (300 and 400 MPa) due to the conformational changes in proteins resulting in reduction of available acidic groups in the meat (McArdle et al. 2010). HPP slightly increase the pH and tenderness at a higher pressure (250 MPa) with lighter colour of the hot-boned beef (Morton et al. 2017). HPP treatment altered the colour (lightness and yellowness) of Sous-Vide cooked beef steaks, with limited variation in the redness of samples (Sun et al. 2019). Higher pressure levels (300 and 400 MPa) resulted in higher L* values ("whitening/brightening" effect) of beef samples compared to that at 200 MPa which can be attributed to heme/haem displacement or release, globin denaturation, and ferrous ion oxidation (McArdle et al. 2010). In recent reports, Rode and Rotabakk (2021) mentioned that HPP (600 MPa for 5 min) treated rehydrated clipfish and saltfish can be stored for >49 days without affecting the affecting the colour parameters, but it increased the drip loss during the storage which can possibly be reduced by the usage of absorbent in the packaging material. It was observed that HPP treatment turned yak meat white, while the combinational treatment (papain + HPP) did not significantly deteriorate the colour parameters (Ma et al. 2019).

Significant effect of pressure induced changes on the colour parameters (browning index, highest colour (ΔE) difference) were observed in carrot juice and pineapple slices (Szczepańska et al. 2020; Rinaldi et al. 2020). The colour values (a* and b*) of HPP treated beetroots were lower than the blanched ones, demonstrating a blue-shift in colour (from red to red-violet) in agreement to high content of betanin (absorbing at 536 nm) in HPP samples (Paciulli et al. 2016). In contrary, it was found that HPP treatment did not affect colour attributes of elephant apple juice compared to untreated ones, while heat treated samples (80 °C for 60 s) had higher L* values and lesser a* and b* values leading to browning (Nayak et al. 2017). It was worth noticing that HPP treated milk had similar colour, casein and fat particle size as the raw and pasteurized milk, indicating the possibility of using HPP as an alternative conventional processes to produce fresh-like milk with microbiological safety (Stratakos et al. 2019).

Hardness and chewiness of the HPP treated beetroot samples were better than the blanched ones, which can be attributed to the release of PME (that form a gel network with divalent ions increasing the hardness) during pressure induced disruption of the tissue (Paciulli et al. 2016). Rinaldi et al. (2020) also noted that pineapple cubes (in sugar syrup) treated with HPP (600 MPa for 3 min) preserved the shape and microstructure as well as the hardness of the cubes. In contrary, texture (hardness) of HPP treated carrots was reduced by 39% compared to untreated ones due to the turgor loss associated with mechanical membrane damage (Vervoort et al. 2012). HPP as a pretreatment for vacuum-fried carrot snacks caused structural modification in the product increasing the hardness and reducing the general acceptability of the snack, while freezing as a pretreatment increased the crispiness and improved organoleptic and nutritional (antioxidant capacity) properties (Albertos et al. 2016). Alsalman and Ramaswamy (2020) reported that HPP treated chickpeas resulted in ~90 to 93% hydration, which were supported by SEM studies showing larger pore sizes and bigger starch granules. In addition, HPP enhanced texture softness and colour parameters of chickpeas thus improving the overall quality compared to conventional soaking overnight (Alsalman and Ramaswamy 2020).

The findings of Zhang et al. (2018) indicated that the application of HPP within the range 100–300 MPa on pork improved the tenderness by inhibiting the rigor process. HPP treatment (250 MPa/15 min) lowered the maximal shear force (by 49%); cooking loss (by 8%) and increased the water holding capacity (by 10%) of the meat during tenderization of the yak meat (Ma et al. 2019). Hot-boned beef treated with HPP at 175 MPa reduced the shear force (by 43%), cooking loss and enhanced the eating quality with higher sensory score compared to that of chill ageing for four weeks (Morton et al. 2017). A negative effect of higher pressure levels (300, 400 MPa) on cook loss were noticed in HPP treated beef samples compared to that at 200 MPa due to the alteration in water binding properties of meat (McArdle et al. 2010). Sous-Vide cooked beef steaks treated with HPP (450 MPa or 600 MPa) received poor sensory scorings with lower tenderness and juiciness compared to control (Sun et al. 2019). HPP treated fish mince showed change in colour and water holding capacity (enhances fluid drain) due to the decrease in protein solubility and carbonylation. The firmness, hardness and cohesiveness of the fish cakes prepared from the HPP treated fish mince decreased, as the pressure increased (Cropotova et al. 2020). From these studies, it can be inferred that HPP retains most of the physical characteristics of the food products, while texture of the food products is altered depending on the conditions employed for processing.

2.4 Flavour and Volatile Compounds

Flavour of food is sensitive to heat and other extrinsic factors; in this regard, there are few reports focusing on the effect of HPP on flavour profile of the food. Hurtado et al. (2015) observed that HPP treatment of fruit smoothies did not alter the flavour profile at 350 and 450 MPa, while thermal processed samples developed cooked-fruit flavour due to the thermal degradation of flavour compounds. HPP treatment also produced fresh-like carrot juice and smoothies with better sensorial attributes, aroma, taste, and overall acceptability than thermally-treated ones (Zhang et al. 2016; Andrés et al. 2016a). Liu et al. (2016) also presented that HPP-treated cucumber juice had a shelf-life of 20 days with high clarity and sensory score, retention of colour and aroma compounds than high temperature short time treated juice. In contrast, HPP treated apple juice did not retain better flavour and colour attributes compared to thermal processed samples (Yi et al. 2017). This could be due to the formation of higher amounts of aldehydes, alcohols, ketones and organosulfur compounds during the thermal processing responsible for cooked notes as a result of Maillard and oxidative reactions.

The HPP treated dry-cured ham had higher total free amino acid and volatile compound content (higher ketone and ester content) compared to control samples (Pérez-Santaescolástica et al. 2019). Martínez-Onandi et al. (2016) noticed that HPP treated Serrano dry-cured ham affected esters and sulfur compounds exhibiting moderate effect on its volatile fraction. On contrary, Martínez-Onandi et al. (2017) reported that HPP had a higher impact on the volatile profile of the Iberian dry-cured ham than the physicochemical characteristics such as intramuscular fat content, salt concentration, salt-in-lean ratio and water activity. The reports suggest that the HPP affects flavour of food depending on the type of the matrix and thus more studies are necessary for in order to broaden the knowledge on effect of HPP on the flavour profile of foods.

2.5 Bioactives

Nutrients/bioactives are vital for maintaining human health and any food process should aim at preserving these health-related components. In this purview, literature reports focus on the efficiency of HPP to maintain the phytonutritive components of the food. A six fold increase in betanin was achieved when beetroot was treated with HHP in a time-dependent manner up to 7 min, because of greater extractability from the broken cells (Paciulli et al. 2016). HPP treatment of smoothie (fruit and vegetable based) and orange juice increased the extractability of bioactive compounds (lycopene, b-carotene, flavonoids, and polyphenols), while retaining ascorbic content (92%) and antioxidants compared to thermal treated ones (Andrés et al. 2016b; de Ancos et al. 2020). Bioactives such as hesperidin, narirutin, phytoene were found to increase by 25, 27 and 40%, respectively in orange juice treated with HPP compared to untreated juice (de Ancos et al. 2020). HPP also improved the quality of chickpeas by reducing tannin and phytic acid content by 27% and 17%, respectively (Alsalman and Ramaswamy 2020). It was reported that HPP treatment preserved ascorbic acid in fruit smoothies and beetroot samples than the thermal treated ones (Hurtado et al. 2015; Paciulli et al. 2016).

Thermal assisted HPP of Indian gooseberry juice significantly retained ascorbic acid (by 85%) and also increased antioxidant and phenolic content of the juice (Raj et al. 2019). The HPP (650 MPa) treatment increased the phenolic content of beetroot samples compared to raw samples, while it was comparable with the blanched (90 °C for 7 min) ones (Paciulli et al. 2016). Szczepańska et al. (2020) reported that the HPP increases the polyphenol content in carrot juice due to the pressure induced release of bound phenolics and by enzymatic degradation and/or condensation. The HPP (400 MPa for 5 min) maintained phenolic content and antioxidant capacity of aronia berry purée and did not affect the physicochemical attributes (colour, pH, acidity, pulp content) (Yuan et al. 2018). The fruit smoothies treated with HPP retained total phenols and flavonoids but resulted in a sucrose hydrolysis (Hurtado et al. 2015). Higher antioxidant activity, phenolic and flavonoid content were noticed in HPP treated elephant apple juice compared to heat treated and untreated samples (Nayak et al. 2017). In contrary, Marszałek et al. (2017) reported that the HPP treatment reduced polyphenols and anthocyanins in strawberry puree by 32% and 73%, respectively during 12-week storage at 6 °C, while thermal pasteurization caused loss by 28% and 54%, respectively.

The HPP treatment minimized the loss of phytochemical content, antioxidant capacity and also had a similar colour profile compared to that of steam-treated purple waxy corn kernels (Saikaew et al. 2018). Hurtado et al. (2015) found that HPP treated fruit smoothies had lower antioxidant capacity (at 48 h post-processing) than the thermally treated ones, indicating a higher tendency of pressurized smoothies to oxidation. Rinaldi et al. (2020) observed that HPP treated pineapple cubes did not show any significant change in antioxidant activity compared to the raw samples. The milk pre-treated with HPP used for the cheddar cheese preparation showed higher antioxidant activity and ACE-inhibitory potential, which could be due to the disruption of casein micelles (Munir et al. 2020), suggesting the application of HPP as pre-treatment to improve the nutritional quality of cheese. Treatment of rice with HPP increased thiamine content and gelatinization capacity, indicating a potential application of HPP for the enhancement of thiamine in white rice in a shorter duration with better organoleptic attributes (Balakrishna and Farid 2020). It was also observed that HPP caused partial gelatinization and reduced the loss of phenolics and flavonoids when corn kernels were pressurized at 250–400 MPa for 30 min, while processing higher than 400 MPa for 30 min resulted in complete gelatinization and loss of phytochemicals (Saikaew et al. 2018). It is mentioned that HPP increases the bioavailability and slows down retrogradation compared to thermal processing (Saikaew et al. 2018).

Processing of beef samples at higher pressures (300 and 400 MPa) resulted in increased rate of lipid oxidation compared to non-treated samples due to the release of free-radicals during pressure-induced protein denaturation (McArdle et al. 2010). The HPP treatment was found to be efficient in reducing the rancidity in brown rice at 200 MPa for 10 min (Wang et al. 2018). It was found that HPP did not have significant effect on the lipid oxidation of fish samples (Cropotova et al. 2020). The thiobarbituric acid reactive substances (TBARS) levels were affected by HPP in cod fish, while no effect was seen in salmon and mackerel samples (Rode and Hovda 2016). It was noticed that HPP at 500 MPa showed negative effect on different physical and chemical parameters of fish compared to that treated at 200 MPa, indicating the effect of pressure on the quality and safety of the food samples. The literature reviews showed that HPP could be a potential non-thermal technology not only for decontaminating foods, but also for enzyme inactivation, without much altering nutritive value and other attributes of food. The commercialization of this technology in a few specific areas showed that it could be explored in many food processing operations.

3 Light-Based Processing

Application of light energy for decontamination of foods is being studied, as it could be one of the potential methods of non-thermal treatment of foods, especially for surface treatments. Although, Pulsed light (PL) technology has been investigated extensively, efficacy of blue light (BL) is being explored in detail in the last few years. Combination of light with other modes of treatment is also gaining popularity to achieve and broaden the knowledge base of hybrid mode processing. The highlights of PL and BL processing in recent years are presented below.

3.1 Pulsed Light Technology

Pulsed light (PL) processing is one among the non-thermal processing techniques, which decontaminate foods in short duration of time using high-intensity light pulses with a wide wavelength range of 180–1100 nm. The term PL was first implemented by Food and Drug Administration (FDA) in 1996 for food processing, even though it was previously recognised in 1980 (FDA 1996). PL technology is also known as intense pulsed light, high intensity light, pulsed white light, etc., which employs inert gas flash lamps to transform high power electric pulses into high-power pulses of radiation (Kramer and Muranyi 2014; Bhavya and Umesh Hebbar 2017). The major advantages of PL are (i) its environment friendly processing with low impact (ii) low processing costs, and (iii) no chemical use (Bhavya and Hebbar 2020). PL will decontaminate foods in a shorter duration and extend their shelf life without affecting the characteristics. PL has a disadvantage of sample heating if treatment duration is longer, however, which has an extra thermal effect on microbial reduction (Bialka and Demirci 2008; Huang and Chen 2014). As an effect of excessive heat generation during PL (more than 10 s), Romaine lettuce showed slight wilting due to photothermal effect (Mukhopadhyay et al. 2021).

Gram-negative bacteria are more susceptible to PL compared to Gram-positive bacteria, which could be attributed to the difference in their cell walls. Viruses are most resistant among the microbes (bacteria, fungi and their spores) and reports suggest that UV region in PL has the most antibacterial effect causing cell death (Ramos-Villarroel et al. 2014; Kramer et al. 2015). In addition, visible and infrared regions having high power also contribute to the inactivation of the micro-organisms (Elmnasser et al. 2007). The cellular inactivation happens when bacteria absorb UV radiation leading to the formation of conjugated carbon–carbon double bonds in proteins and nucleic acids, photochemical dimerization of nucleic acids, and consequent inhibition of transcription and replication (Wang et al. 2005; Gomez-Lopez et al. 2007; Ramos-Villarroel et al. 2012). Few reports also mention that PL caused membrane damage, loss in cell viability, cytoplasm shrinkage, oxidative stress associated DNA damage, rupture of internal organization and also provoked significant disorder in the inner cells (Cheigh et al. 2013; Nicorescu et al. 2013; Kramer and Muranyi 2014; Ferrario and Guerrero 2017). To summarize, PL processing

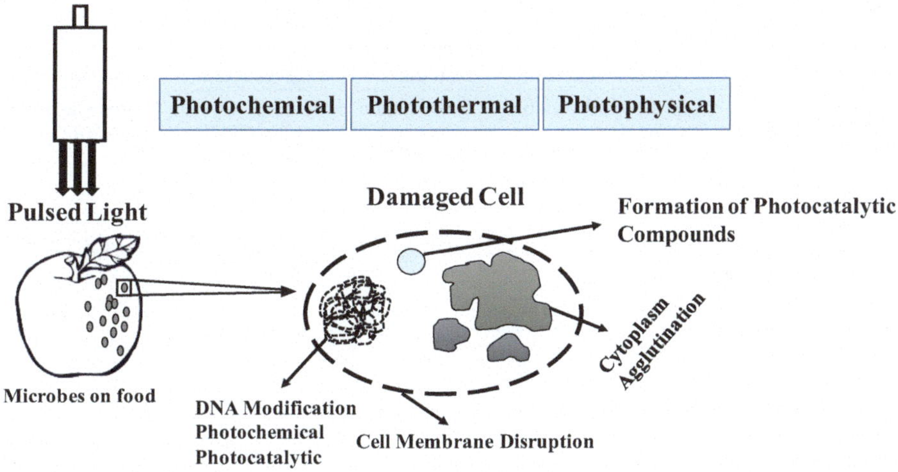

Fig. 6.2 Schematic representation of microbial inactivation by pulsed light. (Bhavya and Hebbar 2020)

inactivates the microbes through photothermal/photophysical and photochemical effects (Fig. 6.2), which cause both alteration of cell membrane integrity and damage to chromosomal DNA (Ramos-Villarroel et al. 2012; Cheigh et al. 2012; Nicorescu et al. 2013). Recent works on effect of light-based processing on microbial inactivation of foods is summarized in Table 6.2.

There are vast literature reports on the decontamination of foods such as fruits, vegetables, meat, fish, milk and their products by PL processing (Pataro et al. 2015; Pinela and Ferreira 2017; Bhavya and Umesh Hebbar 2017; Valdivia-Nájar et al. 2018; Vollmer et al. 2020; Xiaokang et al. 2020; Contigiani et al. 2021). Glimpse of the recent developments in PL processing have been discussed in the present chapter. PL treatments delayed the visual mold development by 2 to 4 days in strawberries (Cao et al. 2019). Tomato slices treated with PL at fluences of 4, 6 or 8 J/cm^2 had lower microbial counts by 2 log CFU/g compared to the control samples throughout the storage period of 20 days at 5 °C (Valdivia-Nájar et al. 2018). Contigiani et al. (2021) reported that mild thermal treatment (2.5 and 5.0 min at 46 ± 0.4 °C) of strawberries (cv. Albion) was more effective in decreasing natural mycobiota/fungal decay and delaying the onset of infection for 6 days compared to that of PL (11.9 J/cm^2) or even combination treatments.

Activities of undesirable enzymes (oxidoreductases) that cause browning in juices like PPO and POD were inactivated by 50% and 42% in PL (1479 J/cm^2) treated pineapple juice, while it was 61% and 81% in thermally pasteurized juice, indicating PPO was more sensitive to PL than POD (Vollmer et al. 2020). While, health-promoting proteolytic enzymes, bromelain, was completely retained by PL compared to thermal treatment which reduced enzyme by 93% (Vollmer et al. 2020). The enzymes PPO and POD were inactivated by 41% and 51%, respectively in mixed fruit beverage treated with PL at 17.42 W/cm^2 for 3 min (Dhar and Chakraborty 2020).

Table 6.2 Microbial inactivation in foods using light-based processing

Treatment	Micro-organism	Matrix/ medium	Log reduction	Reference
30 s at 6 cm (127.2 mJ/cm^2) for UV-C, and 9 s at 3.5 cm (152.6 mJ/ cm^2) for PL	Mesophilic viable counts	Raw salmon	0.9 and 1.3 log CFU/g respectively.	Pedrós-Garrido et al. (2018)
3 J/cm^2 (0.1 J/cm^2 per pulse; ~35 cm from PL lamp; 10 s, high intensity and low dose); 5 J/cm^2 (0.1 J/cm^2 per pulse; ~35 cm from PL lamp; 16 s, high intensity and high dose); 3 J/cm^2 (0.05 J/cm^2 per pulse; ~55 cm from PL lamp; 20 s, low intensity and low dose).	*Salmonella*	Strawberries	0.4 to 0.8 log CFU/g	Cao et al. (2019)
3.6 J/cm^2 of PL	*L. innocua*	Sliced boiled ham, chicken cold cuts and frankfurter sausages	1 log and 3–4 log CFU/ respectively.	Kramer et al. (2019)
4.0–16.8 J/cm^2 of PL	*S. Enteritidis, E. coli* O157:H7, *S. aureus* and *L. monocytogenes*	Fresh-cut lettuce	2.03 to 5.40; 2.20 to 5.08; 2.96 to 6.56 and 2.30 to 4.00, respectively.	Tao et al. (2019)
0.52 to 19.11 J/cm^2 of PL	*Salmonella typhimurium* and *Yersinia enterocolitica*	Pork skin	1.73 to 3.16 and 1.48 to 4.37 log CFU/cm2	Koch et al. (2019)
Pulsed LED (395 nm) for 60 min in a semi-closed system	*Salmonella* cocktail	Wheat flour	2.91 log CFU/g	Du et al. (2020)
31.5 J/cm^2 (30 s) of PL	*Salmonella*	Cherry toma-toes stem scars	2.3 log CFU/g	Leng et al. (2020)
1.8 kV/47 pulses (160 J·cm^{-2}) 2.4 kV/94 pulses (757 J·cm^{-2})	Aerobic mesophiles, yeasts and moulds	Pineapple juice	2.0 and 2.2 log10 CFU/mL, respectively ~5 log cycles	Vollmer et al. (2020)

(continued)

Table 6.2 (continued)

Treatment	Micro-organism	Matrix/ medium	Log reduction	Reference
8.4 J/ cm^2 of PL	*Listeria*	Serrano and Iberian dry-cured ham	1 and 2 log cfu/cm2, respectively	Fernández et al. (2020)
PL of 17.42 W/cm^2 fluence-rate for 3 min (2400 V/2.4 cm/ 3 min).	Aerobic mesophiles, yeast and mold counts	Mixed fruit beverage	>5 log CFU/mL	Dhar and Chakraborty (2020)
Blue LEDs after 7 d of treatment at 4 °C	*L. Monocytogenes* and *P. fluorescens*	Packaged sliced cheese	5.14 and 3.60 log CFU/cm2	Hyun and Lee (2020)
10.5 J/cm^2 PL dose (10 s)	*E. coli* O157:H7	Romaine lettuce	2.68 log CFU/g	Mukhopadhyay et al. (2021)
Fluence of 4.2 J/cm^2 of pulsed UV-C	Total aerobic mesophilic bacterial counts	Beef loin steaks	3.49 log CFU/g	Söbeli et al. (2021)
Continuous flow PL (95.2 J/cm^2) with a fixed flow rate of 100 mL/s	*E. coli* (MTCC 433)	Orange, pineapple juice and tender coconut water	4.0, 4.5 and 5.33 log CFU/mL, respectively	Preetha et al. (2021)
180–1100 nm at 3 pulses/s with a pulse width of 360 μs, treated for 0, 10, 45, and 90 s at a distance of 40 cm	*Cryptosporidium parvum* oocysts	Cilantro, mesclun lettuce, spinach, and tomato	2.4, 4.3, 2.5, and 2.2, respectively.	Craighead et al. (2021)

Shiitake mushrooms pretreated with PL (25 pulses, 400 J) improved surface colour and decreased the browning index, browning degree, PPO activity and 5-hydroxymethylfurfural values by ~43–47% (Zhang et al. 2021b). PL treatment affected the colour (red colour) and odour (fatty and porky smell) in pork skin and loin samples, while reducing *S. Typhimurium* and *Yersinia enterocolitica* significantly (Koch et al. 2019). Colour parameters like a* and b* values were found to decrease in the PL treated beef loin steaks compared to control ones (Söbeli et al. 2021). PL treatment (0.35 J/cm^2–3.6 J/cm^2) did not significantly affect the product colour and appearance of packaged frankfurter sausages, sliced boiled ham and chicken cold cuts during 12 days of chilled storage (Kramer et al. 2019). Velderrain-Rodríguez et al. (2021) noticed that there was no change in colour parameters in PL treated fresh-cut avocados, while reduction in firmness (~ 34%) was observed. PL treatment preserved TSS, colour attributes by delaying browning, and also minimized the weight loss in fresh-cut lettuce stored at 4 °C for 8 days (Tao et al. 2019). Fresh-cut strawberries treated with PL maintained the surface colour on both internal and external surfaces and reduced the softening incidence during cold storage for 14 days without browning the fruit (Avalos-Llano et al. 2018).

Leng et al. (2020) mentioned that PL treatment did not affect the colour and firmness of the cherry tomatoes, but were slightly softened during the latter part of 21 days of storage at 10 °C. Thermal treatment in combination with PL improved the mechanical properties like firmness, stiffness, and resistance to deformation compared to control strawberries (Contigiani et al. 2021). PL treated fresh-cut tomatoes at 8 J/cm^2 demonstrated increased weight loss, slight firmness decrements and changes in the pectinolytic enzymes (PME, polygalacturonase) during 20 days of storage (Valdivia-Nájar et al. 2018). In addition, sensory attributes of PL treated tomatoes were not significantly affected for at least 10 days during the storage (Valdivia-Nájar et al. 2018).

The flavour profile of shiitake mushrooms was affected by both PL treatment and storage conditions (storage temperature and presence of oxygen), especially content of 8 carbons compounds that majorly contribute to mushroom flavour (Xiaokang et al. 2020). PL processing of beef steaks was shown to decrease the level of few compounds arising from microbial metabolism while increased the amounts of some volatile compounds such as hexanal, 2-heptenal, 2-octenal, 2,4-nonadienal, 3-octen-2-one and limonene, indicating the potential of PL in meat industry by enhancing the microbial safety and having minimal effect on the quality characteristics (Söbeli et al. 2021). PL treated ready-to-eat dry-cured ham, Serrano and Iberian (Spanish varieties) did not have rancid notes, but increased concentration of sulphur and metallic note contributing volatile compounds such as methional, dimethyl disulfide and 1-octen-3-one, which disappeared during storage (Fernández et al. 2020). However, PL induced off-flavour formation (burnt and sulphuric notes) in frankfurter sausages and sliced boiled ham (Kramer et al. 2019).

PL treatment did not significantly affect the chemical composition (total moisture, total protein, total lipid and ash content) of the beef loin steaks (Söbeli et al. 2021). In dried shiitake mushrooms, PL pretreatment retained reducing sugar content, umami taste characteristics (monosodium glutamate like components), total phenol, polysaccharides and antioxidant activity compared to untreated ones (Zhang et al. 2021b). PL treated shiitake mushrooms had higher total phenolics content and antioxidant activity compared to untreated ones (Xiaokang et al. 2020). Reports also mention that PL treatment retained ascorbic acid, antioxidant activity, colour parameters and desired enzyme bromelain in pineapple juice compared to thermal pasteurization, indicating PL as a promising novel processing technique for preservation of fresh-like chilled fruit juices (Vollmer et al. 2020). In contrary, PL negatively impacted ascorbic acid content (reduction by 36%) and colour attributes in mixed fruit beverage (Dhar and Chakraborty 2020). While, phenolic content and antioxidant capacity was increased by 14% and 33% in PL treated mixed fruit beverage, which can be attributed to the PL induced de-polymerization of polyphenols or activation of phenolic biosynthesis pathway (Dhar and Chakraborty 2020). Interestingly it was observed that PL without UV portion increased the antioxidant activity thus increasing ascorbic acid, carotenoids and phenolic compounds immediately after irradiation and even during storage period (Velderrain-Rodríguez et al. 2021). It was noticed that PL did not affect the ascorbic acid content and physicochemical properties (colour and TSS) of persimmons at two different maturity stages (unripe yellow-green and ripe orange-red) (Denoya et al. 2020). In contrast, phenolics and antioxidant activity of the unripe persimmons fruit was significantly affected by PL treatment, which could be attributed to the effect of PL on its soluble tannins content (Denoya et al. 2020). The PL treatment did not show any effect on quality attributes (weight loss, firmness, and colour) or the bioactive compounds (total anthocyanin, total phenolics, and total antioxidant activity) of strawberry fruits and also extended the shelf life of the fruits (Avalos-Llano et al. 2018; Cao et al. 2019). Chlorophyll and ascorbic acid content was maintained in PL treated fresh-cut hydroponic lettuce and also prolonged shelf life by 8 days (Tao et al. 2019). Lower doses of PL (4 and 8 J/cm^2) retained the vitamin C and anthocyanin content in strawberries, while higher doses (12 and 16 J/cm^2) decreased the contents by 20% and 30%, respectively (Avalos-Llano et al. 2018). The PL technology is a promising technique for decontamination of foods by maintaining quality, but has a few limitations. When PL technology is combined with other non-thermal technologies, it could be more effective, and hence further hurdle combinations need to be explored for different food matrices.

3.2 *Blue Light Processing*

In recent years, application of visible light is gaining scope for demonstrating as a novel technology for microbial inactivation in spite of its low inactivation efficacy compared to UV light (Luksienė and Zukauskas 2009; Maclean et al. 2009). The visible spectrum is sub-divided into six sub-groups namely, violet (400–450 nm), blue (450–500 nm), green (500–570 nm), yellow (570–590 nm), orange (590–610 nm), and red (610–760 nm), among which red, blue, and green light are commonly used in the food industry (D'Souza et al. 2015; Hyun and Lee 2020; Hinds et al. 2021). Blue light (BL) is high energy wave light possess antimicrobial property that inactivate bacteria, yeast, fungi as well as bacterial spores (Kumar et al. 2017; Murdoch et al. 2013; Moorhead et al. 2016; Trzaska et al. 2017). It is reported that Gram-positive bacteria is more susceptible to BL than Gram-negative bacteria (Maclean et al. 2009; Bhavya and Hebbar 2019a). Antibacterial effect of BL on the bacteria can be attributed to physical damage to cellular membrane rather than DNA (Bhavya and Hebbar 2019a). Srimagal et al. (2016) and Kumar et al. (2016) observed 405 nm had more antimicrobial effect than 460 nm illumination, since 405 nm spectrum fell in UV range which may directly cause the DNA damage. dos Anjos et al. (2020) showed that BL (2 h treatment) significantly reduced *S. aureus*, *E. coli*, *P. aeruginosa*, *S. Typhimurium*, and *Mycobacterium fortuitum* cells by 5 log in whole milk. Guffey et al. (2016) also demonstrated that BL (464 nm, 18 J/cm^2) decontaminated *S. Typhimurium* on cucumbers. At 460 $\pm$ 2 nm illumination, acidic conditions were more detrimental than alkaline conditions for bacteria (Ghate et al. 2015; Bhavya and Hebbar 2019a), suggesting the potential of BL in preserving acidic foods.

Blue Light treatment in presence of exogenous or endogenous photosensitizer (PS) is known as photodynamic treatment. Several studies have demonstrated the use of curcumin-mediated PS against a range of fungi and bacteria such as *S. aureus* (Jiang et al. 2014; Penha et al. 2017; Bhavya and Hebbar 2019a), *S. Typhimurium* (Penha et al. 2017), and *E. coli* (Haukvik et al. 2009; Bhavya and Hebbar 2019a). Haukvik et al. (2009) mentioned 3.0 log reductions in *E. coli* count when treated with 25 μM curcumin and 430 nm illumination in combination. Similarly, Jiang et al. (2014) and Bhavya and Hebbar (2019a) also reported 2.0 and 5.9 log reduction in *S. aureus* population with curcumin (2.5–20 μM) and BL (460–470 nm) treatments *in vitro*. Further, a combination of curcumin (75 μM) and BL (470 nm) reduced the pathogenic bacteria (*E. coli, S. Typhimurium, Aeromonashydrophila, S. aureus, P. aeruginosa*) by 3.5–6.0 log CFU/mL (Penha et al. 2017). Aponiene et al. (2015) observed the reduction of mesophilic bacteria on fresh produce (apricots, plums, and cauliflowers), after a combination treatment with 15 μM hypericin and 585 nm light for 30 min. An exposure of 1199 J/cm^2 light (275, 365, 395, and 455 nm) inactivated *Salmonella* by 1.07 to 3.67 log CFU/g and it was found that 365 and 395 nm had higher inactivating efficacy than 455 nm at an equal energy dose (Subedi et al. 2020). In orange juice and pineapple slices, 3 to 4 log reduction in *E. coli* and *S. aureus* population was observed when curcumin and BL was applied in combination (Bhavya et al. 2021). During photodynamic process, inactivation of bacteria occurs due to the oxidative stress causing disruption of membrane integrity. The BL illumination excites the PS and while returning to ground state, these molecules transfer energy to oxygen molecule, resulting in the production of reactive oxygen species (ROS). ROS such as superoxide anion, hydrogen peroxide, hydroxyl radicals and singlet oxygen may attack cellular components like DNA, lipids, and proteins ultimately causing bacterial death (Luksienė and Zukauskas 2009). Kim et al. (2015), Bhavya and Hebbar (2019a) and Hyun et al. (2020) observed that BL illumination resulted in loss of bacterial membrane permeability and no DNA damage was observed, therefore physical damage might partly be the cause for antibacterial effect during photodynamic inactivation.

Authors reported that colour of pineapple slices was reduced when treated with BL (460 ± 5 nm) due to the absorption of light by β-carotene in pineapple (Ghate et al. 2017; Bhavya et al. 2021). Wheat flour treated with different wavelengths of 365, 395, and 455 nm showed significant colour change, exhibited the drying effect (reduced water activity) and improved functional and rheological properties of wheat flour except for 275 nm illumination (Subedi et al. 2020). Fresh-cut papaya illuminated with 405 ± 5 nm light did not show any change in colour and antioxidant capacity, whereas flavonoid content was increased by 1.5 to 1.9 times than that of untreated fruit (Kim et al. 2017a). Further, there was no significant change in the levels of lycopene, ascorbic acid and β-carotene between BL illuminated fruits and control ones. Similarly, Kim et al. (2017b) observed the preservation of physicochemical and nutritive qualities (ascorbic acid, β-carotene, antioxidant capacity and flavonoids) of fresh-cut mangoes after 405 ± 5 nm illumination. Bhavya and Hebbar (2019b) and Bhavya et al. (2021) also mentioned that there was no significant alteration in colour parameters, phenolics, flavonoids, ascorbic acid and antioxidant activity of the orange juice and pineapple slices treated with BL. Shelf life of BL (405 nm) treated milk was enhanced by 19 h and 9 days at room temperature and refrigerated conditions, respectively with no significant changes in composition and physicochemical properties (Srimagal et al. 2016). From the literature reports, it can be concluded that BL treatment is a novel venture into the food processing that not only decontaminate food but also retain/maintain the nutrient levels of food. There is a need for further investigation to know how process variables dictate the efficiency of the BL process in decontamination and retaining nutrients. This will help to standardize the conditions and analyse the potentials of developing this technique in to a commercially viable technology.

4 Ultrasound

Ultrasound (US) is one of the novel non-thermal technologies that uses sound/acoustic waves above the frequency of 18 kHz. During US processing, the sound waves propagates through the food matrix, causes heating (absorption of acoustic energy), structural changes (compression and rarefaction), cavitation (collapse of micro-bubbles in liquid) and turbulence (acoustic streaming or micro streaming) (Jambrak et al. 2007). Depending on the frequency range, US can be classified in to two types: Low power US (>100 kHz and intensity <1 W/m^2) and high power US (20 kHz to 100 kHz and intensity >1 W/m^2) (Knorr et al. 2004). Low power US has been used for compositional analysis, detection and quality control of foods, while high power US is being used to alter physiochemical and mechanical properties of food (Chemat and Khan 2011). US has been applied to inactivate microbes and enzymes; to extend the shelf life; to tenderize meat; to accelerate the extraction, drying, freezing, crystallization, filtration, and emulsification processes (Kentish and Feng 2014). There are number of literature reports on the application of US in food processing and in the present chapter recent studies focusing on microbial safety and quality of US processed foods are discussed.

4.1 Effect on Micro-organisms and Enzymes

The US has been used abundantly for producing microbiologically safe food and recent reports on the efficiency of US to inactivate micro-organisms has been summarized in Table 6.3. US treatment (78 W, 8 min and 104 W, 4 min) preserved lactic acid bacteria in frozen stored semi-skimmed sheep milk, while inactivating contaminant bacteria, which was more advantageous than high temperature short time (HTST) pasteurization (Balthazar et al. 2019). Temperature of the tomato juice treated with US (20 kHz) was increased from 37 to 52 °C and exhibited 3–4 log reduction in total mesophilic micro-organisms even after 10-day storage and (Starek et al. 2021). The different degree of homogenization of juice and inactivation of

Table 6.3 Microbial inactivation in foods using ultrasound processing

Treatment	Micro-organism	Matrix/ medium	Log reduction	Reference
66.64, 106.19, and 145.74 W/L	Total bacterial count mold and yeast	Cherry tomato	0.42, 0.86, and 1.04 log CFU/g, respectively 0.41, 0.70, and 0.93 log CFU/g, respectively	Wang et al. (2015)
28 kHz for 60–90 min at 20 ° C	Total plate count. Yeast and mold	Grapefruit juice	1.0–2.0 log CFU/mL	Aadil et al. (2015)
33 kHz, 60 W for 40 min	Total bacterial and yeast and mold count	Strawberry	2 and 1.2 log CFU/g respectively	Gani et al. (2016)
33 kHz, 60 W for 10–60 min	Total bacterial count, yeast and molds	Cherry	0.6 to 1.6 log CFU/g	Muzaffar et al. (2016)
40 kHz; 0.5 W/cm2 for 20, 40 or 60 min	Total aerobic bacteria, yeast and mould	Carrot juice	1 log CFU/mL	Zou and Jiang (2016)
600 W (2 vs),	Aerobic mesophilic heterotrophic bacteria, yeasts and molds	Prebiotic whey beverage	2 and 0.4 log CFU/mL, respectively	Guimarães et al. (2018)
23.06 min by UH55°C 400 W, 14.18 min by UH63° C 200 W, 9.59 min by UH63°C 400 W	*B. subtilis*	Chinese bayberry juice	5 log CFU/mL	Li et al. (2019a)
20 kHz for 5–15 min	Total bacterial count mold and yeast	Fresh-cut cucumber	0.49–1.02 log CFU/g 0.41–0.84 log CFU/g	Fan et al. (2019a)
130 W, 42 kHz	Aerobic mesophilic bacteria	Fresh-cut lettuce	0.8 to 1.7 log CFU/g	Irazoqui et al. (2019)
20 kHz, 200 W for 3 min	Total plate count, yeast and molds	Peanut milk	1.12 and 0.7 log CFU/mL, respectively	Salve et al. (2019)
20 kHz, amplitude (10 and 30 μm) for 5 min	*E. coli* and *S. aureus*	Orange juice	3.02 and 0.18 log CFU/ mL, respectively	Bhavya and Hebbar (2019b)
50 °C, 750 W and 36 min	Total aerobic bacteria	Mandarin juice	2.4 log CFU/mL	Cheng et al. (2020)
55 °C (3 min) and 517.1 mW/mL acoustic energy density	Total mesophilic aerobic bacteria and yeasts and molds	Strawberry juice	1–1.5 log CFU/mL	Yildiz et al. (2020)

(continued)

Table 6.3 (continued)

Treatment	Micro-organism	Matrix/ medium	Log reduction	Reference
1000 W, 20 to 25 kHz	Total bacteria count, *E. coli*, yeast and molds	Grape juice	0.1–1.0 log CFU/mL	Ma et al. (2020)
20 kHz and 1 MHz for 10 min	*L. innocua*	Blueberries	2.5 log CFU/g	Zhang et al. (2021a)
40 kHz/110 W	Aerobic mesophiles, yeast and molds, coliforms	Strawberries	0.7–1.0 log CFU/g	Donatti Leão Alvarenga et al. (2021)

micro-organisms depends on the intensity/amplitude of the US (Starek et al. 2021). Throughout the storage period (5 °C for 10 days), significant decrease in the microbial growth was observed in US treated strawberry juice compared to the untreated ones, thus extending the shelf-life of the juice by 7–10 days (Tomadoni et al. 2017). Irazoqui et al. (2019) observed that US affected the wax layer of lettuce, thereby destroying the natural defence mechanism against microbes, which was confirmed by the higher microbial growth in US treated lettuce throughout the storage (8 days at 5 °C). Amaral et al. (2015) and Sulaiman et al. (2017) noticed that US inactivated PPO in strawberry puree and fresh-cut potatoes by 11–18 and 50%, respectively.

4.2 *Effect on Physico-Chemical Attributes*

Effect of US on the physical and chemical characteristics of food products in the recent years have been discussed below. US (40 kHz) treated strawberry juice did not have any alteration in colour parameters, while thermally treated (90 °C) samples had lower L* values and higher hue angles (Tomadoni et al. 2017). US treatment was found to alter the b* values of the mandarin juice, which could be due the isomerization of coloured compounds by hydroxyl ions produced during the US processing (Cheng et al. 2020). Ordóñez-Santos et al. (2017) mentioned that significant increase in total colour difference, phenolics, carotenoids and retinol activity was observed in US treated Cape gooseberry juice compared to control. Meanwhile, a reduction in colour parameters (chroma values, yellowness index) and ascorbic acid content was noticed in US treated gooseberry juice (Ordóñez-Santos et al. 2017). US treatment did not have any effect on the sensory attributes, L* and Hue values and browning of potato strips (Amaral et al. 2015). Beef samples treated with US had higher a*, b* and water holding capacity after 7 days of storage at 4 °C (Carrillo-lopez et al. 2018). Studies showed that US can be applied to enhance the rheological properties/physical attributes of peach juice with increased stability to pulp sedimentation and consistency/turbidity without influencing the colour characteristics of the juice during storage (Rojas et al. 2016).

US treatment did not affect sugar content, vitamin C, pH, TSS and total acidity of strawberry fruit/juice and mandarin juice compared to the untreated samples (Gani et al. 2016; Cheng et al. 2020; Yildiz et al. 2020). Muzaffar et al. (2016) revealed that US (33 kHz) affected the firmness and increased the antioxidant activity of cherry fruit, while maintaining the colour attributes during storage (15 days at 4 °C). It was observed that US did not affect the pH value of tomato juice, while it slightly varied the lycopene and ascorbic content of the juice (Starek et al. 2021). US treatment of carrot juice did not cause any significant change in pH values as compared to the untreated samples, while electrical conductivity, viscosity and colour of the treated juice were increased, as exposure time increased (Zou and Jiang 2016). The pH of the fresh-cut potato strips and strawberry juice was affected significantly by US exposure and storage time (Amaral et al. 2015; Yildiz et al. 2020). US increased the degree of protein hydrolysis, TSS and decreased titratable acidity of the peanut milk compared to control samples (Salve et al. 2019).

US increased the phenolic content and antioxidant capacity of the non-clarified strawberry juice and did not affect the physicochemical attributes (TSS, acidity) compared to that of thermal treatment (Tomadoni et al. 2017). Significant improvement in TSS, total sugars, total carotenoids and ascorbic acid contents were observed in US treated carrot juice compared to that of untreated ones (Zou and Jiang 2016). At the end of the storage (15 days at 4 °C), antioxidant capacity was slightly increased in US treated strawberry fruits compared to control samples (Gani et al. 2016). Fruits juices like mandarin, apple-grape blend (50:50) and strawberry juice treated with US retained the nutritional value (ascorbic acid, total phenolics, total carotenoids, total anthocyanins and radical scavenging activity) compared to that of the thermally pasteurized samples (Cheng et al. 2020; Yildiz et al. 2020; Aadil et al. 2020). Grapefruit treated with US (28 kHz for 90 min) showed an increase in total carotenoids, phenolics, sugar and lycopene content, while it had a negative effect on the viscosity of the juice compared to the control samples (Aadil et al. 2015). Wang et al. (2015) observed that US effectively delayed postharvest ripening by lowering or inhibiting the ethylene production and respiration rates of mature-green cherry tomatoes compared to control ones. It was also reported that US maintained the firmness, flavour, phenolics, flavonoids, antioxidant and enzyme activities of cherry tomatoes (Wang et al. 2015). US induced the bound aroma compounds in mandarin juice which was higher compared to the untreated and pasteurized ones, thus preserving the flavour of mandarin juice (Cheng et al. 2020). Honey treated with US (24 kHz) was found to accelerate liquefaction and decrease HMF content compared to the conventional thermal treatments (Önür et al. 2018).

The beef treated with US (40 kHz) did not have any negative effect on the overall quality during 14 days of storage, even though there was slight increase in lipid oxidation and decrease in shear force (Peña-González et al. 2017). US treated semi-skimmed sheep milk did not affect the protein and amino acid profile, proving the applicability of US in dairy industry for maintaining the product quality (Balthazar et al. 2019). US treatment increased the germination rate, sprout length and preserved the isoflavone, and amino acids including gamma-aminobutyric acid (GABA) content in the soybean sprouts/seeds (Yang et al. 2015). US treatment improved the edibility and nutritional quality of the soybean sprouts by unaltering the protein pattern and decreasing the lipoxygenase isozyme activity and trypsin inhibitor content (Yang et al. 2015). High intensity US improved stability of inulin enriched whey beverage without phase separation, decreasing the particle size, denaturing the whey proteins compared to HTST (Guimarães et al. 2018).

Beef samples stored initially under vacuum (4 °C for 0, 7 or 14 days) and then ultrasonicated (40 kHz, 11 W/cm^2 for 60 min) had lower shear force, intense fresh meat smell and oily flavour than control samples, suggesting a safe method for tenderizing bovine meat (Peña-Gonzalez et al. 2019). US treatment increased the meat tenderness making beef samples juicier compared to control, and sensory analysis revealed that US did not change panellists' perception of beef quality (Peña-González et al. 2017). The US improved the porosity, softness and reduce the particle density, chewiness, gumminess with slight alteration in the colour of blueberries (Nowak et al. 2019). Further, the microstructure of fresh-cut potato was found to be influenced by US, while firmness was not altered (Amaral et al. 2015). During US assisted dough fermentation, US significantly affected the hardness (reduced by 22.4%) and specific volume (increased by 6.7%) of steamed bread compared to the control samples (Luo et al. 2018). US treated peanut milk had better colour attributes, separation index and microstructure with smaller particle and fat globule size (Salve et al. 2019). Guava juice treated with US increased the *in vitro* accessibility of lycopene, physical stability avoiding pulp sedimentation and decreased particle size without any alteration in the colour attributes of the juice (Campoli et al. 2018).

4.3 Ultrasound as Hurdle Technology

In the recent years, many reports on the application of US along with other processing technique as a hurdle are available to maintain food safety and retain quality of food. Hurdle technologies along with US have shown to inactivate microorganisms to safe levels by more than 5 log cells and improved quality characteristics of the food products with enhanced shelf life compared to the untreated and thermally processed foods (Guerrero et al. 2017; Khandpur and Gogate 2016). Combination of US, acetic acid (AC) and gibberellin acid (GA), (US+GA + AC) processing was shown to reduce the microbial load and physiological deterioration, inactivate significantly phenylalanine ammonia lyase (PAL) and POD activities, suppressing the accumulation of lignin of green asparagus (Wang and Fan 2019). Combination of US and ε-polylysine inactivated micro-organisms, retained quality and extend shelf-life of fresh-cut leafy greens like lettuce (Fan et al. 2019b). Combination of ultrasound-pasteurization inactivated the microbes (total plate count, yeast and mould) completely and enzymes (PPO, POD, PME) by 95–98% and retained ascorbic acid, bioactive compounds (phenolics) in pear juice

(Saeeduddin et al. 2015). It was observed that a combination of US (600 W, 10 min)-UV treatment inactivated both pathogenic bacteria and enzymes (PPO, POD, PME) completely in mango juice (Wang et al. 2020). It was noticed that the combination processing (US and ε-polylysine) reduced weight loss and total color difference, inactivated POD and PPO, retained phenolic content, vitamin C and chlorophyll content in fresh-cut lettuce during storage compared to untreated samples (Fan et al. 2019b).

The combinational treatment, US+UV, tripled the carotenoid content (all-trans-β-carotene) and preserved quality/nutritional value (ascorbic acid, phenolics, soluble dietary fiber, antioxidant ability, acidity, TSS, rheological behaviour, volatile aroma compounds, metal elements) of mango juice (Wang et al. 2020). The US+GA + AC processing demonstrated retention of ascorbic acid, TSS, chlorophyll content, phenolic content and sensory characteristics in green asparagus (Wang and Fan 2019). Colour and anthocyanin content was higher in bayberry juice treated with a combination of US and mild heat compared to that with thermal processing, while there were no changes in pH and soluble solids content (Li et al. 2019a). US+mild heat treatment significantly preserved the aroma of the bayberry juice by inhibiting the decline of key volatile compounds (Li et al. 2019a). A combination of US treatment (600 and 1200 W/L for 5 min) and HPP (450 MPa for 5 min) preserved fructo-oligosaccharides (FOS), organic acids (>90% retention) and anthocyanin (increase by 24%) content in prebiotic cranberry juice fortified with FOS (Gomes et al. 2017). US combined with modified/controlled atmosphere packaging preserved the quality of cucumbers in terms of freshness, weight loss, colour, TSS, firmness, flavour and taste compared to that of control and controlled atmospheric package samples (Feng et al. 2018; Fan et al. 2019a). It is noteworthy that US treatment required more energy than HPP and thermal processing; US retained better colour, while HPP preserved better antioxidant activity of strawberry puree during storage (3 °C for 30 days) compared to thermal treatment and control samples (Sulaiman et al. 2017). Rajashri et al. (2020) mentioned that the tender coconut water processed using a combination of US and nisin treatment was unacceptable (change in colour due to oxidation) after 2 weeks of refrigerated storage. US treated blueberries (a combination of US with carvacrol and carbonated water) had an impact on colour and texture, while having no effect on total phenolic content and anthocyanin content of blueberries (Zhang et al. 2021a). In this regard, US technology has been explored on different products by the food industry personnel's to reach quality safe food to the consumer's table.

5 Cold Plasma

Cold plasma (CP) is one of the emerging non-thermal technology which uses plasma, "the fourth state of matter". Plasma is generated when electrical energy is applied to a gas to produce complete or partly ionized gas/reactive species such as charged particles, formed molecules, ultraviolet photons, free radicals, and other reactive elements of nitrogen, oxygen, and hydrogen. Plasma possess neutral charge and exists in either ground or its excited state (Thirumdas et al. 2015; Pankaj et al. 2018). Non-thermal plasma is formed at low pressure and power without a localized thermodynamic equilibrium. There are three different types of CP discharge systems such as glow discharge plasma, radio frequency discharge and discharge barrier type. The reactive species such as reactive oxygen (ROS) and nitrogen (RNS) species that are directly generated within the sealed packages during CP processing is suitable for decontamination of food (Varilla et al. 2020). CP has several factors that dictate the process efficiency such as gas composition, plasma reactor design and structure, plasma energy, frequency, modulation, pulse form and duration of input energy.

5.1 *Effect of CP on Micro-organisms and Enzymes*

The CP has been applied for producing microbiologically safe food and recent reports on the efficiency of CP to inactivate micro-organisms has been presented in Table. 6.4. Tender coconut water treated with CP was microbiologically shelf stable till 8 days at 6 °C in a glass bottle, while a blend of coconut water and orange juice with ascorbic acid exhibited a shelf life of 35 and 18 days on storage at 6 °C and 27 °C, respectively (Chutia and Mahanta 2021). A study by Rana et al. (2020) showed that shelf life of strawberry fruit can be extended by 5 and 9 days at 25 and 4 °C, respectively in sealed atmospheric CP package, while untreated strawberries spoiled in 2 days. Plasma (500 Hz for 10 min) treated Pacific white shrimp extends its shelf life by 4 days compared to control ones (de Souza Silva et al. 2019). CP improved the shelf life of fig samples compared to that of untreated ones and also retained the quality attributes of the samples except for pH and a* values (Abbaszadeh et al. 2018).

It is noteworthy to mention that CP treatment effectively inactivated the PPO and POD by 82.4% and 42.3%, respectively in açai pulp (Dantas et al. 2021). It was noticed that the spark discharge plasma (10.5 kV) inactivated the PPO in cloudy apple juice with a residual activity of 16% and 27.6% after 5 and 4 min of treatment, respectively (Illera et al. 2019). The α-amylase activity in brown rice was found to be increased by 1.21-fold at 5 min CP treatment and even water uptake rate also increased significantly (Lee et al. 2016a). Interestingly, the superoxide dismutase (SOD) activity was increased and decay rates of blueberries were reduced with dielectric-barrier discharge (DBD) plasma compared to control treated (Dong and Yang 2019; Ji et al. 2020). It is noteworthy that CP has been one of the probable technology with promising scope in the future to decontaminate microbes and enzymes that are responsible for food spoilage.

Table 6.4 Microbial inactivation in foods using cold plasma processing

Treatment	Micro-organism	Matrix/ medium	Log reduction	Reference
Atmospheric cold plasma for 0, 15, 30, 45, 60, 90, or 120 s at a working distance of 7.5 cm with a mixture of 4 cubic feet/minute (cfm) of plasma jet and 7 cfm of ambient air	Total aerobic plate count	Blueberries	0.8 to 1.6 log CFU/g and 1.5 to 2.0 log CFU/g c after 1 and 7 days, respectively	Lacombe et al. (2015)
Plasma processed air for 60 min	Native microbial flora	Pepper seeds and paprika powder	>3 log CFU/g	Hertwig et al. (2015)
Cold plasma (250 W, 15 kHz, ambient air) for 20 min	*B. cereus*, *B. subtilis*, *E. coli* O157:H7	Brown rice	2.30 log CFU/g	Lee et al. (2016a)
Atmospheric cold plasma apparatus (air DBD, 15 kV) for 15 and 30 min of treatment (in afterglow at 70 mm from the discharge, at 22 °C and 60% of RH)	*E. coli* O157:H7 and *L. monocytogenes*	Radicchio leaves (red chicory)	1.35 and 2.2 log MPN/cm2	Pasquali et al. (2016)
DBD atmospheric CP at 34.8 kV for 5 min	*E. coli* O157:H7, *Salmonella*, *L. monocytogenes*, and Tulane virus	Lettuce	1.1, 0.4, 1.0 log CFU/g, and 1.3 log PFU/g, respectively	Min et al. (2016)
Atmospheric CP for 15 and 45 s at a working distance of 7.5 cm with 4 cubic feet/ minute (cfm) of CP jet.	Murine norovirus and Tulane virus	Blueberries	0.5 and 1.5 plaque forming units/g, respectively	Lacombe et al. (2017)
High voltage atmospheric CP treatment at 80 kV for 4 min	*Saccharomyces cerevisiae*	Grape juice	7.4 log CFU/mL	Pankaj et al. (2017)
High voltage (80 kV) DBD direct treatment for 20 min	Bacteria and fungi	Barley Wheat	2.4 and 2.1 log CFU/g, respectively 1.5 and 2.5 log CFU/g, respectively	Los et al. (2018)
Plasma treatments – air as a precursor under constant gas flow (3000 L/h) at 650 W for 30, 60, 90, and 120 s	*E. coli*	Apple, orange, tomato juices and sour cherry nectar	4.02, 1.59, 1.43, and 3.34 log CFU/mL, respectively	Dasan and Boyaci (2018)
Air DBD plasma discharge at atmospheric pressure for 10 min	Total bacteria count and fungi	Blueberries	2.01 and 0.58 log CFU/g, respectively	Dong and Yang (2019)

(continued)

Table 6.4 (continued)

Treatment	Micro-organism	Matrix/ medium	Log reduction	Reference
Atmospheric CP treatment of 15 min-dielectric barrier discharge (DBD) at 60 kV with an input voltage of 260 V at 50 Hz.	Total microbial count	Strawberry	2 log CFU/g	Rana et al. (2020)
Atmospheric CP generated with an open-air high-voltage DBD pilot-scale reactor, operated in static (batch) mode	*E. coli* *L. innocua*	Strawberries and spinach	2.0 and 2.2 log CFU/mL, respectively 1.3 and 1.7 log CFU/ml, respectively	Ziuzina et al. (2020)
High voltage atmospheric cold plasma at 100 kV (60 Hz frequency) for 5 min.	Total aerobic mesophiles, and yeast and mold	Carrots	2 log CFU/g	Mahnot et al. (2020)
Atmospheric cold plasma (12 kV, 5 kHz) for 60 s	Total aerobic bacteria and mold	Blueberries	0.34–1.24 and 0.57–0.87 log CFU/g respectively	Ji et al. (2020)
Dielectric barrier discharge plasma (250 W, 15 kHz, ambient air) treatments for 20 min.	*B. cereus* and *E. coli* O157:H7	Rice noodles	4.10 and 2.75 log CFU/g, respectively.	Kim et al. (2020)
High voltage atmospheric cold plasma under direct mode of exposure in dry air and MA65 (65% O_2, 30% CO_2, 5% N_2) gas blends for up to 5 min at 100 kV	*L. innocua* and *E. coli* K-12	Queso fresco (fresh cheese)	1.4 and 3.5 log CFU/g, respectively	Wan et al. (2021)

5.2 *Effect of CP on Physico-Chemical Attributes*

Consumer's choice depends on the physico-chemical attributes of the food product that attracts them at the first instance. In this regard, studies on the CP has also been focused to enhance/maintain quality of the food. Whiteness index and lightness values of brown rice treated with CP was higher compared to control samples, while yellowness and redness were slightly reduced in CP treated samples (Lee et al. 2016a). The colour parameters of CP treated mango flour noodles such as lightness (L*) decreased and redness (a*) and yellowness (b*) values increased compared to the untreated ones (Abidin et al. 2018). CP significantly reduced redness in red paprika powder, while colour parameters of pepper seeds and oregano were maintained (Hertwig et al. 2015). Fresh-cut kiwifruit treated with DBD plasma resulted in less darkened samples compared to untreated ones during storage

(Ramazzina et al. 2015). The colour of the CP treated cloudy apple juice was altered and phenolic content was enhanced by 64–69% (Illera et al. 2019). Pankaj et al. (2017) noticed that CP did not significantly affect the pH, colour parameters, acidity and electrical conductivity of the grape juice, while increased the non-enzymatic browning of the samples.

Lee et al. (2016a) noticed that there was a decrease in pH values of the CP treated brown rice samples, while no significant difference was observed among the plasma treatment times (5, 10 and 20 min). Further, CP treatment did not affect the physical attributes (colour, firmness, TSS, pH and moisture) of strawberry fruit and spinach (Rana et al. 2020; Ziuzina et al. 2020). Mahnot et al. (2020) observed that alteration in pH, texture, colour, and total carotenoids in CP treated carrots were minimal compared to the control samples. CP retained the colour, pH, phenolic content and antioxidant activity and also oligosaccharide content in the prebiotic orange juice (Almeida et al. 2015).

The sugar content in blueberries treated with air DBD plasma at atmospheric pressure was increased initially by 1.5 fold and delayed the reduction of sugar content during storage (Dong and Yang 2019). There was decrease in glucose and fructose content in cashew apple juice treated with CP (700 Hz), while malic acid concentration increased in the CP treated samples compared to the control ones (Leite et al. 2021). Vitamin C content was increased by 1.5 fold by DBD plasma treatment in blueberries compared to the control samples, after 16 days of storage, lengthening the preservation time of vitamin C (Dong and Yang 2019). Blueberries treated with CP increased phenolic and flavonoid content significantly, while decreasing the ascorbic acid at longer plasma doses (Sarangapani et al. 2017). The ascorbic acid content was not affected immediately after the CP treatment, while there was a ~7% loss during 4 days of storage in kiwifruits (Ramazzina et al. 2015). Increment in vitamin C and better bioaccessibility of vitamin C was found in CP treated cashew apple juice (Leite et al. 2021). It was noticed that CP treatment (60 kV) increased the phenolic content of grape pomace extract by 10.9–22.8%, which contains higher amounts of anthocyanins (Bao et al. 2020). It was evident from the study of Kovačević et al. (2016) that CP had positive impact on the colour and anthocyanin stability (increase in content by 21–35%) in cloudy pomegranate juice. Authors noticed that total anthocyanin content was increased by 2.2 fold in DBD plasma treated blueberries (8- and 10-min) compared to that of untreated ones after 4 days of storage and was maintained during prolonged storage period of 20 days (Dong and Yang 2019). Lacombe et al. (2015) noticed that the CP processing reduced the anthocyanins significantly after 90 s, while surface colour (L^* and a^* values) was altered after 120 s. The reports conclude that the CP is apt for inactivating microorganisms on food products, but optimization is necessary to improve the quality of the produce.

CP treatment for 10 min with a flow rate 20 mL/min maintained the TPC, mono and polyunsaturated fatty acids, volatile compounds and ACE-inhibitory activity of chocolate milk drinks (Coutinho et al. 2019). Concentration of bioactive compounds, vitamin C, antioxidant activity and volatile components were increased in guava-flavored whey beverage by CP processing compared to pasteurized beverage, but CP decreased the carotenoid content and favourable fatty acids (Silveira et al. 2019). In addition, CP increased the antioxidant activity and bioaccessibility of bioactive compounds (catechin, epicatechin, epigallocatechin gallate, rutin, caffeic acid, and chlorogenic acid) in açai pulp (Dantas et al. 2021). A recent study by Wan et al. (2021) reported that high voltage atmospheric CP caused minimal quality changes in Queso fresco (fresh cheese) in terms of colour, pH and lipid oxidation, while retaining the texture of the cheese. It was mentioned that CP processing (DBD-60 kV for 15 min) enhanced bioactives such as phenolics, gallic acid, phloretin, chlorogenic acid, hyprin, 4-hydroxybenzaldehyde, vanillin, and rutin in strawberry fruit during storage of 5 days (Rana et al. 2020). There was a decrease in chlorophyll a content immediately after CP treatment by ~15%, but during storage, chlorophyll and carotenoid content was retained better in CP treated kiwifruits compared to the untreated ones (Ramazzina et al. 2015). CP treated germinated brown rice also showed higher levels of vitamin E, phenolics, anthocyanins and phytosterols compared to the control samples (Yodpitak et al. 2019). An investigation conducted by Kim et al. (2020) revealed that colour parameters, peroxide values and TBARS were increased by CP treatment in rice noodles, while it also enhanced the lipid oxidation of the noodles. Yadav et al. (2020) mentioned that ready-to-eat ham samples treated with CP showed significant changes in colour parameters and lipid oxidation after post-treatment storage (1 and 7 days at 4 °C).

CP processing did not have any negative effect on aw, TPC and antioxidant activity of the mango flour noodles (Abidin et al. 2018). CP treatment had no effect on weight, sugar content, colour and surface morphology of banana and CP inactivated mold growth compared to untreated samples (Trivedi et al. 2019). The total antioxidant capacity, phenolic, ascorbic acid and anthocyanin content were found to increase by CP processing in blueberry fruit and the firmness of the fruit were maintained throughout the storage (Ji et al. 2020). There was no significant change in antioxidant content (polyphenols) and activity among the CP treated and control kiwifruits (Ramazzina et al. 2015). It was mentioned that CP (DBD at 45 kV for 1 min) inhibited the microbial growth, promoted the accumulation of total phenolics, flavonoid, and anthocyanin content and in addition textural properties of the fresh-cut strawberries were also maintained (Li et al. 2019b). Pasquali et al. 2016 mentioned that there was no significant change in antioxidant activity and external appearance of the radicchio leaves, immediately after CP treatment, but a few alterations emerged in the quality during storage which need to be considered in future studies.

It is noteworthy that the CP treatment increased the flavonols and decreased the total phenolics, flavonoids and antioxidant activity in white grape juice which was comparable to the thermally pasteurized juice (Pankaj et al. 2017). It was observed that CP had a positive effect on phenolic content of orange, tomato, apple juices, and sour cherry nectar, while there was no effect on the pH and colour parameters of the samples (Dasan and Boyaci 2018). The high voltage atmospheric CP (120 s) processing reduced vitamin C and PME activity by 22% and 74%, respectively and CP was better than heat pasteurization and did not have any negative impact on Brix or pH of the orange juice (Xu et al. 2017). CP increased the antioxidant activity of prickly pear cactus fruit and grape pomace extract by disrupting the epidermal cell structure (Kim et al. 2019; Bao et al. 2020). It was also noticed that CP treatment increased sucrose, vitamin C, phenolic and flavonoid content in cashew apple juice, whilst decreased glucose and fructose content (Rodríguez et al. 2017). Atmospheric CP (500 Hz/5 min) did not affect the anthocyanin content of açai pulp, while increased the phenolic content by 39% (Dantas et al. 2021).

Firmness of blueberries was slightly affected by CP processing, but there was no change in colour parameters (Sarangapani et al. 2017). It was observed that DBD plasma did not affect the hardness/texture and soluble solid content of the kiwifruit slices (Ramazzina et al. 2015). Meanwhile, firmness, colour parameter (b*) and organoleptic characteristic of the fermented vegetable was enhanced by CP processing compared to the pasteurized samples, suggesting that CP could improve the softening and browning of the fermented vegetable (Zhao et al. 2020). These reports suggest that CP not only maintains the microbial safety of food products, but also provides a better textural quality.

Low-pressure CP processing improved the cooking properties by reducing the cooking time upto 8 min and also increased the water absorption of parboiled rice (Sarangapani et al. 2015). CP treatment reduced the cooking time by 8 min in brown rice due to the increase in the degree of gelatinization and water uptake (Thirumdas et al. 2016). As a result of CP treatment, the hardness of the brown rice was reduced significantly compared to the untreated ones (Lee et al. 2016a). Improvement in cooking time, surface energy and hardness was observed in CP (20 kV for 10 min and 15 min) treated fortified white rice (Akasapu et al. 2020). The CP treated fortified rice exhibited higher *in vitro* bioaccessibility of iron compared to control samples and also lessened the rate of oxidation of iron during storage (Akasapu et al. 2020). It was observed that the contact an gle, hardness and stickiness of the parboiled rice was decreased and surface energy was increased by CP treatment, thus improving the texture properties of rice (Sarangapani et al. 2015). Liu et al. (2021) also demonstrated that CP processing (120 W for 20 s) improves the cooking properties of milled rice by decreasing the cooking time and hardness; increasing the water absorption rate, rice dough development, adhesiveness, elasticity, and gruel solid loss of the rice. CP (120 W–20 s) processing also exhibited rough kernel surface, weak protein network, and a higher speed of starch gelatinization in milled rice (Liu et al. 2021). Thirumdas et al. (2016) also showed that the CP treatment altered the textural properties of the brown rice by increasing surface energy and decreasing the hardness and chewiness of the rice. CP treatment (DBD for 5–30 min at 80 kV) increased the pasting, final viscosities and flour hydration properties of wheat flour, thus improving the functionality of wheat flour (Chaple et al. 2020).

CP treatment increased the root length, seedling height and germination percentage of germinated brown rice by 57%, 69%, and 84%, respectively (Yodpitak et al. 2019). Antioxidant activity of CP treated germinated brown rice was not altered, while high content of γ-oryzanols were estimated in CP treated brown rice compared to the untreated ones (Yodpitak et al. 2019). Interestingly, CP treatment did not affect the appearance of milled rice significantly (Liu et al. 2021). CP processing (5 min) was a promising technology in retaining the germination capacity and quality parameters of barley and wheat grains (Los et al. 2018). CP lowered the nitrite content of fermented vegetable by ~48% compared to untreated samples, but did not affect the salt content and amino acid nitrogen of the samples (Zhao et al. 2020). CP processing (644 W for 36 min) slightly improved the water solubility and dispersion stability of prickly pear cactus fruit extract (Kim et al. 2019). Interestingly, DBD plasma (80 kV and 5 min) was found to degrade pesticides such as boscalid and Imidacloprid on blueberries by 80% and 76%, respectively (Sarangapani et al. 2017). To summarize, studies on the CP suggested that it has a promising scope in decontaminating food, while retaining quality of food products and thus showing good probability adopting this by food industry for commercial exploitation.

6 Membrane Processing

Membrane processing is an economical and environment-friendly technique of high significance due to its key role and wide applications in the field of food technology. It is a separation process in which one liquid feed is passed through a selectively permeable solid barrier known as membrane and separated into two liquid product streams, namely permeate and retentate, due to the difference in transmembrane pressure. The target compound of interest during the processing may either exist in permeate or retentate stream.

Factors that affect the functioning of separation process and efficiency of membrane processing are transmembrane pressure difference, temperature, feed flow rate, feed composition, interactions between feed components and membrane, and nature of membrane (Lin et al. 1997). The type of membrane based on pore size, construction material, and driving force required for mass transfer decides its selectivity, diffusivity, and permeability (Cheryan 1998). Concentration polarization and membrane fouling are the significant factors responsible for decreasing the permeate flux and degenerating membrane performance. Concentration polarization occurs at the initial stage when large-sized solute particles accumulate on the surface of membrane, causing more resistance and higher localized osmotic pressure. Fouling, on the other hand, affects at a later stage due to solute deposition and various interactions of solute particles with the membrane (de Morais Coutinho et al. 2009).

The different types of membrane processes used in food processing industries are Microfiltration (MF), Ultrafiltration (UF), Nanofiltration (NF), and Reverse Osmosis (RO).

Membrane processing has various applications in the food processing industries. One of the common applications is the use of microfiltration and ultrafiltration for clarification and concentration of fruits and milk based liquid products. Reverse Osmosis is extensively being used for the purification of drinking quality water. Ultrafiltration is used to concentrate milk for cheese manufacturing, while beer clarification requires microfiltration (Ahmad and Ahmed 2014) (Table. 6.5).

Table 6.5 Membrane classification and applications

Membrane process	Pore size	Molecular weight	Osmotic pressure (MPa)	Applications
Microfiltration (MF)	0.1–10 μm	>100 kDa	0.2	Microbial reduction Clarification Separates high molecular weight proteins
Ultrafiltration (UF)	1–100 nm	0.3–1000 kDa	1.0	Clarification and concentration Separates biomolecules, polymers, and microbes
Nanofiltration (NF)	0.5–2 nm	0.2–2 kDa	1.0–4.0	Concentrate salts and sugars Separates a few bioactive compounds Partial demineralization
Reverse Osmosis (RO)	<0.1 nm	<350 Da	4.0–10.0	Demineralization Concentration

Bhattacharjee et al. (2017); Castro-Muñoz et al. (2020); Nakao (1994)

Recently membrane processing has been explored for its applications in the fats and oils industry. The oil refining process steps can be carried out using membrane processing, such as desolventization, deacidification, degumming, and separation of pigments, antioxidants, and other compounds present in trace amounts in the oil. Refining the oils by membrane processing has several advantages over conventional methods, such as lower processing temperature, low energy consumption, retention of nutrients conventionally damaged by heat, less use of chemical additives, and reduction in the formation of undesirable compounds (Subramanian et al. 2001).

Another significant importance of membrane processing in various sectors of food processing industries is its application for food preservation by the separation of micro-organisms from feed into the retentate. Microfiltration and ultrafiltration are commonly applied to reduce microbial population in permeate stream. MF and UF processing methods not only separate vegetative microbial cells but also the thermostable bacterial spores along with fat, high molecular weight proteins, and other polymers without the use of extensive heat treatment (Goulas and Grandison 2008). Various researchers have explored the application of membrane processing for microbial reduction in different food commodities to improve their shelf life in the last decade.

Milk is a highly perishable food commodity used in nearly every household worldwide. It is an ideal medium for the growth of micro-organisms due to its composition and structure. Microfiltration and ultrafiltration are emerging technologies being applied in the dairy industry to preserve milk and different milk products.

Walkling-Ribeiro et al. (2011) compared the effect of Microfiltration (MF), Pulsed Electric Field (PEF), and their combinations on skim milk for the inactivation of native microbial population with conventional thermal processing. MF (1.4 μm, 35 °C) and PEF (815 kJ/L) resulted in 3.7 and 2.5 log CFU/mL microbial reduction when applied individually, while thermal processing (75 °C /24 s) reduced microbial counts to 4.6 log CFU/mL. MF followed by PEF produced results (4.8 log CFU/mL) comparable with thermal processing, whereas PEF followed by MF caused the highest microbial reduction of 7.1 log CFU/mL. The milk processing by PEF/MF extended the shelf life of milk to 7 days at 4 °C. Wang et al. (2019) extended the shelf life of skim milk to more than 92 days when stored at 6 °C by processing it through a ceramic membrane of 1.4 μm pore size with transmembrane pressure difference of 75.8 kPa and subsequently pasteurizing it at 72 °C for 25 seconds. MF alone decreased the microbial count by 3.4 log cycles and 4 log cycles when combined with pasteurization. MF helped reduce the microbial load significantly without affecting the sensory quality of the skim milk. Panopoulos et al. (2020) studied the effect of microfiltration on microbial load, proteins, enzymes, nutritional composition, and renneting behaviour of defatted bovine and ovine milk. The bovine and ovine milk passed through a ceramic membrane of 1.4 μm pore size at 50 °C with transmembrane pressure of 1.5 bar showed a reduction in mesophilic microbial counts of 2 and 4 log cycles, respectively. It also affects the composition of milk by decreasing the proteins and total solids in permeate stream. There was no significant difference in the renneting properties of milk which inferred that microfiltration can be used for preservation and as a pretreatment for cheese making. Li et al. (2021) carried out microfiltration of skim milk with membranes of 1.4 and 0.8 μm pore size and compared the observations with UV-C irradiation and pasteurization. Microfiltration reduced the bacteria to 3.0 log cycles while completely removing coliforms, microbial spores, and somatic cells from milk samples.

Barukčić et al. (2014) studied the effect of temperature and pore size on microbial reduction, retentate protein, permeate flux, and membrane fouling during the crossflow microfiltration (CFMF) processing of sweet whey. The sweet whey samples with temperatures of 20, 40, and 50 °C were passed for 65 minutes through ceramic ZrO2 membranes having different pore sizes (0.1, 0.5, and 0.8 μm). It was found that microbial reduction of the permeate, calculated in comparison with conventional pasteurization (5.71 log reduction), improved with smaller membrane pore size (0.1 μm) and higher sample temperature (50 °C) up to 5 log reduction. The flux rates increased with the increase in temperature and pore size for all the membranes, maximum for 0.8 μm membrane at 50 °C (about 100 kg m-2 h-1). On the other hand, fouling was minimum when the membrane of 0.5 μm was used at 20 °C. It was concluded that 0.5 μm membrane appeared to be optimal for CFMF process as it provides the lowest fouling with a substantial microbial reduction in the order of 4 log cycles.

Bovine colostrum contains various health-beneficial components giving it the properties of a nutraceutical. Due to the heat-sensitive nature and variable microbial load of the bovine colostrum, it was investigated by Gosch et al. (2014) to obtain a stable product with improved shelf life using Microfiltration (MF) in combination with High Pressure Processing (HPP). The skimmed raw bovine colostrum was processed through a crossflow microfiltration (CFMF) system using the ceramic membranes of pore sizes 0.8 and 1.4 μm and subsequently by HPP system for 10 minutes at 400 and 500 MPa pressures. This hybrid treatment acted as different hurdles for microbial population growth that allowed the preservation of colostrum at temperatures as low as 40 °C. It prevented the extensive denaturation of protein and separation of fat. The findings elucidate that the microfiltration carried out using 1.4 μm membrane reduces the total viable count of microbes to more than 4 log cycles in permeate with 10–30% decrease of IgG in the retentate. The permeate, when further processed through HPP removes the residual microbial population. The results showed that this combination of MF with HPP demonstrates an advantageous process for preserving heat-sensitive products compared to the individual approach.

Fruit and vegetable juices play a significant role in the market for foods and beverages. They are extremely perishable and need to be preserved. Conventionally, they are being processed thermally, which leads to the loss of many heat-sensitive biologically active components. Hence, the emerging need for a non-thermal preservation technique that should retain the antioxidants, vitamins, pigments, and other thermally degrading phytochemicals, can be met by membrane processing.

Rezzadori et al. (2013) compared microbiological, rheological, physicochemical, and sensory properties of sugar cane and passion fruit mixed juice after processing through microfiltration (organic polyimide hollow fiber membrane, 0.4 μm) and pasteurization (90/95 °C for 30 seconds). Microfiltration of the product resulted in low microbial counts, soluble solids, and acidity. Laorko et al. (2013) carried out microfiltration for non-thermal preservation of pineapple juice, retaining the phytochemical and sensory properties. Polysulfone hollow fiber module with a pore size of 0.2 μm used for microfiltration of juice at 4, 27, and 37 °C showed no microbial growth after 6 months of storage. The rate of degradation of nutrients during the storage period was found less at a lower storage temperature of 4 °C. Zhao et al. (2015) developed a hurdle treatment by combining the two non-thermal processing techniques: microfiltration and UV radiation, to preserve apple cider. Microfiltration was carried out with 0.8 and 1.4 μm pore size ceramic membranes at 10 °C having transmembrane pressure of 155 kPa. Filtration of the juice helped in improving the efficacy of the process when subjected to UV radiation by reducing the initial microbial load. The hybrid treatment of microfiltration with 0.8 μm membrane and UV radiation reduced the microbial count by 5 log cycles which was comparable with thermal treatment. Colantuono et al. (2018) subjected pomegranate juice with microfiltration (mixed cellulose ester membranes, 0.45 μm) and compared it with conventional pasteurization (80 °C for 30 seconds). It was found that microfiltration of juice reduced the Lactic Acid Bacteria and yeasts/molds counts from 2.29 and 2.63 log CFU/mL to undetectable levels, respectively. Vieira et al. (2020) evaluated the microbiological quality of an ovine whey-orange juice beverage processed by 0.2 μm ceramic membrane. The microfiltration was done at 20, 30, 40, and 50 °C, and observations were compared with the conventional pasteurization done by heating the product at 63 °C for 30 minutes. Membrane processing at 30 °C reduces the population of Aerobic Mesophilic Bacteria, molds and yeasts, and Lactic Acid Bacteria by 4, 3, and 4 log CFU/mL.

Coconut water contains various nutritive and non-nutritive components that provide many health benefits. It is a highly perishable food commodity susceptible to microbial spoilage due to high water activity. It is conventionally processed by thermal treatments to improve its shelf-life which negatively affects its nutritional and sensory quality characteristics such as the formation of 5-hydroxy methyl furfural (HMF), non-enzymatic browning, changes in color and flavor. Preservation of heat-sensitive coconut water by membrane processing is recently being explored by researchers (Naik et al. 2020).

The stability of coconut water is also affected by enzymes, concerning color and other compositional changes. Debien et al. (2013) applied the ultrafiltration technology to separate enzymes and prevent the formation of products from enzymatic degradation. Polyethersulfone (150 kDa), polyvinylidene fluoride (150 kDa), and cellulose (30 kDa) membranes were studied and evaluated for permeate flux and enzyme separation. It was found that the permeate flux was inversely proportional to the enzyme retention through the membranes. Cellulose membrane was found to separate 100% enzymes in the retentate with the lowest flux. On the other hand, separation through polyethersulfone and polyvinylidene fluoride membranes resulted in 71% and 85% retention of enzymes, respectively. The ultrafiltration technique represents to be an appropriate method for ensuring the enzymatic and microbial stability of coconut water without the use of extensive thermal treatment. Junmee and Tongchitpakdee (2015) carried out microfiltration of coconut water using polysulfone membrane and further treated it at Ultra High Temperature (UHT) of 140 °C for 4 seconds. Microfiltration through 0.2 μm pore size membrane and UHT treatment extended the shelf life of coconut water up to 28 days in ambient conditions. Haze, total color difference (ΔE), browning index (BI), and HMF were the quality parameters monitored throughout the storage period. Microfiltration removed the initial haze in coconut water significantly, and thereafter it uniformly increased with a storage time of 28 days. The samples processed through 0.2 μm polysulfone membranes showed longer storage stability with better quality parameters compared to the unfiltered control sample with lower BI and HMF values at the end of the 28th day.

Concentration polarization and membrane fouling are the critical factors that significantly reduce the permeate flux and efficiency of membrane processing. Lamdande et al. (2020) conducted studies on the effect of acoustic field on transmembrane flux during ultrafiltration of coconut water. Ultrafiltration was carried out by passing the coconut water through polyethersulfone membranes (30 and 100 kDa) with constant stirring at 100 rpm under the effect of an ultrasonic field, producing the transmembrane pressure difference of 1, 2, 3, and 4.5 bar. The combination of ultrasound with ultrafiltration resulted in 30–50% increase in flux and in achieving cold sterilization by reducing the microbial count from 4.16 to 0.0 log CFU/mL. This provided extended stability in the nutritional, sensory, and microbiological quality of coconut water for up to 3 months. It decreased the turbidity of coconut water by 95.81% and also affected the enzymes as the activity of POD and PPO enzymes was lowered in permeate by 95.10% and 97.89%, respectively. This study indicates the commercial feasibility of membrane processing to preserve highly heat-sensitive and perishable liquid food products.

Membrane processing is a separation technique being employed in various food processing applications such as preservation, fractionation, downstream processing, and other unit operations. It is being increasingly used as an alternative to thermal processing to improve food products' storage stability. Characteristics of both feed and membrane affect transmembrane flux during the process. The applications of different types of membranes for filtration to extend the shelf life of various heat sensitive and perishable liquid plant and animal based food products have been studied by researchers. Membrane processing was also investigated as a part of hurdle technology by combining other techniques such as High Pressure Processing and Ultra High Temperature treatment to preserve products containing heat-labile nutrients. This elucidates that membrane processing has a good potential to be used as an alternative to some of the conventional food processing methods and is receiving highly acceptability due to its, energy-efficiency, ease of operation and scalability.

7 Conclusion

Non-thermal technologies have been used as an alternative to thermal processing due to several advantages. Various processing variables dictate the efficiency of the processing technology and optimization is very crucial for the application in real food systems. In future, research on these non-thermal technologies should focus on scalability and expanding their scope in food industries. To conclude, all the above discussed non-thermal technology have a good future to be employed as commercially viable technologies, with promising assurance in maintaining quality.

Acknowledgement Authors wish to thank Director, CSIR-CFTRI, Mysuru for the extending all the support authors to contribute this chapter.

References

Aadil RM, Zeng XA, Wang MS, Liu ZW, Han Z, Zhang ZH, Jabbar S (2015) A potential of ultrasound on minerals, micro-organisms, phenolic compounds and colouring pigments of grapefruit juice. Int J Food Sci Technol 50(5):1144–1150

Aadil RM, Khalil AA, Rehman A, Khalid A, Inam-ur-Raheem M, Karim A, Afraz MT (2020) Assessing the impact of ultra-sonication and thermo-ultrasound on antioxidant indices and polyphenolic profile of apple-grape juice blend. J Food Process Preserv 44(5):e14406

Abbaszadeh R, Alimohammad K, Zarrabi Ekbatani R (2018) Application of cold plasma Technology in Quality Preservation of fresh fig fruit (Siyah): a feasibility study. Int J Hortic Sci Technol 5(2):165–173

Abera G (2019) Review on high-pressure processing of foods. Cogent Food Agric 5(1):1568725

Abidin NSA, Rukunudin IH, Zaaba SK, Wan Omar WA (2018) The effect of atmospheric cold plasma (ACP) treatment on colour, water activity, antioxidant activity and total phenolic content of mango flour noodles during storage. Int Food Res J 25(4):1444–1449

Ahmad S, Ahmed SM (2014) Application of membrane technology in food processing. In: Food processing: strategies for quality assessment, Food Engineering Series. Springer, New York, pp 379–394. https://doi.org/10.1007/978-1-4939-1378-7_15

Akasapu K, Ojah N, Gupta AK, Choudhury AJ, Mishra P (2020) An innovative approach for iron fortification of rice using cold plasma. Food Res Int 136:109599

Albertos I, Martin-Diana AB, Sanz MA, Barat JM, Diez AM, Jaime I, Rico D (2016) Effect of high pressure processing or freezing technologies as pretreatment in vacuum fried carrot snacks. Innovative Food Sci Emerg Technol 33:115–122

Almeida FDL, Cavalcante RS, Cullen PJ, Frias JM, Bourke P, Fernandes FA, Rodrigues S (2015) Effects of atmospheric cold plasma and ozone on prebiotic orange juice. Innovative Food Sci Emerg Technol 32:127–135

Alsalman FB, Ramaswamy H (2020) Reduction in soaking time and anti-nutritional factors by high pressure processing of chickpeas. J Food Sci Technol 57(7):2572–2585

Amaral RDA, Benedetti BC, Pujola M, Achaerandio I, Bachelli MLB (2015) Effect of ultrasound on quality of fresh-cut potatoes during refrigerated storage. Food Eng Rev 7(2):176–184

Andrés V, Villanueva MJ, Tenorio MD (2016a) Influence of high pressure processing on microbial shelf life, sensory profile, soluble sugars, organic acids, and mineral content of milk-and soy-smoothies. LWT-Food Sci Technol 65:98–105

Andrés V, Villanueva MJ, Tenorio MD (2016b) The effect of high-pressure processing on colour, bioactive compounds, and antioxidant activity in smoothies during refrigerated storage. Food Chem 192:328–335

Aponiene K, Paskeviciute E, Reklaitis I, Luksiene Z (2015) Reduction of microbial contamination of fruits and vegetables by hypericin-based photosensitization: comparison with other emerging antimicrobial treatments. J Food Eng 144:29–35

Ates C, Akdemir Evrendilek G, Uzuner S (2021) High pressure processing of shalgam with respect to quality characteristics, microbial inactivation and shelflife extension. J Food Process Preserv 45:e15598

Avalos-Llano KR, Martín-Belloso O, Soliva-Fortuny R (2018) Effect of pulsed light treatments on quality and antioxidant properties of fresh-cut strawberries. Food Chem 264:393–400

Balakrishna AK, Farid M (2020) Enrichment of rice with natural thiamine using high-pressure processing (HPP). J Food Eng 283:110040

Balamurugan S, Gemmell C, Lau ATY, Arvaj L, Strange P, Gao A, Barbut S (2020) High pressure processing during drying of fermented sausages can enhance safety and reduce time required to produce a dry fermented product. Food Control 113:107224

Balthazar CF, Santillo A, Guimarães JT, Bevilacqua A, Corbo MR, Caroprese M, Albenzio M (2019) Ultrasound processing of fresh and frozen semi-skimmed sheep milk and its effects on microbiological and physical-chemical quality. Ultrason Sonochem 51:241–248

Bao Y, Reddivari L, Huang JY (2020) Enhancement of phenolic compounds extraction from grape pomace by high voltage atmospheric cold plasma. LWT 133:109970

Barukčić I, Božanić R, Kulozik U (2014) Effect of pore size and process temperature on flux, microbial reduction and fouling mechanisms during sweet whey cross-flow microfiltration by ceramic membranes. Int Dairy J 39(1):8–15. https://doi.org/10.1016/j.idairyj.2014.05.002

Bhattacharjee C, Saxena VK, Dutta S (2017) Fruit juice processing using membrane technology: a review. Innov Food Sci Emerg Technol 43:136–153). Elsevier. https://doi.org/10.1016/j.ifset.2017.08.002

Bhavya ML, Hebbar HU (2019a) Efficacy of blue LED in microbial inactivation: effect of photosensitization and process parameters. Int J Food Microbiol 290:296–304

Bhavya ML, Hebbar HU (2019b) Sono-photodynamic inactivation of *Escherichia coli* and *Staphylococcus aureus* in orange juice. Ultrason Sonochem 57:108–115

Bhavya ML, Hebbar HU (2020) High-intensity pulsed light for food preservation. In: Handbook of food preservation. CRC Press, pp 779–804

Bhavya ML, Umesh Hebbar H (2017) Pulsed light processing of foods for microbial safety. Food Qual Saf 1(3):187–202

Bhavya ML, Shewale SR, Rajoriya D, Hebbar HU (2021) Impact of blue LED illumination and natural photosensitizer on bacterial pathogens, enzyme activity and quality attributes of fresh-cut pineapple slices. Food Bioprocess Technol 14(2):362–372

Bialka KL, Demirci A (2008) Efficacy of pulsed UV-light for the decontamination of *Escherichia coli* O157: H7 and *Salmonella* spp. on raspberries and strawberries. J Food Sci 73(5):M201–M207

Campoli SS, Rojas ML, do Amaral JEPG, Canniatti-Brazaca SG, Augusto PED (2018) Ultrasound processing of guava juice: effect on structure, physical properties and lycopene in vitro accessibility. Food Chem 268:594–601

Cao X, Huang R, Chen H (2019) Evaluation of food safety and quality parameters for shelf life extension of pulsed light treated strawberries. J Food Sci 84(6):1494–1500

Carrillo-lopez LM, Luna-Rodriguez L, Alarcon-Rojo AD, Huerta-Jimenez M (2018) High intensity ultrasound homogenizes and improves quality of beef longissimus dorsi. Food Sci Technol 39: 332–340

Castro-Muñoz R, Boczkaj G, Gontarek E, Cassano A, Fíla V (2020) Membrane technologies assisting plant-based and agro-food by-products processing: a comprehensive review. Trends Food Sci Technol 95:219–232. https://doi.org/10.1016/j.tifs.2019.12.003

Chai HE, Sheen S (2021) Effect of high pressure processing, allyl isothiocyanate, and acetic acid stresses on *Salmonella* survivals, storage, and appearance color in raw ground chicken meat. Food Control 123:107784

Chaple S, Sarangapani C, Jones J, Carey E, Causeret L, Genson A et al (2020) Effect of atmospheric cold plasma on the functional properties of whole wheat (*Triticum aestivum* L.) grain and wheat flour. Innovative Food Sci Emerg Technol 66:102529

Cheigh CI, Park MH, Chung MS, Shin JK, Park YS (2012) Comparison of intense pulsed light-and ultraviolet (UVC)-induced cell damage in *Listeria monocytogenes* and *Escherichia coli* O157: H7. Food Control 25(2):654–659

Cheigh CI, Hwang HJ, Chung MS (2013) Intense pulsed light (IPL) and UV-C treatments for inactivating *Listeria monocytogenes* on solid medium and seafoods. Food Res Int 54(1): 745–752

Chemat F, Khan MK (2011) Applications of ultrasound in food technology: processing, preservation and extraction. Ultrason Sonochem 18(4):813–835

Cheng CX, Jia M, Gui Y, Ma Y (2020) Comparison of the effects of novel processing technologies and conventional thermal pasteurisation on the nutritional quality and aroma of Mandarin (*Citrus unshiu*) juice. Innovative Food Sci Emerg Technol 64:102425

Cheryan M (1998) Ultrafiltration and microfiltration handbook. CRC Press. https://doi.org/10.1201/9781482278743

Chutia H, Mahanta CL (2021) Influence of cold plasma voltage and time on quality attributes of tender coconut water (*Cocos nucifera* L.) and degradation kinetics of its blended beverage. J Food Process Preserv 45(4):e15372

Colantuono A, Vitaglione P, Manzo N, Blaiotta G, Montefusco I, Marrazzo A, Pizzolongo F, Romano R (2018) Evaluation of microfiltration and heat treatment on the microbiological characteristics, phenolic composition and volatile compound profile of pomegranate (Punica granatum L.) juice. J Sci Food Agric 98(9):3324–3332. https://doi.org/10.1002/jsfa.8836

Contigiani EV, Jaramillo Sánchez GM, Castro MA, Gómez PL, Alzamora SM (2021) Efficacy of mild thermal and pulsed light treatments, individually applied or in combination, for maintaining postharvest quality of strawberry cv. Albion. J Food Process Preserv 45(1):e15095

Coutinho NM, Silveira MR, Fernandes LM, Moraes J, Pimentel TC, Freitas MQ, Cruz AG (2019) Processing chocolate milk drink by low-pressure cold plasma technology. Food Chem 278:276–283

Craighead S, Huang R, Chen H, Kniel KE (2021) The use of pulsed light to inactivate *Cryptosporidium parvum* oocysts on high-risk commodities (Cilantro, mesclun lettuce, spinach, and tomatoes). Food Control 126:107965

Cropotova J, Mozuraityte R, Standal IB, Ojha S, Rustad T, Tiwari B (2020) Influence of high-pressure processing on quality attributes of haddock and mackerel minces during frozen storage, and fishcakes prepared thereof. Innovative Food Sci Emerg Technol 59:102236

D'Souza C, Yuk HG, Khoo GH, Zhou W (2015) Application of light-emitting diodes in food production, postharvest preservation, and microbiological food safety. Compr Rev Food Sci Food Saf 14(6):719–740

Dantas AM, Batista JDF, dos Santos Lima M, Fernandes FA, Rodrigues S, Magnani M, Borges GDSC (2021) Effect of cold plasma on açai pulp: enzymatic activity, color and bioaccessibility of phenolic compounds. LWT 149:111883

Dasan BG, Boyaci IH (2018) Effect of cold atmospheric plasma on inactivation of Escherichia coli and physicochemical properties of apple, orange, tomato juices, and sour cherry nectar. Food Bioprocess Technol 11(2):334–343

de Alba M, Pérez-Andrés JM, Harrison SM, Brunton NP, Burgess CM, Tiwari BK (2019) High pressure processing on microbial inactivation, quality parameters and nutritional quality indices of mackerel fillets. Innovative Food Sci Emerg Technol 55:80–87

de Ancos B, Rodrigo MJ, Sánchez-Moreno C, Cano MP, Zacarías L (2020) Effect of high-pressure processing applied as pretreatment on carotenoids, flavonoids and vitamin C in juice of the sweet oranges 'Navel' and the red-fleshed 'Cara Cara'. Food Res Int 132:109105

de Morais Coutinho C, Chiu MC, Basso RC, Ribeiro APB, Gonçalves LAG, Viotto LA (2009) State of art of the application of membrane technology to vegetable oils: a review. Food Res Int 42(5–6):536–550). Elsevier. https://doi.org/10.1016/j.foodres.2009.02.010

de Souza Silva DA, da Silva Campelo MC, de Oliveira Soares Rebouças LUCAS, de Oliveira Vitoriano J, Alves C, Alves da Silva JB, de Oliveira Lima P (2019) Use of cold atmospheric plasma to preserve the quality of white shrimp (Litopenaeus vannamei). J Food Prot 82(7): 1217–1223

Debien IC d N, Gomes MT d MS, Ongaratto RS, Viotto LA (2013) Ultrafiltration performance of PVDF, PES, and cellulose membranes for the treatment of coconut water (Cocos nucifera L.). Food Sci Technol 33(4):676–684. https://doi.org/10.1590/S0101-20612013000400012

Denoya GI, Pataro G, Ferrari G (2020) Effects of postharvest pulsed light treatments on the quality and antioxidant properties of persimmons during storage. Postharvest Biol Technol 160:111055

Dhar R, Chakraborty S (2020) Influence of voltage and distance on quality attributes of mixed fruit beverage during pulsed light treatment and kinetic modeling. J Food Process Eng 43(11):e13517

Donatti Leão Alvarenga P, Mileib Vasconcelos C, de São José JFB (2021) Application of ultrasound combined with acetic acid and Peracetic acid: microbiological and physicochemical quality of strawberries. Molecules 26(1):16

Dong XY, Yang YL (2019) A novel approach to enhance blueberry quality during storage using cold plasma at atmospheric air pressure. Food Bioprocess Technol 12(8):1409–1421

Du L, Prasad AJ, Gänzle M, Roopesh MS (2020) Inactivation of Salmonella spp. in wheat flour by 395 nm pulsed light emitting diode (LED) treatment and the related functional and structural changes of gluten. Food Res Int 127:108716

dos Anjos C, Sellera FP, de Freitas LM, Gargano RG, Telles EO, Freitas RO et al (2020) Inactivation of milk-borne pathogens by blue light exposure. J Dairy Sci 103(2):1261–1268

Elmnasser N, Guillou S, Leroi F, Orange N, Bakhrouf A, Federighi M (2007) Pulsed-light system as a novel food decontamination technology: a review. Can J Microbiol 53(7):813–821

Evert-Arriagada K, Trujillo AJ, Amador-Espejo GG, Hernández-Herrero MM (2018) High pressure processing effect on different *listeria* spp. in a commercial starter-free fresh cheese. Food Microbiol 76:481–486

Fan K, Zhang M, Jiang F (2019a) Ultrasound treatment to modified atmospheric packaged fresh-cut cucumber: influence on microbial inhibition and storage quality. Ultrason Sonochem 54:162–170

Fan K, Zhang M, Bhandari B, Jiang F (2019b) A combination treatment of ultrasound and ε-polylysine to improve microorganisms and storage quality of fresh-cut lettuce. LWT 113: 108315

Feng L, Zhang M, Adhikari B, Guo Z (2018) Effect of ultrasound combined with controlled atmosphere on postharvest storage quality of cucumbers (*Cucumis sativus* L.). Food Bioprocess Technol 11(7):1328–1338

Fernandez MV, Denoya GI, Agüero MV, Jagus RJ, Vaudagna SR (2018) Innovative Food Sci Emerg Technol 47:170–179

Fernandez MV, Denoya GI, Jagus RJ, Vaudagna SR, Agüero MV (2019) Microbiological, antioxidant and physicochemical stability of a fruit and vegetable smoothie treated by high pressure processing and stored at room temperature. LWT 105:206–210

Fernández M, Hospital XF, Cabellos C, Hierro E (2020) Effect of pulsed light treatment on *Listeria* inactivation, sensory quality and oxidation in two varieties of Spanish dry-cured ham. Food Chem 316:126294

Ferrario M, Guerrero S (2017) Impact of a combined processing technology involving ultrasound and pulsed light on structural and physiological changes of *Saccharomyces cerevisiae* KE 162 in apple juice. Food Microbiol 65:83–94

Food and Drug Administration (1996) Pulsed Light for the Treatment of Food, 21CFR179.4

Gani A, Baba WN, Ahmad M, Shah U, Khan AA, Wani IA et al (2016) Effect of ultrasound treatment on physico-chemical, nutraceutical and microbial quality of strawberry. LWT-Food Sci Technol 66:496–502

Georget E, Sevenich R, Reineke K, Mathys A, Heinz V, Callanan M et al (2015) Inactivation of microorganisms by high isostatic pressure processing in complex matrices: a review. Innovative Food Sci Emerg Technol 27:1–14

Ghate V, Leong AL, Kumar A, Bang WS, Zhou W, Yuk HG (2015) Enhancing the antibacterial effect of 461 and 521 nm light emitting diodes on selected foodborne pathogens in trypticase soy broth by acidic and alkaline pH conditions. Food Microbiol 48:49–57

Ghate V, Kumar A, Kim MJ, Bang WS, Zhou W, Yuk HG (2017) Effect of 460 nm light emitting diode illumination on survival of *Salmonella* spp. on fresh-cut pineapples at different irradiances and temperatures. J Food Eng 196:130–138

Gomes WF, Tiwari BK, Rodriguez Ó, de Brito ES, Fernandes FAN, Rodrigues S (2017) Effect of ultrasound followed by high pressure processing on prebiotic cranberry juice. Food Chem 218: 261–268

Gomez-Lopez VM, Ragaert P, Debevere J, Devlieghere F (2007) Pulsed light for food decontamination: a review. Trends Food Sci Technol 18(9):464–473

Gosch T, Apprich S, Kneifel W, Novalin S (2014) A combination of microfiltration and high pressure treatment for the elimination of bacteria in bovine colostrum. Int Dairy J 34(1):41–46. https://doi.org/10.1016/j.idairyj.2013.06.014

Goulas A, Grandison AS (2008) Applications of membrane separation. In: Advanced dairy science and technology, pp 35–74. https://doi.org/10.1002/9780470697634.ch2

Gouvea FS, Padilla-Zakour OI, Worobo RW, Xavier BM, Walter EH, Rosenthal A (2020) Effect of high-pressure processing on bacterial inactivation in açaí juices with varying pH and soluble solids content. Innovative Food Sci Emerg Technol 66:102490

Guerrero SN, Ferrario M, Schenk M, Carrillo MG (2017) Hurdle technology using ultrasound for food preservation. In: Ultrasound: advances for food processing and preservation. Academic, pp 39–99

Guffey JS, Payne WC, Motts SD, Towery P, Hobson T, Harrell G et al (2016) Inactivation of salmonella on tainted foods: using blue light to disinfect cucumbers and processed meat products. Food Sci Nutr 4(6):878–887

Guimarães JT, Silva EK, Alvarenga VO, Costa ALR, Cunha RL, Sant'Ana AS, Cruz AG (2018) Physicochemical changes and microbial inactivation after high-intensity ultrasound processing of prebiotic whey beverage applying different ultrasonic power levels. Ultrason Sonochem 44: 251–260

Haukvik T, Bruzell E, Kristensen S, Tønnesen HH (2009) Photokilling of bacteria by curcumin in different aqueous preparations. Studies on curcumin and curcuminoids XXXVII. Pharmazie-Int J Pharm Sci 64(10):666–673

Hernández-Hernández HM, Moreno-Vilet L, Villanueva-Rodríguez SJ (2019) Current status of emerging food processing technologies in Latin America: novel non-thermal processing. Innovative Food Sci Emerg Technol 58:102233

Hertwig C, Reineke K, Ehlbeck J, Erdoğdu B, Rauh C, Schlüter O (2015) Impact of remote plasma treatment on natural microbial load and quality parameters of selected herbs and spices. J Food Eng 167:12–17

Hinds LM, Bhavya ML, O'Donnell CP, Tiwari BK (2021) Visible light. In: Electromagnetic technologies in food science. John Wiley & Sons, Inc., pp 181–199

Hite BH (1899) The effect of pressure in the preservation of milk: a preliminary report, vol 58. West Virginia Agricultural Experiment Station

Hoover DG, Metrick C, Papineau AM, Farkas DF, Knorr D (1989) Biological effects of high hydrostatic pressure on food microorganisms. Food Technol (Chicago) 43(3):99–107

Huang Y, Chen H (2014) A novel water-assisted pulsed light processing for decontamination of blueberries. Food Microbiol 40:1–8

Hurtado A, Picouet P, Jofré A, Guàrdia MD, Ros JM, Bañón S (2015) Application of high pressure processing for obtaining "fresh-like" fruit smoothies. Food Bioprocess Technol 8(12): 2470–2482

Hyun JE, Lee SY (2020) Blue light-emitting diodes as eco-friendly non-thermal technology in food preservation. Trends Food Sci Technol 105:284–295

Hyun JE, Choi C, Lee SY (2020) Synergistic effects of blue light-emitting diodes in combination with antimicrobials against *Escherichia coli* O157: H7 and their mode of action. J Photochem Photobiol B Biol 213:112079

Illera AE, Chaple S, Sanz MT, Ng S, Lu P, Jones J, Bourke P (2019) Effect of cold plasma on polyphenol oxidase inactivation in cloudy apple juice and on the quality parameters of the juice during storage. Food Chem: X 3:100049

Irazoqui M, Romero M, Paulsen E, Barrios S, Pérez N, Faccio R, Lema P (2019) Effect of power ultrasound on quality of fresh-cut lettuce (cv. Vera) packaged in passive modified atmosphere. Food Bioprod Process 117:138–148

Jambrak AR, Mason TJ, Paniwnyk L, Lelas V (2007) Accelerated drying of button mushrooms, Brussels sprouts and cauliflower by applying power ultrasound and its rehydration properties. J Food Eng 81(1):88–97

Ji Y, Hu W, Liao J, Jiang A, Xiu Z, Gaowa S, Liu C (2020) Effect of atmospheric cold plasma treatment on antioxidant activities and reactive oxygen species production in postharvest blueberries during storage. J Sci Food Agric 100(15):5586–5595

Jiang Y, Leung AW, Hua H, Rao X, Xu C (2014) Photodynamic action of LED-activated curcumin against *Staphylococcus aureus* involving intracellular ROS increase and membrane damage. Int JPhotoenergy 2014:1–7, 637601

Junmee J, Tongchitpakdee S (2015) Effect of membrane processing on quality of coconut water. Acta Hortic 1088:605–610. https://doi.org/10.17660/ActaHortic.2015.1088.112

Kentish S, Feng H (2014) Applications of power ultrasound in food processing. Annu Rev Food Sci Technol 5:263–284

Khaliq A, Chughtai MFJ, Mehmood T, Ahsan S, Liaqat A, Nadeem M, Ali A (2021) High-pressure processing; principle, applications, impact, and future prospective. In: Sustainable food processing and engineering challenges. Academic, pp 75–108

Khandpur P, Gogate PR (2016) Evaluation of ultrasound based sterilization approaches in terms of shelf life and quality parameters of fruit and vegetable juices. Ultrason Sonochem 29:337–353

Kim MJ, Mikš-Krajnik M, Kumar A, Ghate V, Yuk HG (2015) Antibacterial effect and mechanism of high-intensity 405±5 nm light emitting diode on *Bacillus cereus*, *Listeria monocytogenes*, and *Staphylococcus aureus* under refrigerated condition. J Photochem Photobiol B Biol 153:33–39

Kim MJ, Bang WS, Yuk HG (2017a) 405±5 nm light emitting diode illumination causes photodynamic inactivation of Salmonella spp. on fresh-cut papaya without deterioration. Food Microbiol 62:124–132

Kim MJ, Tang CH, Bang WS, Yuk HG (2017b) Antibacterial effect of 405±5 nm light emitting diode illumination against *Escherichia coli* O157: H7, *Listeria monocytogenes*, and *Salmonella* on the surface of fresh-cut mango and its influence on fruit quality. Int J Food Microbiol 244: 82–89

Kim SY, Lee SY, Min SC (2019) Improvement of the antioxidant activity, water solubility, and dispersion stability of prickly pear cactus fruit extracts using argon cold plasma treatment. J Food Sci 84(10):2876–2882

Kim HJ, Lee BW, Baek KH, Jo C, Kim JK, Lee JY, Lee B (2020) Quality characteristics of rice noodles treated with cold plasma. Korean J Food Sci Technol 52(5):560–563

Knorr D (1993) Effects of high-hydrostatic-pressure processes on food safety and quality. Food Technol (Chicago) 47(6):156–161

Knorr D, Zenker M, Heinz V, Lee DU (2004) Applications and potential of ultrasonics in food processing. Trends Food Sci Technol 15(5):261–266

Koch F, Wiacek C, Braun PG (2019) Pulsed light treatment for the reduction of *Salmonella Typhimurium* and *Yersinia enterocolitica* on pork skin and pork loin. Int J Food Microbiol 292:64–71

Kovačević DB, Putnik P, Dragović-Uzelac V, Pedisić S, Jambrak AR, Herceg Z (2016) Effects of cold atmospheric gas phase plasma on anthocyanins and color in pomegranate juice. Food Chem 190:317–323

Kramer B, Muranyi P (2014) Effect of pulsed light on structural and physiological properties of *Listeria innocua* and *Escherichia coli*. J Appl Microbiol 116(3):596–611

Kramer B, Wunderlich J, Muranyi P (2015) Pulsed light induced damages in *Listeria innocua* and *Escherichia coli*. J Appl Microbiol 119(4):999–1010

Kramer B, Wunderlich J, Muranyi P (2019) Inactivation of *Listeria innocua* on packaged meat products by pulsed light. Food Packag Shelf Life 21:100353

Kumar A, Ghate V, Kim MJ, Zhou W, Khoo GH, Yuk HG (2016) Antibacterial efficacy of 405, 460 and 520 nm light emitting diodes on *Lactobacillus plantarum*, *Staphylococcus aureus* and *Vibrio parahaemolyticus*. J Appl Microbiol 120(1):49–56

Kumar A, Ghate V, Kim MJ, Zhou W, Khoo GH, Yuk HG (2017) Inactivation and changes in metabolic profile of selected foodborne bacteria by 460 nm LED illumination. Food Microbiol 63:12–21

Lacombe A, Niemira BA, Gurtler JB, Fan X, Sites J, Boyd G, Chen H (2015) Atmospheric cold plasma inactivation of aerobic microorganisms on blueberries and effects on quality attributes. Food Microbiol 46:479–484

Lacombe A, Niemira BA, Gurtler JB, Sites J, Boyd G, Kingsley DH, Chen H (2017) Nonthermal inactivation of norovirus surrogates on blueberries using atmospheric cold plasma. Food Microbiol 63:1–5

Lamdande AG, Mittal R, Raghavarao KSMS (2020) Flux evaluation based on fouling mechanism in acoustic field-assisted ultrafiltration for cold sterilization of tender coconut water. Innov Food Sci Emerg Technol 61(February):102312. https://doi.org/10.1016/j.ifset.2020.102312

Laorko A, Tongchitpakdee S, Youravong W (2013) Storage quality of pineapple juice non-thermally pasteurized and clarified by microfiltration. J Food Eng 116(2):554–561. https://doi.org/10.1016/j.jfoodeng.2012.12.033

Lee KH, Kim HJ, Woo KS, Jo C, Kim JK, Kim SH, Kim WH (2016a) Evaluation of cold plasma treatments for improved microbial and physicochemical qualities of brown rice. Lwt 73:442–447

Lee SH, Choi W, Jun S (2016b) Conventional and emerging combination technologies for food processing. Food Eng Rev 8(4):414–434

Leite AK, Fonteles TV, Miguel TB, da Silva GS, de Brito ES, Alves Filho EG, Rodrigues S (2021) Atmospheric cold plasma frequency imparts changes on cashew apple juice composition and improves vitamin C bioaccessibility. Food Res Int 147:110479

Leng J, Mukhopadhyay S, Sokorai K, Ukuku DO, Fan X, Olanya M, Juneja V (2020) Inactivation of *Salmonella* in cherry tomato stem scars and quality preservation by pulsed light treatment and antimicrobial wash. Food Control 110:107005

Li J, Cheng H, Liao X, Liu D, Xiang Q, Wang J et al (2019a) Inactivation of *Bacillus subtilis* and quality assurance in Chinese bayberry (*Myrica rubra*) juice with ultrasound and mild heat. Lwt 108:113–119
Li M, Li X, Han C, Ji N, Jin P, Zheng Y (2019b) Physiological and metabolomic analysis of cold plasma treated fresh-cut strawberries. J Agric Food Chem 67(14):4043–4053
Li Z, Liu D, Xu S, Zhang W, Hemar Y, Regenstein JM, Zhou P (2021) Effects of pasteurization, microfiltration, and ultraviolet-c treatments on microorganisms and bioactive proteins in bovine skim milk. Food Biosci 43:101339. https://doi.org/10.1016/j.fbio.2021.101339
Lin L, Rhee KC, Koseoglu SS (1997) Bench-scale membrane degumming of crude vegetable oil: process optimization. J Membr Sci 134(1):101–108. https://doi.org/10.1016/S0376-7388(97)00098-7
Liu F, Zhang X, Zhao L, Wang Y, Liao X (2016) Potential of high-pressure processing and high-temperature/short-time thermal processing on microbial, physicochemical and sensory assurance of clear cucumber juice. Innovative Food Sci Emerg Technol 34:51–58
Liu J, Wang R, Chen Z, Li X (2021) Effect of cold plasma treatment on cooking, thermomechanical and surface structural properties of Chinese milled rice. Food Bioprocess Technol 14(5): 866–886
Los A, Ziuzina D, Akkermans S, Boehm D, Cullen PJ, Van Impe J, Bourke P (2018) Improving microbiological safety and quality characteristics of wheat and barley by high voltage atmospheric cold plasma closed processing. Food Res Int 106:509–521
Luksienė Z, Zukauskas A (2009) Prospects of photosensitization in control of pathogenic and harmful micro-organisms. J Appl Microbiol 107(5):1415–1424
Luo D, Wu R, Zhang J, Zhang K, Xu B, Li P, Li X (2018) Effects of ultrasound assisted dough fermentation on the quality of steamed bread. J Cereal Sci 83:147–152
Ma Y, Yuan Y, Bi X, Zhang L, Xing Y, Che Z (2019) Tenderization of yak meat by the combination of papain and high-pressure processing treatments. Food Bioprocess Technol 12(4):681–693
Ma T, Wang J, Wang L, Yang Y, Yang W, Wang H, Sun X (2020) Ultrasound-combined sterilization technology: an effective sterilization technique ensuring the microbial safety of grape juice and significantly improving its quality. Foods 9(10):1512
Maclean M, MacGregor SJ, Anderson JG, Woolsey G (2009) Inactivation of bacterial pathogens following exposure to light from a 405-nanometer light-emitting diode array. Appl Environ Microbiol 75(7):1932–1937
Mahnot NK, Siyu LP, Wan Z, Keener KM, Misra NN (2020) In-package cold plasma decontamination of fresh-cut carrots: microbial and quality aspects. J Phys D Appl Phys 53(15):154002
Marszałek K, Woźniak Ł, Skąpska S, Mitek M (2017) High pressure processing and thermal pasteurization of strawberry purée: quality parameters and shelf life evaluation during cold storage. J Food Sci Technol 54(3):832–841
Martínez-Onandi N, Rivas-Cañedo A, Nuñez M, Picon A (2016) Effect of chemical composition and high pressure processing on the volatile fraction of serrano dry-cured ham. Meat Sci 111: 130–138
Martínez-Onandi N, Rivas-Cañedo A, Ávila M, Garde S, Nuñez M, Picon A (2017) Influence of physicochemical characteristics and high pressure processing on the volatile fraction of Iberian dry-cured ham. Meat Sci 131:40–47
McArdle R, Marcos B, Kerry JP, Mullen A (2010) Monitoring the effects of high pressure processing and temperature on selected beef quality attributes. Meat Sci 86(3):629–634
Milani EA, Ramsey JG, Silva FV (2016) High pressure processing and thermosonication of beer: comparing the energy requirements and *Saccharomyces cerevisiae* ascospores inactivation with thermal processing and modeling. J Food Eng 181:35–41
Min SC, Roh SH, Niemira BA, Sites JE, Boyd G, Lacombe A (2016) Dielectric barrier discharge atmospheric cold plasma inhibits *Escherichia coli* O157: H7, *Salmonella*, *Listeria monocytogenes*, and Tulane virus in Romaine lettuce. Int J Food Microbiol 237:114–120

Moorhead S, Maclean M, MacGregor SJ, Anderson JG (2016) Comparative sensitivity of *Trichophyton* and *Aspergillus* conidia to inactivation by violet-blue light exposure. Photomed Laser Surg 34(1):36–41

Morton JD, Pearson RG, Lee HYY, Smithson S, Mason SL, Bickerstaffe R (2017) High pressure processing improves the tenderness and quality of hot-boned beef. Meat Sci 133:69–74

Mukhopadhyay S, Sokorai K, Ukuku DO, Jin T, Fan X, Olanya OM, Juneja V (2021) Effects of direct and in-package pulsed light treatment on inactivation of *E. coli* O157: H7 and reduction of microbial loads in Romaine lettuce. LWT 139:110710

Munir M, Nadeem M, Qureshi TM, Gamlath CJ, Martin GJ, Hemar Y, Ashokkumar M (2020) Effect of sonication, microwaves and high-pressure processing on ACE-inhibitory activity and antioxidant potential of Cheddar cheese during ripening. Ultrason Sonochem 67:105140

Murdoch LE, McKenzie K, Maclean M, Macgregor SJ, Anderson JG (2013) Lethal effects of high-intensity violet 405-nm light on *Saccharomyces cerevisiae*, *Candida albicans*, and on dormant and germinating spores of *Aspergillus niger*. Fungal Biol 117(7–8):519–527

Muzaffar S, Ahmad M, Wani SM, Gani A, Baba WN, Shah U, Wani TA (2016) Ultrasound treatment: effect on physicochemical, microbial and antioxidant properties of cherry (*Prunus avium*). J Food Sci Technol 53(6):2752–2759

Naik M, Sunil CK, Rawson A, Venkatachalapathy N (2020) Tender coconut water: a review on recent advances in processing and preservation. Food Rev Intl 00(00):1–22. https://doi.org/10.1080/87559129.2020.1785489

Nakao S-i (1994) Determination of pore size and pore size distribution. 3. Filtration membranes. J Membr Sci 96(1–2):131–165. https://doi.org/10.1016/0376-7388(94)00128-6

Nayak PK, Rayaguru K, Radha Krishnan K (2017) Quality comparison of elephant apple juices after high-pressure processing and thermal treatment. J Sci Food Agric 97(5):1404–1411

Nicorescu I, Nguyen B, Moreau-Ferret M, Agoulon A, Chevalier S, Orange N (2013) Pulsed light inactivation of *Bacillus subtilis* vegetative cells in suspensions and spices. Food Control 31(1):151–157

Nowak KW, Zielinska M, Waszkielis KM (2019) The effect of ultrasound and freezing/thawing treatment on the physical properties of blueberries. Food Sci Biotechnol 28(3):741–749

Oey I, Lille M, Van Loey A, Hendrickx M (2008) Effect of high-pressure processing on colour, texture and flavour of fruit-and vegetable-based food products: a review. Trends Food Sci Technol 19(6):320–328

Önür İ, Misra NN, Barba FJ, Putnik P, Lorenzo JM, Gökmen V, Alpas H (2018) Effects of ultrasound and high pressure on physicochemical properties and HMF formation in Turkish honey types. J Food Eng 219:129–136

Ordóñez-Santos LE, Martínez-Girón J, Arias-Jaramillo ME (2017) Effect of ultrasound treatment on visual color, vitamin C, total phenols, and carotenoids content in cape gooseberry juice. Food Chem 233:96–100

Paciulli M, Medina-Meza IG, Chiavaro E, Barbosa-Cánovas GV (2016) Impact of thermal and high pressure processing on quality parameters of beetroot (*Beta vulgaris* L.). LWT-Food Sci Technol 68:98–104

Pankaj SK, Wan Z, Colonna W, Keener KM (2017) Effect of high voltage atmospheric cold plasma on white grape juice quality. J Sci Food Agric 97(12):4016–4021

Pankaj SK, Wan Z, Keener KM (2018) Effects of cold plasma on food quality: a review. Foods 7(1):4

Panopoulos G, Moatsou G, Psychogyiopoulou C, Moschopoulou E (2020) Microfiltration of ovine and bovine milk: effect on microbial counts and biochemical characteristics. Foods 9(3):284. https://doi.org/10.3390/foods9030284

Pasquali F, Stratakos AC, Koidis A, Berardinelli A, Cevoli C, Ragni L, Trevisani M (2016) Atmospheric cold plasma process for vegetable leaf decontamination: a feasibility study on radicchio (red chicory, Cichorium intybus L.). Food Control 60:552–559

Pataro G, Sinik M, Capitoli MM, Donsì G, Ferrari G (2015) The influence of post-harvest UV-C and pulsed light treatments on quality and antioxidant properties of tomato fruits during storage. Innovative Food Sci Emerg Technol 30:103–111

Pedrós-Garrido S, Condón-Abanto S, Clemente I, Beltrán JA, Lyng JG, Bolton D et al (2018) Efficacy of ultraviolet light (UV-C) and pulsed light (PL) for the microbiological decontamination of raw salmon (*Salmo salar*) and food contact surface materials. Innov Food Sci Emerg Technol 50:124–131

Peña-González EM, Alarcón-Rojo AD, Rentería A, García I, Santellano E, Quintero A, Luna L (2017) Quality and sensory profile of ultrasound-treated beef. Ital J Food Sci 29(3):463–475

Peña-Gonzalez E, Alarcon-Rojo AD, Garcia-Galicia I, Carrillo-Lopez L, Huerta-Jimenez M (2019) Ultrasound as a potential process to tenderize beef: sensory and technological parameters. Ultrason Sonochem 53:134–141

Penha CB, Bonin E, da Silva AF, Hioka N, Zanqueta ÉB, Nakamura TU, Mikcha JMG (2017) Photodynamic inactivation of foodborne and food spoilage bacteria by curcumin. LWT-Food Sci Technol 76:198–202

Pérez-Santaescolástica C, Carballo J, Fulladosa E, Munekata PES, Campagnol PB, Gómez B, Lorenzo JM (2019) Influence of high-pressure processing at different temperatures on free amino acid and volatile compound profiles of dry-cured ham. Food Res Int 116:49–56

Pinela J, Ferreira IC (2017) Nonthermal physical technologies to decontaminate and extend the shelf-life of fruits and vegetables: trends aiming at quality and safety. Crit Rev Food Sci Nutr 57(10):2095–2111

Pokhrel PR, Toniazzo T, Boulet C, Oner ME, Sablani SS, Tang J, Barbosa-Cánovas GV (2019) Inactivation of *Listeria innocua* and *Escherichia coli* in carrot juice by combining high pressure processing, nisin, and mild thermal treatments. Innovative Food Sci Emerg Technol 54:93–102

Preetha P, Pandiselvam R, Varadharaju N, Kennedy ZJ, Balakrishnan M, Kothakota A (2021) Effect of pulsed light treatment on inactivation kinetics of *Escherichia coli* (MTCC 433) in fruit juices. Food Control 121:107547

Proctor A (2018) Alternatives to conventional food processing, vol 53, 2nd edn. Royal Society of Chemistry

Raghubeer EV, Phan BN, Onuoha E, Diggins S, Aguilar V, Swanson S, Lee A (2020) The use of high-pressure processing (HPP) to improve the safety and quality of raw coconut (*Cocos nucifera* L) water. Int J Food Microbiol 331:108697

Raj AS, Chakraborty S, Rao PS (2019) Thermal assisted high-pressure processing of Indian gooseberry (*Embilica officinalis* L.) juice–impact on colour and nutritional attributes. LWT 99:119–127

Rajashri K, Roopa BS, Negi PS, Rastogi NK (2020) Effect of ozone and ultrasound treatments on polyphenol content, browning enzyme activities, and shelf life of tender coconut water. J Food Process Preserv 44(3):e14363

Ramazzina I, Berardinelli A, Rizzi F, Tappi S, Ragni L, Sacchetti G, Rocculi P (2015) Effect of cold plasma treatment on physico-chemical parameters and antioxidant activity of minimally processed kiwifruit. Postharvest Biol Technol 107:55–65

Ramos-Villarroel AY, Aron-Maftei N, Martín-Belloso O, Soliva-Fortuny R (2012) The role of pulsed light spectral distribution in the inactivation of *Escherichia coli* and *Listeria innocua* on fresh-cut mushrooms. Food Control 24(1–2):206–213

Ramos-Villarroel A, Aron-Maftei N, Martín-Belloso O, Soliva-Fortuny R (2014) Bacterial inactivation and quality changes of fresh-cut avocados as affected by intense light pulses of specific spectra. Int J Food Sci Technol 49(1):128–136

Rana S, Mehta D, Bansal V, Shivhare US, Yadav SK (2020) Atmospheric cold plasma (ACP) treatment improved in-package shelf-life of strawberry fruit. J Food Sci Technol 57(1):102–112

Rezzadori K, Petrus RR, Benedetti S, Carminatti CA, Petrus JCC (2013) Effects of tangential microfiltration and pasteurization on the rheological, microbiological, physico-chemical and sensory characteristics of sugar cane juice. Int J Food Sci Technol 48(1):1–9. https://doi.org/10.1111/j.1365-2621.2012.03124.x

Rinaldi M, Littardi P, Ganino T, Aldini A, Rodolfi M, Barbanti D, Chiavaro E (2020) Comparison of physical, microstructural, antioxidant and enzymatic properties of pineapple cubes treated with conventional heating, ohmic heating and high-pressure processing. LWT 134:110207

Rode TM, Hovda MB (2016) High pressure processing extend the shelf life of fresh salmon, cod and mackerel. Food Control 70:242–248

Rode TM, Rotabakk BT (2021) Extending shelf life of desalted cod by high pressure processing. Innovative Food Sci Emerg Technol 69:102476

Rodríguez Ó, Gomes WF, Rodrigues S, Fernandes FA (2017) Effect of indirect cold plasma treatment on cashew apple juice (*Anacardium occidentale* L.). LWT 84:457–463

Rojas ML, Leite TS, Cristianini M, Alvim ID, Augusto PE (2016) Peach juice processed by the ultrasound technology: changes in its microstructure improve its physical properties and stability. Food Res Int 82:22–33

Saeeduddin M, Abid M, Jabbar S, Wu T, Hashim MM, Awad FN, Zeng X (2015) Quality assessment of pear juice under ultrasound and commercial pasteurization processing conditions. LWT-Food Sci Technol 64(1):452–458

Saikaew K, Lertrat K, Meenune M, Tangwongchai R (2018) Effect of high-pressure processing on colour, phytochemical contents and antioxidant activities of purple waxy corn (Zea mays L. var. ceratina) kernels. Food Chem 243:328–337

Salve AR, Pegu K, Arya SS (2019) Comparative assessment of high-intensity ultrasound and hydrodynamic cavitation processing on physico-chemical properties and microbial inactivation of peanut milk. Ultrason Sonochem 59:104728

Sarangapani C, Devi Y, Thirundas R, Annapure US, Deshmukh RR (2015) Effect of low-pressure plasma on physico-chemical properties of parboiled rice. LWT-Food Sci Technol 63(1): 452–460

Sarangapani C, O'Toole G, Cullen PJ, Bourke P (2017) Atmospheric cold plasma dissipation efficiency of agrochemicals on blueberries. Innovative Food Sci Emerg Technol 44:235–241

Silveira MR, Coutinho NM, Esmerino EA, Moraes J, Fernandes LM, Pimentel TC et al (2019) Guava-flavored whey beverage processed by cold plasma technology: bioactive compounds, fatty acid profile and volatile compounds. Food Chem 279:120–127

Söbeli C, Uyarcan M, Kayaardı S (2021) Pulsed UV-C radiation of beef loin steaks: effects on microbial inactivation, quality attributes and volatile compounds. Innovative Food Sci Emerg Technol 67:102558

Srimagal A, Ramesh T, Sahu JK (2016) Effect of light emitting diode treatment on inactivation of *Escherichia coli* in milk. LWT-Food Sci Technol 71:378–385

Starek A, Kobus Z, Sagan A, Chudzik B, Pawłat J, Kwiatkowski M, Andrejko D (2021) Influence of ultrasound on selected microorganisms, chemical and structural changes in fresh tomato juice. Sci Rep 11(1):1–12

Stratakos AC, Inguglia ES, Linton M, Tollerton J, Murphy L, Corcionivoschi N, Tiwari BK (2019) Effect of high pressure processing on the safety, shelf life and quality of raw milk. Innovative Food Sci Emerg Technol 52:325–333

Subedi S, Du L, Prasad A, Yadav B, Roopesh MS (2020) Inactivation of salmonella and quality changes in wheat flour after pulsed light-emitting diode (LED) treatments. Food Bioprod Process 121:166–177

Subramanian R, Ichikawa S, Nakajima M, Kimura T, Maekawa T (2001) Characterization of phospholipid reverse micelles in relation to membrane processing of vegetable oils. Eur J Lipid Sci Technol 103(2):93–97. https://doi.org/10.1002/1438-9312(200102)103:2<93::aid-ejlt93>3.0.co;2-9

Sulaiman A, Farid M, Silva FV (2017) Strawberry puree processed by thermal, high pressure, or power ultrasound: process energy requirements and quality modeling during storage. Food Sci Technol Int 23(4):293–309

Sun S, Rasmussen FD, Cavender GA, Sullivan GA (2019) Texture, color and sensory evaluation of sous-vide cooked beef steaks processed using high pressure processing as method of microbial control. LWT 103:169–177

Szczepańska J, Barba FJ, Skąpska S, Marszałek K (2020) High pressure processing of carrot juice: effect of static and multi-pulsed pressure on the polyphenolic profile, oxidoreductases activity and colour. Food Chem 307:125549

Tao T, Ding C, Han N, Cui Y, Liu X, Zhang C (2019) Evaluation of pulsed light for inactivation of foodborne pathogens on fresh-cut lettuce: effects on quality attributes during storage. Food Packag Shelf Life 21:100358

Thirumdas R, Sarangapani C, Annapure US (2015) Cold plasma: a novel non-thermal technology for food processing. Food Biophys 10(1):1–11

Thirumdas R, Saragapani C, Ajinkya MT, Deshmukh RR, Annapure US (2016) Influence of low pressure cold plasma on cooking and textural properties of brown rice. Innovative Food Sci Emerg Technol 37:53–60

Tomadoni B, Cassani L, Viacava G, Moreira MDR, Ponce A (2017) Effect of ultrasound and storage time on quality attributes of strawberry juice. J Food Process Eng 40(5):e12533

Tribst AAL, Júnior BRDCL, de Oliveira MM, Cristianini M (2016) High pressure processing of cocoyam, Peruvian carrot and sweet potato: effect on oxidative enzymes and impact in the tuber color. Innovative Food Sci Emerg Technol 34:302–309

Trivedi MH, Patel K, Itokazu H, Huynh NA, Kovalenko M, Nirenberg G, Sekhon JK (2019) Enhancing shelf life of bananas by using atmospheric pressure pulsed cold plasma treatment of the storage atmosphere. Plasma Med 9(1):23–38

Trzaska WJ, Wrigley HE, Thwaite JE, May RC (2017) Species-specific antifungal activity of blue light. Sci Rep 7(1):1–7

Valdivia-Nájar CG, Martín-Belloso O, Soliva-Fortuny R (2018) Impact of pulsed light treatments and storage time on the texture quality of fresh-cut tomatoes. Innov Food Sci Emerg Technol 45: 29–35

Varilla C, Marcone M, Annor GA (2020) Potential of cold plasma technology in ensuring the safety of foods and agricultural produce: a review. Foods 9(10):1435

Velderrain-Rodríguez GR, Salmerón-Ruiz ML, González-Aguilar GA, Martín-Belloso O, Soliva-Fortuny R (2021) Ultraviolet/visible intense pulsed light irradiation of fresh-cut avocado enhances its phytochemicals content and preserves quality attributes. J Food Process Preserv 45(3):e15289

Vervoort L, Van der Plancken I, Grauwet T, Verlinde P, Matser A, Hendrickx M, Van Loey A (2012) Thermal versus high pressure processing of carrots: a comparative pilot-scale study on equivalent basis. Innovative Food Sci Emerg Technol 15:1–13

Vieira AH, Balthazar CF, Guimaraes JT, Rocha RS, Pagani MM, Esmerino EA, Silva MC, Raices RSL, Tonon RV, Cabral LMC, Walter EHM, Freitas MQ, Cruz AG (2020) Advantages of microfiltration processing of goat whey orange juice beverage. Food Res Int 132:109060. https://doi.org/10.1016/j.foodres.2020.109060

Vollmer K, Chakraborty S, Bhalerao PP, Carle R, Frank J, Steingass CB (2020) Effect of pulsed light treatment on natural microbiota, enzyme activity, and phytochemical composition of pineapple (*Ananas comosus* [L.] Merr.) juice. Food Bioprocess Technol 13:1095–1109

Walkling-Ribeiro M, Rodríguez-González O, Jayaram S, Griffiths MW (2011) Microbial inactivation and shelf life comparison of "cold" hurdle processing with pulsed electric fields and microfiltration, and conventional thermal pasteurisation in skim milk. Int J Food Microbiol 144(3):379–386. https://doi.org/10.1016/j.ijfoodmicro.2010.10.023

Wan Z, Misra NN, Li G, Keener KM (2021) High voltage atmospheric cold plasma treatment of *Listeria innocua* and *Escherichia coli* K-12 on queso fresco (fresh cheese). LWT 146:111406

Wang J, Fan L (2019) Effect of ultrasound treatment on microbial inhibition and quality maintenance of green asparagus during cold storage. Ultrason Sonochem 58:104631

Wang T, MacGregor SJ, Anderson JG, Woolsey GA (2005) Pulsed ultra-violet inactivation spectrum of Escherichia coli. Water Res 39(13):2921–2925

Wang W, Ma X, Zou M, Jiang P, Hu W, Li J, Liu D (2015) Effects of ultrasound on spoilage microorganisms, quality, and antioxidant capacity of postharvest cherry tomatoes. J Food Sci 80(10):C2117–C2126

Wang H, Zhu S, Ramaswamy HS, Hu F, Yu Y (2018) Effect of high pressure processing on rancidity of brown rice during storage. LWT 93:405–411

Wang D, Fritsch J, Moraru CI (2019) Shelf life and quality of skim milk processed by cold microfiltration with a 1.4-μm pore size membrane, with or without heat treatment. J Dairy Sci 102(10):8798–8806. https://doi.org/10.3168/jds.2018-16050

Wang J, Liu Q, Xie B, Sun Z (2020) Effect of ultrasound combined with ultraviolet treatment on microbial inactivation and quality properties of mango juice. Ultrason Sonochem 64:105000

Xiaokang W, Brunton NP, Lyng JG, Harrison SM, Carpes ST, Papoutsis K (2020) Volatile and non-volatile compounds of shiitake mushrooms treated with pulsed light after twenty-four hour storage at different conditions. Food Biosci 36:100619

Xu L, Garner AL, Tao B, Keener KM (2017) Microbial inactivation and quality changes in orange juice treated by high voltage atmospheric cold plasma. Food Bioprocess Technol 10(10): 1778–1791

Yadav B, Spinelli AC, Misra NN, Tsui YY, McMullen LM, Roopesh MS (2020) Effect of in-package atmospheric cold plasma discharge on microbial safety and quality of ready-to-eat ham in modified atmospheric packaging during storage. J Food Sci 85(4):1203–1212

Yang H, Gao J, Yang A, Chen H (2015) The ultrasound-treated soybean seeds improve edibility and nutritional quality of soybean sprouts. Food Res Int 77:704–710

Yi J, Kebede BT, Dang DNH, Buvé C, Grauwet T, Van Loey A, Hendrickx M (2017) Quality change during high pressure processing and thermal processing of cloudy apple juice. Lwt 75: 85–92

Yildiz S, Pokhrel PR, Unluturk S, Barbosa-Cánovas GV (2020) Changes in quality characteristics of strawberry juice after equivalent high pressure, ultrasound, and pulsed electric fields processes. Food Eng Rev 30:1–12

Yodpitak S, Mahatheeranont S, Boonyawan D, Sookwong P, Roytrakul S, Norkaew O (2019) Cold plasma treatment to improve germination and enhance the bioactive phytochemical content of germinated brown rice. Food Chem 289:328–339

Yuan B, Danao MGC, Stratton JE, Weier SA, Weller CL, Lu M (2018) High pressure processing (HPP) of aronia berry purée: effects on physicochemical properties, microbial counts, bioactive compounds, and antioxidant capacities. Innovative Food Sci Emerg Technol 47:249–255

Yuste J, Capellas M, Pla R, Fung DY, MOR-MUR MONTSERRAT (2001) High pressure processing for food safety and preservation: a review 1. J Rapid Methods Autom Microbiol 9(1):1–10

Zhang Y, Liu X, Wang Y, Zhao F, Sun Z, Liao X (2016) Quality comparison of carrot juices processed by high-pressure processing and high-temperature short-time processing. Innovative Food Sci Emerg Technol 33:135–144

Zhang H, Pan J, Wu Z (2018) Investigation of the effects of high pressure processing on the process of rigor in pork. Meat Sci 145:455–460

Zhang ZH, Wang LH, Zeng XA, Han Z, Brennan CS (2019) Non-thermal technologies and its current and future application in the food industry: a review. Int J Food Sci Technol 54(1):1–13

Zhang H, Tsai S, Tikekar RV (2021a) Inactivation of Listeria innocua on blueberries by novel ultrasound washing processes and their impact on quality during storage. Food Control 121: 107580

Zhang J, Yagoub AEA, Sun Y, Mujumdar AS, Ma H, Wahia H, Zhou C (2021b) Intensive pulsed light pretreatment combined with controlled temperature and humidity for convection drying to reduce browning and improve quality of dried shiitake mushrooms. J Sci Food Agric 101(13): 5608–5617

Zhao D, Barrientos JU, Wang Q, Markland SM, Churey JJ, Padilla-Zakour OI, Worobo RW, Kniel KE, Moraru CI (2015) Efficient reduction of pathogenic and spoilage microorganisms from apple cider by combining microfiltration with UV treatment. J Food Prot 78(4):716–722. https://doi.org/10.4315/0362-028X.JFP-14-452

Zhao N, Ge L, Huang Y, Wang Y, Wang Y, Lai H et al (2020) Impact of cold plasma processing on quality parameters of packaged fermented vegetable (radish paocai) in comparison with pasteurization processing: insight into safety and storage stability of products. Innovative Food Sci Emerg Technol 60:102300

Ziuzina D, Misra NN, Han L, Cullen PJ, Moiseev T, Mosnier JP, Bourke P (2020) Investigation of a large gap cold plasma reactor for continuous in-package decontamination of fresh strawberries and spinach. Innovative Food Sci Emerg Technol 59:102229

Zou Y, Jiang A (2016) Effect of ultrasound treatment on quality and microbial load of carrot juice. Food Sci Technol 36:111–115

Chapter 7
Recent Developments in Thermal Processing of Foods

K. R. Jolvis Pou and Vijaya Raghavan

1 Introduction

The application of conventional thermal treatment in food processing and preserving is an age-old practice and it is still widely used in processing different kinds of food in the food industry and for domestic purposes. Thermal processing considers the combination of temperature and holding time required to inactivate food spoilage microorganisms and/or to obtain the desired characteristics of the food products. Thermal treatment is most commonly applied to enhance the safety, eating quality, and shelf-life of foods. Thermal processing involves the manufacture, preservation, and transformation of food products (Wang and Sun 2006). Commonly, food products are subjected to thermal treatments (drying, pasteurization, sterilization, cooking, baking, frying, blanching, evaporation, etc.) by heating to a higher temperature using the traditional well-established techniques. During the treatments, the huge amount of energy transferred to the products may generate undesirable reactions in the food system consequently leading to unfavorable modifications in the food products quality parameters. Thermal processing can effectively reduce a wide range of microorganisms but its impact on the quality attributes of food is of great concern. It is; therefore, the development of innovative thermal processing techniques is imperative to address the issues of detrimental quality changes in thermally unstable food products. Currently, there are several novel food processing technologies implemented in the food industry, yet the thermal process remains as one of the most indispensable pathways towards food safety. Thus, the improvement of the existing technologies and development of the advanced methods would play a vital role to replace or complement the conventional practices. Traditionally, foods were

K. R. J. Pou · V. Raghavan (✉)
Department of Bioresource Engineering, McGill University, Sainte-Anne-de-Bellevue, Montreal, QC, Canada
e-mail: vijaya.raghavan@mcgill.ca

H. U. Hebbar et al. (eds.), *Engineering Aspects of Food Quality and Safety*, Food Engineering Series, https://doi.org/10.1007/978-3-031-30683-9_7

processed primarily for taste, convenience, and appearance. The concept of processing pertinent to health benefits in the food products is comparatively a new trend (Choudhury 2017). Consumers tendencies concerning the quality and safety of food have motivated researchers to investigate the development of alternative thermal food processing techniques.

In recent years, electro-heating technologies such as microwave, ohmic, and radio frequency heating of foods have emerged with great potential to replace and/or complement the conventional thermal techniques with a wide range of applications at industrial scale and domestic uses. Electro-heating can be categorized as direct and indirect electro-heating. In the former method, electrical current is directly applied to the food products (ohmic heating), whereas in the latter, the electrical energy is first transformed to electromagnetic radiation and then allow to generate heat within the food system (microwave or radio frequency heating) (Osaili 2012). These innovative electro-heating approaches are volumetric forms of heating that generate thermal energy within the food products. This form of heat generation offers several benefits such as control of excessive processing time, deliver safe food, retention of nutritional and sensory characteristics, and more energetic and heating efficiency as compared to well-established thermal techniques (Atuonwu and Tassou 2019; Osaili 2012). Traditional thermal techniques depend predominantly on the heat generated outside the food products by fuel combustion or by an electric heater, and its transmission into the sample through convection and conduction mechanisms (Osaili 2012). Thermal food processing, it may be conventional or electro-heating, can be grouped into unit operations such as blanching, drying, thawing, cooking, evaporation, pasteurization, sterilization, etc., and involve heating the product to a certain temperature depending on the objective of the thermal treatment (Delgado et al. 2006). There is extensive use of microwave heating of foods both at home and industrial scale while ohmic heating is primarily confined to industrial applications. Microwave and ohmic heating are being currently used substantially for food processing and preservation. In this study, we have outlined the mechanisms and recent applications of novel thermal processing technologies (microwave and ohmic heating) in the food sector to provide some necessary information to the readers. This chapter emphasizes the impacts of electro-heating techniques namely microwave and ohmic heating of food products on the safety and quality attributes during pasteurization, sterilization, cooking, drying, thawing, and tempering of food products.

2 Microwave Heating

Microwave heating of foods has emerged to be an indispensable process in the twenty-first century in countries like Canada, the United States, and many European countries (Orsat et al. 2017). Microwaves are electromagnetic spectrum with frequencies ranging from 300 MHz to 300 GHz. Certain frequencies within this range are set aside by the International Telecommunications Union for domestic, industrial, medical, and scientific applications (Raghavan et al. 2005). The allocation of

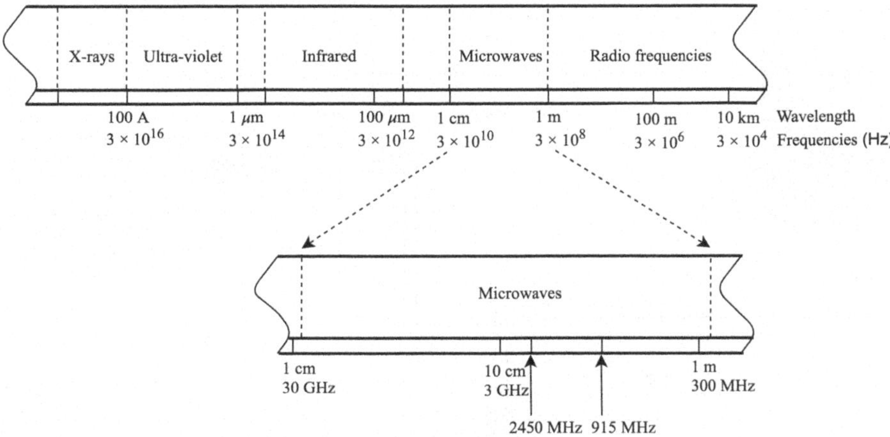

Fig. 7.1 Position of microwave on the electromagnetic spectrum. (Redrawn from Dev et al. 2012)

frequencies for various applications is varied from country to country. The most commonly used frequencies in the Americas for domestic and industrial applications are 2450 and 915 MHz, respectively. The Netherlands, Germany, Austria, Portugal, and Switzerland generally used 433.92 MHz and 896 MHz in the United Kingdom for industrial, scientific, and medical purposes (Orsat et al. 2017). The position of microwave on the electromagnetic spectrum is shown in Fig. 7.1 (Dev et al. 2012).

Generally, microwave heating system consists of a high voltage power source, a cavity magnetron, a high voltage capacitor, control panel, short waveguide, and magnetron control circuit. A typical microwave heating system set up is shown in Fig. 7.2 (Dev et al. 2008). Microwave heating primarily depends on the electric field which is responsible for heat generation by interacting with molecules through two modes viz. dipolar rotation and ionic conduction. The presence of moisture in the food products triggers dielectric heating because of the dipolar nature of water. The dipoles within the food materials align themselves when an external oscillating electric field is applied. This alignment occurs a million times per second when a high frequency electric field is applied and instigates internal friction leading to volumetric heating (Chandrasekaran et al. 2013; Kingston and Haswell 1997; Thuery 1992). Microwave heating may also occur due to ionic conduction in which a free ion or ionic species migrates and attempt to realign with the applied oscillating electric field. The friction produced between these migrating species leads to the generation of heat. The efficiency of the rate of heat generation depends on the polarity and/or ionicity of a species (Raghavan et al. 2005). Microwave heating and distribution of heat depend on several factors. The dielectric properties and the depth of penetration are considered to be the most important factors (Chandrasekaran et al. 2013). When microwaves are applied to a food sample, part of its energy is reflected, and a portion of the energy is transmitted through the surface of the product and part of it is absorbed. The amounts of energy fall into these categories are defined in terms of dielectric properties (Venkatesh and Raghavan 2005). The dielectric constant which is the real part of the dielectric property indicates the ability of electrical

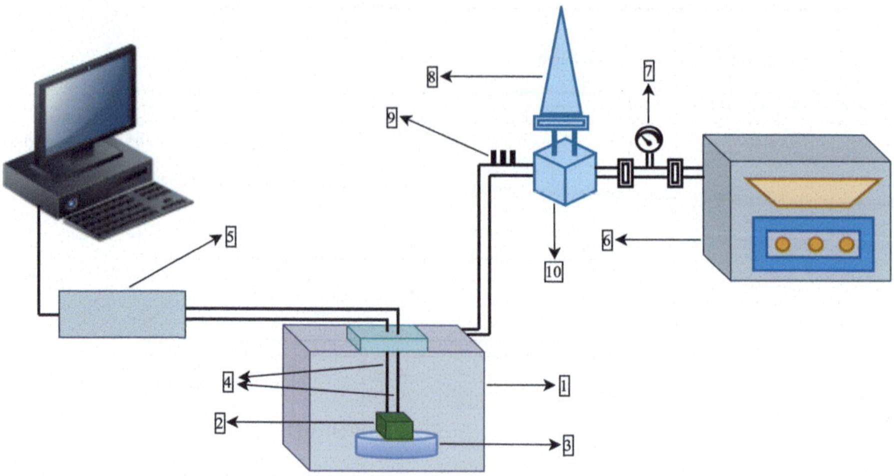

Fig. 7.2 A typical microwave heating system set up, where (1) microwave cavity, (2) food sample, (3) sample holder, (4) fiber optic probes, (5) data collector, (6) microwave generator, (7) microwave meter, (8) microwave absorber, (9) turning screw, and (10) circulator. (Redrawn from Dev et al. 2008)

energy storage and the imaginary part, known as dielectric loss, implies the capability to convert electric energy into heat (Chandrasekaran et al. 2013). The primary dielectric property of a product is the complex relative permittivity of the material. It can be expressed as shown in Eq. 7.1 (Venkatesh and Raghavan 2004, 2005).

$$\varepsilon^* = \varepsilon' - j\varepsilon'' \tag{7.1}$$

Where ε^* is the relative permittivity, ε' and ε'' are dielectric constant and dielectric loss factor, respectively, and $j = \sqrt{-1}$. The absolute permittivity of a vacuum, ε_o, can be expressed as shown in Eq. (7.2) (Venkatesh and Raghavan 2004, 2005).

$$C_o \mu_o \varepsilon_o = 1 \tag{7.2}$$

Where C_o is the speed of light and μ_o is the magnetic constant.

The numerical values of ε_o and μ_o are 8.854×10^{-12} F/m and 1.26×10^{-6} H/m, respectively. The permittivity values of products (solid, liquid, and gaseous) are higher and can be determined relative to the value in vacuum. The relative permittivity of a material, ε_r, is determined as shown in Eq. (7.3) (Venkatesh and Raghavan 2004, 2005).

$$\varepsilon_r = \frac{\varepsilon_{abs}}{\varepsilon_o} \tag{7.3}$$

Where ε_{abs} is the absolute permittivity of the material.

The microwave penetration depth, d_p, from the dielectric surface to inside can be defined as the depth into a product where the microwave power has declined to

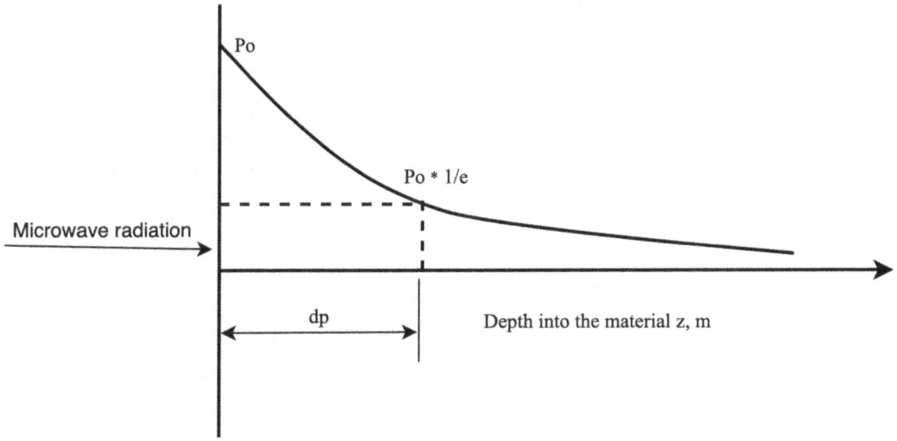

Fig. 7.3 Graphical illustration of the concept of microwave penetration depth. (Redrawn from Dev et al. 2012)

36.8% (or 1/e) of its transmitted amount. It is also expressed as the distance at which the power density has decreased to 50% of transmitted power. A graphical demonstration of the concept of microwave penetration depth is represented in Fig. 7.3 (Dev et al. 2012). The microwave penetration depth can be calculated using Eq. (7.4) (Venkatesh and Raghavan 2005).

$$d_p = \frac{\lambda_o \sqrt{\varepsilon'}}{2\pi\varepsilon''} \tag{7.4}$$

Where λ_o is the free space microwave wavelength ($\lambda_o = 122$ mm for 2450 MHz). In most of the food products, the ε'' value is less than 25, this indicates a penetration depth of 6–10 mm (Venkatesh and Raghavan 2005).

In recent years, there has been much research and development on the applications of microwave technology in the food sector. There is extensive use of microwave heating of foods at home and in the industry. Currently, microwave heating is being applied in several operations such as drying, cooking, pasteurization, sterilization, thawing, reheating, baking, frying, blanching, roasting, etc. (Orsat et al. 2017). Research has demonstrated the promising applications of microwave technology combined with other thermal and non-thermal methods in food processing and preservation. Microwave heating has exhibited to be an efficient alternative and/or complementary technique of conventional thermal treatment of foods. Some of the applications of microwave heating in food processing are summarized in Table 7.1. The advancements of microwave drying, pasteurization, sterilization, cooking, thawing, and tempering of food products are discussed in the following sections.

Table 7.1 Microwave heating of food products

Microwave treatment condition	Purpose	Product	Effect	Reference
Microwave oven (0.75, 1.0, and 2.0 W/g)	Pasteurization	Egg	The results indicated that the eggshell and shell membrane have excellent transparency to microwaves (in their dielectric properties), thus assisting as a suitable container for microwave treatment. The egg yolk heated up faster than the egg white when processed within the shell. The microwave technique emerged to perform better in terms of the product quality and process time when compared with the water bath method.	Dev et al. (2008)
Hot air (60 °C) freeze drying (50 pa and − 40 °C), and microwave-freeze drying (1.6, 2, and 2.3 W/g).	Drying	Sea cucumber	It was observed that the microwave-freeze drying decreased the drying time by about 50% of the freeze drying and resulted in a comparable product quality. The freeze drying method resulted in the best quality yet the long drying time and overall cost of operation is the concern.	Duan et al. (2010)
Microwave (medium and high-power levels for 10, 20, 30, 40, and 50 s)	Sterilization	Grape tomatoes	The bacteria elimination was more evident at a high microwave power level. However, treatment at the medium-power level for 40 s was found to have the overall best effect in terms of nutritional values, pH, color, texture, and reduced *salmonella enterica* by 1.45 log cycles.	Lu et al. (2011)
Microwave (0.5, 1.0, and 2.0 W/g) assisted hot air (40, 50, and 60 °C) and hot air for 3600 s	Drying	Canola seeds	The drying rate increased from 4.3×10^{-5} to 6.8×10^{-5} kg water/kg sample/s during the hot air drying as the drying	Hemis et al. (2015)

(continued)

Table 7.1 (continued)

Microwave treatment condition	Purpose	Product	Effect	Reference
			temperature increased from 40 to 60 °C, however it remained unchanged during microwave-hot air drying. It indicated that the microwave coupled with hot air at low humidity resulted in a higher drying rate.	
Microwave (500 W)-steam (1700 W) and steam for 720 s	Cooking	Purple sweet potatoes	The combined use of microwave and steam resulted in a higher concentration of flavonoids, total phenolics, anthocyanins, phenolic acids, and soluble sugar while significantly shortened the cooking time as compared to steaming alone.	Xu et al. (2016)
Microwave (500 W and 1 kW and between −5 and − 3 °C) and radio frequency (electrode gaps of 160 and 150 mm and between −5 and − 3 °C)	Tempering	Shrimp	The time required to attain the desired temperature (between −5 and − 3 °C) in the product was found to be 4 and 10 min for 1 kW and 500 W, respectively, and 7 and 11 min for 150 and 160 mm, respectively. Among all the methods, microwave (500 W) resulted in the most uniform distribution of internal temperature, although local surface heating was detected at both the power levels considered.	Palazoğlu and Miran (2017)
Microwave (360, 630, and 900 W) and infrared (50, 60 and 70 °C)	Drying	Blueberry leaves	The analysis indicated that the total phenol and antioxidant capacity increased by 12 and 9 times, respectively in the microwave dried sample as compared to the infrared dried sample. Also, catechins and	Borda-Yepes et al. (2019)

(continued)

Table 7.1 (continued)

Microwave treatment condition	Purpose	Product	Effect	Reference
			epicatechin concentrations were observed to be more in microwave drying.	
Ultrasound (400 W, 40 kHz, and 10 °C)-microwave (440 W, 2450 Hz, and 4 °C) and ultrasound-far-infrared (300 W and 10 °C)	Thawing	Red drum	The results showed that the combined application of ultrasound with microwave and far-infrared could retain more muscle fiber structure, higher protein stability as compared to when these methods were applied as alone, and immobilized water had no significant changes compared with the fresh product.	Cai et al. (2019)
Microwave (4, 8, and 10 W/g for 180 s) and conventional cooking	Cooking	Rice starch	The microwave treatment at higher power levels (8 and 10 W/g) induced more order and periodic amorphous crystalline structures. Comparatively, microwave processed starches have higher slowly digestible starch levels and a lower rate of digestion.	Li et al. (2019a)
Magnetic nanoparticles (0.1 mg/mL; Fe_3O_4, 10–30 nm, 25% in H_2O)-microwave (300 W and 2450 MHz at 0 °C) and magnetic nanoparticles-far-infrared (300 W at 0 °C)	Thawing	Red seabream fillets	The quality of freshness and safety after thawing and the thawed sample after 6 h of storage were evaluated based on volatile amino nitrogen content, the total number of viable colonies, pH, formaldehyde, trimethylamine, putrescine, and dimethylamine. The results showed that the combined methods could retain better freshness than far-infrared and microwave alone, and the growth in amines concentration was significantly reduced.	Cai et al. (2020)

(continued)

Table 7.1 (continued)

Microwave treatment condition	Purpose	Product	Effect	Reference
Microwave-infrared radiation (4–12 W/g, 75–375 °C, and 15–75 s)	Sterilization	Parika	It was observed that the combined treatment resulted in 8.849 log reduction of *salmonella* typhimurium and a 7.372 log reduction of *aspergillus flavus.* The short processing time intensive microwave-infrared heating can be used as an effective method of pathogenic bacteria inactivation in the presence of natural microbial flora.	Shirkole et al. (2020)
High pressure (600 MPa at 25 °C for 600 s) and microwave-thermal (70 °C for 120 s)	Pasteurization	Green beans	Microwave assisted treatment resulted in a higher reduction of *listeria innocua* (9.0 log CFU/g) as compared to high pressure processing (3.7 log CFU/g). Both the processes have no significant effect on total chlorophyll content, firmness, and pH; however, vitamin C content was found to be more in the microwave assisted treated product.	Inanoglu et al. (2021)
Microwave-infrared radiation (150, 350, and 390 °C and 10 W/g for 120 s)	Pasteurization	Paprika	Microwave assisted infrared radiation reduced *salmonella* typhimurium and *aspergillus flavus* by 7.389 and 6.182 log CFU/g, respectively. The incubation of the sample for 30 s after the treatment was observed to be effective in improving the bactericidal effect. Less exposure to oxygen during the rapid treatment led to a reduction in quality degradation.	Shirkole et al. (2021)

2.1 Microwave Drying

The methods of drying of agricultural products can be categorized into four generations: (a) first and second-generation drying comprises different convective drying techniques (first generation: kiln, cabinet, belt, and conveyor dryers; second generation: drum and spray dryers), (b) the third generation includes osmotic and freeze drying, and (c) microwave and radio frequency related drying methods are considered as a novel and fourth generation drying (Kumar and Karim 2019). In general, the selection of drying techniques depends on the characteristics of the materials and socio-economic conditions. Energy consumption is of great concern in the selection of drying methods and with the significant rise in fuel price, it becomes a critical issue in all the processing sectors. Electrical energy is an important alternative source of energy for drying food products, particularly where electricity generation is based on renewable energy sources (Orsat et al. 2006). Microwave energy induces an effective gradient of vapor pressure and drives the moisture to the surface of the products (Lv et al. 2019). The moisture can be removed from the surface by passing the air over the surface of the products. Whenever microwave drying is mentioned, in many cases it infers to microwave convective drying. The drying air temperature can be varied depending on the characteristics of the product and the desirability of the end-product. The temperature of the products can be controlled by monitoring either power density (W/g) or duty cycle (power time on/off) (Orsat et al. 2006). In general, microwave drying can accomplish the four key requirements in the drying of food products: energy efficiency, speed of operation, operation cost, and quality of dried products (González-Cavieres et al. 2021; Mousakhani-Ganjeh et al. 2021; Wray and Ramaswamy 2015; Zhang et al. 2006).

Microwave power and the drying temperature are the two main factors in microwave drying which have significant impacts on the drying parameters such as speed, time, efficiency, and the quality of the dried product. In this regard, Li et al. (2010a) have developed a microwave drying system with the facility to control power and temperature automatically for drying of apple. The study suggested that the drying experiment with fixed power without temperature control caused inconsistent temperatures. The combination of feedback temperature control with variable microwave power resulted in the best temperature control and the quality of the final product. On the other hand, variable microwave power levels without feedback temperature control triggered a small variation in temperature but resulted in low repeatability. It is; therefore, control of temperature is always suggested for microwave drying of agricultural products. The change in the electrical properties of the products during microwave drying inferred the alteration in the microwave absorption ability of products, consequently, the product temperature and microwave power were attuned with the change in microwave absorption capacity of the materials. The research suggested that the loss factor and dielectric constant of the potato chips declined during microwave drying, and faster they decreased with the rise in drying temperature. Three-stage different temperature control with prefixed variable microwave power levels was found to be the most appropriate method for drying potato chips (Luo et al. 2019). Xu et al. (2018) have investigated the influence of average temperature gradient (ATG) on the quality attributes and drying rate during microwave combined with hot air drying of fresh carrot cubes. The drying experiments were conducted with three variables namely product temperature (60–80 °C), temperature difference (10–30 °C), and cube size (8–18 mm). The results indicated that the rehydration capacity and color difference decreased and increased in drying time with the decrease of ATG. Both the lowest and highest ATG caused degradation in the quality of the sample. The ATG of 6 °C/mm was found to be optimum and the linear control method developed was observed to have similar quality and drying time with this optimum condition. Li et al. (2010b) have reported that a fixed drying temperature could not accomplish the best drying result when energy efficiency, drying time, the occurrence of charring, and sample volatiles (carrot) were all taken into consideration. To address this issue, they applied fuzzy logic and linear control techniques which enhanced the drying effects, and the acceptable quality was achieved in the final product in terms of taste, color parameters, and overall appearance. The relative humidity is another important factor to be controlled during microwave drying of agricultural products. In an attempt to control and monitor the relative humidity of the convection air, Pu et al. (2016) have developed a new microwave drying system. In this study, two approaches were used to monitor the relative humidity: (a) controlling of air flow rate and (b) addition of complementary water. The ambient air was forced into the sample container and removed the gaseous moisture. The authors proposed nine schemes with different rates of air flow to control the relative humidity. In case of complementary water approach, 15, 25, and 35 mL of water were added in the container and air flow rate was maintained at 4 L/min. During the microwave operation, the heated water caused the vapor to nearly saturation level. Addition of complementary water in the container provided high humidity for different time length during the microwave drying process. Results indicated that the application of a complementary water scheme provides the best quality of the dried product in terms of odor, shape, rehydration ratio, and consistency. On the other hand, the excessive complementary water method is not recommended as it reduced the rehydration capacity. Accelerated and decelerated air flow methods should be avoided if the material quality is of ultimate consideration.

Microwave is commonly combined with vacuum, hot air, freezing, osmotic, fluidized bed, and spouted bed drying to enhance the drying performance (Kumar and Karim 2019). Nordin et al. (2014) have investigated the changes in the quality attributes of red pitaya slices during microwave-hot air, microwave-vacuum, and hot air drying. The results indicated that the microwave-vacuum drying resulted in the best visual appearance. Yet there were no significant differences among the drying methods considered in terms of ascorbic acid content and rehydration ratio. The combination of microwave-vacuum drying and microwave-hot air drying reduced the drying time by 83% and 50%, respectively as compared to hot air drying. The modelling study of microwave assisted heated air drying of high oil content products namely canola, corn, and soybean exhibited reduced rate of drying with the increase of drying temperature inside the microwave cavity due to reduction in differential vapor pressure between the sample and the ambient air. The oilseeds developed small cracks when they were exposed to high microwave power levels. It was observed that the oilseeds can be safely dried without damaging or cracking by controlling the air temperature and at low microwave power levels (Hemis et al. 2016). Dev et al. (2011) have studied the effects of hot air (50, 60, and 70 °C) and microwave (1 W/g) coupled with hot air drying on the drying kinetics and quality characteristics of *Moringa oleifera* pods (drumsticks). The microwave assisted hot air drying at 60 °C was observed to be the optimum condition in terms of the product color parameters and rehydration capacity. On the other hand, microwaves combined with heated air at 50 °C retained the highest amount of volatiles. Further, the research indicated that the use of microwave and hot air as a combined method accomplished the drying process five times faster as compared to hot air drying, hence the drying time was reduced by more than 80%. Similar trends were reported by Md Salim et al. (2017) during the hot air (40, 50, and 60 °C) and microwave combined hot air drying of broccoli stalk slices. The microwave-hot air drying resulted in better color parameters and significantly reduced the drying time by 55% as compared to hot air drying alone. Vacuum assisted microwave drying could remove moisture at a relatively low drying temperature by decreasing the vapor pressure in the microwave cavity. Thus, heat-sensitive agricultural products could retain better quality parameters such as nutritional values, color, texture, rehydration ratio, and density. Moreover, this method is sometimes more effective as compared to freeze drying or hot air drying. The application of osmotic dehydration as a pre-treatment in microwave-vacuum drying of food products results in saving energy with higher nutrients content, better appearance, and flavor (Corrêa et al. 2011). Song et al. (2020) have reported that the osmotic dehydration followed by microwave-vacuum drying of blackberries significantly reduced the drying time by 10–15% and enhanced anthocyanin retention. Similar results were reported by Erle and Schubert (2001) during microwave-vacuum drying (strawberries and apples) combined with osmotic dehydration as a pre-treatment. This hybrid approach can also enhance the molecular structure of the products. Jiang et al. (2017) have conducted a comparative study on freeze-microwave-vacuum drying, freeze drying, hot air drying, microwave-vacuum drying, and microwave-vacuum-hot air drying of functional okra snacks. The results implied that the freeze-microwave-vacuum drying process produced better color and texture, the lesser time required to accomplish the drying, and higher energy saving. Principal component analysis of sensory parameters, antioxidant properties, and efficiency of drying suggested that freeze-microwave-vacuum drying is a promising approach for the drying of functional okra snacks.

One of the main drawbacks of microwave drying is the non-uniform distribution of product temperature owing to the uneven distribution of electromagnetic energy within the drying cavity which leads to non-uniform heat generation. This problem, for the particulate materials, can be addressed by spouting or fluidizing the products. In addition, the fluidizing or spouting air not only removes the surface moisture which enhances the rate of mass transfer but also it acts as a source of heat for moisture evaporation (Jumah and Raghavan 2001). Hu et al. (2017) have studied the optimization of process parameters of microwave assisted fluidized bed drying of fresh carrot slices. The optimized conditions were observed to be at 0.44 W/g and 55 °C for microwave power density and air temperature, respectively, based on the drying time and color components. Further, the pre-treatments (osmotic dehydration, water blanching, and citric acid solution) during drying at the optimized conditions reduced the drying time and improved the color and textural strength of the final products. The research revealed that microwave assisted pulse spouted freeze drying has the benefits of producing better quality products, reducing drying time, saving energy, and improving the capacity of processing which delivered a foundation for industrial scale production. The hybrid microwave and freeze drying of Chinese yam with spouting interval of 10 min were observed to reduce the drying time by 41.38% and save energy consumption up to 34.4% as compared to freeze drying alone (Li et al. 2019b). Microwave spouted bed drying did not considerably change the color, infrared spectrum, solubility, and rheological properties of the powdered dried carrageenan. This indicated that microwave assisted spouted bed drying is an effective technique for the processing of carrageenan powder (Serowik et al. 2018). The combined use of microwave and hot air fluidized bed could be a promising alternative way of processing parboiled rice without steaming. Besides improving the drying rate, microwaves enable gelatinization of the starch within the kernels, which is particularly difficult for the hot air drying process (Saniso et al. 2020). Several researches have indicated the microwave assisted drying concept as an effective technique to reduce drying time and retain the inherent properties of agricultural products.

2.2 *Microwave Pasteurization and Sterilization*

In general, pasteurization refers to mild heat treatment (<100 °C) to inactivate some enzymes and to destroy vegetative pathogenic microorganisms. Its main purposes are to eliminate pathogenic microorganisms to reduce public health hazards and to prolong the shelf life of products for several days or weeks in low acid food products such as milk (pH $>$ 4.5). In high acid foods (pH $<$ 4.5), such as fruit juices, it is aimed to extend the shelf life for several weeks by destroying the food spoilage microorganisms (molds or yeast) and/or inactivation of enzymes. The two key elements to achieve proper pasteurization are the heating temperature and the holding time (Fellows 2009). Pasteurization temperature and residence time vary depending on the target microorganism, pH of the food, and the nature of the product. In most cases, the product is heated up to 60–85 °C and holding for few seconds to an hour, for example, for the destruction of pathogens (*Mycobacterium tuberculosis, Brucella abortis, and Coxiella burnetti*) and inactivation of enzymes and microorganisms in milk (63 °C for 30 min or 71.7 °C for 15 sec or 88.3 °C for 1 sec or 90 °C for 0.5 sec) and for the deactivation of enzymes (polygalacturonase and pectin methylesterase) and destruction of spoilage microorganisms (yeasts and molds) in fruit juice (65 °C for 30 min or 77 °C for 1 min or 80 °C for 10–60 sec) (Fellows 2009; Laguerre and Hamoud-Agha 2020).

Sterilization is considered as one of the most effective food preservation techniques and has been used extensively worldwide. Comparatively, the sterilization process involves the application of high intensity heat (121–140 °C). Sterilization commonly implies a wet heat treatment of about 121 °C for 15 min or its equivalent time-temperature combination to eliminate spores of *Clostridium botulinum* and other spore forming pathogenic microorganisms. In other words, sterilization also means that every food particle should receive a sufficient amount of heat in order to achieve a proper sterilization process. Thus, the rate of heat transfer through the product is necessary to take into account while evaluating the effect of heat destruction (Deák 2014). The main aim of sterilization is to completely eliminate all the food spoilage microorganisms including spores thereby extending the shelf life and enables the products to store and distribute at ambient temperature. However, in practice, a complete sterile product may not be possible. Based on the exponential microbial reduction population, the absolute sterility could not be achieved and there would be a non-zero survivor. The probability of survival can be reduced to an acceptable level which has been set at 10^{-12} part survival of spores of *Clostridium botulinum*, known as the 12D concept. Yet, some high heat resistant spore forming microorganisms such as *Geobacillus stearothermophilus* and *Clostridium thermosaccharolyticum* may still survive but being thermophilic in nature cannot grow under normal storage condition (ambient temperature) which is called as commercial sterility (Deák 2014). Conventional sterilization can effectively reduce microbial hazards but usually suffers serious quality degradation of the products. The standard to assess the commercial sterilization process is the reticence of microbial growth in the product and not their absence or presence. The process is also limited by the packaging condition which could result in recontamination. The efficiency of sterilization is determined based on the destruction of the most heat resistance spore forming bacteria in products (Li and Farid 2016).

In semi-solid or solid foods, during conventional heating, the heat transfer occurs from the surface to the center point commonly referred to as cold point through conduction. This process often requires using severe conditions to achieve the target temperature at the cold point. This may result in overheating at the surface and leads to quality degradation of the products. This concern can be addressed by the application of microwave heating which possessed direct and volumetric interaction between microwaves and food products. Microwave heating has the potential to overcome the drawback imposed by the sluggish thermal diffusion of conventional heating (Laguerre and Hamoud-Agha 2020). Microwaves deactivate food spoilage microorganisms by two mechanisms: non-thermal effects and thermal effects. The non-thermal effects of microwave energy on the inactivation of microorganisms can be explained by four theories: magnetic field coupling, cell-membrane rupture, electroporation, and selective heating (Osaili 2012). The use of microwave energy is rapidly attracting as a feasible approach of pasteurization and sterilization of food products. Investigations on microwave assisted pasteurization and sterilization have been driven by the rapid and efficient heating and the possibility of higher retention of inherent properties of foods as compared to conventional heating methods. Studies have indicated that the use of static mixers at the exit of the microwave applicators could improve the temperature distribution during microwave pasteurization and sterilization of foods (Kumar et al. 2008; Sumnu and Sahin 2012). The uniform distribution of heat and the durability of machinery are crucial in the microwave system. The knowledge of electric field patterns within the microwave assisted heating system cavities would play a vital role in designing effective tray carriers. Jain et al. (2018) have developed a computer simulation model and used it as a tool for designing various food package carriers for the improvement of microwave heating uniformity. Results showed that the pattern of the electric field inside the cavities can be altered by varying the designs of tray carriers. It was observed that for 10 oz. tray carriers, the most effective conditions were achieved when a metal frame around the product package was appended in the tray carrier. The use of a metal plate facilitated the uniform distribution of microwave power in the tray center. The heating pattern in food packages is controlled by the dominant electric field section inside the microwave heating cavity. The heating uniformity could be achieved by adjusting the dimension of the cavity in the direction of the dominant component. Regulating the phase of the standing wave inside the microwave heating cavity can increase the uniform heat distribution in the thickness direction of the product package (Luan et al. 2016).

Microwave pasteurization is preferred over the conventional method for the main reason being fast and involves minimum come-up time to the required temperature (Ahmed and Ramaswamy 2007). Siguemoto et al. (2018) have reported that microwave heating achieved the desired temperature much faster than a tubular heat exchanger during pasteurization of cloudy apple juice. As compared to conventional treatment with similar holding times, the microwave pasteurized cloudy apple juices were observed to have higher retention of volatile compounds and comparable ratios of sugar and acid to that of the fresh sample (Siguemoto et al. 2019). These studies suggest that microwave treatment could reduce the flavor degradation linked to traditional heating since the rate of heating is much higher in microwave techniques. The shelf life of the pasteurized products depends on the processing parameters and storage conditions. Recently, microwave assisted pasteurization of food products has emerged to address the needs for fresh-like and safe food products. Hong et al. (2021) have developed a simplified approach to assist the food industry in the production of microwave assisted pasteurization of ready-to-eat foods. A user-friendly chart was developed based on analytical calculations to easily predict the relationship between package thickness, dielectric loss factor, and heating rates. The developed chart would be helpful for the food processors and product developers in optimizing the product recipes and scheduling the process for the microwave assisted pasteurization of foods. Microwave assisted thermal treatment resulted in superior quality over conventional hot water pasteurization in terms of microbial safety and higher retention of color, ascorbic acid, and chlorophyll during storage of green beans. This suggests that microwave-thermal pasteurization is a promising alternative method to produce quality and safe vegetable products (Qu et al. 2021). Compared with an equivalent hot water method, microwave assisted pasteurization greatly reduced the treatment time, minimize the cooked-like, and enhanced the quality uniformity of the pre-packaged carrots (Peng et al. 2017). Microwave technology showed a promising application in the beverage industry. Microwave assisted pasteurization of beverages could accomplish the process in a relatively short time with the potential of better product quality and microbial safety. González-Monroy et al. (2018) have reported that microwave pasteurization at 490 W and 2450 MHz and holding time of 12 s could achieve the desired temperature (90 °C) for the elimination of pectinmethylesterase in the green beverage and tamarind beverage within 4.25 and 4.67 min, respectively. Further, the results suggested that microwave assisted pasteurization did not influence the sensory and physicochemical characteristics of the tamarind beverage. However, microwave assisted heating affected the sensory and color properties of the green beverage. Microwave pasteurization reduced *E. coli* O157:H7 and *Salmonella* Typhimurium in apple juice by 7 log units, hence meeting the United States Food and Drug Administration guidelines for processed juices. The pasteurization temperatures of 80–90 °C were achieved in 25 s (Mendes-Oliveira et al. 2020). The microwave process inactivated 90% of pectinmethylesterase in orange juice treated at 60 °C for 10 s and 99% reduction at 80 °C holding for 20 s (Brugos et al. 2018). This demonstrated that microwave pasteurization is an excellent alternative for the processing of orange juice at relatively low temperatures in order to avoid over processing and quality degradation. Several researchers have demonstrated the microwave assisted pasteurization to be more efficient in terms of the processing time as well as the product quality than the traditional approach under the same processing conditions, inferring the possibility of non-thermal effects.

A recent study has proposed an innovative hybrid microwave and steam system for palm oil sterilization. The results suggested that the hybrid system increased the mesocarp temperature by 17.5% and the kernel temperature by 25.1%. The post steaming process retained the mesocarp moisture, thereby yielding a better oil quality. The developed system exhibited shorter processing time and consumed lesser power while maintaining the quality parameters (Hock et al. 2020). Studies have shown that microwave assisted sterilization has insignificant impacts on vitamin C, color, and total β-carotene content of vitamin C-fortified sweet potato puree and extended shelf life up to 18 months depending on the storage conditions. This processed product could be a desirable baby food throughout the 18 months. Further research is still required to stabilize vitamin C content for around 3 years for NASA and military missions (Zhang et al. 2019). Sun et al. (2007) have reported that the combined use of microwave and circulated water heating system resulted in higher antioxidant capacities and greener color as compared to hot water and steam retort sterilization of asparagus. This suggests the advantages of microwave sterilization over conventional heating in the processing of vegetables. Tang et al. (2008) have developed a microwave sterilization system for sliced beef in gravy in 7 oz. trays. It was observed that the 915 MHz single mode microwave sterilization was effective for inhomogeneous food treatment. This technique could be applied for sterilization of other packaged inhomogeneous food products such as chicken meat in gravy in trays and fish in gravy in pouches. Microwave treatment at 950 W, 2450 MHz, 63 °C for 25 s resulted in 5.1 log cycles reduction of *Salmonella* Typhimurium without significant degradation in the color parameters of jalapeno peppers (De La Vega-Miranda et al. 2012). Exposure of chicken drumsticks to microwave for 60 s reduced *Listeria monocytogenes* by 6 log cycles (Zeinali et al. 2015). The effectiveness of microwave-based sterilization of food products depends on the food medium, frequency, and the achievement of the desired temperature at various parts of the products. Studies have indicated the promising applications of microwave assisted sterilization of food products and emerged to be an excellent candidate in terms of rapid processing time, energy saving, microbial safety, and retention of the products quality (Soni et al. 2020).

2.3 Microwave Cooking

The cooking of food can lead to a partial or total loss of valuable nutritional values depending on the methods and process parameters followed. Research has shown that many food vitamins are thermally unstable and leached out during thermal treatment. Over the years, microwave technology has exhibited several benefits over well-established conventional methods. The application of microwave technology in the cooking of foods at the industrial level was successfully employed in the processing of meatballs, meat patties, and bacon (Raghavan et al. 2005). Combinations of traditional and microwave ovens are now available for commercial use, which permits the accelerated cooking process, and the surface moisture could be evaporated through convectional airflow over the food surface. Industrial scale microwave bacon cookers are available from Ferrite's Industrial Microwave Systems (Nashua, New Hampshire, USA), Defreeze Corporation (Southborough, Massachusetts, USA), and Microdry Incorporated (Crestwood, Kentucky, USA) (Orsat et al. 2017). Microwave cooking of fresh meat can be challenging due to a lack of browning in the presence of cool air around the meat. The mass transfer from the center to the surface while the products heated up inside may result in dry, flavorless, and tough meat products. Ingredient selection can be a way to solve these issues, such as the use of coloring agent and salt-based surface coating to attract microwave energy to the meat surface, water binding agent (starch) to decrease the loss of moisture, enzyme to preserve tenderness, and addition of flavoring agents (Meda et al. 2017). Generally, it is recommended to maintain the concentration of salt as low as possible to enhance the depth of microwave penetration and gradual surface heating while delivering higher heating uniformity and shunning thermal runaway. However, salt addition may be suggested in some circumstances since it improves the rate of heating at the surface for specific applications (Meda et al. 2017).

The assessment of the food cooking process plays a vital role to comprehend the heat-induced quality modifications and optimize the process parameters. Studies indicated that microwave cooking tends to degrade meat texture, promote N-ε-carboxymethyllysine formation, and reduce essential amino acids by Maillard reaction. The optimization of the process parameters could be a very useful approach to minimize these limitations. The optimum conditions of the effects of different specific powers and cooking times on the tenderness, lysine,

Nε-carboxymethyllysine, polyphenols, and energy consumption during microwave cooking of beef burgundy were found to be at 0.84 W/g specific power and a cooking duration of 84 min (Jouquand et al. 2015). The crayfish tail subjected to microwave (3 W/g) and boiling water treatment observed that the former approach resulted in more cooking uniformity and a lesser degree of overheating. However, the microwave cooked product suffered from central overheating which might be attributed as an important factor for the texture degradation (Fan et al. 2020). Microwave cooking showed a different time-temperature profile compared to traditional cooking. Microwave cooking is characterized by a lower total thermal effect. Although microwave cooking has a less thermal effect, the process resulted in more gelatinized and softer spaghetti than conventional cooking (Cocci et al. 2008). Alajaji and El-Adawy (2006) have recommended cooking chickpea using the microwave technique, since the process improved the nutritional values, reduced the antinutritional level and flatulence factors, increased the digestibility of in-vitro protein, retained minerals and vitamin B, and significantly reduced the cooking time.

Currently, cooking vegetables in microwave bags are emerging to be a prevalent domestic cooking process. Ready-to-cook vegetables packaged in microwave bags are available and fulfilled the needs of modern consumers due to the freshness of the products and ease to cook (Paulsen et al. 2021). A recent study has shown that the microwave bag processed broccoli resulted in higher glucosinolate content retention and higher antioxidant capacity compared to conventional microwave cooking. The process is a rapid, easy, and clean cooking option to satisfy the demand of modern consumers (Paulsen et al. 2021). Zhong et al. (2015) have investigated the influence of microwave bag cooking, traditional microwaving, and steamer steaming on the physical and nutritional characteristics of frozen broccoli. The results suggested that the steamable bag microwaving performed better in terms of antioxidant activity, ascorbic acid retention, tenderness, and color parameters. These findings demonstrated that microwave bag processing of vegetables could be a more appropriate cooking process than conventional cooking methods.

2.4 *Microwave Thawing and Tempering*

Frozen food products are either tempered or thawed before further processing. Tempering is the process of increasing the frozen food temperature from below −18 °C to a temperature of approximately −5 to −2 °C. Tempering is an initial means of raising the temperature of the frozen food which is usually carried forward by complete thawing. Thawing is referred to as the process of raising the temperature of the frozen product to 0 °C and the product is free from ice. At these temperatures (thawing and tempering), the mechanical properties of the product are more suitable for further mechanical processing such as cutting, slicing, milling, and molding (James et al. 2017; Orsat et al. 2017; Sumnu and Sahin 2012). In general, thawing, and tempering processes are achieved conventionally by leaving the frozen food in the air at ambient temperature or by immersing it in water. These ineffective practices are being replaced by microwave, radio frequency, and ohmic heating methods (Raghavan et al. 2005; Swain and James 2005). Microwave thawing and tempering of foods showed several advantages over the conventional methods: rapid process, no temperature abuse, reduction of chemical deterioration, decrease microbial growth, less drip loss, require less space, and low cost (Raghavan et al. 2005; Sumnu and Sahin 2012). The heating uniformity and end temperature control are vital in microwave thawing and tempering, because a localized heating and melting may lead to the thermal runaway effect (Schubert and Regie 2006). The phase change takes place at a defined temperature during microwave thawing and tempering of the pure substance. Conversely, for multicomponent products (foods), the change of phase occurs in a range of temperatures. Thus, during microwave thawing of foods, there occurs three different regions: solid, liquid, and solid-liquid combination. Several numerical methods such as effective heat capacity technique, temperature method, and enthalpy approach have been implemented to analyze the heating dynamics during microwave thawing of food products (Punathil and Basak 2017).

A comparative study has been conducted on the effects of the water bath and microwave thawing of white sauces prepared with modified waxy maize starch and two native starches (corn and potato). The results suggested that the microwave method is better and more effective than water bath heating based on the microstructural and viscoelastic properties and processing time (Arocas et al. 2011). A recent research reported some novel approaches of thawing with potential application in the food industry. Magnetic nanoparticles combined with microwave and magnetic nanoparticles assisted far-infrared thawing of red seabream fillets showed a desirable gelation property, thermal stability, stable secondary and tertiary protein structures, and the immobilized water and free water had insignificant changes compared with the fresh product (Cao et al. 2018). Another study on magnetic nanoparticles plus microwave thawing of largemouth bass fillets was conducted by Cao et al. (2019). The study observed that the stability of the treated product quality from the perspective of proteomics was quite satisfactory. Peng et al. (2021) have reported that the microwave assisted thawing could retain the structural and physicochemical properties of myofibrillar proteins of porcine *longissimus dorsi* and the immobilized water and free water had no significant differences compared with the fresh sample, denoting a firmer interaction between the muscle protein and water in microwave assisted thawing method. Microwave or ultrasound vacuum thawing of red seabream fillets showed no significant differences in free water. Microwave-vacuum thawing resulted in more desirable viscoelasticity of muscle protein, however, ultrasound vacuum treated product has higher stability of the secondary structure. This signifies that the microwave or ultrasound combined with vacuum could be used to increase the proteins physicochemical properties during the thawing of fillets (Cai et al. 2018). A comparative study on the effects of four different tempering methods namely microwave, radio frequency, forced air convection, and water immersion on the quality attributes of cryogenic immersion frozen pork loin was conducted by Choi et al. (2017). It was observed that the electro-heating methods (microwave and radio frequency) significantly reduced the processing time, whereas the forced air convection process was the most time consuming. Comparatively, the radio frequency process was found to have superior performance in terms of drip loss, color, and microstructure.

3 Ohmic Heating

Ohmic heating also referred to as joule heating, direct electrical resistance heating, electrical resistance heating, electroconductive heating, or direct electro-heating is known as a process where heat is generated directly inside the product owing to electrical resistance when an electric current passed through the material (Alwis and Fryer 1990; Kaur and Singh 2016). Ohmic heating is different from other electro-heating techniques since the electrodes are in contact with the products unlike microwave or inductive heating, in which the electrodes are absent. Also, the frequency used is relatively less compared with microwave or radio frequency level and the waveform is unrestricted while typically sinusoidal (Osaili 2012; Salari and Jafari 2020). Several factors such as field strength, electrical conductivity, ionic concentration, particle size, concentration, and electrodes have effects on the ohmic heating of food products. Among these, based on review of literature, the authors suggest that electrical conductivity is observed to be an important parameter in ohmic heating modeling (Kaur and Singh 2016). A product to be ohmically heated must be physically able to conduct electricity. A product is classified as a conductor when electrical charges can move within it from a point to another to complete an electrical circuit. Most of the foods can conduct electricity due to the presence of a huge amount of water and dissolved salts and these solutions can conduct electricity through electrolytic conduction. Ohmic heating obtains its name from Ohm's law, which describes the relationship between current (I), resistance (R), and voltage (V) as shown in Eq. 7.5 (Icier 2012). The food product placed in between the electrodes has a resistance function in the circuit.

$$I = \frac{V}{R} \tag{7.5}$$

The electrical resistance offered by the food material instigates heat generation inside the product. In other words, electrical energy is converted to heat energy. The collision of the moving ions and molecules results in momentum transfer which increases the kinetic energy thereby leading to heat generation. The interaction between the local electrical conductivity and field strength will govern the rate of local heat generation as expressed in Eq. 7.6 (Lyng et al. 2018).

$$Q = E^2 k = \lambda J^2 \tag{7.6}$$

Where Q is the rate of heat generation (W/m^3), k is the electrical conductivity (S/m), E is the electrical field strength (V/cm), J is the current density (A/m^2), and λ is the electrical resistivity (Ωm). The actual rate of heating for the product can then be determined using Eq. (7.7) (Lyng et al. 2018).

$$\frac{dT}{dt} = \frac{Q}{\rho C} \tag{7.7}$$

Where C is the specific heat capacity (kJ/kg °C), ρ is the density (kg/m^3), and ρC is termed as the volumetric heat capacity. Eq. 7.7 indicates that a high Q does not assure a quick rise in the rate of temperature as it also depends on ρC.

The main purpose of ohmic heating is to rise the temperature of food products to a level at which the product is adequately heated. The proper assessment of the temperature profile is required when heat is applied to the food products. The temperature distribution can be determined by considering an energy balance for the products. This requires the information of heat, mass, and momentum transfer during ohmic heating. The mass transfer (presence: continuous mode; absence: batch operation) through the heating unit will change the hydrodynamic conditions. A schematic representation of heat, mass, and momentum transfer during ohmic heating is shown in Fig. 7.4 (Vicente et al. 2006).

The ohmic heating system varies based on the type of process (continuous or batch) and the applications (cooking, thawing, pasteurization, etc.). Commonly, the basic components of a static ohmic heating unit include ohmic cell, electrodes, variable voltage transformer, data acquisition system, temperature measuring device, and computer system as shown in Fig. 7.5 (Salari and Jafari 2020). Ohmic heating was first used in the nineteenth century for milk pasteurization termed as 'electropure'. But its applications in the food sector was not successful at that time due to high cost of electricity, electrolysis related effects, less availability of inert materials for electrode, process regulations, and other practical difficulties (Icier 2012; Kang and Jun 2021). However, in recent years, research has taken great momentum and could address several issues encountered in the past. In present days, with the development of solid-state power supply, the use of ohmic heating in pulse mode is possible in order to economically regulate the electrolytic effects to harmless degrees. Ohmic technology is now well-engineered, less expensive, and more sophisticated than in the past years (Icier 2012). Ohmic heating is being currently explored in various applications in the food processing sector such as thawing, tempering, cooking, evaporation, blanching, baking, pasteurization, sterilization, peeling, distillation, fermentation, tissue softening, modify food texture and structure, and pre-treatment for extraction (Gavahian et al. 2019b; Knirsch et al. 2010; Makroo et al. 2020; Moreno-Vilet et al. 2018; Pereira et al. 2021; Ríos-Ríos et al. 2021). The main advantages of ohmic heating over other novel thermal food heating methods (microwave, radio frequency, and infrared) are more heating uniformity, lower fouling, the higher temperature in particles than liquids, more energy efficiency, and lower cost (Salari and Jafari 2020). Ohmic heating can be suitably combined with other thermal and non-thermal food processing technologies

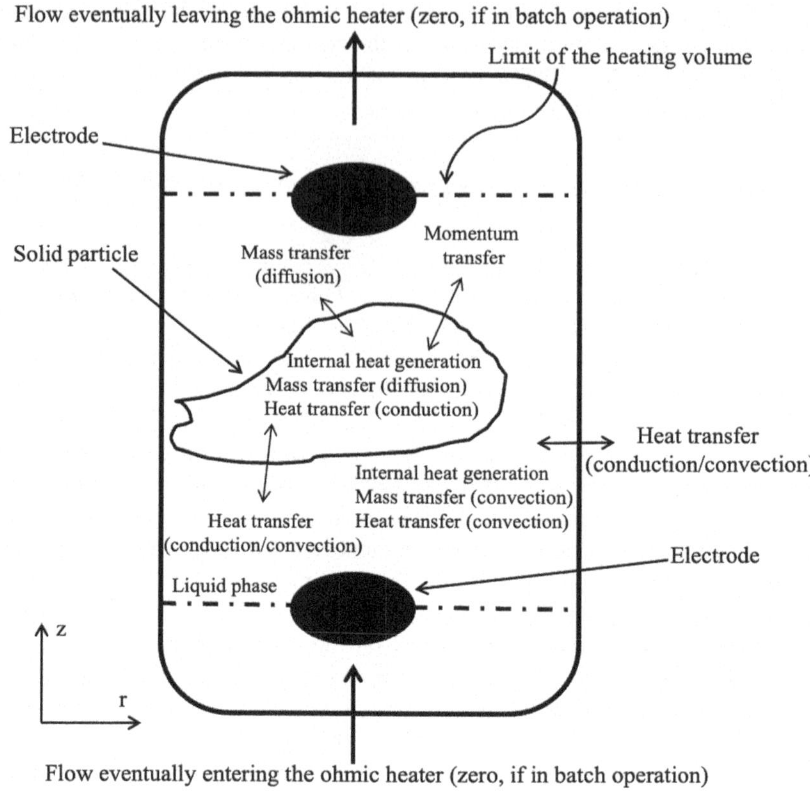

Fig. 7.4 A schematic illustration of heat, mass, and momentum transfer during ohmic heating. (Redrawn from Vicente et al. 2006)

to obtain synergistic effects. Ohmic heating is a novel thermal food treatment method and showed several benefits over the traditional technique in terms of food quality retention, microbial safety, rapid process, clean process, and energy efficiency. Some of the recent applications of ohmic heating in food processing are shown in Table 7.2. The recent developments of ohmic heating in food operations such as sterilization, pasteurization, cooking, thawing, and tempering of food products are summarized in the following sections.

3.1 Ohmic Pasteurization and Sterilization

The microbial inactivation mechanism in relation to ohmic heating is predominantly attributed to thermal effects, much like traditional heating. However, two additional non-thermal effects namely chemical and mechanical mechanisms have also been proposed due to the presence of the electric field. Chemical deactivation of

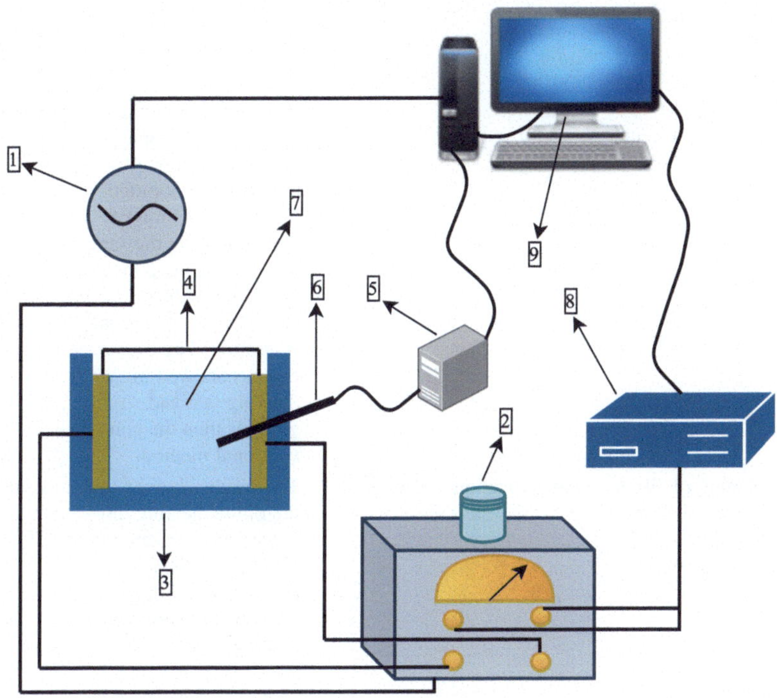

Fig. 7.5 A schematic illustration of an ohmic heating system, where (1) AC power supply, (2) variable voltage transformer, (3) ohmic heating cell, (4) electrodes, (5) thermocouple, (6) thermocouple sensor, (7) food sample (8) data logger, and (9) computer system. (Redrawn from Salari and Jafari 2020)

microorganisms has been ascribed to the development of free hydrogen and oxygen, radicals of hydroperoxyl and hydroxyl, and metal ions which cause a microbial reduction. Mechanical microbial inactivation has been attributed to the disruption of the cell membrane of microorganisms resulting in mild electroporation thereby causing the cell contents leakage (Lyng et al. 2018; Maloney and Harrison 2016; Sun et al. 2008). The reason for the additional impact of ohmic heating on microbial death may be due to its low frequency of 50–60 Hz, which permits the cell wall to store up charges and generate pores. Conversely, in the case of the high frequency heating methods (microwave and radio frequency), before the adequate charge is formed at the cell walls, the electric field reversed back (Kaur and Singh 2016; Knirsch et al. 2010; Osaili 2012).

Ohmic pasteurization was applied to eliminate polyphenol oxidase and peroxidase in coconut water. The electrical conductivity of coconut water is naturally good, which makes it suitable for ohmic heating treatment. Kanjanapongkul and Baibua (2021) have reported that the ohmic heating completely reduced peroxidase at 90 °C for 3 min although about 10% of polyphenol oxidase activity still remained in coconut water. Comparatively, the polyphenol oxidase activity levels in ohmic pasteurized products were considerably lower than in the conventionally processed

Table 7.2 Ohmic heating of food products

Ohmic treatment condition	Purpose	Product	Effect	Reference
Ohmic heating (20 V, 60 Hz, and sample center temperature − 2 °C)	Thawing	Surimi	It was observed that the electrode solution concentration below 4% resulted in a homogeneous distribution of temperature in the sample. The thawing rate increased with the rise in electrode solution concentration. The ohmic process showed higher thawing rate and stronger gels than the conventional method.	Miao et al. (2007)
Ohmic heating (5–30 A, 14.4 V/cm at 72 °C for 15 s) and conventional heating (72 °C for 15 s)	Pasteurization	Goat milk	The results showed that the ohmic heating did not promote modification of free fatty acid contents as compared with the conventional heating process. This suggests that ohmic heating can accomplish goat milk pasteurization without affecting the flavor.	Pereira et al. (2008)
Ohmic heating (10, 20, and 30 V/cm) and conventional thawing (25 °C and 95% RH)	Thawing	Beef cuts	The ohmic treated samples were harder, while the conventional process caused more springiness. The effects of voltage gradient during ohmic heating on gumminess, springiness, hardness, and chewiness were statistically significant. Comparatively, the ohmic thawed sample showed a lesser change in textural and histological properties.	Icier et al. (2010)
Ohmic heating (10 V/cm) and water bath (85 °C)	Cooking	Pork	Ohmic cooking resulted in lower color components, metmyoglobin, and deoxymyoglobin than water bath treatment at the same endpoint temperatures (60–80 °C). Moreover, the ohmic	Dai et al. (2013)

(continued)

Table 7.2 (continued)

Ohmic treatment condition	Purpose	Product	Effect	Reference
			processed sample showed higher sarcoplasmic protein solubility and water holding capacity.	
Ohmic heating (10, 20, and 30 kHz)	Tempering	Potato puree	The ohmic heating substantially shortened the tempering time by 90% of the conventional method. The tempering time was appeared to be reduced with the increase of frequency and salt concentration by increasing the electrical conductivity and heating rate.	Seyhun et al. (2013)
Ohmic heating (40–20,000 Hz and 20–60 V) and conventional heating (100 °C for 480 s)	Pasteurization	Fermented red pepper paste	It was observed that the ohmic method inactivated vegetable cells of Bacillus strains by 99.7% against 81.9% reduction with the conventional process. The physicochemical and organoleptic properties of the ohmic processed samples were comparable with the raw samples.	Cho et al. (2016)
Ohmic heating (15 A, 0–250 V, and 50 Hz) and steam cooking (72 °C)	Cooking	Shrimps	The steam method needed an overall cooking time of 59 and 38 s for large and small shrimps, respectively, while only 40 s were required for the ohmic process irrespective of shrimp size. No significant differences were observed for texture and cooking loss in both methods. This emphasizes the potential of ohmic heating for a quicker and more cooking uniformity while delivering a similar color, texture, and yield as compared to steam cooking.	Lascorz et al. (2016)

(continued)

Table 7.2 (continued)

Ohmic treatment condition	Purpose	Product	Effect	Reference
Ohmic heating (12 V/cm and 72 °C) and water bath cooking (80 °C; until the center temperature reached 72 °C)	Cooking	Beef muscle	Comparatively, the shear force and cooking loss were found to be significantly lower and the redness value was considerably higher in the ohmic treated sample. Proteomics analysis indicated that the ohmic cooked sample had lower protein damage than the water bath cooked meat.	Tian et al. (2016)
Ohmic heating (0.05 duty ratio, 60 Hz, and 12.1 V_{rms}/cm), carvacrol (0.2 mg/g) and ohmic-carvacrol	Sterilization	Salsa	The combined treatment showed a synergistic effect on the inactivation of *Escherichia coli* O157:H7, *listeria monocytogenes*, and *salmonella* typhimurium. Moreover, color components were improved by the ohmic-carvacrol process compared to ohmic heating alone.	Kim and Kang (2017)
Ohmic heating (200 V, 50 Hz-20 kHz and sample center temperature 20 °C)	Thawing	Tuna	Increasing the frequency resulted in a drop in electrical resistance which caused a higher heating rate thereby reduced the thawing time. The electrical conductivity and thawing time were the lowest and longest for the muscle with the highest fat level and lowest moisture content and vice versa. The removal of muscle membranes and the use of parallel electrical circuits led to higher electrical conductivity.	Liu et al. (2017)
Ohmic heating (30, 40, 50, and 60 V/cm)	Sterilization	Apple juice	Results showed that about 5 log reduction in *Escherichia coli* O157:H7, *listeria monocytogenes*, and *salmonella enterica serovar*	Park et al. (2017)

(continued)

Table 7.2 (continued)

Ohmic treatment condition	Purpose	Product	Effect	Reference
			typhimurium were obtained after the treatment at 30 V/cm for 60 s in 36 °brix and this same result was also achieved in 48 °brix sample processed at 60 V/cm for 20 s without degrading the product quality. The optimum process conditions were found to be at 60 V/cm in 48 °brix based on the bactericidal efficacy and system performance coefficients for the processing of apple juice concentrates.	
Ohmic heating (0.22 S/m, 65 °C, for 300 and 420 s, and 70 °C for 180 and 300 s) and conventional (63 °C for 1800 s)	Pasteurization	Pulque	The ohmic treated samples were observed to have a higher content of lactic acid bacteria (5.96 to 6.68 $\log_{10}$ CFU/mL compared with conventional heating (3.54 $\log_{10}$ CFU/mL). The physicochemical properties such as color, pH, and alcoholic content were found to be stable during the 22 days of storage in both methods. However, the ohmic processed samples were observed to have better sensory acceptance.	Alcántara-Zavala et al. (2019)
Ohmic heating (250 V and 10 kHz for 30 s) and water bath cooking (90 °C)	Cooking	Surimi-canned corn mixed gels	The modified cohesiveness and hardness of the mixture decreased with the increase of corn in both the cooking methods. Yet, the ohmic heated product exhibited a lesser reduction in modified cohesiveness and hardness. The shear force of corn in the ohmic heated sample was higher than the water bath method and	Jung et al. (2020)

(continued)

Table 7.2 (continued)

Ohmic treatment condition	Purpose	Product	Effect	Reference
			there was an insignificant difference between the shear force of ohmic heated corn and canned corn without heating. Comparatively, the ohmic cooked mixed gels showed lower moisture loss.	
Ohmic heating (4, 8, or 12 V/cm, 72–75 °C for 15 s)	Pasteurization	Milk	Ohmic heating decreased the hardness, firmness, and elasticity while improved the sensory properties of Minas Frescal cheese prepared from ohmic pasteurized milk. In addition, higher antidiabetic activities, antihypertensive, and antioxidant activities as well as more short, medium, and long chain fatty acids concentrations were observed.	Rocha et al. (2020)
Ohmic heating (15, 17.5, and 20 V/cm, 60 Hz, and 72 °C for 0, 15, 30, 45, 60, and 120 s) and conventional heating (72 °C for 0, 15, 30, 45, 60, and 120 s)	Pasteurization	Mango pulp	Ohmic heating completely inactivated polyphenol oxidase in 15 s. Ohmic treatment promoted insoluble dietary fiber release, increased apparent viscosity, and favored water retention. Ohmic treatment at 20 V/cm reduced polyphenol oxidase 1.5 times faster than the conventional process. This suggests the potential of ohmic heating for fast pasteurization of mango pulp to inactivate polyphenol oxidase while maintaining its physicochemical properties.	Barrón-García et al. (2021)

samples. Additionally, the ohmic heating process significantly reduced the processing time and prevented the discoloration of pink color in coconut water during cold storage. A rapid and uniform ohmic heating of red pepper paste resulted in the inactivation of microorganisms by 2 log reduction while maintaining a shorter treatment time than the conventional method. The consistency index (thickening properties) was appreciably higher in ohmic treated samples, which indicates the suitability of ohmic heating of highly viscous food products (Cho et al. 2017). Ohmic heating was found to reduce completely *Listeria monocytogenes* with more than a 5 log reduction in sausage, whereas conventional heating achieved only 3 to 4 log reduction. Ohmic pasteurized products (sausage) showed no statistical change in chemical composition, lipid oxidation, pH, water holding capacity, and cooking loss and only negligibly changed in color and texture (Inmanee et al. 2019). Ohmic heating shows the potential application in sausage processing, for both process efficiency and product safety while maintaining the quality attributes of the products. The ohmic heating through the conductive package demonstrated a prospect of producing safe solid-liquid products rapidly with a faster rate of heating, more heating uniformity, and higher vitamin C retention compared to traditional heating (Wattanayon et al. 2021). Pasteurization of citrus juices with ohmic heating suggested a potential to maintain the carotenoid profile. Achir et al. (2016) have reported that the ohmic heating of citrus juices resulted in a comparatively lesser loss of xanthophyll (30%) and epoxyxanthophylls (20%) against 70% and 40%, respectively, with conventional heating. Moreover, no non-thermal effects of ohmic treatment were observed on the carotenoid profile. Also, a recent study on the ohmic heating of sugarcane juice showed a total phenolic content comparable to raw fresh samples, denoting the absence of additional non-thermal impacts on the products (Rodrigues et al. 2021). Ohmic heating is a promising approach for the preservation of carrot juice. The ohmic pasteurized carrot juice samples were observed to be microbiologically stable during 60 days at 4 °C while maintaining the quality attributes (Negri Rodríguez et al. 2021).

The control of nutritional value degradation during baby food processing is very important to support an adequate nutritional status in infants. A comparative study on the effects of two ultra-high temperature techniques, direct steam injection, and continuous ohmic sterilization of liquid infant formula on the nutritional parameters under the same processing conditions were conducted by Roux et al. (2016). The results demonstrated that the sample sterilized by the ohmic process was appeared to be richer in Maillard products (carboxymethyllysine and furosine), better color parameters, and higher vitamin C content than by the steam injection method. Continuous ohmic heating used at the pilot scale provided satisfactory results in terms of maintenance of the nutritional values while no fouling was detected after 2.5 h processing at 140 °C. Research indicated that the retorted vegetable baby foods were appeared to have about 20% less total amino acids than in the raw samples (9% non-essential amino acids and 35% essential amino acids). Conversely, ohmic heating did not reduce total amino acids, and the essential and non-essential amino acid contents were comparable with the untreated samples. Ohmic sterilization promotes less degradation in nutritional values of vegetable baby purees, therefore this technology may be successfully applied as an alternative technique to ensure nutritional and microbiological properties of infant foods (Mesías et al. 2016). Park et al. (2013) have exhibited the ohmic assisted process as a feasible approach for sterilization of carrot puree, green pea puree, and tomato juice. It was observed that the ohmic assisted sterilization had sporicidal effects and reduced significantly *Bacillus amyloliquefaciens* and *Geobacillus stearothermophilus* spores. The combined treatment of ohmic, pressure, and thermal (30 V/cm, 105 °C at 600 MPa for 0, 1, 3, and 5 min) of carrot was appeared to have better performance than when applied alone or pressure assisted thermal approach. The ohmic-pressure-thermal sterilized product showed the least textural damage, highest crunchiness index, and required the least thermal come-up time as compared with other methods considered in this study (Park et al. 2014). The combined application of ohmic, thermal, and pressure exhibits to be a potent technique for rapid elimination of pressure-thermal resistant microbial spores while maintaining the quality parameters in low-acid vegetable foods by means of a synergy of the electric field, pressure, and thermal energy.

3.2 *Ohmic Cooking*

Ohmic cooking is an advanced method of food processing based on electromagnetic techniques. Ohmic cooking is distinctive due to its volumetric nature and it is characterized by a nearly linear expansion in temperature of the food product during cooking. Ohmic heating could be applied as a continuous in-line treatment process of pumpable products for cooking and heating of viscous foods and mixtures encompassing particulate foods. It is considered as very beneficial for food products since heat is generated rapidly which is distributed uniformly because of high electrical conductivity, thereby resulting in higher retention of flavor and nutritional values. The important factors to be considered during ohmic cooking are the electrical conductivity inside the medium and the current. The electrical conductivity and ohmic cooking increase with the rise in salt concentration while the fat content reduces the electrical conductivity. Other factors to be taken into consideration include difficulty in controlling time-temperature, electrode corrosion, and complex coupling between the electrical distribution and temperature (Suleman et al. 2020; Van der Sman 2017). These factors would influence the effectiveness of the ohmic cooking and should be taken care during ohmic cooking of food products. Higher frequency reduces the rate of electrode corrosion in ohmic cooking. It was observed that the stainless steel 904 L has a greater resistance to corrosion than 304 and 316 L, but due to the high content of nickel and chromium, the total migrations of ions to the product are comparable. The nickel and chromium ions migration to the product (meat) were found to be lower than iron ion, resulting in within the safe limit of nickel and chromium ions concentration when the high frequency is used (Wang and Farid 2015).

Meat products contain rich nutritional values which serve as a suitable environment for the food spoilage microorganisms and common foodborne pathogenic bacteria. Ohmic heating offers a feasible approach to produce safer meat products by effectively reducing the microbial proliferation through high uniformity heating and cooking quickly and instantly within the product. The difficulties of ohmic heating of meat and meat products are due to solid in nature, presence of fats, and heterogeneous structures of meat (Yildiz-Turp et al. 2013). Usage of plate heating and flat meat patties was recommended in order to improve the contact between the electrode surface and the product. Bozkurt and Icier (2010a) have investigated the electrical conductivity variations of minced beef fat blends (fat levels: 2, 9, and 15%) during ohmic cooking (20, 30, and 40 V/cm). The major factors affecting the electrical conductivity were found to be the blending composition and temperature. The impact of initial fat level on the electrical conductivity was significant, however, voltage variation had no significant influence. It was reported that the electrical conductivity increased with the rise in temperature up to the critical temperature about 60–70 °C depending on the fat content, thereafter, reduced because of the structural changes and bound water increment during cooking. Tian et al. (2019) have conducted a comparative study on the survival of Escherichia coli O157:H7 in pork batter after water bath and ohmic cooking. The results indicated that the time required for the sample to achieve the desired endpoint temperature of 61, 65, and 72 °C was shorter in the ohmic process and the addition of NaCl considerably shortened the ohmic cooking time, however, no significant outcome was observed in water bath process. Ohmic cooked sample appeared to have a lower cooking loss, but the inactivation effect of Escherichia coli O157:H7 was observed to be similar in both the methods. Ohmic cooking of pork was observed to have a shorter cooking time without affecting the cooking loss, water holding capacity, and color parameters and appreciated by consumers and cooks alike when compared with pan cooking (Ángel-Rendón et al. 2019). Ohmic cooked ground beef was firmer and higher volume reduction than the conventional process, yet the yield and retention of fat were similar in both the cooking methods. The voltage variation during ohmic cooking was not linked to the product quality (Bozkurt and Icier 2010b). Ángel-Rendón et al. (2020) have compared four different cooking methods namely ohmic, pan, *sous vide*, and vacuum cooking based on the microstructure, physicochemical, and sensory properties of cooked pork meat. The results indicated that the ohmic process produced firmer meats and myofibrils, vacuum cooked products appeared more drier and loss of structure, and *sous vide* treated meats observed as insipid. Pan cooking yielded softer meat and was described as juicy, tasty, and tender by the consumers. No significant differences were observed for water holding capacity and cooking loss in all the methods. Ohmic treated samples, which needed shorter processing time, exhibited comparable qualities to the pan cooked product, could be used as an efficient alternative and/or complementary technique to conventional cooking methods.

The impacts of volumetric heating (microwave and ohmic) on energy consumption, color, hydration, and texture of rice were examined and the obtained results were compared with the hotplate method (Gavahian et al. 2019a). The investigation revealed that the ohmic cooking negatively degraded the color parameters, however, it produced more softening rates as compared to the conventional process. Furthermore, no fouling was detected in ohmic cooking and consumed 69% of the energy required in microwave heating for a 50% softening of the rice grains. This novel approach could be a potential alternative to the conventional rice cooking process with further developments and improvements. The electrical conductivity of germinated brown rice, brown rice, white rice (Sao Hai), and white rice (KDML105) were found to be 0.485–1.182, 0.617–1.370, 0.375–1.005, and 0.246–0.900 S/m, respectively. The rice cooked by ohmic technique showed a significant difference in textural properties compared with the sample using an electric rice cooker. The ohmic cooking system consumed approximately 73–90% of the energy required for the traditional rice cooker. It is recommended to add 0.1 M salt solution to substitute water in rice cooking to increase the electrical conductivity (Jittanit et al. 2017). The temperature come-up time to 100 °C during ohmic cooking of instant noodles at 10, 12.5, 15, and 17.5 V/cm of electric fields were found to be 3.9, 2.5, 2.1, and 1.3 min, respectively. The highest heat transfer ratio was observed at 15 V/cm and the system performance coefficient was found to be highest at 15 V/cm and holding time of 90 s and produced the most desirable textural qualities (Jo and Park 2019). This indicates the potential of the ohmic process for the rapid cooking of instant noodles while providing good textural qualities and saving energy consumption. The texture softening rates were appeared to be increasing with the increase of input power during ohmic cooking of potatoes and about 8 times faster in softening the texture as compared to the conventional method. Comparatively, treatment at 300 kHz produced a better uniform texture than 12 kHz. The textural properties were enhanced and reduced weight loss by adjusting the input power level and liquid conductivity during the ohmic cooking of potatoes (Gratz et al. 2021). Ohmic cooking was observed to soften the texture of red beet, carrot, and golden carrot at higher rates than microwave and conventional cooking (Farahnaky et al. 2012). This indicates the potential of ohmic heating for the rapid processing of some root vegetables.

3.3 Ohmic Thawing and Tempering

During thawing and tempering, food products are subjected to impairment by physical, chemical, and microbiological modifications. Rapid thawing at a lower temperature to avoid a significant increase in temperature and high dehydration of product is desirable to ensure food quality (Seyhun et al. 2014). Modern consumers' dependent on frozen foods is ever increasing worldwide, therefore there is a demand for the development of advanced thawing methods. Electro-heating (volumetric method) provides solutions to traditional thawing concerns. Volumetric heating methods do not require a huge amount of water and the process is fast as the thawing rate does not depend on food product thermal conductivity. Ohmic heating is one of the emerging volumetric heating processes. Employing ohmic heating to temper and thaw frozen products is a novel approach. This innovative method utilizes the electrical resistance of frozen food to generate heat volumetrically inside the product itself when the electrical current is passed through the product (Li and Sun 2002; Seyhun et al. 2014). The proper contact between the electrodes and the frozen food is crucial for the uniform heating of the frozen food product.

Research indicated that thawing of frozen meat and meat-derived products is observed to have greater impacts on the quality of the products than freezing and frozen storage. Thus, the selection of appropriate thawing methods is crucial to retain the quality attributes of food products. Excessive weight loss and reduction in the nutritional values of meat products depending on the growth of water infiltration ratio are found to be the main problems due to inappropriate thawing processes (Duygu and Ümit 2015). Bozkurt and Icier (2012) have reported that the applied voltage gradient (10, 20, and 30 V/cm) did not cause significant influence on the drip loss, some color parameters (lightness and hue angle), and thawing homogeneity, whereas it had a substantial effect on thawing rate, thawing time, and energy utilization ratio during ohmic thawing of beef cuts. A study on ohmic assisted thawing of frozen beef showed that the pressure-ohmic thawing (200 MPa and 40 V/cm) enabled fast thawing of 0.8 min against 11.5, 5.5, and 43.3 min for pressure, ohmic, and conventional thawing, respectively. The combined treatment of pressure and ohmic retained better textural properties as compared to other methods. The quick ice-to-water transition during pressure assisted ohmic thawing reduced the meat muscle breakdown and minimized any modifications in structural integrity (Min et al. 2016). Jia et al. (2019) have demonstrated that the thermal properties of pork changed with the addition of salt and caused to increase in the thawing rate during ohmic heating. The pork tenderloin thawed using the ohmic technique had significantly reduced the drip loss and higher water holding capacity as compared to air thawing. Additionally, the ohmic process was found to be effective for increasing salt diffusion particularly at a high electric field strength (Jia et al. 2019). Research showed that ohmic treatment significantly shortened the thawing time (5.95 times) and reduced cooking losses during the ohmic thawing of frozen tuna as compared to the conventional process (Fattahi and Zamindar 2020). The ohmically and conventionally thawed shrimp showed no statistically significant difference in sensory parameters, total aerobic microbial loads, and moisture content. This suggests a potential application of ohmic heating in the thawing of shrimp blocks (Roberts et al. 2002). Cokgezme and Icier (2019) have investigated the ohmic thawing (10, 15, and 20 V/cm) and conventional thawing (4 °C) of frozen cherry juice (15, 30, 40, and 50% of total soluble solid). The results suggested that the thawing time was reduced by about 90% with the ohmic process compared to the conventional method. The applicability of ohmic heating in the thawing of frozen fruit juice was verified.

4 Some Limitations of Microwave and Ohmic Heating of Food Products

Currently, the higher cost input limits the applications of ohmic heating at a larger scale. The ohmic heating of food products containing oils and fats are not effective due to low electrical conductivity. In addition, microorganisms present inside the fats are exposed less to heat as compared to those present outside (Kaur and Singh 2016). The electrical conductivity increases with the rise in food products temperature due to motion of electrons. Studies reported that the narrow band frequencies of ohmic is another disadvantage for food treatment. Furthermore, electrodes corrosion due to electrochemical reactions is a problem and this corrosion increases with the ohmic heating operation (Alkanan et al. 2021; Patel and Singh 2018). The main drawback of microwave heating is non-uniform distribution of temperature resulting in cold and hot spots in food products (Jain et al. 2018; Muñoz-Almagro et al. 2021). Another problem of microwave heating is the difficulty in controlling the temperature of the final product particularly in drying of food products. High microwave power may trigger undesirable changes in the food matrix due to elevated temperature. The lack of actual temperature profile availability can cause difficulty in monitoring the heating process; thus, it may result in unfavourable changes in the end products (Michalak et al. 2020). However, the present limitations of these novel thermal technologies can be overcome by bridging the knowledge gaps and overall process design. The intelligently combined use of electro-heating technology with other processing technologies can address some of the drawbacks of its application in food treatment. It is anticipated to have a stronger existence of this electro-heating technology in the food industry during the years ahead.

5 Conclusions

In recent years, several innovative food processing methods were developed and implemented in the food industry, however, the thermal technique remains as one of the most indispensable pathways towards food safety. Therefore, the improvement of the existing approaches and new development of advanced thermal techniques is crucial to address the concerns of well-established traditional methods. Currently, electro-heating technologies such as microwave heating and ohmic heating have emerged as an efficient alternative and/or complementary methods to conventional processes. These novel electro-heating methods are volumetric forms of heating that generate heat within the food products. This form of heat generation provides several benefits such as reduction of heating time, retention of sensory and nutritional values, microbial safety, and more energetic and heating efficiency as compared to conventional heating. There is extensive use of microwave heating of foods both at home and industrial scale while ohmic heating is primarily confined to industrial applications. Microwave and ohmic heating are being currently used substantially in some major food operations such as drying, pasteurization, sterilization, cooking, thawing, and tempering of food products. Applications of microwave and ohmic heating processes combined with other thermal and non-thermal methods were found to be effective in terms of food quality retention, microbial safety, reduction of processing time, and energy efficiency with the process sometimes been synergistic. The novel heating technology applications in food processing and preservation are expected to grow due to increasing demand for safe and quality food products. The problem becomes more complex when meat products are used in the novel approaches. The future application of electro-heating in the food sector is promising with the advancement in the development of process technology. The successful implication of electro-heating relies on fully understanding the science involves in the interaction of foods and electro-heating and the efficient design of electro-heating systems and processes. With the development of powerful computer systems and the availability of advanced simulation and modeling packages, it will be possible to design innovative electro-heating units and processes or improve the existing systems. Collaboration between academia and industry has been instrumental in progressing these food heating techniques. Such continual partnership is important for the research, development, and implementation of these technologies.

Acknowledgments This study was supported by the National Overseas Scholarship, Government of India.

References

Achir N, Dhuique-Mayer C, Hadjal T, Madani K, Pain J-P, Dornier M (2016) Pasteurization of citrus juices with ohmic heating to preserve the carotenoid profile. Innovative Food Sci Emerg Technol 33:397–404

Ahmed J, Ramaswamy HS (2007) Microwave pasteurization and sterilization of foods. In: Rahman MS (ed) Handbook of food preservation. CRC Press, pp 691–711

Alajaji SA, El-Adawy TA (2006) Nutritional composition of chickpea (*Cicer arietinum* L.) as affected by microwave cooking and other traditional cooking methods. J Food Compos Anal 19(8):806–812

Alcántara-Zavala AE, de Dios Figueroa-Cárdenas J, Morales-Sánchez E, Aldrete-Tapia JA, Arvizu-Medrano SM, Martínez-Flores HE (2019) Application of ohmic heating to extend shelf life and retain the physicochemical, microbiological, and sensory properties of pulque. Food Bioprod Process 118:139–148

Alkanan ZT, Altemimi AB, Al-Hilphy ARS, Watson DG, Pratap-Singh A (2021) Ohmic heating in the food industry: developments in concepts and applications during 2013–2020. Appl Sci 11(6):2507

Alwis AAP, Fryer PJ (1990) The use of network theory in the finite-element analysis of coupled thermo-electric fields. Commun Appl Numer Methods 6(1):61–66

Ángel-Rendón SV, Filomena-Ambrosio A, Cordon-Diaz S, Benítez-Sastoque ER, Sotelo-Díaz LI (2019) Ohmic cooking: application of a novel technology in pork and influences on water holding capacity, cooking loss and colour. Int J Gastron Food Sci 17:100164

Ángel-Rendón SV, Filomena-Ambrosio A, Hernández-Carrión M, Llorca E, Hernando I, Quiles A, Sotelo-Díaz I (2020) Pork meat prepared by different cooking methods. A microstructural, sensorial and physicochemical approach. Meat Sci 163:1–9

Arocas A, Sanz T, Hernando MI, Fiszman SM (2011) Comparing microwave-and water bath-thawed starch-based sauces: infrared thermography, rheology and microstructure. Food Hydrocoll 25(6):1554–1562

Atuonwu J, Tassou S (2019) Energy issues in microwave food processing: a review of developments and the enabling potentials of solid-state power delivery. Crit Rev Food Sci Nutr 59(9): 1392–1407

Barrón-García OY, Gaytán-Martínez M, Ramírez-Jiménez AK, Luzardo-Ocampo I, Velazquez G, Morales-Sánchez E (2021) Physicochemical characterization and polyphenol oxidase inactivation of Ataulfo mango pulp pasteurized by conventional and ohmic heating processes. LWT-Food Sci Technol 143:111113

Borda-Yepes VH, Chejne F, Daza-Olivella LV, Alzate-Arbelaez AF, Rojano BA, Raghavan VG (2019) Effect of microwave and infrared drying over polyphenol content in *Vaccinium meridionale* (Swartz) dry leaves. J Food Process Eng 42(1):e12939

Bozkurt H, Icier F (2010a) Electrical conductivity changes of minced beef–fat blends during ohmic cooking. J Food Eng 96(1):86–92

Bozkurt H, Icier F (2010b) Ohmic cooking of ground beef: effects on quality. J Food Eng 96(4): 481–490

Bozkurt H, Icier F (2012) Ohmic thawing of frozen beef cuts. J Food Process Eng 35(1):16–36

Brugos AF, Gut JA, Tadini CC (2018) Inactivation kinetics of pectin methyl esterase in the microwave-assisted pasteurization of orange juice. LWT-Food Sci Technol 97:603–609

Cai L, Cao M, Cao A, Regenstein J, Li J, Guan R (2018) Ultrasound or microwave vacuum thawing of red seabream (*Pagrus major*) fillets. Ultrason Sonochem 47:122–132

Cai L, Zhang W, Cao A, Cao M, Li J (2019) Effects of ultrasonics combined with far infrared or microwave thawing on protein denaturation and moisture migration of *Sciaenops ocellatus* (red drum). Ultrason Sonochem 55:96–104

Cai L, Dai Y, Cao M (2020) The effects of magnetic nanoparticles combined with microwave or far infrared thawing on the freshness and safety of red seabream (*Pagrus major*) fillets. LWT-Food Sci Technol 128:109456

Cao M, Cao A, Wang J, Cai L, Regenstein J, Ruan Y, Li X (2018) Effect of magnetic nanoparticles plus microwave or far-infrared thawing on protein conformation changes and moisture migration of red seabream (*Pagrus major*) fillets. Food Chem 266:498–507

Cao M, Cao A, Li Y, Wang W, Wang Y, Cai L (2019) Effects of magnetic nanoparticles plus microwave on the thawing of largemouth bass (*Micropterus salmoides*) fillets based on iTRAQ quantitative proteomics. Food Chem 286:506–514

Chandrasekaran S, Ramanathan S, Basak T (2013) Microwave food processing—a review. Food Res Int 52(1):243–261

Cho W-I, Yi JY, Chung M-S (2016) Pasteurization of fermented red pepper paste by ohmic heating. Innovative Food Sci Emerg Technol 34:180–186

Cho W-I, Kim E-J, Hwang H-J, Cha Y-H, Cheon HS, Choi J-B, Chung M-S (2017) Continuous ohmic heating system for the pasteurization of fermented red pepper paste. Innovative Food Sci Emerg Technol 42:190–196

Choi EJ, Park HW, Chung YB, Park SH, Kim JS, Chun HH (2017) Effect of tempering methods on quality changes of pork loin frozen by cryogenic immersion. Meat Sci 124:69–76

Choudhury MH (2017) Thermal processing in food technology: latest trends. In: Meghwal M, Goyal MR, Kaneria MJ (eds) Food technology. Apple Academic Press, pp 63–94

Cocci E, Sacchetti G, Vallicelli M, Angioloni A, Dalla Rosa M (2008) Spaghetti cooking by microwave oven: cooking kinetics and product quality. J Food Eng 85(4):537–546

Cokgezme OF, Icier F (2019) Effect of voltage gradient on ohmic thawing characteristics of sour cherry juice concentrates for the temperature range of– 18°C to+ 4°C. Food Sci Technol Int 25(8):659–670

Corrêa J, Dev S, Gariepy Y, Raghavan G (2011) Drying of pineapple by microwave-vacuum with osmotic pretreatment. Dry Technol 29(13):1556–1561

Dai Y, Miao J, Yuan S-Z, Liu Y, Li X-M, Dai R-T (2013) Colour and sarcoplasmic protein evaluation of pork following water bath and ohmic cooking. Meat Sci 93(4):898–905
De La Vega-Miranda B, Santiesteban-Lopez N, Lopez-Malo A, Sosa-Morales M (2012) Inactivation of *Salmonella* Typhimurium in fresh vegetables using water-assisted microwave heating. Food Control 26(1):19–22
Deák T (2014) Food technologies: sterilization. In: Encyclopedia of food safety, vol 3. Elsevier B. V., pp 245–252
Delgado AE, Sun D-W, Rubiolo AC (2006) Thermal physical properties of foods. In: Sun D-W (ed) Thermal food processing: new technologies and quality issues. CRC Press, pp 3–31
Dev S, Raghavan G, Gariepy Y (2008) Dielectric properties of egg components and microwave heating for in-shell pasteurization of eggs. J Food Eng 86(2):207–214
Dev S, Geetha P, Orsat V, Gariépy Y, Raghavan G (2011) Effects of microwave-assisted hot air drying and conventional hot air drying on the drying kinetics, color, rehydration, and volatiles of *Moringa oleifera*. Dry Technol 29(12):1452–1458
Dev SRS, Birla SL, Raghavan GSV, Subbiah J (2012) Microbial decontamination of food by microwave (MW) and radio frequency (RF). In: Demirci A, Ngadi MO (eds) Microbial decontamination in the food industry. Woodhead Publishing, pp 274–299
Duan X, Zhang M, Mujumdar AS, Wang S (2010) Microwave freeze drying of sea cucumber (*Stichopus japonicus*). J Food Eng 96(4):491–497
Duygu B, Ümit G (2015) Application of ohmic heating system in meat thawing. Procedia Soc Behav Sci 195:2822–2828
Erle U, Schubert H (2001) Combined osmotic and microwave-vacuum dehydration of apples and strawberries. J Food Eng 49(2–3):193–199
Fan H, Fan D, Huang J, Zhao J, Yan B, Ma S, Zhou W, Zhang H (2020) Cooking evaluation of crayfish (*Procambarus clarkia*) subjected to microwave and conduction heating: a visualized strategy to understand the heat-induced quality changes of food. Innovative Food Sci Emerg Technol 62:102368
Farahnaky A, Azizi R, Gavahian M (2012) Accelerated texture softening of some root vegetables by ohmic heating. J Food Eng 113(2):275–280
Fattahi S, Zamindar N (2020) Effect of immersion ohmic heating on thawing rate and properties of frozen tuna fish. Food Sci Technol Int 26(5):453–461
Fellows PJ (2009) Pasteurization. CRC Press
Gavahian M, Chu Y-H, Farahnaky A (2019a) Effects of ohmic and microwave cooking on textural softening and physical properties of rice. J Food Eng 243:114–124
Gavahian M, Tiwari BK, Chu Y-H, Ting Y, Farahnaky A (2019b) Food texture as affected by ohmic heating: mechanisms involved, recent findings, benefits, and limitations. Trends Food Sci Technol 86:328–339
González-Cavieres L, Pérez-Won M, Tabilo-Munizaga G, Jara-Quijada E, Díaz-Álvarez R, Lemus-Mondaca R (2021) Advances in vacuum microwave drying (VMD) systems for food products. Trends Food Sci Technol 116:626–638
González-Monroy AD, Rodríguez-Hernández G, Ozuna C, Sosa-Morales ME (2018) Microwave-assisted pasteurization of beverages (tamarind and green) and their quality during refrigerated storage. Innovative Food Sci Emerg Technol 49:51–57
Gratz M, Schottroff F, Gall L, Zejma B, Simon F, Jaeger H (2021) Advantages of ohmic cooking in the kilohertz range–part II: impact on textural changes and approaches to improve heating uniformity. Innovative Food Sci Emerg Technol 67:1–13
Hemis M, Choudhary R, Gariépy Y, Raghavan VG (2015) Experiments and modelling of the microwave assisted convective drying of canola seeds. Biosyst Eng 139:121–127
Hemis M, Choudhary R, Becerra-Mora N, Kohli P, Raghavan V (2016) Modelling of microwave assisted hot-air drying and microstructural study of oilseeds. Int J Agric Bio Eng 9(6):167–177
Hock TK, Chala GT, Cheng HH (2020) An innovative hybrid steam-microwave sterilization of palm oil fruits at atmospheric pressure. Innovative Food Sci Emerg Technol 60:102289

Hong Y-K, Liu F, Tang Z, Pedrow PD, Sablani SS, Yang R, Tang J (2021) A simplified approach to assist process development for microwave assisted pasteurization of packaged food products. Innov Food Sci Emerg Technol 68:102628

Hu X, Kurian J, Gariepy Y, Raghavan V (2017) Optimization of microwave-assisted fluidized-bed drying of carrot slices. Dry Technol 35(10):1234–1248

Icier F (2012) Ohmic heating of fluid foods. In: Taylor SL (ed) Novel thermal and non-thermal Technologies for Fluid Foods. Academic, pp 305–367

Icier F, Izzetoglu GT, Bozkurt H, Ober A (2010) Effects of ohmic thawing on histological and textural properties of beef cuts. J Food Eng 99(3):360–365

Inanoglu S, Barbosa-Cánovas GV, Patel J, Zhu M-J, Sablani SS, Liu F, Tang Z, Tang J (2021) Impact of high-pressure and microwave-assisted thermal pasteurization on inactivation of *Listeria innocua* and quality attributes of green beans. J Food Eng 288:110162

Inmanee P, Kamonpatana P, Pirak T (2019) Ohmic heating effects on *Listeria monocytogenes* inactivation, and chemical, physical, and sensory characteristic alterations for vacuum packaged sausage during post pasteurization. LWT-Food Sci Technol 108:183–189

Jain D, Tang J, Liu F, Tang Z, Pedrow PD (2018) Computational evaluation of food carrier designs to improve heating uniformity in microwave assisted thermal pasteurization. Innovative Food Sci Emerg Technol 48:274–286

James S, James C, Purnell G (2017) Microwave-assisted thawing and tempering. In: Regier M, Knoerzer K, Schubert H (eds) The microwave processing of foods. Woodhead Publishing, pp 252–272

Jia G, Sha K, Meng J, Liu H (2019) Effect of high voltage electrostatic field treatment on thawing characteristics and post-thawing quality of lightly salted, frozen pork tenderloin. LWT-Food Sci Technol 99:268–275

Jiang N, Liu C, Li D, Zhang Z, Liu C, Wang D, Niu L, Zhang M (2017) Evaluation of freeze drying combined with microwave vacuum drying for functional okra snacks: antioxidant properties, sensory quality, and energy consumption. LWT-Food Sci Technol 82:216–226

Jittanit W, Khuenpet K, Kaewsri P, Dumrongpongpaiboon N, Hayamin P, Jantarangsri K (2017) Ohmic heating for cooking rice: electrical conductivity measurements, textural quality determination and energy analysis. Innovative Food Sci Emerg Technol 42:16–24

Jo Y-J, Park SH (2019) Evaluation of energy efficacy and texture of ohmically cooked noodles. J Food Eng 248:71–79

Jouquand C, Tessier FJ, Bernard J, Marier D, Woodward K, Jacolot P, Gadonna-Widehem P, Laguerre J-C (2015) Optimization of microwave cooking of beef burgundy in terms of nutritional and organoleptic properties. LWT-Food Sci Technol 60(1):271–276

Jumah RY, Raghavan G (2001) Analysis of heat and mass transfer during combined microwave convective spouted-bed drying. Dry Technol 19(3–4):485–506

Jung H, Moon JH, Park JW, Yoon WB (2020) Texture of surimi-canned corn mixed gels with conventional water bath cooking and ohmic heating. Food Biosci 35:100580

Kang T, Jun S (2021) Fundamentals of ohmic processing: modeling and commercial applications. Innov Food Process Technol 1:689–698

Kanjanapongkul K, Baibua V (2021) Effects of ohmic pasteurization of coconut water on polyphenol oxidase and peroxidase inactivation and pink discoloration prevention. J Food Eng 292: 1–8

Kaur N, Singh A (2016) Ohmic heating: concept and applications—a review. Crit Rev Food Sci Nutr 56(14):2338–2351

Kim S-S, Kang D-H (2017) Synergistic effect of carvacrol and ohmic heating for inactivation of *E. coli* O157: H7, *S.* Typhimurium, *L. monocytogenes*, and MS-2 bacteriophage in salsa. Food Control 73:300–305

Kingston HM, Haswell SJ (1997) Microwave-enhanced chemistry: fundamentals, sample preparation, and applications. American Chemical Society

Knirsch MC, Dos Santos CA, de Oliveira Soares AAM, Penna TCV (2010) Ohmic heating–a review. Trends Food Sci Technol 21(9):436–441

Kumar C, Karim M (2019) Microwave-convective drying of food materials: a critical review. Crit Rev Food Sci Nutr 59(3):379–394

Kumar P, Coronel P, Truong VD, Simunovic J, Swartzel K, Sandeep K, Cartwright G (2008) Overcoming issues associated with the scale-up of a continuous flow microwave system for aseptic processing of vegetable purees. Food Res Int 41(5):454–461

Laguerre J-C, Hamoud-Agha MM (2020) Microwave heating for food preservation. In: Socaci SA, Fărcaş AC, Aussenac T, Laguerre J-C (eds) Food preservation and waste exploitation, pp 1–18. IntechOpen

Lascorz D, Torella E, Lyng JG, Arroyo C (2016) The potential of ohmic heating as an alternative to steam for heat processing shrimps. Innovative Food Sci Emerg Technol 37:329–335

Li X, Farid M (2016) A review on recent development in non-conventional food sterilization technologies. J Food Eng 182:33–45

Li B, Sun D-W (2002) Novel methods for rapid freezing and thawing of foods–a review. J Food Eng 54(3):175–182

Li Z, Raghavan G, Orsat V (2010a) Temperature and power control in microwave drying. J Food Eng 97(4):478–483

Li Z, Raghavan GV, Wang N (2010b) Carrot volatiles monitoring and control in microwave drying. LWT-Food Sci Technol 43(2):291–297

Li N, Cai Z, Guo Y, Xu T, Qiao D, Zhang B, Zhao S, Huang Q, Niu M, Jia C (2019a) Hierarchical structure and slowly digestible features of rice starch following microwave cooking with storage. Food Chem 295:475–483

Li L, Zhang M, Wang W (2019b) A novel low-frequency microwave assisted pulse-spouted bed freeze-drying of Chinese yam. Food Bioprod Process 118:217–226

Liu L, Llave Y, Jin Y, Zheng D-y, Fukuoka M, Sakai N (2017) Electrical conductivity and ohmic thawing of frozen tuna at high frequencies. J Food Eng 197:68–77

Lu Y, Turley A, Dong X, Wu C (2011) Reduction of *Salmonella enterica* on grape tomatoes using microwave heating. Int J Food Microbiol 145(1):349–352

Luan D, Tang J, Pedrow PD, Liu F, Tang Z (2016) Analysis of electric field distribution within a microwave assisted thermal sterilization (MATS) system by computer simulation. J Food Eng 188:87–97

Luo G, Song C, Hongjie P, Li Z, Xu W, Raghavan G, Chen H, Jin G (2019) Optimization of the microwave drying process for potato chips based on the measurement of dielectric properties. Dry Technol 37(11):1329–1339

Lv W, Li D, Lv H, Jin X, Han Q, Su D, Wang Y (2019) Recent development of microwave fluidization technology for drying of fresh fruits and vegetables. Trends Food Sci Technol 86: 59–67

Lyng JG, McKenna BM, Arroyo C (2018) Ohmic heating of foods. In: Proctor A (ed) Alternatives to conventional food processing. The Royal Society of Chemistry, pp 95–137

Makroo H, Rastogi N, Srivastava B (2020) Ohmic heating assisted inactivation of enzymes and microorganisms in foods: a review. Trends Food Sci Technol 97:451–465

Maloney N, Harrison M (2016) Advanced heating technologies for food processing. In: Leadley CE (ed) Innovation and future trends in food manufacturing and supply chain technologies. Woodhead Publishing, pp 203–256

Md Salim NS, Gariépy Y, Raghavan V (2017) Hot air drying and microwave-assisted hot air drying of broccoli stalk slices (*Brassica oleracea* L. Var. Italica). J Food Process Preserv *41*(3):e12905

Meda V, Orsat V, Raghavan V (2017) Microwave heating and the dielectric properties of foods. In: Regier M, Knoerzer K, Schubert H (eds) The microwave processing of foods. Woodhead Publishing, pp 23–43

Mendes-Oliveira G, Deering AJ, San Martin-Gonzalez MF, Campanella OH (2020) Microwave pasteurization of apple juice: modeling the inactivation of *Escherichia coli* O157: H7 and *Salmonella* typhimurium at 80–90 C. Food Microbiol 87:103382

Mesías M, Wagner M, George S, Morales FJ (2016) Impact of conventional sterilization and ohmic heating on the amino acid profile in vegetable baby foods. Innovative Food Sci Emerg Technol 34:24–28

Miao Y, Chen JY, Noguchi A (2007) Studies on the ohmic thawing of frozen surimi. Food Sci Technol Res 13(4):296–300

Michalak J, Czarnowska-Kujawska M, Klepacka J, Gujska E (2020) Effect of microwave heating on the acrylamide formation in foods. Molecules 25(18):1–21

Min S-G, Hong G-P, Chun J-Y, Park SH (2016) Pressure ohmic thawing: a feasible approach for the rapid thawing of frozen meat and its effects on quality attributes. Food Bioprocess Technol 9(4): 564–575

Moreno-Vilet L, Hernández-Hernández H, Villanueva-Rodríguez S (2018) Current status of emerging food processing technologies in Latin America: novel thermal processing. Innovative Food Sci Emerg Technol 50:196–206

Mousakhani-Ganjeh A, Amiri A, Nasrollahzadeh F, Wiktor A, Nilghaz A, Pratap-Singh A, Mousavi Khaneghah A (2021) Electro-based technologies in food drying – a comprehensive review. LWT 145:111315

Muñoz-Almagro N, Morales-Soriano E, Villamiel M, Condezo-Hoyos L (2021) Hybrid high-intensity ultrasound and microwave treatment: a review on its effect on quality and bioactivity of foods. Ultrason Sonochem 80:105835

Negri Rodríguez LM, Arias R, Soteras T, Sancho A, Pesquero N, Rossetti L, Tacca H, Aimaretti N, Rojas Cervantes ML, Szerman N (2021) Comparison of the quality attributes of carrot juice pasteurized by ohmic heating and conventional heat treatment. LWT 145:111255

Nordin M, Puspasari I, Tasirin S, Daud W, Gariépy Y, Talib M (2014) Quality changes of red pitaya (*Hylocereus undatus*) slices dried in hot air, microwave-hot air and microwave-vacuum dryers. Iranian (Iranica) J Energy Environ 5(3):313–322

Orsat V, Changrue V, Raghavan VG (2006) Microwave drying of fruits and vegetables. Stewart Postharvest Rev 2(6):1–7

Orsat V, Raghavan GSV, Krishnaswamy K (2017) Microwave technology for food processing: an overview of current and future applications. In: Regier M, Knoerzer K, Schubert H (eds) The microwave processing of foods. Woodhead Publishing, pp 100–116

Osaili TM (2012) Developments in the thermal processing of food. In: Bhat R, Alias AK, Paliyath G (eds) Progress in food preservation. John Wiley & Sons Ltd., pp 211–230

Palazoğlu TK, Miran W (2017) Experimental comparison of microwave and radio frequency tempering of frozen block of shrimp. Innovative Food Sci Emerg Technol 41:292–300

Park SH, Balasubramaniam V, Sastry SK, Lee J (2013) Pressure–ohmic–thermal sterilization: a feasible approach for the inactivation of *Bacillus amyloliquefaciens* and *Geobacillus stearothermophilus* spores. Innovative Food Sci Emerg Technol 19:115–123

Park SH, Balasubramaniam V, Sastry SK (2014) Quality of shelf-stable low-acid vegetables processed using pressure–ohmic–thermal sterilization. LWT-Food Sci Technol 57(1):243–252

Park I-K, Ha J-W, Kang D-H (2017) Investigation of optimum ohmic heating conditions for inactivation of *Escherichia coli* O157: H7, *Salmonella enterica serovar* Typhimurium, and *Listeria monocytogenes* in apple juice. BMC Microbiol 17(1):1–8

Patel A, Singh M (2018) Ohmic heating for food products- a review. Current J Appl Sci Technol 27: 1–7

Paulsen E, Moreno DA, Periago PM, Lema P (2021) Influence of microwave bag vs. conventional microwave cooking on phytochemicals of industrially and domestically processed broccoli. Food Res Int 140:110077

Peng J, Tang J, Luan D, Liu F, Tang Z, Li F, Zhang W (2017) Microwave pasteurization of pre-packaged carrots. J Food Eng 202:56–64

Peng Z, Zhu M, Zhang J, Zhao S, He H, Kang Z, Ma H, Xu B (2021) Physicochemical and structural changes in myofibrillar proteins from porcine longissimus dorsi subjected to microwave combined with air convection thawing treatment. Food Chem 343:1–8

Pereira R, Martins RC, Vicente A (2008) Goat milk free fatty acid characterization during conventional and ohmic heating pasteurization. J Dairy Sci 91(8):2925–2937

Pereira RN, Rodrigues RM, Machado L, Ferreira S, Costa J, Villa C, Barreiros MP, Mafra I, Teixeira JA, Vicente AA (2021) Influence of ohmic heating on the structural and immunoreactive properties of soybean proteins. LWT 148:111710

Pu H, Li Z, Hui J, Raghavan GV (2016) Effect of relative humidity on microwave drying of carrot. J Food Eng 190:167–175

Punathil L, Basak T (2017) Microwave food processing applications: modeling investigations. Ref Module Food Sci 1:1–18

Qu Z, Tang Z, Liu F, Sablani SS, Ross CF, Sankaran S, Tang J (2021) Quality of green beans (*Phaseolus vulgaris* L.) influenced by microwave and hot water pasteurization. Food Control 124:107936

Raghavan GV, Orsat V, Meda V (2005) Microwave processing of foods. Stewart Postharvest Rev 3(2):1–7

Ríos-Ríos KL, Gaytán-Martínez M, Rivera-Pastrana DM, Morales-Sánchez E, Villamiel M, Montilla A, Mercado-Silva EM, Vázquez-Barrios ME (2021) Ohmic heating pretreatment accelerates black garlic processing. LWT 151:112218

Roberts JS, Balaban MO, Luzuriaga DA (2002) Comparison of quality attributes of ohmic and water immersion thawed shrimp. J Aquat Food Prod Technol 11(2):3–11

Rocha RS, Silva R, Guimarães JT, Balthazar CF, Pimentel TC, Neto RP, Tavares MIB, Esmerino EA, Freitas MQ, Cappato LP (2020) Possibilities for using ohmic heating in Minas Frescal cheese production. Food Res Int 131:1–9

Rodrigues NP, Brochier B, de Medeiros JK, Marczak LDF, Mercali GD (2021) Phenolic profile of sugarcane juice: effects of harvest season and processing by ohmic heating and ultrasound. Food Chem 347:1–11

Roux S, Courel M, Birlouez-Aragon I, Municino F, Massa M, Pain J-P (2016) Comparative thermal impact of two UHT technologies, continuous ohmic heating and direct steam injection, on the nutritional properties of liquid infant formula. J Food Eng 179:36–43

Salari S, Jafari SM (2020) The influence of ohmic heating on degradation of food bioactive ingredients. Food Eng Rev 12:191–208

Saniso E, Prachayawarakorn S, Swasdisevi T, Soponronnarit S (2020) Parboiled rice production without steaming by microwave-assisted hot air fluidized bed drying. Food Bioprod Process 120:8–20

Schubert H, Regie M (2006) Novel and traditional microwave applications in the food industry. In: Willert-Porada M (ed) Advances in microwave and radio frequency processing. Springer, pp 259–270

Serowik M, Figiel A, Nejman M, Pudlo A, Chorazyk D, Kopec W, Krokosz D, Rychlicka J (2018) Drying characteristics and properties of microwave− assisted spouted bed dried semi− refined carrageenan. J Food Eng 221:20–28

Seyhun N, Ramaswamy HS, Zhu S, Sumnu G, Sahin S (2013) Ohmic tempering of frozen potato puree. Food Bioprocess Technol 6(11):3200–3205

Seyhun N, Sumnu SG, Ramaswamy H (2014) Ohmic heating as thawing and tempering technology. In: Ramaswamy HS, Marcotte M, Sastry S, Abdelrahim K (eds) Ohmic heating in food processing. CRC Press, pp 369–378

Shirkole SS, Jayabalan R, Sutar PP (2020) Dry sterilization of paprika (*Capsicum annuum* L.) by short time-intensive microwave-infrared radiation: establishment of process using glass transition, sorption, and quality degradation kinetic parameters. Innov Food Sci Emerg Technol 62: 102345

Shirkole SS, Sadashiv Mujumdar A, Jayabalan R, Prakash Sutar P (2021) Dry pasteurization of paprika (*Capsicum annuum* L.) by short time intensive microwave-infrared radiation: inactivation of *Salmonella* Typhimurium and *Aspergillus flavus* considering quality degradation kinetics. Food Chem 338:128012

Siguemoto ÉS, dos Santos Funcia E, Pires MN, Gut JAW (2018) Modeling of time-temperature history and enzymatic inactivation of cloudy apple juice in continuous flow microwave assisted pasteurization. Food Bioprod Process 111:45–53

Siguemoto ÉS, Purgatto E, Hassimotto NM, Gut JA (2019) Comparative evaluation of flavour and nutritional quality after conventional and microwave-assisted pasteurization of cloudy apple juice. LWT-Food Sci Technol 111:853–860

Song C, Ma X, Li Z, Wu T, Raghavan GV, Chen H (2020) Mass transfer during osmotic dehydration and its effect on anthocyanin retention of microwave vacuum-dried blackberries. J Sci Food Agric 100(1):102–109

Soni A, Smith J, Thompson A, Brightwell G (2020) Microwave-induced thermal sterilization-a review on history, technical progress, advantages and challenges as compared to the conventional methods. Trends Food Sci Technol 97:433–442

Suleman R, Wang Z, Aadil RM, Hui T, Hopkins DL, Zhang D (2020) Effect of cooking on the nutritive quality, sensory properties and safety of lamb meat: current challenges and future prospects. Meat Sci 167:1–12

Sumnu SG, Sahin S (2012) Microwave heating. In: Sun D-W (ed) Thermal food processing: new technologies and quality issues. CRC Press, pp 555–582

Sun T, Tang J, Powers JR (2007) Antioxidant activity and quality of asparagus affected by microwave-circulated water combination and conventional sterilization. Food Chem 100(2): 813–819

Sun H, Kawamura S, Himoto J-i, Itoh K, Wada T, Kimura T (2008) Effects of ohmic heating on microbial counts and denaturation of proteins in milk. Food Sci Technol Res 14(2):117–123

Swain M, James S (2005) Thawing and tempering using microwave processing. In: Schubert H, Regier M (eds) The microwave processing of foods. Woodhead Publishing, pp 174–191

Tang Z, Mikhaylenko G, Liu F, Mah J-H, Pandit R, Younce F, Tang J (2008) Microwave sterilization of sliced beef in gravy in 7-oz trays. J Food Eng 89(4):375–383

Thuery J (1992) Microwaves: industrial, scientific, and medical applications. Artech House

Tian X, Wu W, Yu Q, Hou M, Jia F, Li X, Dai R (2016) Quality and proteome changes of beef M. longissimus dorsi cooked using a water bath and ohmic heating process. Innovative Food Sci Emerg Technol 34:259–266

Tian X, Shao L, Yu Q, Yang H, Li X, Dai R (2019) Comparative study of survival of Escherichia coli O157: H7 inoculated in pork batter after ohmic cooking and water bath cooking. Int J Food Microbiol 304:11–18

Van der Sman R (2017) Model for electrical conductivity of muscle meat during ohmic heating. J Food Eng 208:37–47

Venkatesh M, Raghavan G (2004) An overview of microwave processing and dielectric properties of agri-food materials. Biosyst Eng 88(1):1–18

Venkatesh M, Raghavan G (2005) An overview of dielectric properties measuring techniques. Can Biosyst Eng 47(7):15–30

Vicente AA, Castro ID, Teixeira JA, Pereira RN (2006) Ohmic heating for food processing. In: Sun DW (ed) Thermal food processing: new technologies and quality issues. CRC Press, pp 459–499

Wang R, Farid MM (2015) Corrosion and health aspects in ohmic cooking of beef meat patties. J Food Eng 146:17–22

Wang L, Sun D-W (2006) Heat and mass transfer in thermal food processing. In: Sun D-W (ed) Thermal food processing: new technologies and quality issues. CRC Press, pp 33–63

Wattanayon W, Udompijitkul P, Kamonpatana P (2021) Ohmic heating of a solid-liquid food mixture in an electrically conductive package. J Food Eng 289:110180

Wray D, Ramaswamy HS (2015) Novel concepts in microwave drying of foods. Dry Technol 33(7): 769–783

Xu Y, Chen Y, Cao Y, Xia W, Jiang Q (2016) Application of simultaneous combination of microwave and steam cooking to improve nutritional quality of cooked purple sweet potatoes and saving time. Innovative Food Sci Emerg Technol 36:303–310

Xu W, Song C, Li Z, Song F, Hu S, Li J, Zhu G, Raghavan GV (2018) Temperature gradient control during microwave combined with hot air drying. Biosyst Eng 169:175–187

Yildiz-Turp G, Sengun I, Kendirci P, İçier F (2013) Effect of ohmic treatment on quality characteristic of meat: a review. Meat Sci 93(3):441–448

Zeinali T, Jamshidi A, Khanzadi S, Azizzadeh M (2015) The effect of short-time microwave exposures on *Listeria monocytogenes* inoculated onto chicken meat portions. Vet Res Forum 6(2):173–176

Zhang M, Tang J, Mujumdar AS, Wang S (2006) Trends in microwave-related drying of fruits and vegetables. Trends Food Sci Technol 17(10):524–534

Zhang H, Patel J, Bhunia K, Al-Ghamdi S, Sonar CR, Ross CF, Tang J, Sablani SS (2019) Color, vitamin C, β-carotene and sensory quality retention in microwave-assisted thermally sterilized sweet potato puree: effects of polymeric package gas barrier during storage. Food Packag Shelf Life 21:100324

Zhong X, Dolan KD, Almenar E (2015) Effect of steamable bag microwaving versus traditional cooking methods on nutritional preservation and physical properties of frozen vegetables: a case study on broccoli (*Brassica oleracea*). Innovative Food Sci Emerg Technol 31:116–122

Chapter 8
Advanced Computational Tools for Enhanced Food Quality and Safety

S. Sreevathsan, Bhavana B. K., Sukumar Debnath, and Sandeep N. Mudliar

Abbreviations

AI	Artificial Intelligence
ANN	Artificial Neural Network
CAC	Codex Alimentarius Commission
CFD	Computational Fluid Dynamics
FAO	Food and Agriculture Organization
FCM	fuzzy c-means
GA	Genetic Algorithm
GMO	Genetically Modified Organism
HACCP	Hazard Analysis and Critical Control Point
IoT	Internet of Things
ISO	International Organization for Standardization
KNN	k-Nearest Neighbour

Authors Sreevathsan S. and Bhavana B.K. have equally contributed to this chapter.

S. Sreevathsan · S. N. Mudliar (✉)
Plant Cell Biotechnology Department, CSIR – Central Food Technological Research Institute, Mysuru, Karnataka, India

Academy of Scientific and Innovative Research (AcSIR), Ghaziabad, Uttar Pradesh, India
e-mail: sandeepm@cftri.res.in

Bhavana B. K.
Food Engineering Department, CSIR – Central Food Technological Research Institute, Mysuru, Karnataka, India

S. Debnath
Academy of Scientific and Innovative Research (AcSIR), Ghaziabad, Uttar Pradesh, India

Food Engineering Department, CSIR – Central Food Technological Research Institute, Mysuru, Karnataka, India

H. U. Hebbar et al. (eds.), *Engineering Aspects of Food Quality and Safety*, Food Engineering Series, https://doi.org/10.1007/978-3-031-30683-9_8

ML	Machine Learning
OUOD	One Up One Down
QR	Quick Response
RFID	Radio Frequency Identification
SVM	Support Vector Machine
USDA	United States Department of Agriculture
US-FDA	United States-Food and Drug Administration
IR	4th Industrial Revolution
HSV	Hue, Saturation and Value
SOCIP	Self-Operating Clean-in-Place
P2P	peer-to-peer
M2M	Machine to Machine
IPFS	Interplanetary File System
PPB	parts per billion
NIR	Near Infra-Red
BSI	Biometric Signature Identification
HVAC	Heating, Ventilation and Air Conditioning
RANS	Reynolds Averaged Navier-Stokes
SST	Shear Stress Transport
SIMPLE	Semi-Implicit Method for Pressure-Linked Eq.
VCC	Virtual Cold Chain
PISO-SIMPLE	Pressure Implicit with Splitting of Operators
CT	Computed Tomography
MRI	Magnetic Resonance Imaging
CAD	Computer-Aided Design
SDGs	Sustainable Development Goals

1 Food Quality and Safety

Food quality and safety have turned out to be the challenging aspects in the food sector, due to the ever-increasing global flows of goods, and needs to be assessed by a 'farm to fork' approach. Food quality and safety can be addressed from many perspectives at different stages of the food value chain, from agricultural practice and post-harvesting followed by supply chain, processing, and transport. Agricultural practices involve the irrational use of pesticides and fertilizers, which may be taken up by the agricultural produce. During post-harvest, improper maintenance of storage conditions deteriorates the quality of food products and decreases their shelf life. Supply chain is mostly recognized by a "one up, one down" (OUOD) approach, where the food supply chain participants know only the immediate supplier (one link up the chain) and the immediate customer (one link down the chain) for a product, thereby affecting food traceability (Pearson et al. 2019). During food processing, various unit operations such as drying, baking, mixing, modified atmosphere packaging, pasteurization, irradiation, and high-pressure processing are adopted to manufacture a wide range of food products like dairy products, frozen foods, meat, seafood, beverages, and processed foods. The selection of food processing methods depends on several internal and external quality and safety assessment factors. Internal factors affecting food quality such as appearance, nutrients, flavour, origin; and food safety involves assessment of pathogen, artificial colorants, pesticide residuals, toxins and heavy metal contaminants (Gao et al. 2020; Talaviya et al. 2020; Lin et al. 2018; Zhang et al. 2014; Xu et al. 2020). The selection of appropriate food processing methods is crucial for retaining the nutritional value and quality of the food product, which also depends on external factors such as temperature, relative humidity, light, gas concentration, sanitation procedure and shelf-life of food products.

1.1 Current Practices to Ensure Food Quality and Safety

In order to address these above-mentioned food quality and safety problems, several measures have already been taken by governments, regulatory bodies and food manufacturers. Quality assurance after the production step has been made mandatory at all food processing and storage units. Additionally, in order to regulate risks associated with food processing, ISO 22000 – food safety management operational guidelines are being followed. ISO also ensures best practices for managing risks in all areas of food production (Cho and Kang 2011). Similarly, Hazard Analysis and Critical Control Point (HACCP) management system, based on United States – Food and Drug Administration (US-FDA) guidelines, involves the evaluation and control of physical, chemical and biological hazards from procurement and handling of raw material to production, manufacturing, supply and consumption of the final product (https://www.iso.org/iso-22000-food-safety-management.html). The Codex Alimentarius is a collection of guidelines and standards established cooperatively by the Food and Agriculture Organization (FAO) and the World Health Organization (WHO), that defines the basic conditions and practices required to safeguard consumer health and ensure fair practices in food industries (https://www.fda.gov/food/hazard-analysis-critical-control-point-haccp/haccp-principles-application-guidelines; http://www.fao.org/fao-who-codexalimentarius/en/).

Meanwhile, the increasing demand for organic food and certification to verify its quality and origin, needs efficient traceability. Traceability of organic food is essential in the food supply chain as it can indicate any usage of pesticides, genetically modified organisms (GMOs), and track the environmental /carbon footprint. The reasons for the authenticity of organic foods include partnering with certified organic farmers, meeting global testing standards as per United States Department of Agriculture (USDA), substituting chemical fumigation with food grade CO_2 fumigation and meeting global quality and hygiene standards in state-of-the-art facilities.

Conventional methods of food quality assessment by physical, chemical, microbiological, nutritional, and sensory parameters depend on specific attributes such as sensory properties, based on colour, flavour, aroma, taste, texture and quantitative properties of the food product. The traditional microbiological detection and identification methods for food borne pathogens are time consuming and is unable to meet demands for food more rapid food testing. Further, the evaluation of internal factors as mentioned above is vital especially to fresh foods such as dairy, meat, fruits and vegetables, and processed foods. As it is difficult to quantitatively assess the factors related to the quality and safety of food products, technologies developed will require high sensitivity, high specificity, low detection limits and portable usage (Cho and Kang 2011). These demanding characteristics can be tackled by the use of robust, diverse and cost-effective computational tools, which will facilitate a significant improvement in the quality and safety of food products.

1.2 Key Issues Controlling Performance

According to the Codex Alimentarius Commission (CAC), "food safety is the guarantee that the corresponding food will not harm the consumer when prepared or eaten as per its intended use" (Trafialek 2019). However, there are several factors/parameters responsible for food contamination which needs to be resolved by viable methods. The factors causing spoilage of food threaten its safe consumption and make it harmful to human health. Contamination of food can often occur via pathogens such as bacteria, viruses and parasites, chemical agents and toxins at almost any stage of the food supply chain, which eventually trigger foodborne diseases. These diseases sometimes lead to mortality in developing countries, due to poor personal and food hygiene. These outbreaks can be averted by maintenance of appropriate storage conditions thereby preventing microbial contamination (Uçar et al. 2016). In other cases, fallacious agricultural practices such as irrational pesticide use can be kept under control by supply-chain traceability. For example, traceability will allow a consumer to acquire data on ripening methods (natural/chemical) and storage conditions of the mangoes purchased at a local supermarket.

As discussed earlier, post-harvest storage is a critical step for safe handling of all foods. Post-harvest losses are sometimes as high as 38% and in developing countries, the percentage may rise up to 60% due to lack of dedicated storage areas with proper temperature maintenance (Gustavsson et al. 2011). Cold-chains are employed to meet the requirements of post-harvest storage, slow down the ripening process, minimize respiratory heating, avoid moisture loss and microbial contamination. Temperature is the most critical parameter to be monitored during storage. However, analysis and improvement of these operating conditions in cold chains requires model-based evaluation by testing different designs and concepts, thereby leading to optimization of the complete process (Ambaw et al. 2013). The following subsections discuss the advent of modern numerical modeling and computational tools in the area of food quality and safety.

1.3 Computational Paradigms

Modeling and simulation tools have proven to be very effective in designing and optimizing a large variety of processes and systems across various industrial sectors. Additionally, food manufacturers have discovered and applied some of these techniques in the past decade (Padhi 2020; Misra et al. 2020; Tian 2016; Delele et al. 2010). A wide variety of factors are involved in the food supply-cum-processing chain and therefore traceability and optimization are key paradigms to effectively manage food quality and safety. With the rapidly growing concern of food safety, reliable scientific data-driven computational tools and databases are the need of the hour. Automated detection of food parameters by modern techniques generates large amount of data and hence requires efficient computational tools. Some of the tools are artificial intelligence (AI) based on artificial neural network (ANN), support vector machine (SVM), random forest, k-nearest neighbour (KNN), decision tree, fuzzy c-means (FCM), genetic algorithm (GA), and so on (Zhu et al. 2021; Zhou et al. 2019). Techniques like blockchain technology are essential for food traceability. On the other hand, computational fluid dynamics (CFD), an advanced modeling and simulation tool, can be extensively used for optimizing fluid-based problems throughout the food processing chain, especially in food storage. It is to be noted that food safety is a function of food processing and supply chain. There are very minimal reports on the application of novel and emerging computational tools for food quality and safety. In view of the above, the chapter discusses the use of AI and blockchain from a traceability and supply chain perspective along with the application of CFD from a food processing/ storage angle.

The massive generation of data by Internet of Things (IoT) devices, sensing systems, web applications and social media have contributed to the rise of Artificial Intelligence (AI) (Koch 2018). AI is a technology in which computers and machines are rendered to mimic the problem-solving and decision-making abilities of the human mind. Most of the machine learning and deep learning methods of AI are applied in various food industries with automated food processing units via sensors, Radio Frequency Identification (RFID) tags, Quick Response (QR) codes. These are primarily employed for data collection followed by data optimization using the respective AI models, which is the key driver to avoid food hazards. Further, food traceability issues can be addressed by the use of blockchain technology, which may be described as a decentralized, distributed digital ledger that records transactions across the entire network. Blockchain can be used for efficiently tracking and authenticating the agri-food products in the supply chain thereby ensuring transparency. CFD is a robust design and analysis technique that involves the simulation of fluid engineering systems. As many processes in the food industry involve fluid flow and thermal systems, CFD simulation has traditionally been employed in food processing industries for the past two decades to provide a powerful early-stage qualitative and quantitative evaluation of the performance of food processing operations, thereby allowing modification and optimization of design parameters or operating conditions for a better workflow. Therefore, CFD can also be utilized for designing and optimizing post-harvest storage systems, airflow chambers and transportation systems in order to extend shelf life and maintain quality and safety of the foods.

This chapter discusses the comprehensive applications of advanced computational tools such as AI, blockchain and CFD to enhance food quality and safety. Brief introductions on AI, blockchain, food supply management and CFD with schematic illustrations are furnished to understand the applicability of these emerging tools. Additionally, case studies pertaining to the application of AI for the prediction of food contaminants, blockchain for traceability of agri-food products, food supply chain management and CFD for cold chains, along with their challenges have been summarized. We believe this chapter will be a significant addition to the existing knowledgebase on food quality, safety and sustainability, through a better understanding of AI-enabled blockchain technology, food supply chain management, and CFD.

2 Modeling Approaches for Food Quality and Safety

2.1 Artificial Intelligence

Recent advancements in AI, machine learning (ML), big data and the era of 4th Industrial Revolution (4.0 IR) are highly influencing the methods of crop farming, cultivation, production and food processing by adhering to the regimens of food quality and safety (Younus 2017). Artificial intelligence works in integration with digital data, sensors and robotics, by analyzing large amounts of data with quick iterative processing and intelligent algorithms thus allowing the AI model to learn automatically from features or patterns in the data. The role of AI is to analyse the data from both inputs and outputs, to derive an output with an enhanced degree of precision and efficiency without the human interference.

2.1.1 Architecture and Working of AI

The basic structure of AI is the artificial neural network (ANN), a theoretical mathematical model which comprises of a number of linear or nonlinear processing elements called nodes, which are interconnected through weighted connections. ANN usually includes 3 parts – input layer, the hidden layer and the output layer, which are usually specified by architecture, learning algorithm and neuron model. The architecture characterizes the interconnection pattern between the different layers of neurons, the learning algorithm updates the weightage to model a particular task appropriately, and the neuron model which is defined by its activation function transforms a neuron's weighted input to its activated output (Njikam and Zhao 2016; Thike et al. 2020). Once the data is collected, AI performs a prediction from the data sets based on a centralized model, in which a group of servers run a specific model against training and validating datasets. The AI model will be validated and optimized for minimum error based on the experimental and predicted values. ML-based AI is more appropriate for agri-food systems, where higher accuracy is expected and frequent training of the system is not a limitation as compared to rule-based AI (model based on pre-defined outcomes).

AI-based applications are gaining attention in the food industry with respect to food safety and quality assurance due to their ability to analyze large amounts of data generated from various monitoring techniques such as X-rays, lasers, spectroscopy or cameras. The data obtained is then examined by both intrinsic and extrinsic characteristics of the produce from harvest to packaging and the output is predicted in few seconds. Some novel methods such as computer vision are used to evaluate the quality of beverages, where the usage of automated analysis techniques integrated with AI is beneficial. The advantages of AI-integrated analysis of food quality parameters include cost-effectiveness and lower time-consumption, which in turn allow accurate, consistent, and reliable results. Further, the use of drones and robots on the fields makes it possible to thoroughly assess the sorting of foods, ripening status of fruits and to monitor the use of herbicides and pesticides during agricultural practices. The following applications exemplify the use of AI in revolutionizing the process mechanism in food industry in order to maintain high standards of food quality and safety:

AI for sorting of foods

Conventional food sorting systems which work on 'programmed as "acceptable" from the "rejected" lot of food products' is being replaced with advanced systems that can make optimized decisions based on AI. AI based food sorting uses image processing technologies such as cameras and near-infrared sensors for sorting and grading according to the size and colour of food product.

The application of image processing for sorting of red and green tomatoes and red and green grapes based on colour and size was analysed for an agricultural product packaging system. The colour recognition was carried out by HSV (Hue, Saturation and Value) analysis, and the size recognition was calculated by measuring the diameter of the object/fruit in the grayscale image and setting the thresholding. The data is of fruit size and colour is fed to the microcontroller to sort the fruit by moving it to the box of red/green tomato or red/green grape (Dewi et al. 2020). To arrive at an optimized decision, effective data collection and monitoring of food processes using AI-based optical sensors, to measure and regulate temperature, humidity, pressure and time along with other areas of improvements are essential (Kosior et al. 2017).

Enhanced traceability with precision

The conventional practice of tracing the food product depended entirely on simply gathering the data from a specific region and interpreting it accurately with respect to that specific region. However, traceability across the global food supply chain requires execution of strategic safety interventions to gather data from various regions, interpretation and validation of the collected data with minimal time consumption. In this regard, AI systems have made it possible to correlate past data and predict certain events across multiple timelines from different regions (Ramírez et al. 2019).

Automation for Self-Operating Clean-in-Place systems (SOCIP)

'Smart agriculture' is a classic example of how automation is expanding the production, processing, and packaging of food products using clean-in-place (CIP) systems, which involves periodic cleaning of the equipments to maintain hygiene. The advantage of a self-operating CIP system is that it avoids human intervention, which in turn limits the risks of cross-contamination via foodborne pathogens (Garbie 2010). Martec of Whitwell Ltd. is now examining its self-optimizing CIP system which utilizes optical fluorescence and ultrasonic imaging technologies to feed data to the designed AI program for the measurement of microbial debris and residual food within the equipments. Based on the output parameters i.e., extent of fouling from the debris, the intelligent decision-making tool for CIP will stop the cleaning phase. Some of the applications of AI in food processing is summarized in Table 8.1.

Table 8.1 AI applications in the agri-food supply chain and processing industries

Artificial intelligence	Process	Reference
Bee project hive network-Oracle	Hive uses IoT sensors to remotely collect data and provide insights into the nature of relationships bees share with their environments and data is stored in Oracle's Cloud.	https://worldbeeproject.org/, https://indianexpress.com/article/cities/pune/a-sweet-success-story-in-12-years-indias-honey-production-grows-by-200-exports-by-207-5736611/
TOMRA	X-ray, NIR (Near Infra-Red) spectroscopy, lasers, cameras and a unique machine-learning algorithm for fruit and vegetable sorting	https://www.tomra.com/en/sorting/food/food-technology
Detox™	Laser sorter utilizes an optical design to identify aflatoxin contamination in nuts such as almonds, peanuts, hazelnuts and figs. The Laser technique captures the low intensity of light reflected by the fungus enabling the detection and removal of aflatoxin contaminated nuts.	https://www.tomra.com/en/sorting/food/food-technology
Genius™	Genius™ sorter offers a variety of inspection technologies in different inspection zones with high resolution cameras and lasers. Further the state-of-art guns reject the foreign materials and defective vegetables, fresh cut fruits and nuts within milliseconds and process the approved products to the processing line, ensuring food safety.	https://www.tomra.com/en/sorting/food/sorting-equipment/genius
Google's TensorFlow	Cameras, X-rays, Near Infra-Red (NIR) spectroscopy and lasers to measure and automatically detect variances in diced potatoes, applicable in baby food for ensuring safety standards. Potatoes can be	https://blog.google/products/google-cloud/how-ai-can-help-make-safer-baby-food-and-other-products/

(continued)

Table 8.1 (continued)

Artificial intelligence	Process	Reference
	sorted based on various end products such as French fries or potato chips. Additionally, it optimises and predicts the end-product which will produce the least waste when cut for specific products	
ICatador	Artificial neural network-based virtualization and analysis of organoleptic attributes of cheese using NIR spectrometry as input data for the quality control process.	García-Esteban et al. (2018)

The data stored in AI may be altered or manipulated by hacking due to storage in a centralized manner (Dinh and Thai 2018). Besides, the origin of data and the authenticity of the sources generating the data are not certain and guaranteed (Qi and Xiao 2018). This may lead to an inaccurate and risky prediction. In order to address these risks and issues in AI-based technology, the concept of blockchain was introduced.

2.2 *Blockchain Technology*

Blockchain was introduced in 2008 as a distributed ledger technology to perform digital transactions via Bitcoin, avoiding the need for intermediaries such as payment gateways, banks, etc. (Tripoli and Schmidhuber 2018). As the vigorous decentralized functionality of the blockchain technology is established and proven for global financial systems, it can also be extended to food safety management systems to resolve food traceability issues across the global food supply chain. Further, the application of blockchain to enhance quality and safety of agri-foods can be addressed by data transparency and traceability, improving food safety and quality monitoring, and reducing the cost of financial transactions.

2.2.1 Working Principle of Blockchain

The architecture of blockchain includes a decentralized model which enables secure peer-to-peer (P2P) transactions based on a cryptographically protected approach such as Hash function (#) (Steiner et al. 2015). A unique hash value (combination of numbers and strings) specific to each transaction (similar to a fingerprint) is generated and the hash algorithm is constructed. Instantaneously, each transaction which is stored in a block has to be validated and approved by computing systems following the blockchain protocol (also referred to as nodes). The nodes can detect if there is any change in transaction by reading the hash value. Each block includes its own hash value and the hash value of the immediately preceding block, thereby forming a chain of blocks. A minor change in the transaction of any previously recorded block will change its associated hash value and consequently breaks the chain. However, the information stored in the blocks are no longer changeable once they are uploaded to the blockchain. Additionally, there is an added advantage that new blocks can be integrated with the blockchain for more transactions as the blockchain can update itself periodically (Xu et al. 2020). Some of the popular decentralized storage technologies of blockchain are Filecoin, BigChainDB, Interplanetary File System (IPFS), Storj, Swarm (Protocol Labs 2017; McConaghy et al. 2016; Benet 2014; Wilkinson 2014; Hartman et al. 1999).

With advanced features of transparency and traceability, blockchain represents an emerging technology in the field of agri-foods, which may transform many aspects of the food industry and enable the improved quality and safety of agri-foods. The traceability chain from 'farm-to-fork' via blockchain technology is depicted in Fig. 8.1a (Borah et al. 2020). Food products that have been monitored by blockchain technology in earlier reports are summarized in Table 8.2.

2.3 Blockchain-Enabled AI Applications in Agri-food Industry for Improved Food Security

The emerging concept of decentralized AI, which is basically a combination of AI and blockchain, enables processing and analysis of the data on trusted, digital platforms (Team 2018). The data can also be secured and shared on the blockchain, in a decentralized model. The key features and benefits of block chain integration with AI are enhanced data security, improved trust on robotic decisions, collective decision making, decentralised intelligence and high efficiency (Salah et al. 2019). AI-enabled blockchain can therefore be applied in food supply chain management for simultaneous food traceability and data protection of the supply chain.

2.3.1 AI-Enabled Blockchain in Rice Supply Chain Management

The application of AI-enabled blockchain in agri-food industry to maintain food quality and safety is discussed using the rice supply chain management (Fig. 8.1b, c) (Kumar and Iyengar 2017). The rice supply chain which begins at farm land requires detailed understanding of artificial intelligence models for the optimization of agricultural applications such as smart irrigation, farming, plant data analysis, next generation farming, food processing, use of pesticides etc. The technology of smart irrigation was developed to increase the agricultural produce without the involvement of manpower by using sensors that detect the level of water, temperature of the soil, nutrient content, operation of irrigation pumps and forecast weather. Machine to Machine technology (M2M) was developed to ease the data sharing and

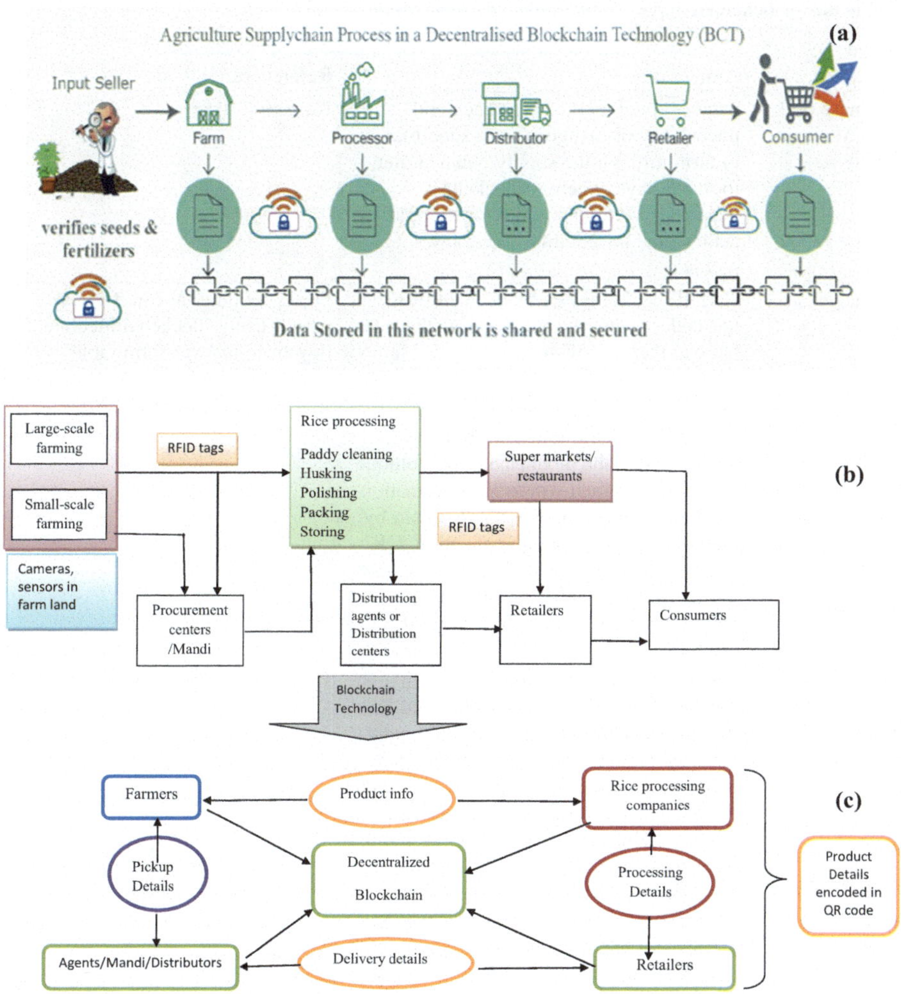

Fig. 8.1 (**a**) The traceability chain from 'farm-to-fork' by implementation of blockchain technology (Reproduced from Borah et al. (2020) with permission from Springer Nature) (**b**) Rice supply chain in India and (**c**) Blockchain framework. (Adapted and modified from Antonucci et al. (2019) and Kumar and Iyengar (2017))

communication gaps with the server or the cloud connected by a central network to all the nodes of the agricultural field (Shekhar et al. 2017). Image analysis of rice cultivation will be captured by computer vision and data collection through sensors, which will be trained, analysed in the developed model based on AI. Further, the harvested rice will be tracked during its shipping which is tagged by RFID to avoid the food fraud affecting the food safety. The traceability of rice from farm to processing unit to procurement centres followed by retailers and consumers will be encoded in a QR code, where the data collected will be stored in a decentralized blockchain.

Table 8.2 Blockchain applications in the agri-food supply chain

Food products	Aim	References
Chinese pork and Mexican mangoes	Conventional methods for back-end traceability of mangoes from supermarket to farm through the supply chain earlier took 6.5 days, whereas blockchain adapted by Walmart and Kroger could achieve the traceability within few seconds	Kamath (2018)
Turkeys	Cargill Inc. harnesses blockchain using QR code to trace the turkeys from the store back to the farm with its Honeysuckle White brand	https://fooddigital.com/food/cargill-using-blockchain-technology-trace-turkeys-farm-table
Coffee	Data (text, images, videos etc) from plantation/ processing/ transportation was uploaded into the blockchain and a unique code was assigned where the information can be shared within the supply chain by web platform or mobile application. QR code on the packaging of the product that transforms normal labels into "smart labels" provides information about the product.	San Domenico coffee (2018)
Tuna fish	Blockchain Supply Chain Traceability Project (WWF 2018) by World Wildlife Foundation (WWF) was initiated to eradicate the illegal tuna fishing, in which the fishermen can register their fish caught, on the blockchain through RFID tags and scanners. Fish caught will be registered by fishermen via SMS and will be stored in blockchain using a permanent, unique identifier such as QR, RFID or barcode. The caught fish transferred to supplier will also be registered on the blockchain and can be accessible to everyone with the unique identifier.	Provenance (2016)
Beef	Blockchain based Hyperledger where the data is stored. Scanning the QR code on the beef product using a mobile application decodes the information regarding the farm that raised the cow, section of the cow, where it was processed, and its package date, beef steak's serial number and 64-digit alphanumeric code referring to the transaction.	Huang (2017)
Beer	DOWNSTREAM beer product is a pioneer using blockchain technology with uniquely marked and authenticated smart QR code, enabling full traceability of every bottle through the brewery and the supplier network to the consumer,	https://www.down-stream.io/

(continued)

Table 8.2 (continued)

Food products	Aim	References
	providing access to information on the premium raw materials – malt, hops, yeast and water and brewing methods	
Bio and DOCG product	To ensure traceability in the agri-food chain by quality and digital identity [for bio and DOCG (Designation of Origin Controlled and Guaranteed) products]	AgriOpenData (2016)
Pork/ Beef	Arc-net is a cloud-based product authentication and traceability service to avoid food fraud involving horse meat labelled and sold as pork/ beef.	http://arc-net.io/
Fresh food	Information on origin of product including the data collection from sensors, permitting data transparency from farm to fork	https://ripe.io/
Milk	To eliminate food fraud in the dairy supply chain by automating the procurement and the registration of information	Milk CyberSecurity (2018)
Pasta	To identify the whole supply chain (i.e., manufacturer, products and flours used, type of drying, transport)	Pasta supply chain (2018)

Adapted and modified from Antonucci et al. (2019)

2.4 Computational Fluid Dynamics

Computational fluid dynamics is a refined modeling and simulation that uses powerful computers to design, analyze and predict fluid flow, phase change, heat transfer, mass transfer, chemical reactions, mechanistic movements, and solid-fluid interaction. CFD facilitates the evaluation of several different design constraints of a physical system using a specifically constructed computational model representing the system. The yardstick of success is observed by how well the results of the numerical simulations correspond with the results of the experiment, in circumstances where laboratory experiments could be performed and how proficiently the simulations can envisage highly complicated processes that cannot be established in the laboratory (Sun 2019).

CFD is a branch of knowledge primarily based on the theory of fluid mechanics. Fluid flows are defined by the conservation laws of mass, momentum and energy in the form of partial differential equations. These equations are replaced by respective algebraic equations during model development in CFD and are then numerically solved. These algebraic equations portray the relation between pressure, temperature, velocity and liquid density with respect to the fluid problem under study. CFD offers a qualitative prediction of fluid flows using: (i) precise mathematical modeling (using Navier-Stokes transport equations) (ii) numerical methods and (iii) software packages (solvers, pre- and post-processing functions) (Tu et al. 2018; Zawawi et al. 2018). The outcome of a CFD study is a function of the physics and numerics available within the software package and can deliver a comprehensive in-depth analysis of a flow system for specific application (Norton and Sun 2006).

Originally, CFD was solely associated with industries involving aerospace and mechanical activities allowing simulation of combustion processes in rocket engines and other physicochemical reactions in and around the rocket airframe. Successively, chemical engineers started employing CFD tools primarily for design of reaction vessels and fluid flow systems (Ranade 2002). Today, CFD is applied in various disciplines across several industries such as aerospace, automotive, chemical, manufacturing, food processing, biomedical, power generation, petroleum exploration, polymer processing, pulp and paper processing, meteorology, astrophysics, medical research and so on (Bayatian et al. 2021; Toparlar et al. 2017; Bakalis et al. 2015). The upcoming sub-sections discuss the equations governing CFD and its applications for enhanced food quality and safety.

2.4.1 CFD in Food Quality and Safety

The food processing sector is unique due to the wide range of distinct constraints that affect process design and development. The food products have to be desirable to the consumer in terms of visual appeal, nutrition, quality and safety. Food processing involves physicochemical and biological interactions of material and non-material components followed by storage and transportation of the manufactured product. This occurs by the means of physical, chemical and biological reactions through transformation of mass, momentum and energy. Robust modeling and simulation tools are employed to diligently study the various transformations during food processing thereby facilitating improved quality and safety (Bakalis et al. 2015).

CFD has grown to be an indispensable part of the design and analysis ecosystem of many food manufacturers due to its capability to predict and analyze the performance of new processes or designs prior to their implementation or fabrication (Padhi 2020; Parpas et al. 2017). One of the undeniable advantages of CFD modeling of food processing operations is the evaluation of the multiphase flow behaviour and prediction of the effect of different operating conditions on the overall process. Process engineers, equipment designers and researchers use CFD to improve the execution and performance of food processing operations, such as drying (Malekjani and Jafari 2018; Ramachandran et al. 2018), baking (Salish et al. 2021; Zhou and Therdthai 2019), sterilization and thermal processing (Park and Yoon 2018; Anandharamakrishnan 2013), refrigeration (Pham et al. 2021; Hoang et al. 2015), mixing (Gomes et al. 2019; Pires et al. 2017) and so on. CFD can be utilized as a robust technique for modeling and simulation of a new food processing facility or retrofitting existing facilities, leading to better performance, quality, safety and energy-optimized operation thus saving up on time, economic cost and manpower.

2.4.2 Governing Equations

The governing equations for CFD of fluid dynamics, heat and mass transfer can be regarded as mathematical expressions of the conservation laws of fluid mechanics and are referred to as the Navier-Stokes equations (Norton and Sun 2006; Farid 2010). These laws of conservation correlate the rate of change of a specific fluid property and associated external forces, when applied to a fluid continuum and can be labelled as:

1. The law of conservation of mass (continuity equation), which states that the mass flows entering a fluid element must balance exactly with those leaving.

$$\frac{\partial \rho}{\partial t} + \frac{\partial}{\partial xi}(\rho uj) = 0 \tag{8.1}$$

2. The law of conservation of momentum (Newton's second law of motion), which states that the sum of the external forces acting on a fluid particle is equal to its rate of change of linear momentum.

$$\frac{\partial}{\partial t}(\rho ui) + \frac{\partial}{\partial xj}(\rho uiuj) = \frac{\partial}{\partial xj}\left[-p\delta ij + \mu\left(\frac{\partial ui}{\partial xj} + \frac{\partial uj}{\partial xi}\right)\right] + \rho gi \tag{8.2}$$

3. The law of conservation of energy (the first law of thermodynamics), which states that the rate of change of energy of a fluid particle is equal to the heat addition and work done on the particle.

$$\frac{\partial}{\partial t}(\rho CaT) + \frac{\partial}{\partial xj}(\rho ujCaT) - \frac{\partial}{\partial xj}\left(\lambda\frac{\partial T}{\partial xj}\right) = sT \tag{8.3}$$

By applying these conservation laws over discrete spatial volumes in a fluid domain, a systematic interpretation of the changes in mass, momentum and energy can be obtained. The aforementioned equations and their governance on CFD can be further explored in detail from "Mathematical modeling of food processing by Farid M.M. (Ed.). 2010" (Farid 2010).

2.4.3 Numerical Analysis and Visualization

CFD code developers and researchers use a selection of diverse numerical methods to discretize the modelled fluid domain. The most significant ones are finite difference, finite element and finite volume methods. Finite difference methods are seldom used for engineering flows due to their limitations in processing of complex geometries. Finite elements are used to model arbitrary geometries such as wind and building interactions in agricultural building designs (Wang et al. 2016). However, due to the technical difficulties involved in the programming and understanding of

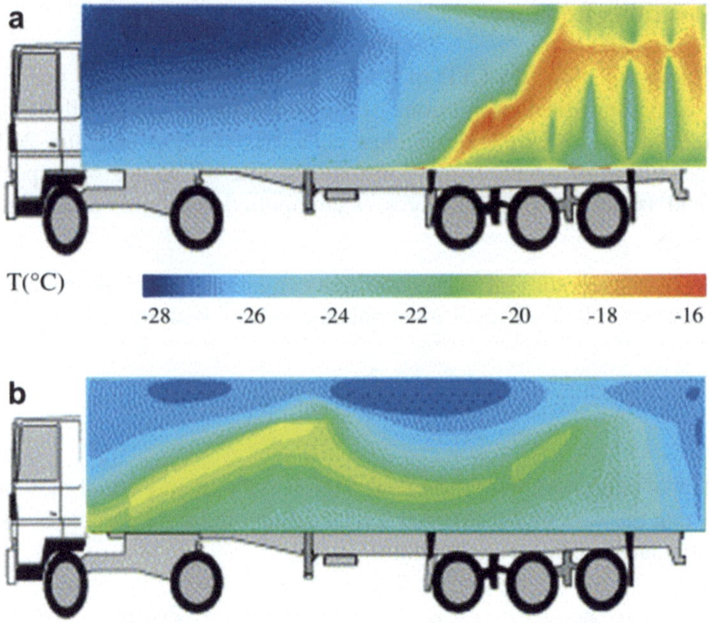

Fig. 8.2 Contours of temperature field within a refrigerated truck (a) with air and (b) without air. (Reproduced from Moureh and Flick (2004) with permission from Elsevier)

this method, a very limited number of commercial finite element packages exist. The finite volume method provides an organized report of the changes in mass, momentum and energy of the fluid across the boundaries within the computational domain (Versteeg and Malalasekera 2007). Therefore, finite volumes are most commonly used due to their ease of understanding, programming and flexibility. The model to be selected for CFD simulation may be laminar or turbulent – based on the type of fluid flow. In a turbulent model, which generally denotes the realistic fluid flow in most applications, the effect of driven forces, eddies and vorticities are incorporated.

Visualization of a CFD problem is almost always necessary to represent its accurate solution. Contour plots, vector and line plots augment the validity of simulation results and have also been employed in several studies to support design and fabrication (Ciano et al. 2021; Inostroza et al. 2021; Demissie et al. 2019). Figure 8.2 illustrates an example of the use of visualization methods to enhance the design process (Moureh and Flick 2004). Today, there are a number of commercial CFD codes and packages that are capable of addressing different problems faced in different areas of engineering (Boysan et al. 2009). Some of the most commonly used commercial CFD codes are ANSYS FLUENT, CFD-ACE, ANSYS CFX, PHOENICS, FLOTHERM and OpenFOAM. These codes incorporate almost all functionalities, employ intuitive graphical user interfaces and support Windows, UNIX and Linux platforms.

3 Applications of Advanced Computational Tools in Food Quality and Safety – Case Studies

3.1 Food Supply Chain Management

Food supply chain is a complex system involving the uninterrupted food flow from farm to consumers. However, there is a lack of transparency and traceability due to the interference of various middlemen. This interference may have significantly negative effects on the freshness of agricultural produce and food quality, and hence the information of all the processes and stages involved in the food supply chain needs to be tracked. The main stages characterizing an agri-food supply chain are agricultural production, post-harvest handling, processing, distribution, retailing and consumption (Caro et al. 2018).

The supply chain in the food industry depends on automated machines which is based on AI, machine vision, navigation technologies and sensor technologies to record the data of temperature, microbiological information and other food quality parameters over the lifecycle of the food products (Jedermann et al. 2014; Heising et al. 2014; Abad et al. 2009). For instance, Radiofrequency Identification (RFID), a sensor-based technology is being applied in food processes, agri-food supply chain industry as one of the efficient traceability systems. The use of RFID and storage of data using blockchain technology in the agri-food supply chain reduces the losses during the logistics process and also maintains food quality and safety. Various food chain traceability initiatives have been taken up and one such initiative by BigchainDB, is based on HACCP, blockchain and IoT (McConaghy et al. 2016). The system provides an information platform for all the supply chain members ensuring transparency and safety (Tian 2016). Also, a recent initiative by Walmart to track and trace the supply of leafy green vegetables by uploading the data to the blockchain using IBM Food Trust Network, can be applicable to other suppliers to ensure food safety along the supply chain. This ensured quick traceability of products back to the farms and harvest sites.

3.1.1 Foodborne Outbreaks in Supply Chain

Food sector faces additional challenges apart from food processing, such as the foodborne outbreaks in the supply chain, which needs immediate attention in the food regulatory framework. The insufficient measures adopted by food regulatory agencies to trace the origin of contamination in foodborne outbreaks affects the public health thereby losing trust of the food product in the supply chain indefinitely. Some of the foodborne outbreaks are detailed in Table 8.3. Therefore, an understanding of the significant impact of foodborne outbreaks, technological advancements and integrated measures to reduce the risks associated with future outbreaks and therefore its prevention by identification and control of hazards at critical stages of the supply chain within a limited time period should be of utmost priority.

Table 8.3 Foodborne outbreaks in agri-food supply chain

Agricultural food products	Reason	Reference
E. coli outbreak in spinach, 2006	Outbreak originated from Yuma, Arizona, and it took almost 7 days for the Walmart food safety division to identify the source of contamination across the supply chain of lettuce	Kamath (2018)
Salmonella contamination in papayas, 2017	Salmonella outbreak in US market in papayas required almost 3 weeks to trace the source to a single farm in Mexico. Papaya farmers from unaffected areas suffered economic losses because of the inability to rapidly track and trace food products	Kamath (2018)
Traceability of sliced mangoes in USA	Walmart and IBM utilized blockchain to trace sliced mangoes from South and Central America to North America highlighting the significant gap in the traceability. The conventional method of traceability took 7 days to connect the supply chain from consumer to the origin of the mangoes, however, blockchain delivered information within 2.2 seconds	Kamath (2018)
Aflatoxin detection in crops	Monitoring of biotic and abiotic conditions in crop field facilitates the identification of the areas of increased incidence of aflatoxins before the harvested crop could enter the food chain	Pillmann et al. (2006)
Poultry meat supply chain	Poultry meat supply chain needs to be restructured to prevent the spread of the avian influenza A virus (H7N9), as the government authorities enforced the closure of live bird markets (LBM) in disease-affected areas of China	Khokhar et al. (2015)

3.2 Aflatoxin Detection in Nuts: Detox™ by TOMRA, a Case Study of AI

Food contamination by aflatoxins, which can cause cancer is considered a major health risk to consumers and also a commercial risk to food businesses. *Aspergillus* (fungi) species present in the environment as *A. flavus* and *A. parasiticus*, are commonly found in orchards of nuts. The parasite grows when suitable conditions arise and produces a chemical compound known as aflatoxin. Aflatoxin is a secondary metabolite and a potent carcinogen. The concentration of aflatoxins is measured in parts per billion (PPB), which is equivalent to a pinch of salt to a 10-ton bag of potato chips. As aflatoxins are known to cause fever, malaise, abdominal pain, vomiting, hepatitis and cancer, they need to be addressed for food safety. Aflatoxin B1, occurs naturally in a wide range of foods and primarily infects cereal crops, spices, figs, dried fruits, cocoa beans, rice and nuts such as walnut, peanuts and tree nuts. Further, Aflatoxin M1 is sometimes present in milk taken from animals that have consumed feed contaminated by aflatoxin B1. Development of technologies for detection of aflatoxin in milk is essential as pasteurisation of milk does not protect against aflatoxin infections (https://food.tomra.com/blog/aflatoxins-in-foods).

The advanced modeling technology Detox™ using AI sensor-based sorting by TOMRA, uses fluorescent lighting, advanced lasers and near infra-red (NIR) spectroscopy to examine and identify the extremely low intensity of light reflected by the aflatoxin moulds in a variety of food types, thereby detecting aflatoxin contamination (Fig. 8.3). TOMRA's sorting machines works on the principle of unique biometric signature identification (BSI) technology. BSI works by scanning and detecting the biometric characteristics of the nuts, which is compared with the database to determine whether the items should be accepted or rejected (https://www.fareasternagriculture.com/technology/infrastructure/laser-sorting-machines-can-eliminate-the-risk-aflatoxins-in-foods). This technology detects and removes smaller defects as compared to conventional spectral technology, as false-rejection rates are exceptionally low with high yields.

(a) Aflatoxin contamination in brazil nuts and figs

(b) TOMRA Nimbus 640 with double Laser in DETOX ™ Configuration

(c) Aflatoxin detection using Detox™ developed by TOMRA

Fig. 8.3 Aflatoxin detection in nuts using Detox™ model (**a**) Aflatoxin contamination in brazil nuts and figs (**b**) TOMRA Nimbus 640 with double Laser in DETOX ™ Configuration (**c**) Aflatoxin detection using Detox™ developed by TOMRA

3.3 Case Studies of Blockchain Technology

3.3.1 Dairy Industry: A Pioneer in the Adoption of Blockchain Technology

Milk with acceptable quality is processed into a variety of dairy products such as yoghurt, liquid milk products, butter, ice creams, cheese and ghee by food processing operations such as homogenization, pasteurization, packaging and storage (https://www.fao.org/publications/sofa/2016/en/). However, although milk is considered as an essential nutritional supplement for human beings, the quality of dairy products has been a major issue affecting the health, especially that of infants and older people. Therefore, traceability of milk and milk products is essential as it provides access to critical information about its origin and processing methods (Francisco and Swanson 2018).

Figure 8.4 illustrates the traceability of milk supply chain, which begins from the milking process at the farm and labelled with an activated QR code for each churn. The data will be updated on a blockchain and further the QR code of each milk churn can be scanned to validate the quality before combining them together in the giant tank. The transport tank will be attached with another activated QR code security seal to trace the truck movement. The activated QR code provides the information on the location of the farm where the milk was sourced, procurement time and volumes, results of onsite tests, and details of the transportation vehicle, which will be stored in a distributed ledger using blockchain technology.

Further, the raw milk arriving at the dairy industry will be scanned to access the data of milk procurement and the shipment will be accepted post-verification. The information encrypted in the QR code will be logged and the status of the milk batch will be updated on the blockchain server. If any discrepancies regarding the milk quality should arise at the processing unit, the milk batch will be traced back to the particular farm through the unique QR code. The batch of milk with inferior quality can then be rejected as it is no longer safe to be processed further.

The data of the entire dairy supply chain is automatically recorded and uploaded to the blockchain system using the various sensors installed. Finally, at the packaging/ bottling stage, a QR code will be printed on each product which provides complete information of the dairy supply chain to the end users. Each QR code, encrypted by a hash value, provides the information from farm to finished product along with additional data consisting of processing details, batch number and expiry date (Tan and Ngan 2020).

Considering the multiple processes involved in dairy supply chain, a smart contract integrated with IoT sensors will be efficient and time-saving for the transaction between the dairy industry and retailers. Smart contracts offer information on shipment details, process parameters such as temperature and humidity during shipment and shelf-life of the product. The product after delivery will be certified for maintenance of temperature and humidity during the food product shipment, so that the quality is maintained throughout the supply chain. The consumers, who are

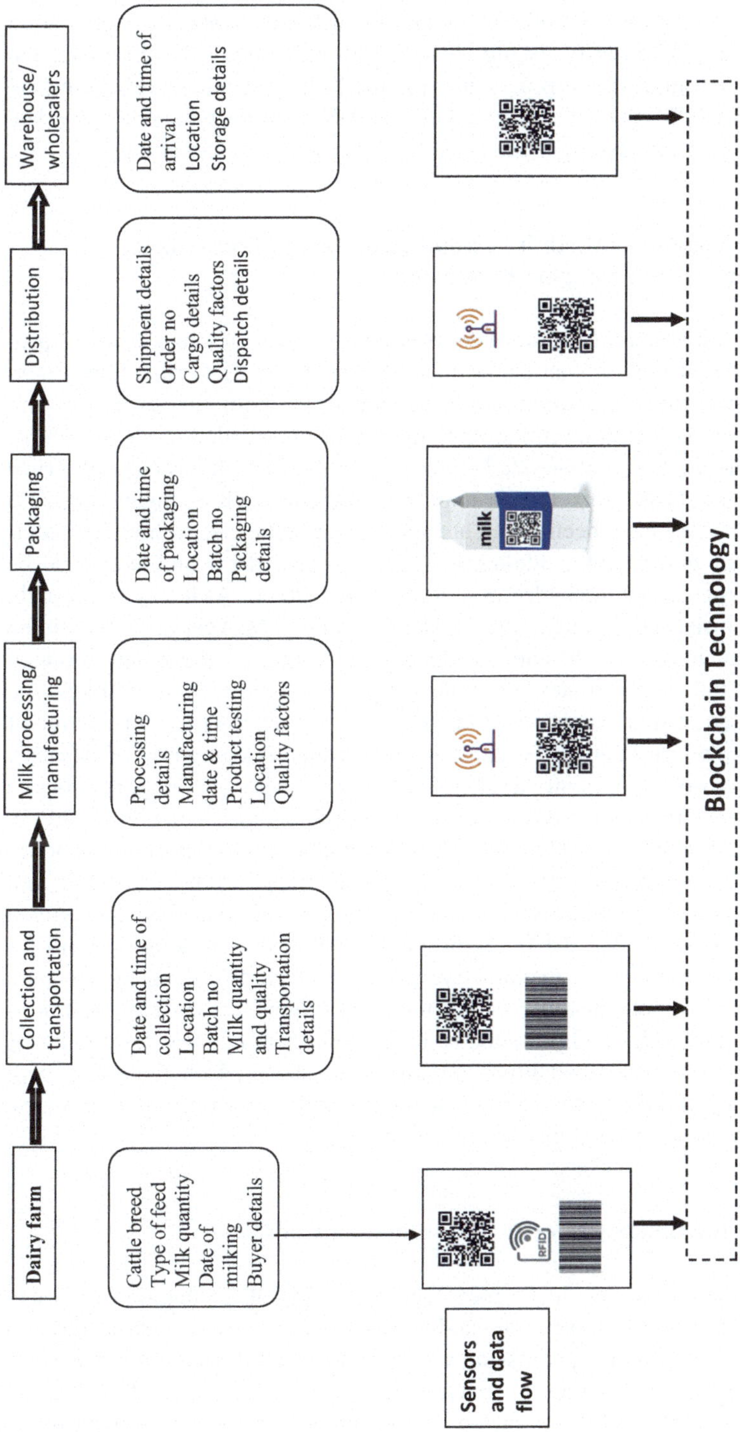

Fig. 8.4 Framework for blockchain-based traceability system in the dairy supply chain. (Adapted and modified from Tan and Ngan (2020))

the end users, can scan the QR code on the product using the respective application on their smartphones to verify its provenance and authenticity. Furthermore, the details of dairy farming, type of processing, the food safety standards regulating the supply chain (HACCP, ISO, national standards) and the manufacture and expiry date of the dairy product can also be made available by blockchain technology.

3.3.2 Hive to Tongue Traceability: Implementation of Blockchain Technology (an Emerging Case Study)

Globally, there are about 94 million beehives with an average annual honey production of about 1.77 million metric tons in 2020 (Statista 2018). Quality honey is an immunity booster, and its traceability is beneficial for both the producers and consumers. Honey is often one of the most impure or mislabelled foods as 76% of honey sold in the market is adulterated (Everstine et al. 2013). It has been observed in recent years that honey mixed with impure ingredients such as sugar, salt, corn syrup and even toxins have been found at suppliers and retailers. Although, analytical methods may detect honey adulteration, they are expensive and tedious to be performed on a regular and scalable basis (Strayer et al. 2014). As honey is currently a highly unorganised sector, due to the constant shifting of beekeepers based on the bee's movement, making it difficult to identify the source of the honey product. Also, to reduce the cost of honey, the honey suppliers and traders often blend the honey from different sources of low-grade honey with a relatively small volume of high-quality, expensive honey. The natural honey which is composed of fructose, glucose, and sucrose is commonly adulterated with corn or sugar syrup, results in a homogeneous mixture which makes it difficult to differentiate (Wang et al. 2015). The syrup or sugar residues in adulterated honey are identical to the natural residues in pure honey. Therefore, the detection of these adulterants becomes difficult, and hence new methods to distinguish the differences between pure and adulterated honey (Drummond and Sun 2010). Additionally, with the increasing demand for organic honey, the suppliers or manufacturers need to meet the criteria such as pure organic honey should not constitute any GMO based content, presence of hazardous substances like antibiotics and pesticide residue, and other chemical residue levels should be below the prescribed limits (Wang et al. 2015). With the increase of adulterated honey flooding the market, the origin and traceability of honey (via labelling) can be powered by blockchain technology.

3.3.3 Unique Identifiers to Verify the Authenticity of Honey

The sale of impure honey could be prevented by testing of adulteration in honey using blockchain-enabled supply chain and traceability. For instance, ultra-filtration is an operation used in honey processing which removes all the naturally beneficial pollen from the honey in order to extend the shelf life of the product thereby rendering it impossible to tell the origin of the honey or the location of beehive.

The collection of data comprising the unique identifiers such as health of bees, temperature of bees and the amount of honey produced can be stored in the blockchain and shared throughout the supply chain between different parties. These unique identifiers can then be identified by the label on the product and can be traced for its origin to verify the authenticity of the honey.

3.3.4 Future of Honey Verification with Blockchain

The tampering of honey with adulterants can be validated by contemplating the locality of the bee hives where the honey was harvested. The analysis of the locality and type of farms surrounding the hives can uncover evidence on the adulteration of honey. For example, if the sample of honey contains maple pollen but the hive is at a location where there are no maple trees, it is understood that the honey is probably adulterated. It is more likely and economically practical that any "standard" applied to honey would be based on how the farm is being managed – for example, the farmer is setting aside a portion of the land to grow pollinator friendly plants all year round. The same can be monitored by capturing aerial images of the land and analysing the division of the lands.

3.4 CFD in Cold Food Chains

3.4.1 Cold Food Chains

A large variety of raw, partially cooked or ready-to-eat chilled and frozen foods are available to consumers all over the world. As the demand for innovative food products and ready-to-eat meals increases, the need for enhanced food safety and improved, optimized cold chain facilities also increases. In order to maintain the quality and extend shelf life of a large number of food products, temperature control is crucial at almost every stage of the food processing chain (Tassou et al. 2015). A chain that comprises the production and distribution of frozen and chilled food products is usually described as a 'cold food chain'. Cold food chains involve heating, ventilation and air conditioning (HVAC) systems, frozen and chilled food display cabinets and refrigerated storage systems: low-temperature (−30 to −40 °C) for frozen foods and medium/ high temperature (−8 to −15 °C) for chilled foods storage. It also includes refrigerated transportation that refers to the conveyance of food by land, sea and air via refrigerated containers (Brown 2008). In the subsequent sections, the effects of freezing and chilling on the quality and safety of fresh and processed foods are explored along with special reference to three case studies, wherein CFD tool is used to enable significant reduction of post-harvest losses and facilitate improved food quality and safety.

3.4.2 Food Quality and Safety Issues at Low Temperatures

Food quality and safety is essential to all fresh and processed foods especially meat, fruits and vegetables. The main objective of chilling or freezing is to extend the shelf life of the food product thereby maintaining its quality. However, certain variation in the characteristics of the food product are often expected due to mismanagement of the cold storage. Oxidation and enzymatic biochemical reactions, especially in fruits, vegetables and sea-food lead to off-flavours and odours. In other cases, the effect of unoptimized cold room conditions is responsible for the degradation of product appearance, palatability and texture (Tassou et al. 2015; Darwesh and Elmetwalli 2015). Additionally, several factors influence the cold chain process efficacy such as temperature distribution, air flow rates, cooling method, storage conditions, thermal-physical interactions, physiological properties, packing design and stacking pattern (Delele et al. 2013a; Chourasia and Goswami 2007a; Pathare et al. 2012).

In order to analyze and improve the existing process and design of the cold chain, modeling of fluid flow, heat and mass transfer is essential during cooling, storage and transportation operations (Ambaw et al. 2013; Delele et al. 2010). CFD solves the governing equations (Eqs. 8.1, 8.2, and 8.3) for cold chain operations to a high degree of accuracy using a very fine mesh of the geometry under study. CFD is also reported to be the most suitable and commonly used technique to analyze the aerodynamics and thermal distributions inside cold storages (Brown 2008; de Albuquerque et al. 2019). Table 8.4 recapitulates the recent use of CFD for design and evaluation of post-harvest cold storages.

3.4.3 Case Study I – CFD Based Optimization of Precooling Conditions of Dates in Cold Room

Ghiloufi and Khir (2019) studied the modeling and optimization of precooling process for dates using CFD. The cold storage (5.6 × 3.62 × 3.3) was designed to preserve about 11 tons of dates at 5 °C. Compression vapor refrigeration system, with 3 axial fans of 45 cm diameter and rotating velocity of 1500 RPM, was used to deliver the necessary cooling capacity. The dates were packed in 432 plastic bins (62 × 32 × 15 cm) and arranged in 3 rows; each row containing 84 bins with 2 m height and circulation corridors were provided. Simulation was performed considering three designs of the cold storage: (1) normal cold room, (2) cold room with single deflector and (3) cold room with three aerodynamic deflectors.

Model Development

The Reynolds Averaged Navier-Stokes (RANS) equations were resolved in three dimensions to determine the air flow profiles. The turbulent model used was the k-ω (SST) developed by Menter (Liu et al. 2012) to study the airflow inside the cold room. This model is best suited to describe the air spinning patterns in a cold room (Nahor et al. 2005). The governing equations for the study have already been

Table 8.4 Recent applications of CFD in design of post-harvest storage facilities

Food product/ replica	Modelled system	CFD Modeling			Reference
		Turbulence model	Numerical technique	Validation	
Beef	Microbial inactivation of pasteurized foods	No	Finite element	3D-heat transfer model, thermal and microbial inactivation kinetics	Delele et al. (2019a)
Beef carcasses	Industrial cooling system Heat and Mass transfer during Industrial cooling	k-ω SST	Finite element	Temperature profile, heat and mass transfer coefficient	Delele et al. (2019b)
Dates	Precooling system and cold storage design	k-ω SST	Finite volume	Design validated with airflow and temperature distribution profiles	Ghiloufi and Khir (2019)
Apple	Heat transfer between product and air	k-ω SST	Finite volume	Anemometry – air velocity and temperature measurements	Hoang et al. (2015)
Citrus fruit	Forced convective cooling conditions, energy consumption	k-ω SST	Finite volume	Cooling rate and heat transfer distribution across packaging	Delele et al. (2012)
Apple	Thermonebulisation fungicide fogging system	SST	Finite volume	Particle deposition as a function of stack positioning	Ambaw et al. (2011)
Apple	Diffusion and adsorption of gas during cold storage	SST	Finite volume	1-MCP concentration measurements	Ferrua and Singh (2011)
Strawberry	Forced-air cooling process of fresh strawberry packages	No (laminar)	Finite volume	Particle image velocimetry, temperature measurements	Arêdes Martins et al. (2011)
Apple	Forced air cooling of fruits	No (laminar)	Finite volume	Comparison of model calculated Nusselt number against literature	Ho et al. (2006)
Pear	Gas permeation, diffusion and respiration kinetics	No	Finite element	Gas concentration profiles and fluxes	Ho et al. (2008, 2010); Cuesta and Lamúa (2009)

(continued)

Table 8.4 (continued)

Food product/ replica	Modelled system	CFD Modeling Turbulence model	Numerical technique	Validation	Reference
Not specific	Heat conduction during chilling of fruit and vegetables	No	Empirical correlation (Fourier series)	No validation	Delele et al. (2009)
Chicory root	Evaluation of chicory root cold store humidification	SST	Finite volume	Mean air velocity	Moureh et al. (2009a)
Cheese and meat products	Mist flow process in refrigerated display cabinets	RNG k-ε	Empirical correlation	Velocity and temperature measurements	Moureh et al. (2009b); Amara et al. (2008)
Not specific	Flow field inside domestic refrigerator	No (laminar)	Finite volume	Particle image velocimetry	Alvarez and Flick (2007)
PVC spheres	Cooling of stack of food products	Macroscopic	Empirical correlation + finite volume	Air velocity and temperature measurements	Chourasia and Goswami (2007b)
Potato	Airflow, heat and mass transfer	No	Finite volume	Time-temperature history and weight loss of product	Allais et al. (2006)
Gel-filled celluloid spheres	Mist-chilling of a stack of spheres	No	Empirical correlation + finite volume	Air velocity and water mass flow rate measurement	Menter (1993)

discussed under Sect. 2.4.2 (Eqs. 8.1, 8.2, and 8.3) of this chapter. Precooling process for dates takes place immediately after harvest from 35 to 40 h depending on initial temperature and refrigeration conditions. Several earlier reports have indicated that the heat respiration of the product during this precooling period is extremely minimal and hence can be neglected (Delele et al. 2012; Gowda et al. 1997; Qiu and Wang 2015). Therefore, the heat respiration and mass losses of dates were not considered for the actual simulation. Assuming each bin of dates to be a solid block, the convective heat transfer coefficient h_{cv} related to the convective flux $q_{dates\text{-}air}$ can be obtained by:

$$hcv = \frac{qdates - air}{Td - TS} \tag{8.4}$$

where T_d is the initial temperature of dates and T_S is the storage temperature.

The model was established using CFD code ANSYS FLUENT 17 using second-order discretization method and SIMPLE algorithm for pressure-velocity coupling. Steady-state simulations are performed to study the airflow distribution, temperature distribution and convective heat transfer coefficient at date-air interface. The cold room space was divided into 16 million cells using the ANSYS Mesher with additional finned meshing near the interfaces. Further details on the model, boundary conditions, geometry and meshing can be found in Ghiloufi and Khir (2019).

Simulation Results

The established CFD model was validated considering earlier studies performed on cold storage of apples (Hoang et al. 2015). The simulation results for the three different cold storage designs are detailed below.

Normal cold room

Figure 8.5a displays the air flow circulation. The air velocity has a swirling profile due to the rotation of the fans. The velocity is maximum (7.5 m s^{-1}) at the extremities and almost zero at the centre. A recirculation zone is observed due to the air turbulences near the walls. This air flow distribution affects the cooling behaviour of the product in the cold storage. In order to obtain high and uniform air velocity, a steady temperature distribution throughout the cold storage needs to be established. The temperature distribution after 25 h of cooling was observed in three different planes. A large variation in the cooling zones can be observed in the normal room due to the heterogeneous cooling rate causing poor air flux distribution. This will lead to increased biological and mass losses and deteriorate the quality of the stored product.

Cold room *with single deflector*

The air flow after 25 h of cooling inside the cold room using a single deflector (baffle) is depicted in Fig. 8.5b. Although most of the zones are optimally cooled, several bins remain at temperatures >15 °C. It can be observed that although the usage of a single baffle improves the product refrigeration, the arrangement was not able to ensure uniform cooling of the product.

Cold room with three aerodynamic deflectors

A better airflow distribution was obtained with the three deflector design (Fig. 8.5c). Several recirculation zones were observed above the bins due to the increased turbulence developed by the deflectors. A uniform temperature distribution was observed inside the cold room with three aerodynamic deflectors. All the product bins are cooled to temperature < 10 °C in all the planes which validates the effectiveness of the arrangement. Additionally, results indicate that the new design significantly reduces the precooling period by 6 h compared to cold room with single deflector and by 10 h compared to normal cold room. The new arrangement with three deflectors also considerably minimizes the energy consumption and therefore the operating cost of the cold storage.

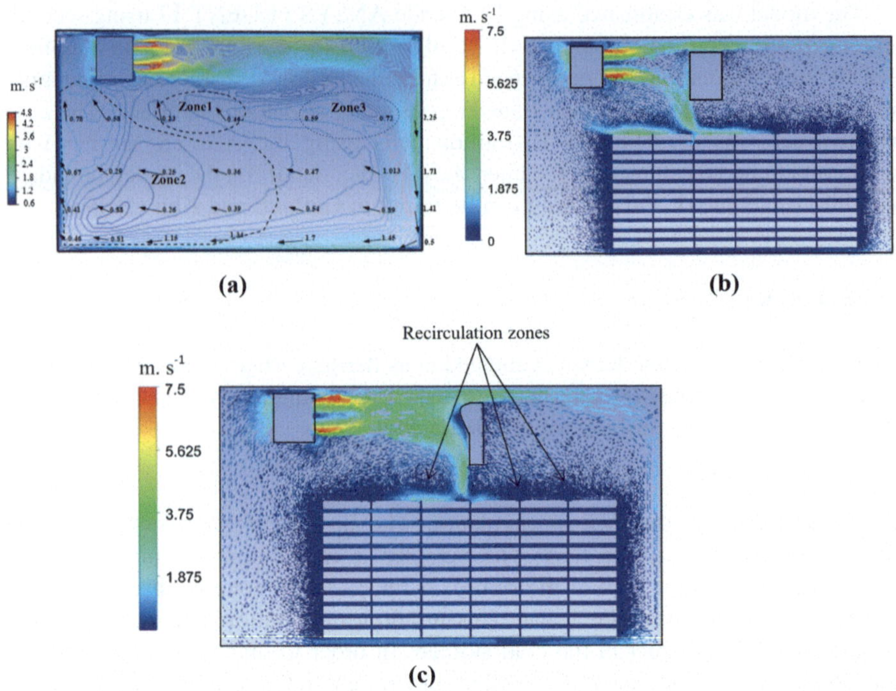

Fig. 8.5 Temperature distribution in (**a**) normal cold room (**b**) cold room with single deflector (**c**) cold room with three deflectors. (Reproduced from Ghiloufi and Khir (2019) with permission from Springer Nature)

3.4.4 Case Study II – Virtual Cold Chain (VCC) Method to Model Post-harvest Temperature History and Quality Loss of Citrus Fruit

Loss of quality is a growing concern in the fruit industry. Post-harvest losses are often as high as 13–38% (Gustavsson et al. 2011). During the preservation and storage of fresh produce, temperature is the most important parameter affecting their quality, shelf life and spoilage rates. It is essential to maintain optimum temperature throughout the supply chain (Pelletier et al. 2011). Therefore, quality observation throughout the cold chain is crucial to track the temperature history of the fruit. Conventionally, temperature monitoring is performed using point probes at specific locations on the pallets or cartons and it does not accurately represent overall fruit quality (Delele et al. 2013b). A better alternative would be to apply numerical modeling via computational fluid dynamics and calculate the core temperatures of each fruit by modeling the heat transfer within the fruit and the surrounding airflow (Wu et al. 2018). This can be accomplished by the use of a virtual cold chain (VCC) strategy wherein, temperatures of packed fruit can be tracked throughout different unit operations of a cold chain. By definition, a virtual cold chain employs a CFD-based workflow to obtain the thermal history and predict the quality evolution of every product going through the different operations of a cold chain. With the help of VCCs, post-harvest losses in the fruit and vegetable industry can be significantly reduced (Menter 1994).

Wu et al. (2018) developed a VCC method for tracking the temperature of packed citrus fruit through various unit operations in different cold chain scenarios using CFD modeling. The temperature information of individual fruits are collated to predict fruit quality loss throughout the entire post-harvest supply chain. The VCC method for citrus fruit (oranges) is discussed in the case study for a conventional cold chain entailing three operations – precooling, refrigerated transport and cold storage. Although only a single carton is used for this study, it is adequate to show the efficacy of the VCC method and can be later extended to other complex models.

Cold Chain Strategies

The VCC method in the present study is used to comparatively assess the temperature history and fruit quality loss throughout the five cold chain strategies. The VCC method recreates five different cold chain strategies used in the fruit industry using different combinations of the above-mentioned unit operations. The different cold chain strategies and their operating parameters are illustrated in the form of a flow chart in Fig. 8.6a. The baseline cold chain (I) is chosen as the control/ representative scenario as it is the most widely used cold chain in the fruit industry; it involves partial precooling to remove fruit heat followed by further heat removal during transport. The cold-disinfestation precooling (II) consists of lower temperatures required for markets demanding a disinfestation protocol to kill insect larvae in fruit. The third cold chain – ambient cooling (III) does not include a precooling step. The fruits are instead stored in static cold storage for 5 days before shipment, establishing a slow cooling process. In the ambient loading (IV) cold chain model, the fruits are directly loaded to the refrigerated container for transport after packing. The final cold chain strategy – holding after precooling (V) contains an additional cold storage before shipment and simulates the scenario where fruits are stored for few days after precooling before being transported.

Model Development

A fibreboard carton (0.4 × 0.3 × 0.27 m) filled with 64 orange fruits according to a predetermined pattern was selected as a model for the study. The fruits are discretely modelled as spheres (d = 7.5 cm) with the total weight of the carton being 13.6 kg. Three separate computational models are developed for precooling, refrigerated transport and storage (Fig. 8.6b). The air flow rates are 1 L kg^{-1} s^{-1}, 0.02 L kg^{-1} s^{-1} and 0.002 L kg^{-1} s^{-1} for precooling, transport and storage, respectively. The initial temperature of the fruit and cardboard is assumed to be 21 °C.

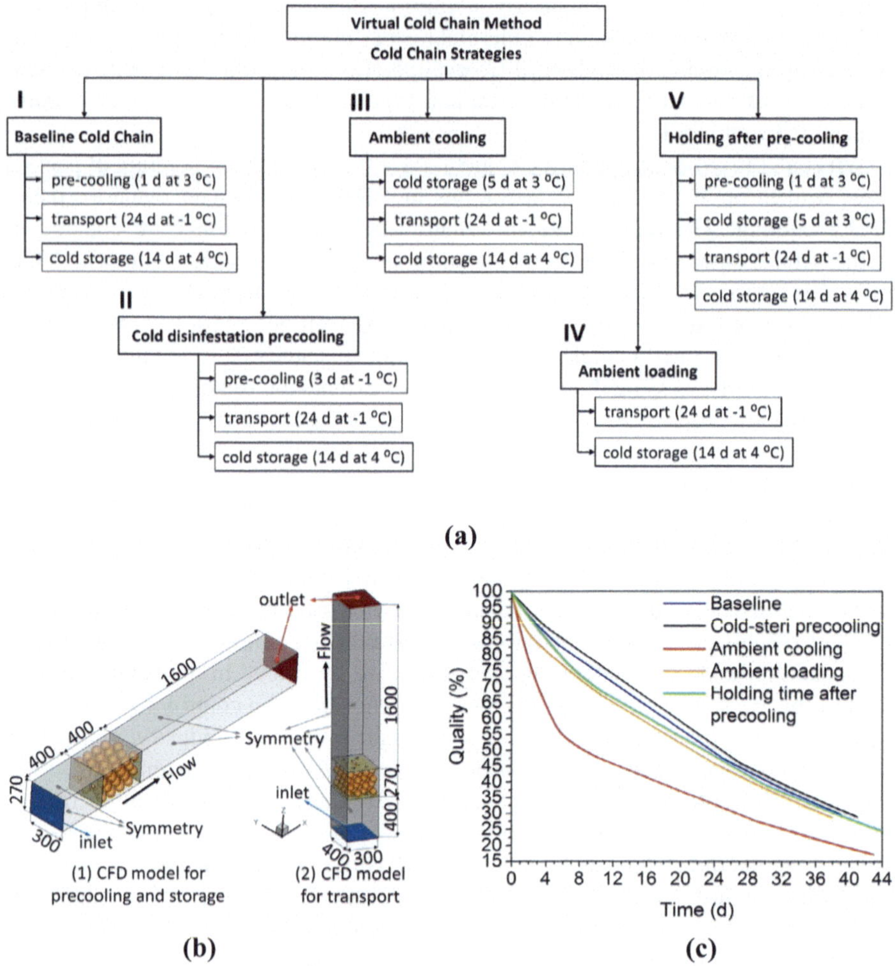

Fig. 8.6 (**a**) Virtual cold chain strategies (**b**) computational models for the VCC method (**c**) average quality loss of fruits in the five cold chains. (Reproduced from Wu et al. (2018) with permission from Elsevier)

The simulations are performed using RANS equations with the shear stress (SST) k-ω turbulence model on the open source CFD code OpenFOAM 2.4.0 (Robertson 2016). All CFD models are meshed using tetrahedral cells. The latent heat of evaporation and heat of respiration have negligible effect on the cooling rate of the oranges and hence are not incorporated in the model. SIMPLE and merged PISO-SIMPLE (PIMPLE) algorithms are used for steady state and time-dependent simulations, respectively. A standard model for predicting the disparity in overall fruit quality over time is developed based on the kinetic rate law (Dermesonlouoglou et al. 2008):

$$-\frac{dA}{dt} = kAn \tag{8.5}$$

where t is the time [s], k is the rate constant [s^{-1}], n is the order of the reaction and A is the change in overall fruit quality estimated by the simulated virtual cold chains. First order reactions are assumed for the overall quality change (A) of oranges. Therefore, Eq. (8.5) can be integrated as:

$$A = A0e - kt \tag{8.6}$$

Arrhenius equation gives the temperature dependency of overall fruit quality loss (Delele et al. 2019b):

$$k(T) = k0e^{\frac{-Ea}{RT}} \tag{8.7}$$

where k_0 is a constant [d^{-1}], E_a is the activation energy [J mol^{-1}], R is the ideal gas constant [8.314 J mol^{-1} K^{-1}], T is the absolute temperature [K].

Simulation Results

The average temperature history of all fruits reveals the differences amongst the five cold chains in a comprehensive manner. The cold disinfestation precooling (II) cold chain is observed to be the most efficiency strategy to lower the fruit temperature. Due to the effects of precooling, the baseline cold chain (I) and holding time after precooling (V) chain are also efficient to cool the fruit quickly to 3 °C. On the contrary, the ambient cooling (III) and loading (IV) cold chains take longer duration to cool the fruit due to lower air flow rates.

The quality degradation of all fruits, based on first order kinetics is illustrated in Fig. 8.6c. The remaining quality after baseline (I), cold-disinfestation precooling (II), ambient cooling (III), ambient loading (IV) and holding time after precooling (V) cold chains are 30%, 29%, 17%, 29%, and 25%, respectively. The ambient cooling cold chain has the lowest residual quality, almost 23% elevated quality loss as compared to the baseline cold chain. This may be attributed to the longer duration of initial cold storage (5 d) before shipment, leading to a significant quality loss. A maximum variation of 2% in quality loss is estimated between individual fruits in the cold-disinfestation after precooling (II) cold chain making it the most effective strategy. Conversely, a maximum difference in quality loss of 5% is observed among the fruits in ambient cooling (III) cold chain, making it the least effective strategy.

3.4.5 Case Study III – Performance Improvement of the Industrial Cooling Process of Large Beef Carcasses by CFD Modeling

Beef is usually stored in cold rooms as carcasses, before being processed to the consumer's preferred size and shape, thereby maintaining hygiene, quality and visual appeal. During industrial air blast cooling in the cold rooms, the carcasses are exposed to the cooling air at a specific temperature, velocity, flow direction and relative humidity which determine the cooling rate of the beef; which in turn affects weight loss, moisture loss, shelf life, rate of biochemical reactions within the carcass and final meat quality (Hamoen et al. 2013). As both very fast and very slow cooling rates have side effects on final meat quality, modern slaughterhouses use controlled cooling methods with an aim to produce meat with desired quality in the shortest possible cooling time. The following case study is discussed in brief as it is fairly similar to the case study discussed in Sect. 3.3.3. on pre-cooling operation during cold storage.

Delele et al. (2019a) evaluated the optimization of industrial cooling of large beef carcasses using CFD. The CFD model allows combination of airflow, temperature, moisture transport, energy transport, weight loss and quality aspects to optimize the cooling efficiency (Kuffi et al. 2016). The study was performed using a validated model of beef carcass developed by Kuffi et al. [142] and solved using commercial CFD code ANSYS-CFX. The cooling problem was solved using the commonly used RANS equations based on the governing equations using a turbulence model. Further details on the model, mesh and computational boundary conditions can be found in Delele et al. (2019a). The study analyzed the effect of pre-cooling temperature and air flow rate on the cooling time quality loss of the beef carcass.

Simulation results indicated that precooling air temperature and duration have a major effect on the cooling time and weight loss, whereas air velocity was not considered a critical factor. Low air velocity was found to be optimal for both energy consumption and retaining overall meat quality. Additionally, low temperature was noted to be the optimum operating condition for the pre-cooling process. If quick drying of the product is desired, high velocity with high temperature is more appropriate with limitations of higher weight and colour loss. However, as low weight loss was aimed in the current study, low velocity and low temperature pre-cooling were deemed ideal. For further discussions on the simulation results, Delele et al. (2019a) can be referred. Therefore, the study summarises that comprehensive CFD models are capable of predicting the optimum operating conditions of industrial cooling of beef carcasses and the approach can be extended to similar cooling operations in the cold chain industry.

4 Challenges, Summary and Outlook

4.1 Challenges

AI and blockchain have been gaining increasing significance due to the recent change in food safety regulations. AI and blockchain focus on the prediction and prevention model rather than the reaction and response model for early prediction of future foodborne outbreaks and other food safety issues. Although the data of food traceability being stored in a decentralized blockchain ledger looks promising, it does not possess a verification system to check the authenticity of the data collected from RFID tags or barcodes. Additionally, as blockchain involves P2P interactions, there is a need to standardize the protocols used in blockchain and its integration with AI for ensuring transparency and data security.

The application of CFD simulations reveals that products can be processed and stored in more efficient systems thereby improving its quality and shelf life. Though CFD provides significant data on the performance of food processing operations and storage environments, the simulation process is time-consuming and requires powerful supercomputers with large storage databases. In recent years, 3D image data from computed tomography (CT) and magnetic resonance imaging (MRI) provides realistic high quality 3D models. However, most existing meshing tools can only operate with models developed by computer-aided design (CAD) tools and have difficulty meshing 3D images. Image-based meshing is expected to open new avenues of modeling in food processing and storage that were earlier challenging due to unavailability of suitable models thereby enabling enhanced food safety.

4.2 Summary

In this chapter, an overview on the current state-of-the-art of AI and blockchain technology in food supply chain management, and CFD for optimization of post-harvest cold storage rooms is reviewed. The applicability of AI and blockchain in gathering the data of agricultural practices and tracing of food products from 'farm to plate approach' is discussed. The role of CFD in food processing as an enabling tool for enhanced food quality and safety is outlined. As a case study, the application of AI in aflatoxin detection in nuts is presented using the Detox™ model by TOMRA. Additionally, the implementation of blockchain technology in the dairy industry is discussed along with an emerging case study of honey traceability. A case study on the application of CFD for optimization of pre-cooling conditions for dates storage in a cold room is reviewed, wherein it was observed that cold room with three deflectors proved to be the most suitable configuration with the best airflow and temperature distribution. In another case study, the development of a virtual cold chain method for tracking the temperature of individual fruits in the cold chain and its simulation using CFD is evaluated. The optimization of industrial pre-cooling of beef carcasses using a comprehensive CFD model is also reviewed as another case study.

AI-enabled blockchain, with its added advantage of decentralized, distributed data storage, enables better agri-food management with enhanced traceability and transparency. The food industry is also exploring the benefits of blockchain technology and next generation genome sequencing for traceability of pathogen outbreaks to ensure food safety. Even though CFD has been a conventional modeling tool in food processing for the last decade, recent computational advancements will enable CFD to be an obligatory tool for optimization of food quality and safety operations. Further, CFD can also facilitate food industries to expand and develop new process strategies corresponding to the demands in the market, while also maintaining high levels of product quality. Therefore, quantitative assessment of the quality and safety of food products is deemed to be highly essential for both the manufacturers and consumers.

4.3 Outlook

Future food safety guidelines to ensure food quality and safety insist on embracing advanced computational tools as an essential part of the global food supply-cum-processing chain. AI-enabled blockchain has the potential to address the challenges of sustainable development goals (SDG) to achieve food security, improve nutrition and promote sustainable agriculture by predicting the environmental effects on nutritional quality and safety of food. In this regard, there is a need to focus on the emerging concern of micropollutants such as heavy metals, pesticides and its end-to-end traceability in vegetables and seafoods across the food chain, in order to predict any foodborne outbreaks. Further, from a policy perspective, a stringent regulatory framework has to be mandated for providing the quality standards information for primary processed food products such as rice, cereals, fruits, vegetables etc. sold by the retail chain along with back-end traceability to locate the agricultural practices of fruits and vegetables. Additionally, usage of big data analytics, though currently in its infancy, will be highly useful in food safety and quality in near future, as it can improve the efficiency of the entire food processing chain by providing predictive insights, correlations, hidden patterns and taking real-time decisions.

CFD aids food manufacturers to stay in complete control of every aspect of food production from the initial to the final stage, thereby delivering a high quality product that is most desirable to the consumer. The continuous advancements in computational technology, parallel processing, advances in numerical methods and persistent effort from CFD researchers are the key factors that will make integration of CFD in food processing conceivable in the near future. Deep machine learning will also be an enabling tool to improve the speed, accuracy, applicability and user-friendliness of the CFD software. Although the modeling platforms of AI, blockchain and CFD have been rigorously adopted in other domains, it is still in its infancy in the agri-food supply chain sector and requires implementation supported by strong research. The advancements in traceability of agri-food products, effective maintenance of operating conditions, and monitoring the quality and integrity of foods pre- and post- processing using emerging advanced computational tools are key to minimize quality issues in the food industry to ensure overall food safety.

Acknowledgments The authors are immensely thankful to Prof. V.V. Ranade for his insights, suggestions and comments on the chapter.

References

Abad E, Palacio F, Nuin M, De Zarate AG, Juarros A, Gómez JM, Marco S (2009) RFID smart tag for traceability and cold chain monitoring of foods: demonstration in an intercontinental fresh fish logistic chain. J Food Eng 93(4):394–399. https://doi.org/10.1016/j.jfoodeng.2009.02.004

AgriOpenData (2016) https://www.youtube.com/watch?v=7o3245dZ_gk. Accessed 24 June 2021

Allais I, Alvarez G, Flick D (2006) Modelling cooling kinetics of a stack of spheres during mist chilling. J Food Eng 72(2):197–209. https://doi.org/10.1016/j.jfoodeng.2004.11.010

Alvarez G, Flick D (2007) Modelling turbulent flow and heat transfer using macro-porous media approach used to predict cooling kinetics of stack of food products. J Food Eng 80(2):391–401. https://doi.org/10.1016/j.jfoodeng.2005.12.052

Amara SB, Laguerre O, Charrier-Mojtabi MC, Lartigue B, Flick D (2008) PIV measurement of the flow field in a domestic refrigerator model: comparison with 3D simulations. Int J Refrig 31(8): 1328–1340. https://doi.org/10.1016/j.ijrefrig.2008.04.005

Ambaw A, Beaudry R, Bulens I, Delele MA, Ho QT, Schenk A, Nicolai BM, Verboven P (2011) Modeling the diffusion–adsorption kinetics of 1-methylcyclopropene (1-MCP) in apple fruit and non-target materials in storage rooms. J Food Eng 102(3):257–265. https://doi.org/10.1016/j.jfoodeng.2010.08.028

Ambaw A, Delele MA, Defraeye T, Ho QT, Opara LU, Nicolaï BM, Verboven P (2013) The use of CFD to characterize and design post-harvest storage facilities: past, present and future. Comput Electron Agric 93:184–194. https://doi.org/10.1016/j.compag.2012.05.009

Anandharamakrishnan C (2013) Applications of computational fluid dynamics in the thermal processing of canned foods. In: Computational fluid dynamics applications in food processing. Springer, New York, NY, pp 27–36

Antonucci F, Figorilli S, Costa C, Pallottino F, Raso L, Menesatti P (2019) A review on blockchain applications in the Agri-food sector. J Sci Food Agric 99(14):6129–6138. https://doi.org/10.1002/jsfa.9912

Arêdes Martins M, Soares de Oliveira L, Osorio Saraz JA (2011) Numerical study of apple cooling in tandem arrangement. Dyna 78(166):158–165

Bakalis S, Knoerzer K, Fryer PJ (eds) (2015) Modeling food processing operations. Elsevier

Bayatian M, Azari MR, Ashrafi K, Jafari MJ, Mehrabi Y (2021) CFD simulation for dispersion of benzene at a petroleum refinery in diverse atmospheric conditions. Environ Sci Pollut Res 1-12: 32973–32984. https://doi.org/10.1007/s11356-020-12254-1

Benet J (2014) IPFS-content addressed, versioned, P2P file system. [Online]. Available: https://arxiv.org/abs/1407.3561. Accessed 24 June 2021

Borah MD, Naik VB, Patgiri R, Bhargav A, Phukan B, Basani SG (2020) Supply chain management in agriculture using blockchain and IoT. In: Advanced applications of blockchain technology. Springer, Singapore, pp 227–242. https://doi.org/10.1007/978-981-13-8775-3_11

Boysan HF, Choudhury D, Engelman MS (2009) Commercial CFD in the service of industry: the first 25 years. In: 100 volumes of 'notes on numerical fluid mechanics'. Springer, Berlin, Heidelberg, pp 451–461. https://doi.org/10.1007/978-3-540-70805-6_35

Brown M (ed) (2008) Chilled foods: a comprehensive guide. Elsevier

Caro MP, Ali MS, Vecchio M, Giaffreda R (2018, May) Blockchain-based traceability in Agri-food supply chain management: a practical implementation. In: 2018 IoT vertical and topical summit on agriculture-Tuscany (IOT Tuscany). IEEE, pp 1–4. https://doi.org/10.1109/IOT-TUSCANY.2018.8373021

Cho YJ, Kang S (eds) (2011) Emerging technologies for food quality and food safety evaluation. CRC Press. https://doi.org/10.1201/b10710

Chourasia MK, Goswami TK (2007a) Simulation of effect of stack dimensions and stacking arrangement on cool-down characteristics of potato in a cold store by computational fluid dynamics. Biosyst Eng 96(4):503–515. https://doi.org/10.1016/j.biosystemseng.2006.12.010

Chourasia MK, Goswami TK (2007b) Steady state CFD modeling of airflow, heat transfer and moisture loss in a commercial potato cold store. Int J Refrig 30(4):672–689. https://doi.org/10.1016/j.ijrefrig.2006.10.002

Ciano T, Ferrara M, Babanezhad M, Khan A, Marjani A (2021) Prediction of velocity profile of water based copper nanofluid in a heated porous tube using CFD and genetic algorithm. Sci Rep 11(1):1–12. https://doi.org/10.1038/s41598-021-90201-x

Cuesta FJ, Lamúa M (2009) Fourier series solution to the heat conduction equation with an internal heat source linearly dependent on temperature: application to chilling of fruit and vegetables. J Food Eng 90(2):291–299. https://doi.org/10.1016/j.jfoodeng.2008.06.036

Darwesh M, Elmetwalli AH (2015) Effect of cold storage conditions on physical properties of potato tubers. Misr J Agric Eng 32(2):789–806. https://doi.org/10.21608/mjae.2015.98640

de Albuquerque CD, Curet S, Boillereaux L (2019) A 3D-CFD-heat-transfer-based model for the microbial inactivation of pasteurized food products. Innovative Food Sci Emerg Technol 54: 172–181. https://doi.org/10.1016/j.ifset.2019.04.007

Delele MA, Schenk A, Ramon H, Nicolaï BM, Verboven P (2009) Evaluation of a chicory root cold store humidification system using computational fluid dynamics. J Food Eng 94(1):110–121. https://doi.org/10.1016/j.jfoodeng.2009.03.004

Delele MA, Verboven P, Ho QT, Nicolai B (2010) Advances in mathematical modelling of postharvest refrigeration processes. Stewart Postharvest Rev 2(1):1–8. https://doi.org/10.2212/spr.2010.2.1

Delele MA, Vorstermans B, Creemers P, Tsige AA, Tijskens E, Schenk A, Opara UL, Nicolaï BM, Verboven P (2012) CFD model development and validation of a thermonebulisation fungicide fogging system for postharvest storage of fruit. J Food Eng 108(1):59–68. https://doi.org/10.1016/j.jfoodeng.2011.07.030

Delele MA, Ngcobo ME, Opara UL, Meyer CJ (2013a) Investigating the effects of table grape package components and stacking on airflow, heat and mass transfer using 3-D CFD modelling. Food Bioprocess Technol 6(9):2571–2585. https://doi.org/10.1007/s11947-012-0895-5

Delele MA, Ngcobo MEK, Getahun ST, Chen L, Mellmann J, Opara UL (2013b) Studying airflow and heat transfer characteristics of a horticultural produce packaging system using a 3-D CFD model. Part I: model development and validation. Postharvest Biol Technol 86:536–545. https://doi.org/10.1016/j.postharvbio.2013.08.014

Delele MA, Kuffi KD, Geeraerd A, De Smet S, Nicolai BM, Verboven P (2019a) Optimizing precooling of large beef carcasses using a comprehensive computational fluid dynamics model. J Food Process Eng 42(4):e13053. https://doi.org/10.1111/jfpe.13053

Delele MA, Kuffi KD, Nicolai B, Verboven P (2019b) CFD modeling to improve the performance of industrial cooling of large beef carcasses. In: Computational fluid dynamics in food processing. CRC Press, p 61

Demissie P, Hayelom M, Kassaye A, Hailesilassie A, Gebrehiwot M, Vanierschot M (2019) Design, development and CFD modeling of indirect solar food dryer. Energy Procedia 158: 1128–1134. https://doi.org/10.1016/j.egypro.2019.01.278

Dermesonlouoglou EK, Pourgouri S, Taoukis PS (2008) Kinetic study of the effect of the osmotic dehydration pre-treatment to the shelf life of frozen cucumber. Innovative Food Sci Emerg Technol 9(4):542–549. https://doi.org/10.1016/j.ifset.2008.01.002

Dewi T, Risma P, Oktarina Y (2020) Fruit sorting robot based on color and size for an agricultural product packaging system. Bull Electr Eng Inform 9(4):1438–1445. https://doi.org/10.11591/eei.v9i4.2353

Dinh TN, Thai MT (2018) AI and blockchain: a disruptive integration. Computer 51(9):48–53. https://doi.org/10.1109/MC.2018.3620971

Drummond L, Sun DW (2010) Effects of chilling and freezing on safety and quality of food products. In: Processing effects on safety and quality of foods. CRC Press, pp 295–322

Everstine K, Spink J, Kennedy S (2013) Economically motivated adulteration (EMA) of food: common characteristics of EMA incidents. J Food Prot 76(4):723–735. https://doi.org/10.4315/0362-028X.JFP-12-399

Farid MM (ed) (2010) Mathematical modeling of food processing. CRC Press. https://doi.org/10.1201/9781420053548

Ferrua MJ, Singh RP (2011) Improved airflow method and packaging system for forced-air cooling of strawberries. Int J Refrig 34(4):1162–1173. https://doi.org/10.1016/j.ijrefrig.2011.01.018

Francisco K, Swanson D (2018) The supply chain has no clothes: technology adoption of blockchain for supply chain transparency. Logistics 2(1):2. https://doi.org/10.3390/logistics2010002

Gao Z, Shao Y, Xuan G, Wang Y, Liu Y, Han X (2020) Real-time hyperspectral imaging for the in-field estimation of strawberry ripeness with deep learning. Artif Intell Agric 4:31–38. https://doi.org/10.1016/j.aiia.2020.04.003

Garbie IH (2010) Enhancing the performance of industrial firms through implementation of lean techniques. In: IIE annual conference. Proceedings. Institute of Industrial and Systems Engineers (IISE), p 1

García-Esteban JA, Curto B, Moreno V, González-Martín I, Revilla I, Vivar-Quintana A (2018) A digitalization strategy for quality control in food industry based on artificial intelligence techniques. In: 2018 IEEE 16th international conference on industrial informatics (INDIN). IEEE, pp 221–226. https://doi.org/10.1109/INDIN.2018.8471994

Ghiloufi Z, Khir T (2019) CFD modeling and optimization of pre-cooling conditions in a cold room located in the south of Tunisia and filled with dates. J Food Sci Technol 56(8):3668–3676. https://doi.org/10.1007/s13197-019-03812-8

Gomes LC, Miranda J, Mergulhão FJ (2019) Operation of biofilm reactors for the food industry using CFD. In: Computational fluid dynamics in food processing. CRC Press, pp 561–590

Gowda BS, Narasimham GSVL, Murthy MK (1997) Forced-air precooling of spherical foods in bulk: a parametric study. Int J Heat Fluid Flow 18(6):613–624. https://doi.org/10.1016/S0142-727X(97)00028-3

Gustavsson J, Cederberg C, Sonesson U, van Otterdijk R, Meybec A (2011) Global food losses and food wastes–extent, causes and prevention. FAO, Rome. http://www.fao.org/3/mb060e/mb060e00.htm. Accessed 25 June 2021

Hamoen JR, Vollebregt HM, Van Der Sman RGM (2013) Prediction of the time evolution of pH in meat. Food Chem 141(3):2363–2372. https://doi.org/10.1016/j.foodchem.2013.04.127

Hartman JH, Murdock I, Spalink T (1999) The swarm scalable storage system. In: Proceedings 19th IEEE International Conference on Distributed Computing Systems. IEEE, pp 74–81

Heising JK, Dekker M, Bartels PV, Van Boekel MAJS (2014) Monitoring the quality of perishable foods: opportunities for intelligent packaging. Crit Rev Food Sci Nutr 54(5):645–654. https://doi.org/10.1080/10408398.2011.600477

Ho QT, Verlinden BE, Verboven P, Nicolaï BM (2006) Gas diffusion properties at different positions in the pear. Postharvest Biol Technol 41(2):113–120. https://doi.org/10.1016/j.postharvbio.2006.04.002

Ho QT, Verboven P, Verlinden BE, Lammertyn J, Vandewalle S, Nicolaï BM (2008) A continuum model for metabolic gas exchange in pear fruit. PLoS Comput Biol 4(3):e1000023. https://doi.org/10.1371/journal.pcbi.1000023

Ho QT, Verboven P, Verlinden BE, Nicolaï BM (2010) A model for gas transport in pear fruit at multiple scales. J Exp Bot 61(8):2071–2081. https://doi.org/10.1093/jxb/erq026

Hoang HM, Duret S, Flick D, Laguerre O (2015) Preliminary study of airflow and heat transfer in a cold room filled with apple pallets: comparison between two modelling approaches and experimental results. Appl Therm Eng 76:367–381. https://doi.org/10.1016/j.applthermaleng.2014.11.012

http://arc-net.io/. Accessed 24 June 2021
http://www.fao.org/fao-who-codexalimentarius/en/. Accessed 24 June 2021
https://blog.google/products/google-cloud/how-ai-can-help-make-safer-baby-food-and-other-products/. Accessed 25 June 2021
https://food.tomra.com/blog/aflatoxins-in-foods. Accessed 24 June 2021
https://fooddigital.com/food/cargill-using-blockchain-technology-trace-turkeys-farm-table. Accessed 24 June 2021
https://ripe.io/. Accessed 24 June 2021
https://worldbeeproject.org/. Accessed 25 June 2021, https://indianexpress.com/article/cities/pune/a-sweet-success-story-in-12-years-indias-honey-production-grows-by-200-exports-by-207-5736611/
https://www.down-stream.io/. Accessed 24 June 2021
https://www.fao.org/publications/sofa/2016/en/. Accessed 24 June 2021
https://www.fareasternagriculture.com/technology/infrastructure/laser-sorting-machines-can-eliminate-the-risk-aflatoxins-in-foods. Accessed 24 June 2021
https://www.fda.gov/food/hazard-analysis-critical-control-point-haccp/haccp-principles-application-guidelines. Accessed 24 June 2021
https://www.iso.org/iso-22000-food-safety-management.html. Accessed 24 June 2021
https://www.tomra.com/en/sorting/food/food-technology. Accessed 25 June 2021
https://www.tomra.com/en/sorting/food/sorting-equipment/genius. Accessed 25 June 2021
Huang E (2017) Blockchain could fix a key problem in China's food industry: the fear of food made in China. Quartz. Available: https://qz.com/1031861/blockchain-could-fix-a-key-problemin-chinasfood-industry-the-fear-of-food-made-in-china/. Accessed 24 June 2021
Inostroza C, Solimeno A, García J, Fernández-Sevilla JM, Acién FG (2021) Improvement of real-scale raceway bioreactors for microalgae production using computational fluid dynamics (CFD). Algal Res 54:102207. https://doi.org/10.1016/j.algal.2021.102207
Jedermann R, Pötsch T, Lloyd C (2014) Communication techniques and challenges for wireless food quality monitoring. Philos Trans R Soc A Math Phys Eng Sci 372(2017):20130304. https://doi.org/10.1098/rsta.2013.0304
Kamath R (2018) Food traceability on blockchain: Walmart's pork and mango pilots with IBM. J Br Blockchain Assoc 1(1):3712. https://doi.org/10.31585/jbba-1-1-(10)2018
Khokhar SG, Min Q, Su C (2015) Bird flu (H7N9) outbreak and its implications on the supply chain of poultry meat in China. J Appl Poult Res 24(2):215–221. https://doi.org/10.3382/japr/pfv007
Koch M (2018) Artificial intelligence is becoming natural. Cell 173(3):531–533. https://doi.org/10.1016/j.cell.2018.04.007
Kosior E, Mitchell J, Davies K, Kay M, Ahmad R, Billiet E, Silver J (2017) Plastic packaging recycling using intelligent separation technologies for materials. In: Proceedings of the annual technical conference, Hyatt Regency, Santa Clara, CA, USA, 12–14 July 2017, pp 500–506
Kuffi KD, Defraeye T, Nicolai BM, De Smet S, Geeraerd A, Verboven P (2016) CFD modeling of industrial cooling of large beef carcasses. Int J Refrig 69:324–339. https://doi.org/10.1016/j.ijrefrig.2016.06.013
Kumar MV, Iyengar NCS (2017) A framework for Blockchain technology in rice supply chain management. Adv Sci Technol Lett 146:125–130. https://doi.org/10.14257/ASTL.2017.146.22
Lin J, Shen Z, Zhang A, Chai Y (2018, July) Blockchain and IoT based food traceability for smart agriculture. In: Proceedings of the 3rd international conference on crowd science and engineering, pp 1–6. https://doi.org/10.1145/3265689.3265692
Liu CC, Ferng YM, Shih CK (2012) CFD evaluation of turbulence models for flow simulation of the fuel rod bundle with a spacer assembly. Appl Therm Eng 40:389–396. https://doi.org/10.1016/j.applthermaleng.2012.02.027
Malekjani N, Jafari SM (2018) Simulation of food drying processes by computational fluid dynamics (CFD); recent advances and approaches. Trends Food Sci Technol 78:206–223. https://doi.org/10.1016/j.tifs.2018.06.006

McConaghy T, Marques R, Müller A, De Jonghe D, McConaghy T, McMullen G, Henderson R, Bellemare S, Granzotto A (2016) Bigchaindb: a scalable blockchain database. White paper, BigChainDB

Menter F (1993) Zonal two equation kw turbulence models for aerodynamic flows. In: *23rd fluid dynamics, plasmadynamics, and lasers conference*, p 2906. https://doi.org/10.2514/6.1993-2906

Menter FR (1994) Two-equation eddy-viscosity turbulence models for engineering applications. AIAA J 32(8):1598–1605. https://doi.org/10.2514/3.12149

Milk CyberSecurity (2018) Available: https://www.greatitalianfoodtrade.it/consum-attori/blockchain-nella-filiera-alimentare-il-prototipo-di-bari/. Accessed 24 June 2021

Misra NN, Dixit Y, Al-Mallahi A, Bhullar MS, Upadhyay R, Martynenko A (2020) IoT, big data and artificial intelligence in agriculture and food industry. IEEE Internet Things J 9:6305–6324. https://doi.org/10.1109/JIOT.2020.2998584

Moureh J, Flick D (2004) Airflow pattern and temperature distribution in a typical refrigerated truck configuration loaded with pallets. Int J Refrig 27(5):464–474. https://doi.org/10.1016/j.ijrefrig.2004.03.003

Moureh J, Tapsoba S, Derens E, Flick D (2009a) Air velocity characteristics within vented pallets loaded in a refrigerated vehicle with and without air ducts. Int J Refrig 32(2):220–234. https://doi.org/10.1016/j.ijrefrig.2008.06.006

Moureh J, Létang G, Palvadeau B, Boisson H (2009b) Numerical and experimental investigations on the use of mist flow process in refrigerated display cabinets. Int J Refrig 32(2):203–219. https://doi.org/10.1016/j.ijrefrig.2008.06.005

Nahor HB, Hoang ML, Verboven P, Baelmans M, Nicolai BM (2005) CFD model of the airflow, heat and mass transfer in cool stores. Int J Refrig 28(3):368–380. https://doi.org/10.1016/j.ijrefrig.2004.08.014

Njikam ANS, Zhao H (2016) A novel activation function for multilayer feed-forward neural networks. Appl Intell 45(1):75–82. https://doi.org/10.1007/s10489-015-0744-0

Norton T, Sun DW (2006) Computational fluid dynamics (CFD)–an effective and efficient design and analysis tool for the food industry: a review. Trends Food Sci Technol 17(11):600–620. https://doi.org/10.1016/j.tifs.2006.05.004

Padhi MR (2020) A review on applications of computational fluid dynamics (CFD) in the food industry. PalArch's J Archaeol Egypt/Egyptol 17(7):10159–10169

Park HW, Yoon WB (2018) Computational fluid dynamics (CFD) modelling and application for sterilization of foods: a review. PRO 6(6):62. https://doi.org/10.3390/pr6060062

Parpas D, Amaris C, Tassou SA (2017) Experimental investigation and modelling of thermal environment control of air distribution systems for chilled food manufacturing facilities. Appl Therm Eng 127:1326–1339. https://doi.org/10.1016/j.applthermaleng.2017.08.134

Pasta supply chain (2018) Available: https://www.aldocozzi.it/news/lablockchain-sbarca-nel-settore-alimentare-la-confezione-di-pastaporta-con-se-tutta-la-sua-storia/. Accessed 24 June 2021

Pathare PB, Opara UL, Vigneault C, Delele MA, Al-Said FAJ (2012) Design of packaging vents for cooling fresh horticultural produce. Food Bioprocess Technol 5(6):2031–2045. https://doi.org/10.1007/s11947-012-0883-9

Pearson S, May D, Leontidis G, Swainson M, Brewer S, Bidaut L, Frey JG, Parr G, Maull R, Zisman A (2019) Are distributed ledger technologies the panacea for food traceability? Glob Food Sec 20:145–149. https://doi.org/10.1016/j.gfs.2019.02.002

Pelletier W, Brecht JK, do Nascimento Nunes MC, Emond JP (2011) Quality of strawberries shipped by truck from California to Florida as influenced by postharvest temperature management practices. HortTechnology 21(4):482–493. https://doi.org/10.21273/HORTTECH.21.4.482

Pham AT, Moureh J, Belaidi M, Flick D (2021) CFD modelling of a pallet of heat-generating product applied to a cheese product. Int J Refrig 128:163–176. https://doi.org/10.1016/j.ijrefrig.2021.03.011

Pillmann W, Geiger W, Isenmann R (2006) Informatics for environmental protection, sustainable development, and risk management 20 years technical committee on environmental informatics. In: Managing Environmental Knowledge. Shaker Verlag. http://enviroinfo.eu/sites/default/files/pdfs/vol114/0001.pdf. Accessed 24 June 2021

Pires JC, Alvim-Ferraz MC, Martins FG (2017) Photobioreactor design for microalgae production through computational fluid dynamics: a review. Renew Sust Energ Rev 79:248–254. https://doi.org/10.1016/j.rser.2017.05.064

Protocol Labs (2017) Filecoin: a decentralized storage network. [Online]. Available: https://filecoin.io/filecoin.pdf. Accessed 24 June 2021

Provenance (2016) Available: https://www.provenance.org/trackingtuna-on-the-blockchain#pilot-phase-2/. Accessed 24 June 2021

Qi Y, Xiao J (2018) Fintech: AI powers financial services to improve people's lives. Commun ACM 61(11):65–69. https://doi.org/10.1145/3239550

Qiu S, Wang J (2015) Effects of storage temperature and time on internal quality of Satsuma mandarin (Citrus unshiu marc.) by means of E-nose and E-tongue based on two-way MANOVA analysis and random forest. Innovative Food Sci Emerg Technol 31:139–150. https://doi.org/10.1016/j.ifset.2015.08.005

Ramachandran RP, Akbarzadeh M, Paliwal J, Cenkowski S (2018) Computational fluid dynamics in drying process modelling—a technical review. Food Bioprocess Technol 11(2):271–292. https://doi.org/10.1007/s11947-017-2040-y

Ramírez EC, Albarrán JC, Salazar LAC (2019, October) The control of water distribution systems as a Holonic system. In: International workshop on service orientation in Holonic and multi-agent manufacturing. Springer, Cham, pp 352–365. https://doi.org/10.1007/978-3-030-27477-1_27

Ranade VV (2002) Computational flow modeling for chemical reactor engineering, vol 5. Academic

Robertson GL (ed) (2016) Food packaging: principles and practice, 3rd edn. CRC Press. https://doi.org/10.1201/b21347

Salah K, Rehman MHU, Nizamuddin N, Al-Fuqaha A (2019) Blockchain for AI: review and open research challenges. IEEE Access 7:10127–10149. https://doi.org/10.1109/ACCESS.2018.2890507

Salish K, Pushpadass HA, Franklin MEE, Mitra H, Muniandy S, Ghosh BC (2021) Three-dimensional computational fluid dynamics modeling of baking of chhana podo (milk cake). J Food Process Eng 44(1):e13587. https://doi.org/10.1111/jfpe.13587

San Domenico coffee (2018) Available: https://www.blockchain4innovation.it/mercati/agrifood/dalla-piantagionealla-tazzina-la-blockchain-entra-nel-caffe/. Accessed 24 June 2021

Shekhar Y, Dagur E, Mishra S, Sankaranarayanan S (2017) Intelligent IoT based automated irrigation system. Int J Appl Eng Res 12(18):7306–7320

Statista (2018) Honey Market Worldwide and in the US. Retrieved Nov 15, 2019 from https://www.statista.com/topics/5090/honey-market-worldwide/

Steiner J, Baker J, Wood G, Meiklejohn S (2015) Blockchain: the solution for transparency in product supply chains. https://www.provenance.org/whitepaper. Accessed 24 June 2021

Strayer SE, Everstine K, Kennedy S (2014) Economically motivated adulteration of honey: quality control vulnerabilities in the international honey market. Food Prot Trends 34(1):8–14. https://www.cropin.com/blogs/following-the-hive-to-honey-trail-using-digital-traceability/

Sun D-W (ed) (2019) Computational fluid dynamics in food processing, 2nd edn. CRC Press. https://doi.org/10.1201/9781351263481

Talaviya T, Shah D, Patel N, Yagnik H, Shah M (2020) Implementation of artificial intelligence in agriculture for optimisation of irrigation and application of pesticides and herbicides. Artif Intell Agric 4:58–73. https://doi.org/10.1016/j.aiia.2020.04.002

Tan A, Ngan PT (2020) A proposed framework model for dairy supply chain traceability. Sustain Futures 2:100034. https://doi.org/10.1016/j.sftr.2020.100034

Tassou SA, Gowreesunker BL, Parpas D, Raeisi A (2015) Modelling cold food chain processing and display environments. In: Modeling food processing operations. Elsevier, pp 185–208. https://doi.org/10.1016/B978-1-78242-284-6.00007-6

Team NA (2018) NEBULA AI (NBAI) – decentralized AI blockchain whitepaper. https://neironix.io/documents/whitepaper/4082/NBAI_whitepaper_EN.pdf. Accessed 24 June 2021

Thike PH, Zhao Z, Shi P, Jin Y (2020) Significance of artificial neural network analytical models in materials' performance prediction. Bull Mater Sci 43(1):1–22. https://doi.org/10.1007/s12034-020-02154-y

Tian F (2016) An agri-food supply chain traceability system for China based on RFID & blockchain technology. In: 2016 13th international conference on service systems and service management (ICSSSM). IEEE, pp 1–6. https://doi.org/10.1109/ICSSSM.2016.7538424

Toparlar Y, Blocken B, Maiheu B, Van Heijst GJF (2017) A review on the CFD analysis of urban microclimate. Renew Sust Energ Rev 80:1613–1640. https://doi.org/10.1016/j.rser.2017.05.248

Trafialek J (2019) The role of suppliers of raw materials in ensuring food safety. Glob J Nutr Food Sci 1(4):1–2. https://doi.org/10.33552/GJNFS.2019.01.000517

Tripoli M, Schmidhuber J (2018) Emerging opportunities for the application of Blockchain in the agri-food industry. FAO and ICTSD: Rome and Geneva. Licence: CC BY-NC-SA, 3. http://www.fao.org/publications/card/en/c/CA9934EN/. Accessed 24 June 2021

Tu J, Yeoh GH, Liu C (2018) Computational fluid dynamics: a practical approach. Butterworth-Heinemann. https://doi.org/10.1016/C2015-0-06135-4

Uçar A, Yilmaz MV, Cakiroglu FP (2016) Food safety–problems and solutions. Significance, prevention and control of food related diseases. IntechOpen, pp 1–26. https://doi.org/10.5772/63176

Versteeg HK, Malalasekera W (2007) An introduction to computational fluid dynamics: the finite volume method. Pearson Education

Wang S, Guo Q, Wang L, Lin L, Shi H, Cao H, Cao B (2015) Detection of honey adulteration with starch syrup by high performance liquid chromatography. Food Chem 172:669–674. https://doi.org/10.1016/j.foodchem.2014.09.044

Wang L, Quant R, Kolios A (2016) Fluid structure interaction modelling of horizontal-axis wind turbine blades based on CFD and FEA. J Wind Eng Ind Aerodyn 158:11–25. https://doi.org/10.1016/j.jweia.2016.09.006

Wilkinson S (2014) Storj a peer-to-peer cloud storage network. http://citeseerx.ist.psu.edu/viewdoc/summary?doi=10.1.1.693.785. Accessed 24 June 2021

Wu W, Cronjé P, Nicolai B, Verboven P, Opara UL, Defraeye T (2018) Virtual cold chain method to model the postharvest temperature history and quality evolution of fresh fruit–a case study for citrus fruit packed in a single carton. Comput Electron Agric 144:199–208. https://doi.org/10.1016/j.compag.2017.11.034

Xu J, Guo S, Xie D, Yan Y (2020) Blockchain: a new safeguard for Agri-foods. Artif Intell Agric 4:153–161. https://doi.org/10.1016/j.aiia.2020.08.002

Younus UM (2017) Fourth industrial revolution. https://www.dawn.com/news/1331909. Accessed 25 June 2021

Zawawi MH, Saleha A, Salwa A, Hassan NH, Zahari NM, Ramli MZ, Muda ZC (2018, November) A review: fundamentals of computational fluid dynamics (CFD). In: AIP conference proceedings, vol. 2030, no. 1. AIP Publishing LLC, pp 020252. https://doi.org/10.1063/1.5066893

Zhang B, Huang W, Li J, Zhao C, Fan S, Wu J, Liu C (2014) Principles, developments and applications of computer vision for external quality inspection of fruits and vegetables: a review. Food Res Int 62:326–343. https://doi.org/10.1016/j.foodres.2014.03.012

Zhou W, Therdthai N (2019) Three-dimensional CFD modeling of continuous industrial baking process. In: Computational fluid dynamics in food processing. CRC Press, pp 193–224

Zhou L, Zhang C, Liu F, Qiu Z, He Y (2019) Application of deep learning in food: a review. Compr Rev Food Sci Food Saf 18(6):1793–1811. https://doi.org/10.1111/1541-4337.12492

Zhu L, Spachos P, Pensini E, Plataniotis KN (2021) Deep learning and machine vision for food processing: a survey. Curr Res Food Sci 4:233–249. https://doi.org/10.1016/j.crfs.2021.03.009

Chapter 9
Recent Trends in Materials and Coatings for Food Packaging and Storage

Rajeshwar S. Matche and Yashika Singh

1 Introduction

Throughout the history, there has been a need to store and preserve all kinds of food products. In the modern world, food industry is one of the major users of packaging materials and has a great impact on our lives. The basic principle of the packaging is to preserve, protect and prolong the shelf-life, marketing of the food product. The changes in need of the consumer habits and preferences leads to a response for development of better materials for food packages. More recently, food packages have been developed which offer biodegradability, more protection in terms of improving shelf life of the food products, packaging that can be consumed along with the food itself. Often the food product and its package are developed to be an integrated unit such edible coating on vegetables, fruits and meat. Technological breakthroughs in bio-polymers have provided a multitude of opportunities for improved food packaging. It is well known that plastic has been the most widely used material for packaging in food industry. The plastics packaging industry has shown the tremendous growth for many reasons. The reasons for plastics to become so popular over the years are many such as they have a wide range of physical and barrier properties; can be moulded and transformed into various shapes and sizes; can be drawn into thin sheets; have good printability; is stronger and many other features that are not available with other materials. But ecological and environmental concerns are however growing rapidly, with the regulations and laws which has affected the entire food packaging industry. The idea of using biopolymers (from

R. S. Matche (✉) · Y. Singh
Department of Food Packaging Technology, CSIR-Central Food Technological Research Institute, Mysore, Karnataka, India
e-mail: rsmatche@cftri.res.in

H. U. Hebbar et al. (eds.), *Engineering Aspects of Food Quality and Safety*, Food Engineering Series, https://doi.org/10.1007/978-3-031-30683-9_9

renewable resources and biodegradable materials) in packaging, to contribute to sustainable development is recognized, since it is possible to dispose of the waste to get degraded in nature. This solution is particularly interesting for food packaging since these kinds of materials are usually contaminated by food residues that constitute a health hazard in sorting and mechanical recycling. Biopolymers are manufactured using biodegradable and renewable sources and have a great potential to reduce the environmental hazards caused by the usage of fossil fuel-based polymers such as plastics. The major advantage of replacing and using biopolymers is the reduced carbon footprint and biodegradability. The process of creating conventional packaging material degrades the environment, as toxic gases and effluents are introduced into the air, water and land. The non-biodegradable packaging material stays in the environment for an immensely prolonged duration. So, the alternative use of biopolymers for making food packaging materials in the form of primary packaging, membranes and films, and edible coating will help to reduce waste on the planet, and save resources as well.

2 Biodegradable Membranes and Coatings

Distinguishing between biopolymer and biodegradable materials helps to avoid the confusion and gives a clear idea about both. As per American Society for Testing and Materials (ASTM) Standard D-5488-94d, a biodegradable material is defined as "material capable of undergoing decomposition into carbon dioxide, methane, water, inorganic compounds, or biomass in which the predominant mechanism is the enzymatic action of microorganisms, that can be measured by standardized tests, in a specified period of time, reflecting available disposal condition".

Biopolymers are polymerics substances which are exclusively obtained from renewable raw materials. However, it must be noted that biopolymers are biodegradable but not all biodegradable materials are biopolymers (i.e., derived from renewable sources). Some biodegradable materials are obtained from non-renewable resources and thus cannot be regarded as biopolymers like polyglycolide (PGA), polycaprolactone (PCL) and polybutylene succinate adipate (PBSA). Biopolymers are usually classified on the basis of their origin:

1. Natural polymers are directly extracted from biomass such as polysaccharides like starch, cellulose, and galactomannans; and proteins like casein and gluten.
2. Synthetic polymers are prepared by renewable materials like polylactic acid (PLA) which are biodegradable. PLA is made from fermented starch from corn, cassava, maize, sugarcane or sugar beet pulp.
3. Microbial polymers are produced by microorganisms, like xanthan gum, gellan gum and pullulan, polyhydroxyalkanoates (PHA) etc.

Polysaccharides or polycarbohydrates are the most abundant carbohydrates found in nature. It is made up of monomeric units which are linked together covalently by glycosidic linkages. Depending on their structure, polysaccharides have various functions such as storing energy in plants (e.g., starch), providing structural support in plants (e.g., cellulose) and animal (e.g., chitin in exoskeleton) (Thakur and Thakur 2016). Polysaccharides have been extensively explored and have found usage to produce biodegradable films and membranes in the recent years which can also serve usage as edible coatings in various industries such as food and pharmaceutical (Mohamed et al. 2020). Polysaccharide membranes are generally of great interest because of their good mechanical properties, barrier properties for carbon dioxide and oxygen. The significant advantage of polysaccharide films and membrane is their edibility, non-toxicity, availability and low cost. The poor thermal and chemical stability accounts for brittleness and shrinking. Presence of hydroxyl, amino or carboxyl group is responsible for imparting high hydrophilicity and hence easy swelling and poor vapour barrier at higher humidity and is a major drawback for making packaging for food storage (Otoni et al. 2018). The coatings made from polysaccharides tend to be colourless, have an oily-free appearance and minor calorific content and thus can be applied on fruits, vegetables, shellfish or meat products to prolong their shelf-life by substantially reducing dehydration, darkening of the surface and oxidative rancidity.

2.1 Cellulose

Cellulose is a naturally occurring and abundant polymer present is cell walls of all plants in which β-1, 4 glycosidic bonds are present among repeating D-glucose units (Gupta et al. 2019). Cellulose is used for making films and coatings because of its specific properties such as low density, high mechanical strength, durability, good film-forming properties, chemical stability, biocompatibility, ease of availability, low cost, non-toxicity, and biodegradability (Gupta et al. 2019; Duan et al. 2016). Wood and cotton fibres are the most used raw material for producing cellulose-based biodegradable films and coatings. Some of the drawbacks of cellulose to be used as a food packaging material is its hydrophilic nature, insolubility in water and crystallinity (Thakur and Thakur 2016). Since the cellulosic material show hard mechanical properties, the use of plasticisers alone or in combination can be used to improve its film-forming abilities. Cellulose and cellophane film packaging are both plant-based, biodegradable and without a doubt 'green' product. For making cellophane, cellulose is dissolved in an alkali for several days (mercerization), followed by addition of disulphide (to form a solution of cellulose xanthate) and mixture of sodium sulphate and dilute sulphuric acid. Lastly, glycerine is added for durability after several washings. This makes it possible to produce a hydrophilic layer with good mechanical properties but without having thermoplastic properties and cannot be heat sealed.

2.2 Starch

Starch is an abundant reserve polysaccharide in plants. It is insoluble in water and mostly comprised of amylose, a linear chain polymer and amylopectin, a polymer of glucose with branched chain structure (Hassan et al. 2018). Starch is a renewable resource, biodegradable, abundantly available at low cost, non-toxic and shows thermoplastic behaviour. It can be extracted from tubers (like potato, tapioca or manioc), cereals (like corn, wheat, rice), grain (like amaranth) or even nuts (like cashew). Commercially starches are extracted from corn, potato and tapioca. Starches offer transparency to films and are odourless, tasteless, show thermoplasticity and are biodegradable.

Starches have been extensively utilised in making biodegradable films and coatings because they are transparent, odourless, tasteless and have good barrier against CO_2 and O_2 (Acosta et al. 2015). One of the limitations to use starch as a food packaging material is its brittleness, highly hydrophilic nature (retrogradation), poor processing quality, low physiological resistance (Ortega-Toro et al. 2015). The hydrophilicity is also responsible for the starch-based films to exhibit water solubility and poor water vapour barrier. These drawbacks can be overcome by using plasticizers such as glycerol, glycol and sorbitol, which in turn will increase the chain mobility and improve the flexibility to create starch-based films. The combination of starch with other materials will reduce the weaknesses of the film & produce a biodegradable film with improved properties (mechanical strength, flexibility and water barrier). Corn starch comprises of high amylose offering a massive delivery for the synthesis of edible films and coatings.

2.3 Galactomannans

Galactomannan is a heterogenous polysaccharide consisting of a β-(1–4)-D-mannan backbone with single D-galactose branches which are α-linked to mannose C-6 sites (Chouana et al. 2017). The major source of galactomannan is plants and mainly the seeds of family *Leguminosae*. It is the seeds endosperm acting as a glucose reserve for germination. Guar bean, cassia tora bean, sesbania seeds are some bean family members from which it is extracted. The galactomannans are extracted from the seeds through a combination of extraction and purification processes. In the recent years, galactomannan films have gained considerable interest due to the sustainability, their biocompatibility and biodegradability. All the gums like Guar gum, Tara gum and Locust bean gum differ by their composition of mannose/galactose ratio of 2, 3 & 4 respectively (Wu et al. 2017). Their capability to form viscous solutions at rather low concentration and their resistance to pH alterations, ionic strength and heat processing are their main distinct features to be used as films and coatings to improve the shelf-life, safety and quality of food products (Galgano 2015). The M/G ratio, degree of substitution and the degree of polymerisation are the main parameters affecting the membrane forming properties of the galactomannans. Liu et al., (2020a), studies the film forming properties of galactomannans with M/G ratio of 2,3 & 4 by characterising the film forming solution and their formed films. Their findings suggested that the galactomannans with proper M/G ratios have better film forming properties for application in food packaging. Edible membranes and coatings of galactomannans have found application in cheese (Lima et al. 2021).

2.4 Chitin and Chitosan

Chitin is a structural polysaccharide like cellulose but with a replacement of a hydroxyl group at C-2 with an acetylated amino group. Chitin is the second most abundant polysaccharide after cellulose & the biopolymer is found in a plethora of living organisms. It is the main component in cell walls of fungi and yeast & found in the exoskeleton of many arthropods (Thakur and Thakur 2016). It is a safe, natural, allergen free and biocompatible polymer with health benefits. The functional properties along with good O_2 & CO_2 barrier properties makes it a burgeoning bioactive polymer for applications in food industry. The films and coatings made from chitosan have good antimicrobial properties. Chitin is insoluble in water & most organic solvents whereas chitosan is readily soluble in dilute acidic solutions. Chitosan has been found to be semipermeable to gases like oxygen, essential for some food products preservation, and a medium water vapour barrier (Van den Broek et al. 2015). Chitosan has excellent film-forming properties allowing the production of films and membranes (thickness > 30 μm) and coatings (<30 μm). Chitosan films in pure form are made by casting method. Literature suggests that chitosan has antimicrobial properties which are influenced by pH, temperature as well as the age of bacterial cell (also known as growth phase). Chitosan has antibacterial and antifungal properties inhibiting growth of bacteria, fungi and yeast in conditions of in-vivo and in-vitro interactions.

The alkyl deacetylation of chitin forms a random copolymer 'Chitosan' consisting of units of D-glucosamine and N-acetyl-D-glucosamine, linked by β-1,4 glycosidic linkages. The degree of deacetylation (approximated by the ratio between two units) reaches 50%, chitosan becomes soluble in acidic aqueous media (Verlee et al. 2017) (Fig. 9.1).

Addition of glycerol followed by thermo-mechanical treatment to chitosan membrane obtains a thermoplastic material with better mechanical properties (Thakur and Thakur 2016). The combination of chitosan and anionic polymers have improved mechanical and barrier properties over those made of chitosan only. Also, when chitosan is combined with other polysaccharides such as starch, pectin, or alginate (Luchese et al. 2018; Hastuti and Hadi 2019) and proteins like gelatin (Cai et al. 2019), and whey proteins (Zhai et al. 2021), improvement was seen in mechanical properties, lower water vapour permeability and lower water solubility. Lipid addition (such as waxes, resins, fatty acids and vegetable oils) to films impart hydrophobicity and thus a reduction in moisture transfer.

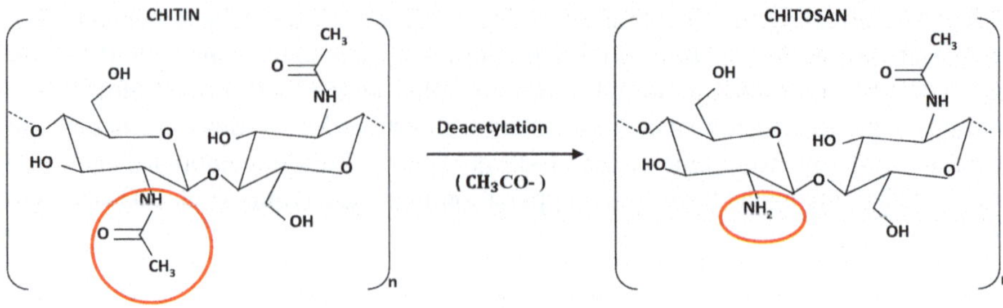

Fig. 9.1 Conversion of Chitin to Chitosan under alkyl deacetylation (Younes and Rinaudo 2015)

2.5 Carrageenan

Carrageenan is a polysaccharide directly extracted from the biomass of seaweeds, red algae. It is a naturally occurring hydrophilic, anionic sulfated linear polysaccharide (Hanani 2017; Tavassoli-Kafrani et al. 2016). It is a linear sulphated polysaccharide composed of d-galactose units. Carrageenans are classified based on their solubility in potassium chloride, into different types (λ, κ, ι, ε and μ). The degree of sulphation in lowest in κ-carrageenan; hence they have excellent properties to form gels and membrane and exhibit better mechanical properties as compared to λ- and ι-carrageenan which (Rhein-Knudsen et al. 2015). Carrageenan is an approved food grade additive by European Union (EU) and is used to produce edible films and coatings to preserve food items by lowering moisture loss, decreasing gas exchange, preventing change in colour, maintaining texture (Younes et al. 2018; Tavassoli-Kafrani et al. 2016).

2.6 Alginate

Alginate is extracted from brown seaweeds and some soil bacteria and has film forming properties. It is linear polysaccharide and can form water insoluble polymers useful for coating of fresh fruits and vegetables. The good film-forming properties of alginate gives a transparent, glossy and uniform look to the films. The membranes have been found to be possess impermeability to fat and oil, and high permeability to water vapour. Some desirable properties of alginate are reduction in shrinkage of film, colour, odour, good oxygen barrier and moisture retention. The interest of using alginates for films and coatings lies in their non-toxicity, biodegradability, biocompatibility and low cost. The capacity of alginates to crosslink with Calcium ions influences the mechanical properties of alginate films with different (M/G) ratios such as increased tensile strength (Costa et al. 2018) (Table 9.1).

Table 9.1 Different polysaccharides with their composition and applications

Polysaccharide obtained from animals				
Polysaccharide	**Composed of**	**Properties of Membrane**	**Application in food**	**References**
Chitin	N-Acetylglucosamine	Biodegradable, resistant to bacteria and fungus, biocompatible, non-toxic, highly transparent	Coffee capsules, Food bags, packaging films.	Priyadarshi (2020) Souza et al. (2020)
Chitosan	D-glucosamine, N-acetyl-d-glucosamine	Biodegradable, biocompatible, non-toxic, antifungal, antibacterial, good mechanical properties, gas barrier, increased water vapour permeability, brittle	Edible membranes & coatings (strawberries, mongo, guava, cherries), packaging membranes to fruits and vegetables.	Galgano (2015)
Polysaccharide obtained from plants				
Starch	Glucose	Biodegradable, tasteless & odourless, non-toxic, transparent, Retrogradation, high elongation and tensile strength	Flexible packaging: Extruded bags, nets for vegetables & fruits, thermoformed trays & containers for packaging fresh commodities	Sadeghizadeh-Yazdi et al. (2019) Lauer and Smith (2020)
Galactomannans	Mannose, galactose	Biodegradable, semi permeable, resistant to gases, edible	Edible membranes and coatings: Fruits & vegetables	Galgano (2015)
Cellulose	Glucose	Decomposable, good mechanical properties, transparent, highly sensitive to water, resistance to lipids, use of plasticizer & blends	Cellophane membranes	Cazon and Vazquez (2020)
Carrageenan	Galactose	Biodegradable, Brittle & Ductile behaviour, commonly blended with other polymers	Coating of fruits, meat, Encapsulation of aroma compounds	Martiny et al. (2020) Bhat et al. (2020)

(continued)

Table 9.1 (continued)

Alginate	Mannuronic and glucuronic acid	Biodegradable, high water vapour permeability, poor water resistance, strong & brittle membranes, cross link with calcium	Coatings Prevents water loss in fresh cut fruit (apple, papaya, pear & melon), Inhibition of microbial growth, Microwavable foods	Galgano (2015) Nair et al. (2020) Gutt and Amariei (2020)
Polysaccharide obtained from microorgansims				
Polysaccharide & Microorganism	**Composition**	**Membrane properties**	**Main food application**	**Reference**
Pullulan (*Aureobasidium pullulans*)	Maltotriose	Biodegradable, transparent, edible, oil & fat resistant, heat sealable, high water solubility, barrier to oxygen	Coating material, Wrapping material Blended with other polymers to improve mechanical properties, Inner package: Seasoning bag of instant noodles Instant coffee	Kraśniewska et al. (2019)
Xanthan gum (*Xanthomonas compestris*)	Glucose, mannose, glucuronic acid, acetate pyruvate	Biodegradable, edible	Edible coating: Fruit (extend shelf-life) Meat (prevents moisture migration during frying)	Quoc et al. (2015)
FucoPol (*Enterobacter A47*)	Fucose, galactose, glucose, glucuronic acid, acetate, succinate, pyruvate	Biodegradable, transparent, high gas barrier, poor water resistance	Possible application as inner layer in multilayer packaging	Ferreira et al. (2016)

3 Edible Coating

An edible coating is a thin layer of edible material applied to the surface of the food to form a protective coating and can be ingested along with the food products. These layers are mostly applied in liquid form on to the food surface, usually by immersing the product in a film-forming solution and sometimes by spraying the film forming solution on to the product (Hassan et al. 2018). These edible coatings can be prepared by using polysaccharides (cellulose, starch, carrageenan, gums, etc.), proteins (whey protein, soy protein, casein, gelatin, etc) from various plant, animal and microbial sources (Figs. 9.2 and 9.3).

Edible films and coatings are presented with number of advantages on top of traditional non-edible polymeric packaging. Edible packaging lessens the complexity of the food package and even uneaten with the packaged product, and reduce pollution due to their biodegradability (Bashir et al. 2017; Moghadam et al. 2020). Edible films and coatings have been applied to humongous range of food products such as fresh/frozen or cured/processed fruits and vegetables, meat, poultry, grains and confectionary products (Aguirre-Joya et al. 2018; Senturk Parreidt et al. 2018). It helps in reducing the post-harvest losses of fruits and vegetables (Radev and Pashova 2020) by protecting against various microbial contaminants (Bajpai et al. 2021), thus enhancing the shelf life (Sapper and Chiralt 2018; Grosso et al. 2020; Hasan et al. 2020). Edible coatings reduce the deterioration effect (Dong et al. 2020), minimizing oxidation of lipid (Kazemian-Bazkiaee et al. 2020) and moisture loss of food products (De Pilli 2020). The most important functionalities of an edible film or coating includes control of mass transfer, provide protection mechanically and without compromising sensory attributes. In this case, control of mass transfer requires prevention of foods from drying up, regulation of gases around the food

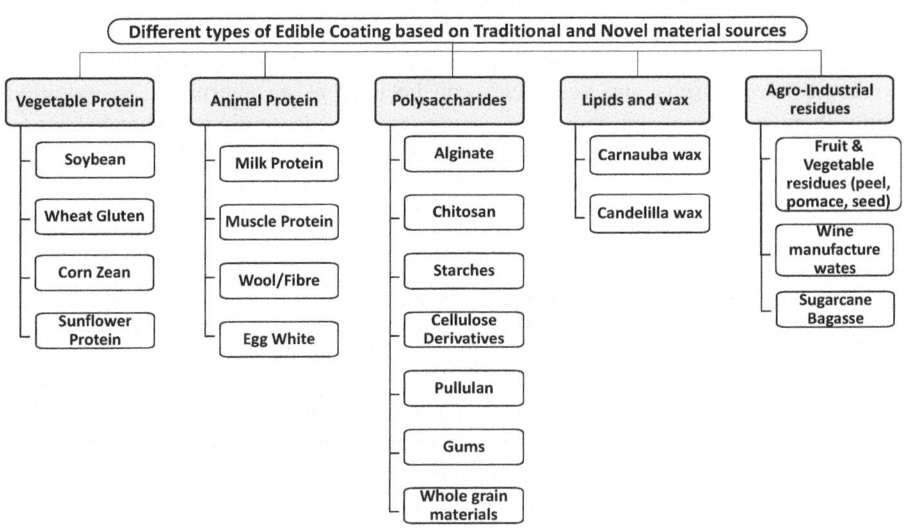

Fig. 9.2 Different types of edible coatings. (Reproduced from Galus et al. 2020)

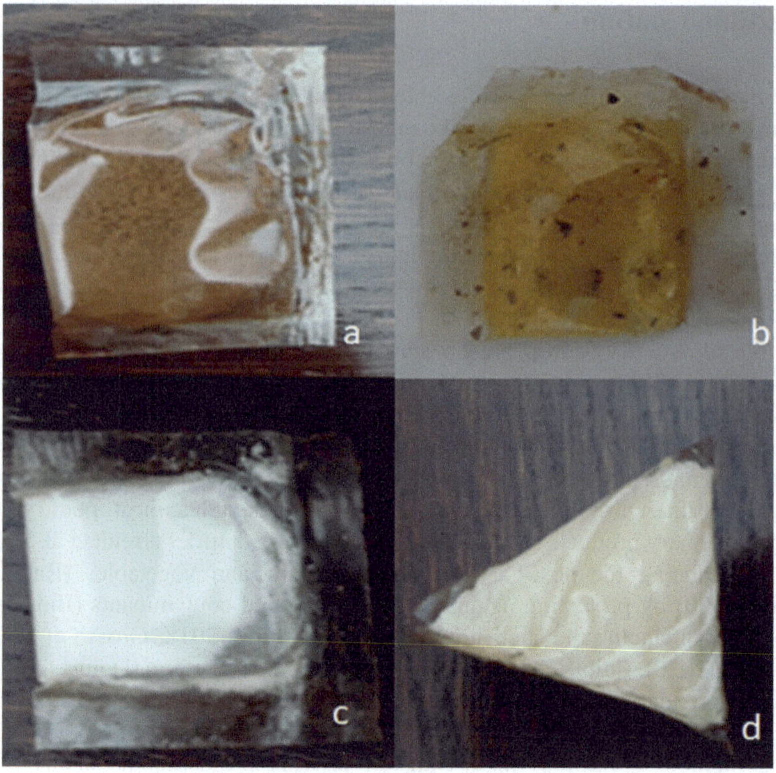

Fig. 9.3 Applications of sodium-alginate-based edible films integrated with *Stevia rebaudiana* for (**a**) soluble coffee, (**b**) dried vegetables, (**c**) medicines in powder form, (**d**) cheese slice; and plasticized with glycerol (Puscaselu et al. 2019)

product and restraining migration of substances in and out of food systems. Edible coating on fresh fruits and vegetables can be used as an alternative to modified atmosphere packaging by serving as means to provide semipermeable barrier to gases and water vapour. This in turn reduces any changes in quality and quantity (Fig. 9.4).

3.1 Method of Film Formation

There are two methods of film formation, wet and dry processes; also called as solvent casting and extrusion processes respectively. The structural chemistry of the biopolymers greatly influences the film forming procedure, whereas physical chemistry determines the functional characteristics such as mechanical strength, elasticity, rheology, moisture and gas permeation, colour, transmittance of light, etc. (Sharma and Ghoshal 2018).

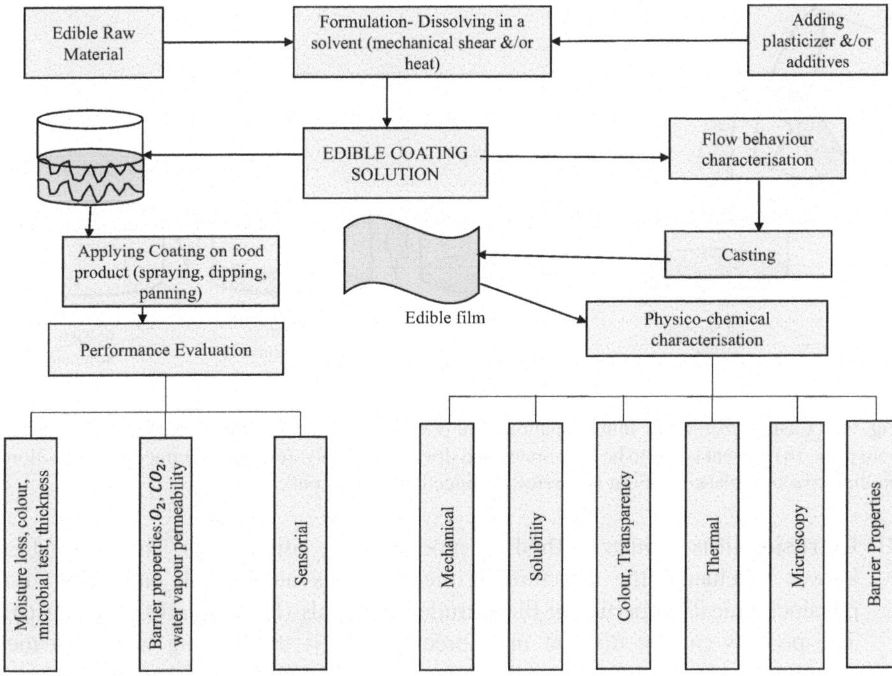

Fig. 9.4 Work flow of edible films and coatings

I. **Casting/Solvent casting**–this method of film-formation is used very often at both laboratory and industrial levels. It comprises majorly of three steps: (i) dissolution of polymer in appropriate solvent, (ii) casting of dissolved polymer onto casting tray, (iii) dehydration of the casted solution. The prepared film should be devoid of any inconsistencies, inclusion or wear and tear. The most important parameters possessed by edible films are thickness, transparency, opacity, swelling degree, thermal stability, mechanical strength, oxygen transmission rate (OTR), water vapor permeability (WVP), and biological characteristics (Khanzadi et al. 2015). The degree of transparency decreases upon adding plasticizers, but other properties such as barrier properties, mechanical strength and thermal properties of the film are enhanced (Sanyang et al. 2015). Solvent casting method does not require the use of complicated machineries and equipment cost, the film preparation method is also simple. The heat sensitive ingredients in the casting solution are also not degraded as it does not involve high temperature processing. The process takes longer time as compared to extrusion and thus is not feasible for mass production of films (Suhag et al. 2020) (Fig. 9.5).

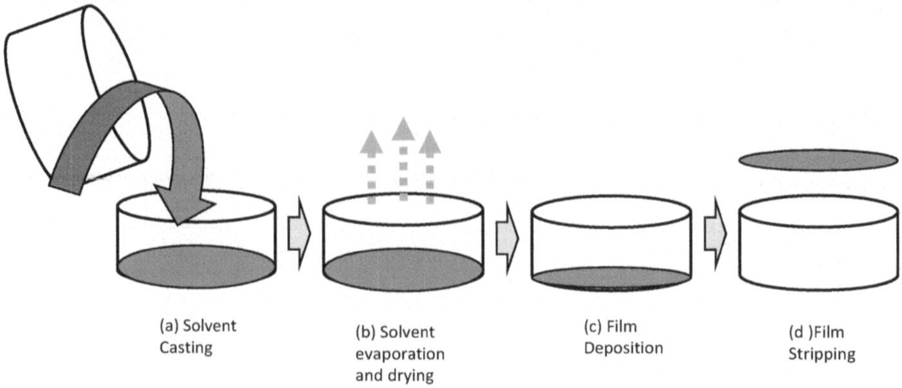

Fig. 9.5 Casting method of film formation. The prepared solvent for film formation is (**a**) Casted onto plate, (**b**) solvent is left to be evaporated and dried completely, (**c**) Film is formed by deposition on the surface of plate, (**d**) Film is carefully stripped and taken out

II. **Extrusion**–It is another method for producing the edible films. This method is known to change the structure of the materials along with enhancing the physiochemical properties of the extruded materials (Fitch-Vargas et al. 2016). The process can be divided into three zones: (i) the feeding zone, (ii) the kneading zone, and (iii) the heating zone at the final part/ exit from the machine (Fitch-Vargas et al. 2019). Mutilayer films can be formed by using co-extrusion that can give rise to flexibility in obtaining the desirable properties of the film. The multilayer film improves the functionality and processibility of the developed film along with helping in the designing structure of the multilayer film (Winotapun et al. 2019). The extruders can easily handle high viscosity biopolymers (like starch), offering high processing pressure for constant flow of biopolymer through the die which is major advantage of extrusion method for film formation using biopolymers. Individual barrel zone temperature control, multiple feeding, and screw configuration for a different degree of mixing gives TSE a large operational flexibility for intensive mixing and compounding of components into plasticized starch. The various advantage of using extruder for biopolymeric films are low cost, ensuring better mixing, heat transfer and control of the molten material (Oksman et al. 2016). It also enables operational flexibility for rigorous mixing and compounding of the biopolymer pellets (Ashter 2016). Sometimes extrusion of biodegradable polymers can be high cost, blowing of film could be difficult because of toughness of thermoplastic starch, difficult to control temperature of biopolymer resulting in leading to moisture losses (Huntrakul et al. 2020) (Fig. 9.6).

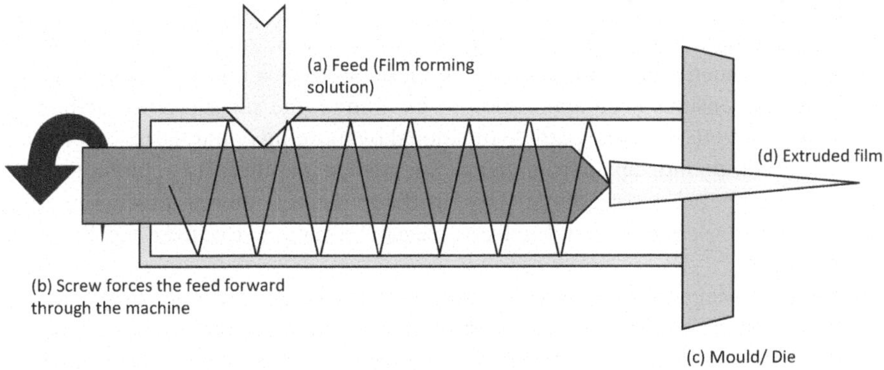

Fig. 9.6 Extrusion method of film formation through an extruder. The film forming solution (**a**) is introduced into the extruder through feed, (**b**) the screw forces the film-forming solution through the flat die system in forward direction, (**c**) flat mould/die to give shape the film, (**d**) film is formed through extrusion

3.2 Methods of Application

There are different methods of applying edible coating. The most common method amongst other is dipping, in which the fruits & vegetables are immersed into coating solution for a certain time & then allowed to air dry (Iniguez-Moreno et al. 2020; Marino et al. 2018). The other commonly used method is spraying, in which the coating solution is forced through a spray machine onto the surface of fruits and vegetables (Cenobio-Galindo et al. 2019). The methods of application are chosen keeping in view of characteristics of both the food product (dimensions and shape) and the coating solution (viscosity) (Atieno et al. 2019).

(a) **Dipping**–It is the most common method of applying edible coating. The product to be coated is dipped in coating solutions and followed by draining of any excess solution which subsequently dries and hardens. Dipping has been time and again used for coating fruits, vegetables, cheeses, meat products (Senturk Parreidt et al. 2018). The fruits and vegetables are submerged for about 5–30 sec in the formulation of edible coating, and ready to store or consume after drying (Raghav et al. 2016). Dipping has not been much effective for direct applications on food surfaces with antimicrobial agents, as leaching on the food, enzymatic activity causes the loss of activity. The simplicity and low cost of the dipping method, makes it the main laboratory method for food coating applications (Atieno et al. 2019).
(b) **Dripping**–This method of application of edible coating is most economic. Dripping is capable of delivering the coating either directly to the surface of food material or to the brushes. The large size of droplets is a snag for a nice and consistent coverage. This can be overcome by tumbling the food commodity adequate enough over several brushes that are saturated with the coatings. It has been commonly used for coating fruits and vegetables.

(c) **Fluidized-bed coating**–is a technique that can be used to apply a very thin layer onto dry particles of very low density or small size. A typical particle size is 100 micrometres to 3 millimetres. The food surface is optimally coated as this technique ensures even application of the film on the surface. This method can also be used to make the particle functional by masking odour, taste, or release of active compounds apart from enhancing storage stability. It is commonly used for coating bakery products. This method requires shorter processing time, imparts total coverage, and reduces cluster formation (Bertuzzi and Slavutsky 2016).
(d) **Panning**–Regardless of being a slow process, panning evenly distributes the coating material., The product is introduced into a large rotating bowl (pan) and then the coating solution is ladled or sprayed onto the product. The rotation of the pan evenly distributes the coating material onto the food surface. Panning is a commonly used in pharmaceutical industry, confectionary, nuts and chocolates.
(e) **Spraying**–This method is useful when a thin and uniform coating is required for certain surfaces. Spraying is the most common method of applying edible coating (Senturk Parreidt et al. 2018). With spray coating larger surface areas can be coated with uniform thin or thick layers. Spraying technique has been reported to enhance the food quality and appearance by (Zhong et al. 2019).
(f) **Electrostatic coating**–the most important factors to be considered while applying electrostatic powder coating is the particle size of the powder, density, charge as well as the surface characteristics and properties of food material to be coated. Electrostatic coating provides better transfer efficiency of the biopolymer on the food surface and also reduces the amount of dust produced during coating by conventional methods. These coating have been promising in few applications, including the infusion of bread with edible vegetable oil, coating of confectionary and chocolate products. Some other factors that impact the performance of electrostatic coating are flowability, resistivity, and target surface properties for powder coating, along with applied voltage, electrical resistivity and liquid's viscosity for liquid coating (Barringer and Sumonsiri 2015).

4 Multilayer Packaging

In recent years, multilayer packaging has gained significant interest in food industry. The requirements of films for food packaging include flexibility, strength and barrier to gases, aroma and water vapour permeation. This can be achieved by combining different polymers together as multi-layered structure. It is an emerging technique aiming at developing a package with improved performance with respect to protection and durability (Kaiser et al. 2018). A multilayer film is better than monolayer of polymer as it can overshadow the limitations of the latter in exhibiting all the functions of food packaging, like containment, protection/preservation, machinability, promotion and convenience along with being cost-effective (Kaiser et al. 2018).

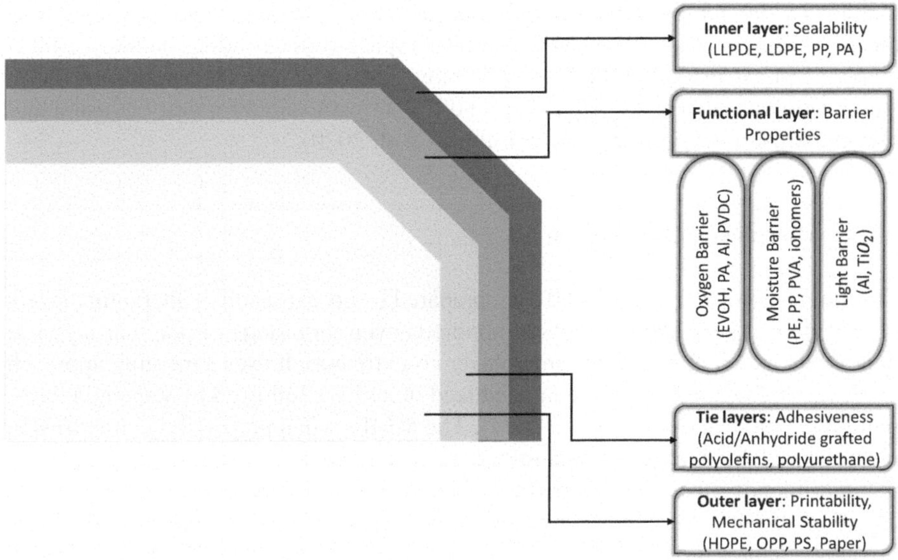

Fig. 9.7 Various layer of multi-layer packaging

Multiple layers when combined together significantly enhances the shelf-life by restricting the transmission rate of oxygen, carbon dioxide and moisture as well as concentration of oxygen inside the package which is necessary for preserving the freshness of the fresh commodities. In case of food packaging the multilayer packages consist of three to seven layers instead of three to twelve layers for general packaging (Butler and Morris 2016). Such multilayer polymeric films offer additional benefits of combined functional and barrier properties of individual layers along with thinner, lighter and compact packages (Butler and Morris 2016; Ramos et al. 2015). The multilayer films are commonly produced by blow extrusion, co-extrusion and various coating techniques (Ajitha et al. 2016). The multilayer packaging uses lesser raw materials along with reducing carbon emissions in the packaging during transport as well (Fig. 9.7).

4.1 Multilayer Film Production Techniques

4.1.1 Coextrusion

It is a process in which two or more than two polymer materials normally in the form of granules or pellets undergo heat treatment and molten separately (Ajitha et al. 2016). The feedblock serves as a medium to shape and combine multiple polymers entering it followed by melt spreading in the die, known as feedblock melt co-extrusion. In an alternative method, the polymer melt can be extruded in a multicavity die, where respective polymer melts are first spread and then combined together (Messin et al. 2017). The methods employed in extruding multilayer films involve cast extrusion/flat die extrusion, blow extrusion, and sheet extrusion. Polymeric materials such as PP, PET, PA; and PE are well suited for cast extrusion and blown extrusion respectively (Anukiruthika et al. 2020).

4.1.2 Microlayer Coextrusion

The multi-microlayer films (MMF) are prepared by co-extrusion of alternating layers of polymers. As the name suggests, hundreds even thousands of polymer layers at microscale are combined using microlayer co-extrusion. It uses a merging approach of processing where two to five layered feed blocks are followed by sequential layer multiplication dies (Messin et al. 2017). The MMFs when prepared by co-extrusion are followed by biaxial orientation & heat treatment of two or more different polymer blends inherit the properties of individual polymer layer. Such properties are compatibility, mechanical properties and water vapour permeation which are favourable characteristics for use in food packaging (Chiu et al. 2020). The overall thickness of a microlayer co-extruded film is no greater than that of a conventional co-extruded film and structure contains same amount of raw material., Such films are better than conventional single or double layered films as they enhance the barrier properties, rigidity and flexibility.

4.1.3 Lamination

It is a process of joining two or more flexible packaging webs using bonding agents. The substrate that makes up the web may consist of films. Web lamination is useful for improving the appearance and barrier properties of the packaging material in order to protect the packaged item and increase its shelf-life. The laminates can be compounded with different microns per layer. Lamination also prevents loss of the freshness and aroma of the packaged product. Some common processes used for lamination are extrusion lamination, adhesive lamination and wax & hot melt lamination. The bonding of extrudate to the webs depends upon chemical compatibility of the two (polymer and web), line speed, temperature of the polymer, thickness of the extrudate. Adhesive laminates often use lamination with aluminium. The selected adhesive must be compatible in terms of chemical compatibility with the webs & with the conditions to which the packaging will be introduced.

4.1.4 Coating

A single layer biopolymer film may not be sufficiently adequate to provide enough barrier. As a solution, multi-layered coated structures can be used to improve the barrier properties. A biopolymer can be applied as a coating on to the external layer on multilayer packaging film. Coated multilayer films provide high barrier to gas without adding to mass and thickness of the polymeric film. More and thinner layers allow efficient usage of materials. For thin coatings like antimicrobial layers, coating via atomization is used. Vartiainen et al. 2016 produced a bio-based multi-layered barrier film by combining 3 different techniques namely dispersion coating, atomic layer deposition and extrusion coating involving materials (cellulose nanofibrils, aluminium oxide and polyglycolic acid respectively). The film containing cellulose nanofibrils and polyglycolic acid were found to be good barrier to oxygen at varying humidity.

4.1.5 Nanostructured Multi-layered Films

Advancements in the use of nanotechnology in multilayer packaging by incorporating nanomaterials and formation of nano layers can help in improving the barrier properties of the packaging matrial further. The nanomaterial can be incorporated into one or more layers. The use of nanomaterial can also avoid the use of adhesives. Multilayer coextrusion at nanoscale allows producing films with up to thousands of layers with layer thickness in both microscale as well as nanoscale. Polymers with different properties can be combined into multilayer structures with up to 4096 layers and with individual layer thickness < 20 nm (Liu et al. 2003). Techniques such as layer by layer (LbL) and Electrohydrodynamic processing (EHDP) can be used to produce such nanostructured multi-layered films. LbL involves adsorption of charged molecules in an aqueous solution on a solid support. The oppositely charged particles (nano particles/ nano crystals) are sequentially adsorbed on the solid support giving rise to a multi-layered structure from 1–100 nanometres. The barrier properties of the polymer nanocomposites are improved by the enhanced tortuosity along with addition of fillers that alters the diffusion rate of the gases (O_2 and CO_2). The addition of nano fillers to the polymer to prepare nanocomposites can overcome the limitations occurred by the use of biopolymers only. The inorganic clay minerals like montmorillonite (MMT) are used as nanofillers in between silicate layers for enhancing barrier properties and provide promising active materials (Huang et al. 2015). A biocomposite film of polylactic acid and hallosite nanotube was found to improve the shelf-life of cherry tomatoes in a study conducted by Risyon et al. (2020). Another example is of a biocomposite film prepared using *Ocimum basilicum* reinforced with MMT as nanofiller. The film was found to be suitable for use in food packaging with improved tensile strength and reduced water vapour permeability (Rohini et al. 2020). Therefore, the principle behind production of multilayer packaging is to develop a solo packing structure possessing various functional properties which can meet all the complex functional necessities of food packaging (Tables 9.2 and 9.3).

Table 9.2 Different Biopolymer-based nanocomposite films with properties

Sl. No.	Biopolymer	Active Ingredient/ Functional compound	Properties	References
1.	Starch-based cellulose nanocomposite film	Cellulose Nanofibres extracted from Banana peel	Increased tensile strength, Young's modulus, water-resistance, opacity and crystallinity	Pelissari et al. (2017)
2.	Soy-protein based antibacterial nanocomposite film with cellulose nanocrystals and zinc nanoparticles	Cellulose nanocrystals, Zinc nanoparticles	Improved tensile strength, barrier properties, thermal stability, Potential of increasing shelf-life of fresh pork	Xiao et al. (2020)
3.	Hydrogel films prepared with agar, alginate and collagen	Silver nanoparticles and grapefruit seed extract	Strong antimicrobial activity, anti-fogging films, effective in preventing potatoes greening during storage	Wang and Rhim (2015)
4.	Smart colour films based on wheat gluten/ chlorophyll/ polypyrrole nanocomposite	Chlorophyll and Polypyrrole	Excellent mechanical, conducting, antibacterial and antioxidant properties, colour change of film can help to assess food quality	Chavoshizadeh et al. (2020b)
5.	Teranary nanocomposite films prepared for packaging application	Gellan gum, xanthan gum and zinc oxide	Increased thermal and mechanical properties, excellent antimicrobial activity, increased hydrophobic nature of film	Rukmanikrishnan et al. (2020)
6.	Bio-nanocomposite films reinforced with magnesium hydroxide	Pectin-based, laboratory made magnesium hydroxide	Improved physical, thermal and barrier properties, better retention of bioactive compound of stored cherry tomatoes.	Kumar et al. (2020)
7.	Semi-refined kappa carrageenan based biocomposite film for food packaging	Zinc oxide	Increase in solubility properties of film	Saputri et al. (2018)

Table 9.3 Different multilayer packaging films with their functions and applications

Type of packaging	Method of preparation	Function	Food sample	References
PE-based multilayer film	Layer by layer spraying	Antimicrobial activity of the films	Chicken meat	Alkan Tas et al. (2019)
Polymer-nanoclay hybrid multilayers	Layer by layer assembly	Improved tensile strength, water vapour, and oxygen barrier property, preventing weigh losses	Strawberries	Li et al. (2019)
Multilayer Zein/gelatin edible film	Casting	Inhibiting bacterial activity, inhibiting quick browning	Freshly cut fruits (kiwi, avocado, banana)	Xia et al. (2019)
Polyethylene terephthalate (PET)-low-density polyethylene (LDPE)-based multilayer films	Coating	Inhibition of lipid oxidation	Fried potatoes	Oudjedi et al. (2019)

5 Intelligent and Antimicrobial Films

Intelligent packaging has proven to be a valuable asset in the maintaining the food quality and safety. However, it is still an emerging technology in the food packaging sector (Sohail et al. 2018). A packaging containing a component that allowing to monitor the state of packaged food or the surrounding microenvironment of food during transportation and storage is called as intelligent packaging. Thus, intelligent packaging is a system providing the user with trustworthy and right information on the conditions of the food, its environment or the package's integrity (Sohail et al. 2018). In broader terms intelligent packaging is an expansion of the communication function of the conventional food packaging, and has the ability to sense, detect, or record changes in the product or its environment. One of the advantages of the intelligent packaging is that it plays a part in improving Hazard Analysis and Critical Control Points (HACCP) and Quality Analysis and Critical Control Points (QACCP) systems (Siracusa and Lotti 2019). The active and intelligent packaging market was valued at USD 17.5 billion in 2020. It is expected to reach a value of USD 25.16 billion by 2026, registering a compound annual growth rate (CAGR) of 6.78% by 2026. During this forecast period, the highest compound annual growth rate (CAGR) is expected to be recorded by Asia Pacific, so it is expected to be the fastest-growing region for the development and utilization of smart packaging (Mirza Alizadeh et al. 2020). Intelligent films are a novel concept which aims at developing films and membranes that are biodegradable and non-toxic and carries the ability to perform the function of sensors, indicators and data carriers. There have been several researches going on for the development of such films (Fig. 9.8).

In recent research, a biopolymer-based pH-responsive colour indicator integrated with natural colorants was developed for real-time monitoring of packaged food

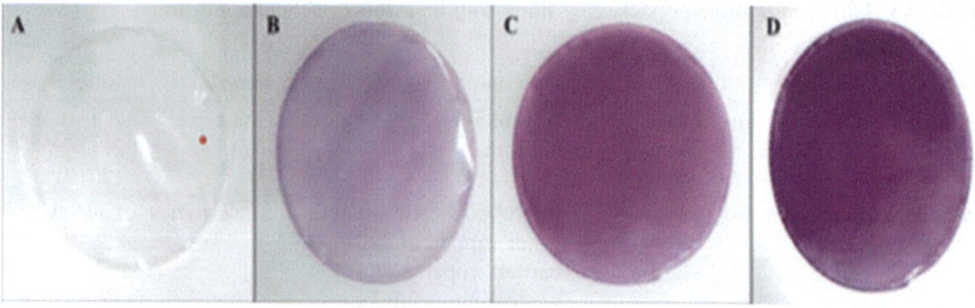

Fig. 9.8 Images showing methylcellulose (MC) intelligent films comprised of jambolão skins extract. (**a**) MC film (control); (**b**) MC film +10% jambolão extract; (**c**) MC film +30% jambolão extract; (**d**) MC film +50% jambolão extract (da Silva et al. 2020)

quality. The pH-responsive colour indicator films were prepared by immobilizing natural food colorants (anthocyanin, curcumin, alizarin, shikonin, betalains) from various plant sources into different biopolymer substrates (Priyadarshi et al. 2019). In a similar study, an intelligent starch/poly-vinyl alcohol (PVA) film capable of monitoring pH changes and inhibiting undesired microbial growth of food was developed. Anthocyanins and Limonene were used to attain simultaneous colorimetric indication and antimicrobial activity. The film possessed a smooth surface, high mechanical strength, showed distinguishable colour changes with changing pH, and also showed excellent antimicrobial activity for microorganisms like *Bacillus subtilis, Aspergillus niger* and *Staphylococcus aureus* (Liu et al. 2017) (Table 9.4).

These compounds can be used to develop active, intelligent, and antimicrobial biodegradable films and be applied to various food types, including meat, seafood, fresh fruits and vegetables and dairy products.

6 Advantages

Trends in the use of active envelopes include oil consumption reduction in deep fried fat food products, transport of bioactive compounds and shelf-life extension of highly perishable products. Incorporation of natural, biodegradable substances and compounds into food packaging system allows to reduce food losses by extending the shelf-life of the product. They are much safer and non-toxic than synthetic compounds and polymers and aren't harmful for health and environment. The incorporation of natural pigments in the biodegradable films increase their thickness due to effective cross-linking between pigment and matrix. Also, the edible films are good carrier of essential nutrients which is intaken via food. Thus, they have the potential to bring revolution in the food packaging industry.

Table 9.4 Various natural compounds incorporated with biodegradable films to develop intelligent and antimicrobial packaging systems

Properties	Source	Biodegradable films	Food Application	References
Compound: Anthocyanins				
Bioactive phenolic compound, water-soluble pigment (red, orange, blue & purple), strong antimicrobials, antioxidant, pH sensitive	Red cabbage anthocyanin Purple sweet potato anthocyanin	Chitin + PVA film, Agar + potato starch film	Pork & various meat products	Zhang et al., 2019; Kurek et al. 2018; Wang et al. 2019a; Vo et al. 2019; Choi et al. 2017
Compound: Curcumin				
Diphenolic, hydrophobic compound, antioxidant, good barrier properties, thermal stability ammonia (NH_3) sensing Change in colour	*Curcuma longa* roots (turmeric)	Gelatin based film, Carrageenan based film, Tara gum+ PVA film, cellulose film, pectin-Sulphur nanocomposite film, casein/zein nanocomposite film	Shrimp, pork, Seafood, meat, Edible packaging	Liu et al. 2018; Ma and Wang. 2016; Tichoniuk et al. 2017; Ezati and Rhim 2020; Wang et al. 2019b
Compound: Betalains				
Water-soluble, phenolic pigment, antimicrobial, antioxidant, sensitivity to NH_3 (change in colour), enhanced mechanical & antioxidant properties, colour change in film	Red pitaya, Amaranth, prickly pear, beet root, Red pitaya peel	Glucomannan + PVA film, starch based film, chitosan/PVA film	Fish, shrimp	Polturak and Aharoni 2019; Apriliyanti et al. 2018; Qin et al. 2020; Yao et al. 2020
Compound: Chlorophyll				
Antioxidant properties, time temperature indicators in temp range 50 °C–70 °C, detection of volatile amines, good antimicrobial, change in colour with varying pH	Plants, cyanobacteria, mosses, Amaranth leaf extract, green tea & basil, Pu erh-tea (fermented tea) & green tea	Gelatin/PVA films, cassava starch film, gluten films, Furcellaran + gelatin film	Chicken, fish Oxidation of oils	Kanatt 2020; Medina-Jaramillo et al. 2017; Chavoshizadeh et al. 2020a; Jamroz et al. 2019

(continued)

Table 9.4 (continued)

Properties	Source	Biodegradable films	Food Application	References
Compound: Carotenoid				
Lipophilic, bioactive pigment (orange, yellow, red), antioxidant, photosensitizers, change in colour with different oxidation conditions, change in colour with thermooxidation & weathering, change in colour with varying pH	Variety of fruits, flowers, green plants, lycopene, cantaloupe, apricots, carrot, sweet potato, green leafy vegetables Green tea & basil	PLA + titanium dioxide, biodegradable film, cassava starch film	NA	Asadi and Pirsa 2020; Latos-Brozio and Masek 2020; Medina-Jaramillo et al. 2017
Compound: Tannins				
Complex bioactive phenols, secondary metabolites, antimicrobial, antioxidant, pH sensitive, enhanced mechanical properties	Edible & inedible plant parts like bark, leaves, nuts, seeds, fruits, spices, legumes Cashew nut extract, pomegranate flesh & peel extract	Cellulose based film, k carrageenan-based film		Liu et al. 2020b
Compound: Brazilin				
Bioactive polyphenol, sensitivity to pH, water soluble, antibacterial, antioxidant, change in colour with varying pH	Sappan wood	PVA/ Glucomannan film	Climacteric fruits like banana	Athinarayana et al. 2017; Kurnianto et al. 2020
Compound: Quercetin				
Antioxidant, flavonoid compound, change in colour with time, change in colour with thermooxidation & weathering	Tea, fruits (grapes, berries, apple, citrus), vegetables (onion, broccoli), grains (quinoa)	Cyclo olefin copolymer film		Masek et al. (2018)

7 Hurdles

The challenges in the use of edible coatings or films and membranes is the selection of suitable material which should be safe for food contact use. Edible films are mostly made from such components that are known to have allergic effects such as gluten (wheat), whey protein and casesin (milk), etc. In such case, allergen information should be clearly declared to the consumer. The use of nanocomposites in food packaging is concerned with migration of the nano-particles in the food itself, though such toxicity hasn't been well reported yet. The other factor that limits the use of edible coatings are off-flavour developments (due to restricted oxygen exchange when the coating is thicker), alteration in normal ripening process of fruits and vegetables (due to restricted gaseous exchange), microbial spoilage (due to the hygroscopic nature of the edible films). To make an antimicrobial compound to be commercially acceptable, several factors have to be considered. The first hurdle is that the antimicrobial compound and the processing surface material must be approved for safe food contact use. The second is the cost of both the coating matrix and the antimicrobial compound, if used any. Since the antimicrobial compounds lose or show reduced activity after immobilisation in a coating matrix, hence, it becomes necessary to understand the relationship between structure and function of the both antimicrobial compound and the coating material. Another challenge is the recycling and degradability if the multilayer packaging. Another great challenge is the recycling of the multilayer packaging since these are harder to separate and degrade naturally. In films with antimicrobial properties due to addition of essential oils the biggest challenge is the safe consumption of these essential oils. Though they have been recognised as GRAS by United States, can still show allergic effects. In fact, the high dose consumption of these natural compounds can be orally toxic, so care must be given that these should not migrate from packaging film to food and a balance between the effectivity, the dose to be consumed (in case of edible coating) and the associated toxicity risk should be established (Ribeiro-Santos et al. 2017). Similarly, the electrostatically charged surfaces (such as cationic polymers) are likely to cause fouling by food matrix component (such as anionic components present naturally in food), which may adversely affect their antimicrobial activity (Bastarrachea et al. 2015).

8 Future of Biodegradable Coating and Films

In future, the application of biodegradable edible coatings and films is going to boost up in the food industry because of the rising environmental concerns regarding pollution and use of non-renewable material. Although, several uses of such coatings are being made in the current times, the future is much brighter. The future trend will focus on effective delivery systems of bioactive compounds from packaging (coating material) to the food contained within such as vitamins, nutraceuticals, prebiotics, probiotics, preservatives to improve the nutritional qualities other than extending shelf life. The use of novel solvents for recycling of polyamide containing multilayer packaging have the potential to change the future of recycling processes. Further, chemical recycling will allow the packaging waste to be re-used as a virgin-like polymer after processing. Researchers are currently designing such bio-polymers that are more user-friendly according to future scenarios.

9 Conclusion

Coatings and membranes can be considered as an active or non-active form of packaging based upon the materials being used & the intended functional properties. They can be either be made from biopolymers which can either be sourced from biodegradable or non-biodegradable materials. These coatings have been shown to lessen the problem of fat uptake in fried products. Natural polymers include polysaccharides (starch, cellulose, galactomannans), proteins (whey, casein); Synthetic polymers include PLA and Microbial polymers such as PHA. Cellulose and cellophane films are biodegradable in nature. Starch films and coatings are transparent, odourless, flexible, possessing barrier against CO_2 and O_2. The mannose/galactose ratio affects the film forming properties of galactomannans. Chitosan is a polymer resulted from alkyl deacetylation of chitin and formed films have antimicrobial and antifungal properties. Carrageenan is a food grade additive approved by European Union and used to form edible coating with good gas barrier properties. Linear structure and good film forming properties of alginates results in stronger film. Edible coatings are environmentally friendly and can serve as carriers of food additives enhancing safety, nutritional and sensory attributes of fruit & vegetables. Edible coatings are helpful in preserving the food by preventing moisture loss, may provide protection against microbes without affecting sensory attributes and also helpful in delivering drugs orally. Edible films can be produced by two methods namely solvent casting and extrusion. The edible coatings can be applied on to surface of foods via methods such as dipping, dripping, fluidized bed coating, panning, spraying and electrostatic coating. Of these, electrostatic coating both in liquid and powder form can improve the quality of food like taste, appearance, aroma and shelf-stability. Electrostatic spraying method for applying edible coatings can be adopted for potential liquid coating application so as to improve the shelf-life of perishable fruits and vegetables as well as minimising the waste of coating materials. Multilayer films are produced by methods such as co-extrusion, microlayer co-extrusion, coating and lamination. Multilayer films have better barrier properties than conventional single layered. Addition of nano-fillers to biopolymers forming nanostructured multi-layered films can be helpful to overcome limitations by use of only biopolymers. Several nanoparticles such as cellulose nanofibers, cellulose nanocrystals, silver nanoparticles, zinc oxide have been used to make bio nanocomposite films. Nanocomposites enhance the properties of biodegradable polymer while still being eco-friendly. Currently, intelligent packaging is being used in food industries more than any other industry offering numerous advantages such as food spoilage alert by functioning as sensor, indicators and data carriers. They aid in ensuring food quality and safety. An antimicrobial compound must retain its activity after being incorporated in films and membranes with expected life of the food product.

References

Acosta S, Jiménez A, Cháfer M, González-Martínez C, Chiralt A (2015) Physical properties and stability of starch-gelatin based films as affected by the addition of esters of fatty acids. Food Hydrocoll 49:135–143

Active and Intelligent Packaging Market- Growth Trends, Covid-19 Impact, and Forecasts (2021–2026). https://www.researchandmarkets.com/reports/4745433/active-and-intelligent-packaging-market-growth

Aguirre-Joya JA, De Leon-Zapata MA, Alvarez-Perez OB, Torres-León C, Nieto-Oropeza DE, Ventura-Sobrevilla JM, Aguilar MA, Ruelas-Chacón X, Rojas R, Ramos-Aguiñaga ME, Aguilar CN (2018, Jan 1) Basic and applied concepts of edible packaging for foods. In: Food packaging and preservation. Academic Press, pp 1–61

Ajitha AR, Aswathi MK, Maria HJ, Izdebska J, Thomas S (2016) Multilayer polymer films. In: Multicomponent polymeric materials. Springer, Dordrecht, pp 229–258

Anukiruthika T, Sethupathy P, Wilson A, Kashampur K, Moses JA, Anandharamakrishnan C (2020) Multilayer packaging: advances in preparation techniques and emerging food applications. Compr Rev Food Sci Food Saf 19(3):1156–1186

Apriliyanti MW, Wahyono A, Fatoni M, Poerwanto B, Suryaningsih W (2018, Nov 1) The potency of betacyanins extract from a peel of dragon fruits as a source of colorimetric indicator to develop intelligent packaging for fish freshness monitoring. In: IOP conference series: earth and environmental science, vol 207, no. 1. IOP Publishing, p 012038

Asadi S, Pirsa S (2020 Feb) Production of biodegradable film based on polylactic acid, modified with lycopene pigment and TiO 2 and studying its physicochemical properties. J Polym Environ 28(2):433–444

Ashter SA (2016) 8—extrusion of biopolymers. Introduction to Bioplastics Engineering; Elsevier, Oxford, UK, pp 211–225

Athinarayanana G, Ranjitsingh AJ, Padmalatha C (2017, Mar 1) Toxicological studies of Caesalpinia sappan wood derived dye in Wister albino rats. Food Sci Human Wellness 6(1): 34–38

Atieno L, Owino W, Ateka EM, Ambuko J (2019, Feb 4) Influence of coating application methods on the postharvest quality of cassava. Int J Food Science 2019:2148914

Bajpai S, Chand N, Lodhi R (2021, Jan 4) Water sorption properties and antimicrobial action of zinc oxide nano particles loaded sago starch film. J Microbiol Biotechnol Food Sci 2021:2368–2387

Barringer SA, Sumonsiri N (2015, Apr 10) Electrostatic coating technologies for food processing. Annu Rev Food Sci Technol 6:157–169

Bashir K, Jan K, Aggarwal M (2017, Apr) Thermo-rheological and functional properties of gamma-irradiated wholewheat flour. Int J Food Sci Technol 52(4):927–935

Bastarrachea LJ, Denis-Rohr A, Goddard JM (2015, Apr 10) Antimicrobial food equipment coatings: applications and challenges. Annu Rev Food Sci Technol 6:97–118

Bertuzzi MA, Slavutsky AM (2016, Sep 19) Standard and new processing techniques used in the preparation of films and coatings at the lab level and scale-up. In: Edible films and coatings. CRC Press, pp 21–42

Bhat KM, Jyothsana R, Sharma A, Rao NN (2020) Carrageenan-based edible biodegradable food packaging: a review. Int J Food Sci Nutr 5(4):69–75

Butler TI, Morris BA (2016, Jan 1) PE-based multilayer film structures. In: Multilayer flexible packaging. William Andrew Publishing, pp 281–310

Cai L, Shi H, Cao A, Jia J (2019) Characterization of gelatin/chitosan ploymer films integrated with docosahexaenoic acids fabricated by different methods. Sci Rep 9(1):1–11

Cazón P, Vázquez M (2020, Dec 10) Bacterial cellulose as a biodegradable food packaging material: a review. Food Hydrocoll 113(11):106530

Cenobio-Galindo AD, Ocampo-López J, Reyes-Munguía A, Carrillo-Inungaray ML, Cawood M, Medina-Pérez G, Fernández-Luqueño F, Campos-Montiel RG (2019, Oct) Influence of bioactive compounds incorporated in a nanoemulsion as coating on avocado fruits (Persea americana) during postharvest storage: antioxidant activity, physicochemical changes and structural evaluation. Antioxidants 8(10):500

Chavoshizadeh S, Pirsa S, Mohtarami F (2020a, Mar) Sesame oil oxidation control by active and smart packaging system using wheat gluten/chlorophyll film to increase shelf life and detecting expiration date. Eur J Lipid Sci Technol 122(3):1900385

Chavoshizadeh S, Pirsa S, Mohtarami F (2020b) Conducting/smart color film based on wheat gluten/chlorophyll/polypyrrole nanocomposite. Food Packag Shelf Life 24:100501

Chiu HT, Yu CY, Huang CM, Lin BS (2020, Nov 5) Coextrusion and biaxial orientation of multimicrolayer films. J Appl Polym Sci 137(41):49245

Choi I, Lee JY, Lacroix M, Han J (2017, Mar 1) Intelligent pH indicator film composed of agar/potato starch and anthocyanin extracts from purple sweet potato. Food Chem 218:122–128

Chouana T, Pierre G, Vial C, Gardarin C, Wadouachi A, Cailleu D, Le Cerf D, Boual Z, El Hadj MO, Michaud P, Delattre C (2017, Nov 1) Structural characterization and rheological properties of a galactomannan from Astragalus gombo Bunge seeds harvested in Algerian Sahara. Carbohydr Polym 175:387–394

Costa MJ, Marques AM, Pastrana LM, Teixeira JA, Sillankorva SM, Cerqueira MA (2018) Physicochemical properties of alginate-based films: effect of ionic crosslinking and mannuronic and guluronic acid ratio. Food Hydrocoll 81:442–448

da Silva FG, Romani VP, Martins VG (2020, Dec 1) Biodegradable and active-intelligent films based on methylcellulose and jambolão (Syzygium cumini) skins extract for food packaging. Food Hydrocoll 109:106139

De Pilli T (2020, Aug 1) Development of a vegetable oil and egg proteins edible film to replace preservatives and primary packaging of sweet baked goods. Food Control 114:107273

Dong C, Wang B, Li F, Zhong Q, Xia X, Kong B (2020, Jan 1) Effects of edible chitosan coating on Harbin red sausage storage stability at room temperature. Meat Sci 159:107919

Duan J, Reddy KO, Ashok B, Cai J, Zhang L, Rajulu AV (2016, Mar 1) Effects of spent tea leaf powder on the properties and functions of cellulose green composite films. J Environ Chem Eng 4(1):440–448

Ezati P, Rhim JW (2020, Feb 15) pH-responsive pectin-based multifunctional films incorporated with curcumin and sulfur nanoparticles. Carbohydr Polym 230:115638

Ferreira AR, Torres CA, Freitas F, Sevrin C, Grandfils C, Reis MA et al (2016) Development and characterization of bilayer films of FucoPol and chitosan. Carbohydr Polym 147:8–15

Fitch-Vargas PR, Aguilar-Palazuelos E, de Jesús Zazueta-Morales J, Vega-García MO, Valdez-Morales JE, Martínez-Bustos F, Jacobo-Valenzuela N (2016, Sep) Physicochemical and microstructural characterization of corn starch edible films obtained by a combination of extrusion technology and casting technique. J Food Sci 81(9):E2224–E2232

Fitch-Vargas PR, Aguilar-Palazuelos E, Vega-García MO, Zazueta-Morales JJ, Calderón-Castro A, Montoya-Rodríguez A, Delgado-Nieblas CI, Camacho-Hernández IL (2019, Jun 1) Effect of a corn starch coating obtained by the combination of extrusion process and casting technique on the postharvest quality of tomato. Mex J Chem Eng 18(3):789–801

Galgano F (2015, Mar 9) Biodegradable packaging and edible coating for fresh-cut fruits and vegetables. Ital J Food Sci 27(1):1–20

Galus S, Arik Kibar EA, Gniewosz M, Kraśniewska K (2020) Novel materials in the preparation of edible films and coatings—a review. Coatings 10(7):674

Grosso AL, Asensio CM, Grosso NR, Nepote V (2020, Jan 1) Increase of walnuts' shelf life using a walnut flour protein-based edible coating. LWT 118:108712

Gupta PK, Raghunath SS, Prasanna DV, Venkat P, Shree V, Chithananthan C, Choudhary S, Surender K, Geetha K (2019) An update on overview of cellulose, its structure and applications. In: Cellulose. Intechopen, London, UK, pp 846–1297

Gutt G, Amariei S (2020, Feb) The use of edible films based on sodium alginate in meat product packaging: an eco-friendly alternative to conventional plastic materials. Coatings 10(2):166

Hanani ZN (2017) Physicochemical characterization of kappa-carrageenan (Euchema cottoni) based films incorporated with various plant oils. Carbohydr Polym 157:1479–1487

Hasan SK, Ferrentino G, Scampicchio M (2020, Jan) Nanoemulsion as advanced edible coatings to preserve the quality of fresh-cut fruits and vegetables: a review. Int J Food Sci Technol 55(1): 1–0

Hassan B, Chatha SA, Hussain AI, Zia KM, Akhtar N (2018, Apr 1) Recent advances on polysaccharides, lipids and protein based edible films and coatings: a review. Int J Biol Macromol 109:1095–1107

Hastuti B, Hadi S (2019, Sep) Synthesis and characterization of pectin-chitosan as candidate materials for slow release system. In: IOP conference series: materials science and engineering, vol 617, no 1. IOP Publishing, p 012002

Huang JY, Chieng YY, Li X, Zhou W (2015, Feb) Experimental and mathematical assessment of migration from multilayer food packaging containing a novel clay/polymer nanocomposite. Food Bioprocess Technol 8(2):382–393

Huntrakul K, Yoksan R, Sane A, Harnkarnsujarit N (2020) Effects of pea protein on properties of cassava starch edible films produced by blown-film extrusion for oil packaging. Food Packag Shelf Life 24:100480

Iñiguez-Moreno M, Ragazzo-Sánchez JA, Barros-Castillo JC, Sandoval-Contreras T, Calderón-Santoyo M (2020, May 1) Sodium alginate coatings added with Meyerozyma caribbica: postharvest biocontrol of Colletotrichum gloeosporioides in avocado (Persea americana mill. cv. Hass). Postharvest Biol Technol 163:111123

Jamróz E, Kulawik P, Krzyściak P, Talaga-Ćwiertnia K, Juszczak L (2019, Feb 1) Intelligent and active furcellaran-gelatin films containing green or pu-erh tea extracts: characterization, antioxidant and antimicrobial potential. International Journal of Biolgical Macromolecules 122:745–757

Kaiser K, Schmid M, Schlummer M (2018, Mar) Recycling of polymer-based multilayer packaging: a review. Recycling 3(1):1

Kanatt SR (2020, Jun 1) Development of active/intelligent food packaging film containing Amaranthus leaf extract for shelf life extension of chicken/fish during chilled storage. Food Packag Shelf Life 24:100506

Kazemian-Bazkiaee F, Ebrahimi A, Hosseini SM, Shojaee-Aliabadi S, Farhoodi M, Rahmatzadeh B, Sheikhi Z (2020, Apr) Evaluating the protective effect of edible coatings on lipid oxidation, fatty acid composition, aflatoxins levels of roasted peanut kernels. J Food Meas Charact 14(2):1025–1038

Khanzadi M, Jafari SM, Mirzaei H, Chegini FK, Maghsoudlou Y, Dehnad D (2015, Mar 15) Physical and mechanical properties in biodegradable films of whey protein concentrate–pullulan by application of beeswax. Carbohydr Polym 118:24–29

Kraśniewska K, Pobiega K, Gniewosz M (2019, Sep 11) Pullulan–Biopolymer with potential for use as food packaging. Int J Food Eng 15(9):20190030

Kumar N, Kaur P, Devgan K, Attkan AK (2020) Shelf life prolongation of cherry tomato using magnesium hydroxide reinforced bio-nanocomposite and conventional plastic films. J Food Process Preserv 44(4):e14379

Kurek M, Garofulić IE, Bakić MT, Ščetar M, Uzelac VD, Galić K (2018, Nov 1) Development and evaluation of a novel antioxidant and pH indicator film based on chitosan and food waste sources of antioxidants. Food Hydrocoll 84:238–246

Kurnianto MF, Poerwanto B, Wahyono A, Apriliyanti MW, Lestari IP (2020) Monitoring of banana deteriorations using intelligent-packaging containing brazilien extract (Caesalpina sappan L.). In: IOP conference series: earth and environmental science, vol 411, no 1. IOP Publishing, p 012043

Latos-Brozio M, Masek A (2020, Jan 1) The application of natural food colorants as indicator substances in intelligent biodegradable packaging materials. Food Chem Toxicol 135:110975

Lauer MK, Smith RC (2020, Nov) Recent advances in starch-based films toward food packaging applications: physicochemical, mechanical, and functional properties. Compr Rev Food Sci Food Saf 19(6):3031–3083

Li H, He Y, Yang J, Wang X, Lan T, Peng L (2019, May 1) Fabrication of food-safe superhydrophobic cellulose paper with improved moisture and air barrier properties. Carbohydr Polym 211:22–30

Lima AEF, Andrade PL, de Lemos TLG, Uchoa DEDA, Siqueira MCA, do Egito AS, Braga RC, da Costa JN, Teixeira Sa DMA (2021) Development and application of galactomannan and essential oil-based edible coatings applied to "coalho" cheese. J Food Process Preserv 45(1): e15091

Liu B, Xu H, Zhao H, Liu W, Zhao L, Li Y (2017, Feb 10) Preparation and characterization of intelligent starch/PVA films for simultaneous colorimetric indication and antimicrobial activity for food packaging applications. Carbohydr Polym 157:842–849

Liu F, Chang W, Chen M, Xu F, Ma J, Zhong F (2020a, Jan 1) Film-forming properties of guar gum, tara gum and locust bean gum. Food Hydrocoll 98:105007

Liu J, Wang H, Wang P, Guo M, Jiang S, Li X, Jiang S (2018, Oct 1) Films based on κ-carrageenan incorporated with curcumin for freshness monitoring. Food Hydrocoll 83:134–142

Liu RY, Jin Y, Hiltner A, Baer E (2003, Nov) Probing nanoscale polymer interactions by forced-assembly. Macromol Rapid Commun 24(16):943–948

Liu Y, Zhang X, Li C, Qin Y, Xiao L, Liu J (2020b, Mar 15) Comparison of the structural, physical and functional properties of κ-carrageenan films incorporated with pomegranate flesh and peel extracts. Int J Biol Macromol 147:1076–1088

Luchese CL, Pavoni JMF, Dos Santos NZ, Quines LK, Pollo LD, Spada JC, Tessaro IC (2018) Effect of chitosan addition on the properties of films prepared with corn and cassava starches. J Food Sci Technol 55(8):2963–2973

Ma Q, Wang L (2016, Nov 1) Preparation of a visual pH-sensing film based on tara gum incorporating cellulose and extracts from grape skins. Sensors Actuators B Chem 235:401–407

Marino AK, Junior JS, Magalhães KM, Mattiuz BH (2018, Aug 23) Chitosan-propolis combination inhibits anthracnose in 'Hass' avocados. Emir J Food Agri:681–687

Martiny TR, Pacheco BS, Pereira CM, Mansilla A, Astorga-España MS, Dotto GL, Moraes CC, Rosa GS (2020, Jul) A novel biodegradable film based on κ-carrageenan activated with olive leaves extract. Food Sci Nutr 8(7):3147–3156

Masek A, Latos M, Piotrowska M, Zaborski M (2018, Jun 1) The potential of quercetin as an effective natural antioxidant and indicator for packaging materials. Food Packag Shelf Life 16: 51–58

Medina-Jaramillo C, Ochoa-Yepes O, Bernal C, Famá L (2017, Nov 15) Active and smart biodegradable packaging based on starch and natural extracts. Carbohydr Polym 176:187–194

Messin T, Follain N, Guinault A, Miquelard-Garnier G, Sollogoub C, Delpouve N, Gaucher V, Marais S (2017, Mar 1) Confinement effect in PC/MXD6 multilayer films: impact of the microlayered structure on water and gas barrier properties. J Membr Sci 525:135–145

Mirza Alizadeh A, Masoomian M, Shakooie M, Zabihzadeh Khajavi M, Farhoodi M (2020, Sep 9) Trends and applications of intelligent packaging in dairy products: a review. Crit Rev Food Sci Nutr:1–5

Moghadam M, Salami M, Mohammadian M, Khodadadi M, Emam-Djomeh Z (2020, Jul 1) Development of antioxidant edible films based on mung bean protein enriched with pomegranate peel. Food Hydrocoll 104:105735

Mohamed SA, El-Sakhawy M, El-Sakhawy MAM (2020) Polysaccharides, protein and lipid-based natural edible films in food packaging: a review. Carbohydr Polym 238:116178
Morris BA (2016, Sep 1) The science and technology of flexible packaging: multilayer films from resin and process to end use. William Andrew
Nair MS, Tomar M, Punia S, Kukula-Koch W, Kumar M (2020, Jul 16) Enhancing the functionality of chitosan-and alginate-based active edible coatings/films for the preservation of fruits and vegetables: a review. Int J Biol Macromol 164:304–320
Oksman K, Aitomäki Y, Mathew AP, Siqueira G, Zhou Q, Butylina S, Tanpichai S, Zhou X, Hooshmand S (2016) Review of the recent developments in cellulose nanocomposite processing. Compos A: Appl Sci Manuf 83:2–18
Ortega-Toro R, Morey I, Talens P, Chiralt A (2015, Aug 20) Active bilayer films of thermoplastic starch and polycaprolactone obtained by compression molding. Carbohydr Polym 127:282–290
Otoni CG, Lorevice MV, de Moura MR, Mattoso LH (2018) On the effects of hydroxyl substitution degree and molecular weight on mechanical and water barrier properties of hydroxypropyl methylcellulose films. Carbohydr Polym 185:105–111
Oudjedi K, Manso S, Nerin C, Hassissen N, Zaidi F (2019, Apr 1) New active antioxidant multilayer food packaging films containing Algerian sage and bay leaves extracts and their application for oxidative stability of fried potatoes. Food Control 98:216–226
Pelissari FM, Andrade-Mahecha MM, do Amaral Sobral PJ, Menegalli FC (2017) Nanocomposites based on banana starch reinforced with cellulose nanofibers isolated from banana peels. J Colloid Interface Sci 505:154–167
Polturak G, Aharoni A (2019, Dec) Advances and future directions in betalain metabolic engineering. New Phytol 224(4):1472–1478
Priyadarshi R, Ezati P, Rhim JW (2019, Feb 2) Recent advances in intelligent food packaging applications using natural food colorants. ACS Food Sci Technol 1(2):124–138
Priyadarshi R, Rhim JW (2020, Jun 1) Chitosan-based biodegradable functional films for food packaging applications. Innovative Food Sci Emerg Technol 62:102346
Puscaselu R, Gutt G, Amariei S (2019, Jun) Biopolymer-based films enriched with stevia rebaudiana used for the development of edible and soluble packaging. Coatings 9(6):360
Qin Y, Liu Y, Zhang X, Liu J (2020, Mar 1) Development of active and intelligent packaging by incorporating betalains from red pitaya (Hylocereus polyrhizus) peel into starch/polyvinyl alcohol films. Food Hydrocoll 100:105410
Quoc LP, Hoa DP, Ngoc HT, Phi TT (2015, Sep 1) Effect of xanthan gum solution on the preservation of acerola (Malpighia glabra L.). Cercetari Agronomice in Moldova 48(3):89–97
Radev R, Pashova S (2020) Application of edible films and coatings for fresh fruit and vegetables. In: Qual Access Success, vol 21, pp 108–112
Raghav PK, Agarwal N, Saini M (2016) Edible coating of fruits and vegetables: A review. Education 1(2):188–204
Ramos M, Valdés A, Mellinas AC, Garrigós MC (2015, Dec) New trends in beverage packaging systems: a review. Beverages 1(4):248–272
Rhein-Knudsen N, Ale MT, Meyer AS (2015) Seaweed hydrocolloid production: an update on enzyme assisted extraction and modification technologies. Mar Drugs 13(6):3340–3359
Ribeiro-Santos R, Andrade M, de Melo NR, Sanches-Silva A (2017, Mar 1) Use of essential oils in active food packaging: recent advances and future trends. Trends Food Sci Technol 61:132–140
Risyon NP, Othman SH, Basha RK, Talib RA (2020, Mar 1) Characterization of polylactic acid/halloysite nanotubes bionanocomposite films for food packaging. Food Packag Shelf Life 23:100450
Rohini B, Ishwarya SP, Rajasekharan R, VijayaKumar AK (2020, Jul 1) Ocimum basilicum seed mucilage reinforced with montmorillonite for preparation of bionanocomposite film for food packaging applications. Polym Test 87:106465
Rukmanikrishnan B, Ismail FRM, Manoharan RK, Kim SS, Lee J (2020) Blends of gellan gum/xanthan gum/zinc oxide based nanocomposites for packaging application: rheological and antimicrobial properties. Int J Biol Macromol 148:1182–1189

Sadeghizadeh-Yazdi J, Habibi M, Kamali AA, Banaei M (2019, Dec 25) Application of edible and biodegradable starch-based films in food packaging: a systematic review and meta-analysis. Curr Res Nutr Food Sci J 7(3):624–637

Sanyang ML, Sapuan SM, Jawaid M, Ishak MR, Sahari J (2015, Jun) Effect of plasticizer type and concentration on tensile, thermal and barrier properties of biodegradable films based on sugar palm (Arenga pinnata) starch. Polymers 7(6):1106–1124

Sapper M, Chiralt A (2018) Starch-based coatings for preservation of fruits and vegetables. Coatings 8(5):152

Saputri AE, Praseptiangga D, Rochima E, Panatarani C, Joni IM (2018, February). Mechanical and solubility properties of bio-nanocomposite film of semi refined kappa carrageenan/ZnO nanoparticles. In: AIP conference proceedings, vol 1927, no. 1. AIP Publishing LLC, p 030040

Senturk Parreidt T, Müller K, Schmid M (2018, Oct) Alginate-based edible films and coatings for food packaging applications. Foods 7(10):170

Sharma R, Ghoshal G (2018, Sep 10) Emerging trends in food packaging. Nutr Food Sci 48(5): 764–779

Siracusa V, Lotti N (2019) Intelligent packaging to improve shelf life. In: Food Quality and Shelf Life, pp 261–279

Sohail M, Sun DW, Zhu Z (2018) Recent developments in intelligent packaging for enhancing food quality and safety. Crit Rev Food Sci Nutr 58(15):2650–2662

Souza VG, Pires JR, Rodrigues C, Coelhoso IM, Fernando AL (2020, Feb) Chitosan composites in packaging industry—current trends and future challenges. Polymers 12(2):417

Suhag R, Kumar N, Petkoska AT, Upadhyay A (2020) Film formation and deposition methods of edible coating on food products: a review. Food Res Int 136:109582

Tas BA, Sehit E, Tas CE, Unal S, Cebeci FC, Menceloglu YZ, Unal H (2019, Jun 1) Carvacrol loaded halloysite coatings for antimicrobial food packaging applications. Food Packag Shelf Life 20:100300

Tavassoli-Kafrani E, Shekarchizadeh H, Masoudpour-Behabadi M (2016) Development of edible films and coatings from alginates and carrageenans. Carbohydr Polym 137:360–374

Thakur VK, Thakur MK (eds) (2016, Jan 5) Handbook of sustainable polymers: processing and applications. CRC Press

Tichoniuk M, Radomska N, Cierpiszewski R (2017) The application of natural dyes in food freshness indicators designed for intelligent packaging. Studia Oeconomica Posnaniensia 5(7): 19–34

van den Broek LA, Knoop RJ, Kappen FH, Boeriu CG (2015, Feb 13) Chitosan films and blends for packaging material. Carbohydr Polym 116:237–242

Vartiainen J, Shen Y, Kaljunen T, Malm T, Vähä-Nissi M, Putkonen M, Harlin A (2016, Jan 10) Bio-based multilayer barrier films by extrusion, dispersion coating and atomic layer deposition. J Appl Polym Sci 133(2). https://doi.org/10.1002/app.42260

Verlee A, Mincke S, Stevens CV (2017) Recent developments in antibacterial and antifungal chitosan and its derivatives. Carbohydr Polym 164:268–283

Vo TV, Dang TH, Chen BH (2019, Jul) Synthesis of intelligent pH indicative films from chitosan/poly (vinyl alcohol)/anthocyanin extracted from red cabbage. Polymers 11(7):1088

Wang X, Yong H, Gao L, Li L, Jin M, Liu J (2019a, Apr 1) Preparation and characterization of antioxidant and pH-sensitive films based on chitosan and black soybean seed coat extract. Food Hydrocoll 89:56–66

Wang L, Xue J, Zhang Y (2019b, Apr 1) Preparation and characterization of curcumin loaded caseinate/zein nanocomposite film using pH-driven method. Ind Crop Prod 130:71–80

Wang LF, Rhim JW (2015) Preparation and application of agar/alginate/collagen ternary blend functional food packaging films. Int J Biol Macromol 80:460–468

Winotapun C, Phattarateera S, Aontee A, Junsook N, Daud W, Kerddonfag N, Chinsirikul W (2019, Aug) Development of multilayer films with improved aroma barrier properties for durian packaging application. Packag Technol Sci 32(8):405–418

Wu Y, Ding W, He Q (2017, Dec 2) Molecular characteristics of tara galactomannans: effect of degradation with hydrogen peroxide. Int J Food Prop 20(12):3014–3022

Xia C, Wang W, Wang L, Liu H, Xiao J (2019, Mar 1) Multilayer zein/gelatin films with tunable water barrier property and prolonged antioxidant activity. Food Packag Shelf Life 19:76–85

Xiao Y, Liu Y, Kang S, Wang K, Xu H (2020) Development and evaluation of soy protein isolate-based antibacterial nanocomposite films containing cellulose nanocrystals and zinc oxide nanoparticles. Food Hydrocoll 106:105898

Yao X, Hu H, Qin Y, Liu J (2020, Sep 1) Development of antioxidant, antimicrobial and ammonia-sensitive films based on quaternary ammonium chitosan, polyvinyl alcohol and betalains-rich cactus pears (Opuntia ficus-indica) extract. Food Hydrocoll 106:105896

Younes I, Rinaudo M (2015) Chitin and chitosan preparation from marine sources. Structure, properties and applications. Mar Drugs 13(3):1133–1174

Younes M, Aggett P, Aguilar F, Crebelli R, Filipič M, Frutos MJ, Galtier P, Gott D, Gundert-Remy U, Kuhnle GG (2018) Re-evaluation of carrageenan (E 407) and processed Eucheuma seaweed (E 407a) as food additives. EFSA J 16(4):e05238

Zhai X, Zhang X, Ao H, Yin Y, Li X, Ren D (2021) Preparation and characterization of whey protein isolate/chitosan/microcrystalline cellulose composite films. Packag Technol Sci 34(9): 589–599

Zhang J, Zou X, Zhai X, Huang X, Jiang C, Holmes M (2019, Jan 30) Preparation of an intelligent pH film based on biodegradable polymers and roselle anthocyanins for monitoring pork freshness. Food Chem 272:306–312

Zhong Y, Zhuang C, Gu W, Zhao Y (2019, May 15) Effect of molecular weight on the properties of chitosan films prepared using electrostatic spraying technique. Carbohydrate Polymer 212:197–205

Chapter 10
A Holistic Approach to Sustainable Food Waste Management and Residue Utilization

Julia Rabelo Vaz Matheus, Thaiza Serrano Pinheiro de Souza, Andreza Salles Barone, and Ana Elizabeth Cavalcante Fai

1 Introduction

Worldwide food waste reaches 1.3 billion tons annually throughout the entire food supply chain from farm to fork. This has not only caused environmental and health impacts but has also led to an economic crisis (Arun et al., 2020; Teigiserova et al. 2019a).

It is important to highlight that in addition to the problems related to food loss, food processing results in the generation of inedible food residue that, despite the efforts being made to mitigate quantities, will persist. The good news: these residues may represent a stable renewable material for the future biobased value chains and products (Teigiserova et al. 2019a).

This is possible due to the chemical composition and bioactive molecules found in food residues. If food residues are used to recover macro and micronutrients,

J. R. V. Matheus · A. S. Barone
Food and Nutrition Graduate Program, Federal University of Rio de Janeiro State (UNIRIO), Rio de Janeiro, Brazil

T. S. P. de Souza
Food and Nutrition Graduate Program, Federal University of Rio de Janeiro State (UNIRIO), Rio de Janeiro, Brazil

Basic and Experimental Nutrition, Institute of Nutrition, Rio de Janeiro State University (UERJ), Rio de Janeiro, Brazil

School of Agriculture, Food and Ecosystem Sciences, Faculty of Science, The University of Melbourne, Melbourne, Australia

A. E. C. Fai (✉)
Food and Nutrition Graduate Program, Federal University of Rio de Janeiro State (UNIRIO), Rio de Janeiro, Brazil

Basic and Experimental Nutrition, Institute of Nutrition, Rio de Janeiro State University (UERJ), Rio de Janeiro, Brazil

H. U. Hebbar et al. (eds.), *Engineering Aspects of Food Quality and Safety*, Food Engineering Series, https://doi.org/10.1007/978-3-031-30683-9_10

bioactives, biopolymers, among others, it is feasible to reduce the cost of these valuable compounds as well as to reduce waste (Arun et al. 2020). Thus, advanced methodologies to characterize, extract and transform those compounds through technology and biotechnology routes have been explored in the last years. Obtaining new sustainable materials for different industrial purposes has also been extensively investigated (Brito et al. 2021, 2020a, b; Matheus et al. 2021, 2020b; Teigiserova et al. 2019a).

In light of this situation, a holistic approach is urgently desirable to reduce food loss and waste, as well as for the management of food residue inherent in food production. Additionally, technological and biotechnological research is required to maximize the economic incomes, by transforming food residues into valuable bioproducts, with low-cost investments.

The sustainable utilization of food residues is aligned with the precepts established in the circular economy, which adopts the concept of re-using and recycling biowastes, leading interested parties and producers to embrace more environmentally-friendly processes (Ng et al. 2020). The circular bioeconomy adopts the use of renewable biological resources to transform them into high-value products, such as bioactive components, bioplastic, and bioenergy, and to preserve the resources for a longer period of time with the objective of producing zero waste and reducing greenhouse gases (GHG) emissions (Sharma et al. 2021). This model is also aligned with the Sustainable Development Goal (SDG) Target 12.3, which aims to halve global food residues at the retail and consumer levels and reduce food losses, including postharvest losses, along supply chains by 2030, by ensuring sustainable consumption and production patterns (Lipinski 2020).

Thus, a holistic approach to sustainable food waste management and residue utilization is proposed in this chapter.

2 Economic and Nutritional Aspects of Food Loss and Waste: An Old Issue with Modern Relevance

Food supply chains include primary agricultural production, manufacturing, retail, and household consumption. Throughout its life cycle, food may be lost or wasted due to technological, economic and/or social reasons (Otles et al. 2015). Food loss or waste is defined as the masses of food lost or wasted in some part of the food chain between the producer and consumer, from edible products intended for human consumption (FAO 2011). Food loss is an unintended loss in edible food quantity or quality before consumption during harvest, post-harvest handling, processing, and distribution. Food waste is when safe and nutritious food for human consumption is discarded or not consumed, primarily in the retail and consumption stages (including services) (Fig. 10.1) (de Brito Nogueira et al. 2020; Teigiserova et al. 2019b; FAO 2011). Table 10.1 shows several reasons for food loss and waste at the distinct stages of the food supply chain.

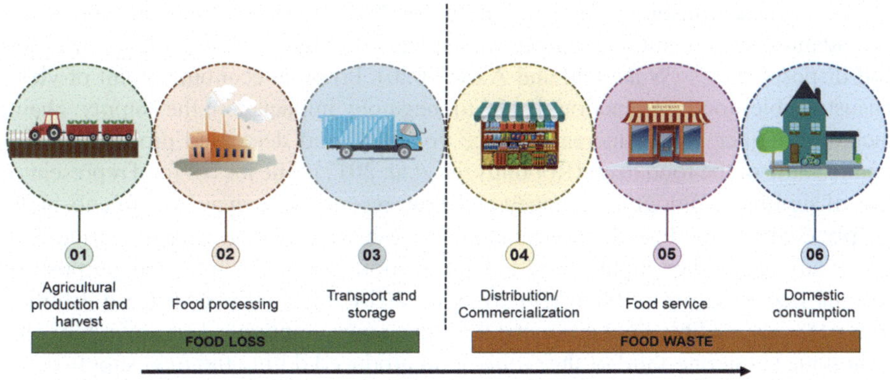

Fig. 10.1 Food loss and food waste: where each of them is in the stages of the food supply chain

Table 10.1 Possible causes of food loss and waste throughout the food supply chain

Stage of food supply chain	Causes of food loss and waste
Production	Infrastructural limitation
	Overproduction
	Harvesting timing and method (manual/mechanical)
	Pesticides and fertilizers
	Economic problems
	Quality standards and norms
Postharvest handling and storage	Degradation and spoilage of product composition
	Loss during transportation from farm to distribution
	Storage infrastructure
Processing and packaging	Unavoidable losses
	Technical malfunctions
	Methods and changes in processing lines
	Contamination in processing lines
	Legislation restrictions
	Packaging system
	Overproduction
Distribution and marketing	Inappropriate transport conditions (temperature-controlled aircrafts and ships)
	Contamination of transportation
	Transportation and market facilities
	Road and distribution vehicles
	Packaging management
	Commercial conditions
	Consumer preference
Consumption	Composition unit and size of household
	Income group
	Demographics and culture
	Individual attitude
	Cooking practices and methods

Adapted from: Singh (2020)

Food waste constitutes a main driver of both economic and environmental degradation, since natural resources are depleted during the production, preparation and disposal stages (Wakefield and Axon 2020). From an economic point of view, unsustainable food production leads to negative impacts on the supply chain, including higher prices, increased price volatility, and decreased profits and economic value of the food itself (Roodhuyzen et al. 2017). The loss of food represents a loss of significant amounts of money and other resources, such as investments in the supply chain in food production, including fresh water, labor, energy, agricultural chemicals, and other inputs for food production, when the intended purpose of feeding people is not fulfilled (Buzby et al. 2014). It is estimated that 1.3 billion tons per year of edible food produced for human consumption is lost or wasted. This represents about one-third of the edible food produced during the food supply chain (FAO 2011). A more recent study estimates that about 14% of food is lost in the pre-retail levels (e.g., agriculture, post-harvest, slaughter, and catch) (FAO 2019). Regarding the global economy, the value of food lost or wasted annually is estimated at US$ 1 trillion (FAO 2015a). The economic cost of waste management comprises the maintenance of landfills, transportation, treatment plant operation and separation and segregation of the waste. The total annual economic, environmental and social cost of food waste for the global economy considering food waste that is not collected separately and disposed of in landfills is US$ 2.6 trillion (Singh 2020).

The quantity of food loss and waste is influenced by several drivers, including the modernization of society, increased globalization of trade, urbanization, cultural changes, dietary transitions, and sociodemographic factors (Wakefield and Axon 2020). Regarding per capita income, developed countries waste six times more food by weight than developing countries (Chen et al. 2020). However, food losses in high-income and low-income nations are estimated to be on the same level. In developed countries, more than 40% of food is lost during the marketing and consumption stages, while in developing countries about 40% is lost during the postharvest and processing stages (FAO 2011). Figure 10.2 presents the quantity of worldwide food loss and waste generation and its distribution along the food supply chain, based on data reported by FAO (2011) and by Lipinski et al. (2013). At the regional level, food loss ranges from 5.8% in Australia and New Zealand, to 20.7% in Central and Southern Asia (Fig. 10.3a) (FAO 2019). Food commodities, roots, tubers and oil-bearing crops report the highest losses – 25.3%, followed by fruit and vegetables – 21.6%, mostly owing to their highly perishable nature (Fig. 10.3 b) (FAO 2019).

It is important to add the nutritional panorama that has been drawn to this economic data. This outlook portrays a global food system that is failing to meet the nutritional needs of the world population, and focuses on health issues related to insufficient micronutrient consumption (Ritchie et al. 2018). Food insecurity, defined as "inefficient access to safe, sufficient and nutritious food to meet a person's dietary needs and food preferences in order to pursue an active and healthy lifestyle" (Tester et al. 2020) is a worldwide public health problem. This concept has been associated with hunger, malnutrition, obesity, and inadequate access to nutritionally satisfying foods (Brown et al. 2019).

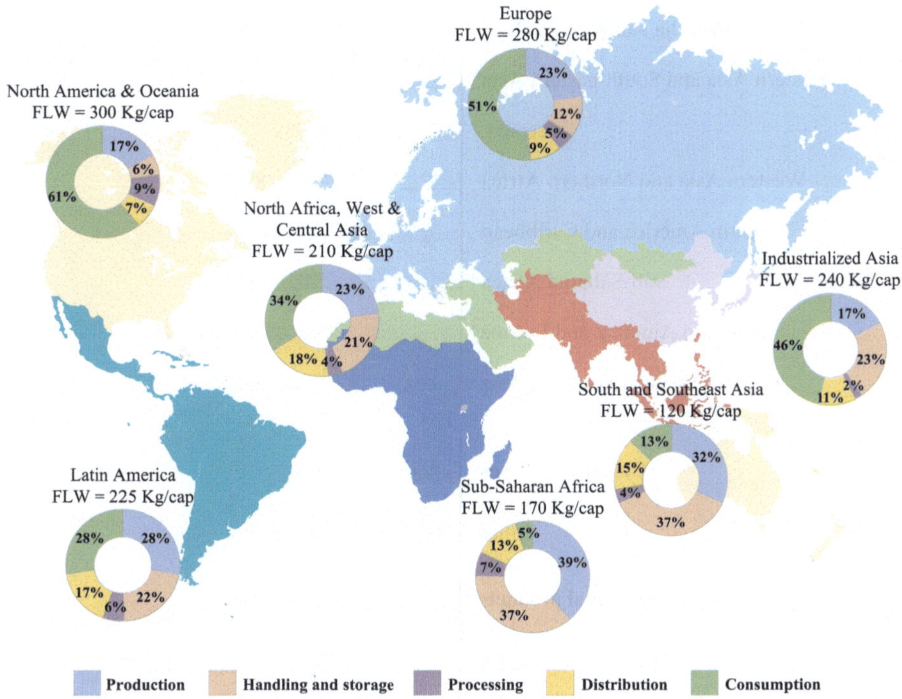

Fig. 10.2 Worldwide food loss and waste and its distribution in the different stages (production; handling and storage; processing; distribution; and consumption) of the food supply chain in 2009. FLW refers to the total food loss and waste in kg per capita/year

Although food insecurity can be obesogenic, linked to unhealthy eating patterns, it can also lead to weight loss, especially in its most severe form – hunger (Morales and Berkowitz 2016). It is estimated that in 2020 between 720 and 811 million people suffered from hunger all over the world (FAO et al. 2021), contributing to the failure to achieve the UN's Sustainable Development Goals (SDG), in particular SDG2, which aims to eradicate hunger, achieve food security, improve nutrition and promote sustainable agriculture by 2030. This fact is observed and reported mainly in low- and middle-income countries, where the population does not have regular access to safe, nutritious and sufficient food (FAO et al. 2019), accompanied by the disincentive of public and social policies that contribute to addressing the problem (Watson et al. 2020). Food insecurity may also be associated to dietary patterns characterized by a lower consumption of healthy food groups and a poor quality of diet, particularly regarding the intake of fruit and vegetables (Morales and Berkowitz 2016). In addition, it is known that in low-income neighborhoods, there is a dichotomy between the limited supply of fresh and nutritious food and an increase in fast-food outlets, which are "low cost" per calorie compared to healthier foods (Tester et al. 2020). Understanding these relationships and the social impacts linked to them is particularly useful, requiring a profound overhaul of food systems in order to provide healthy and sustainably produced diets for a constantly growing world population (FAO et al. 2019; Torres-León et al. 2018).

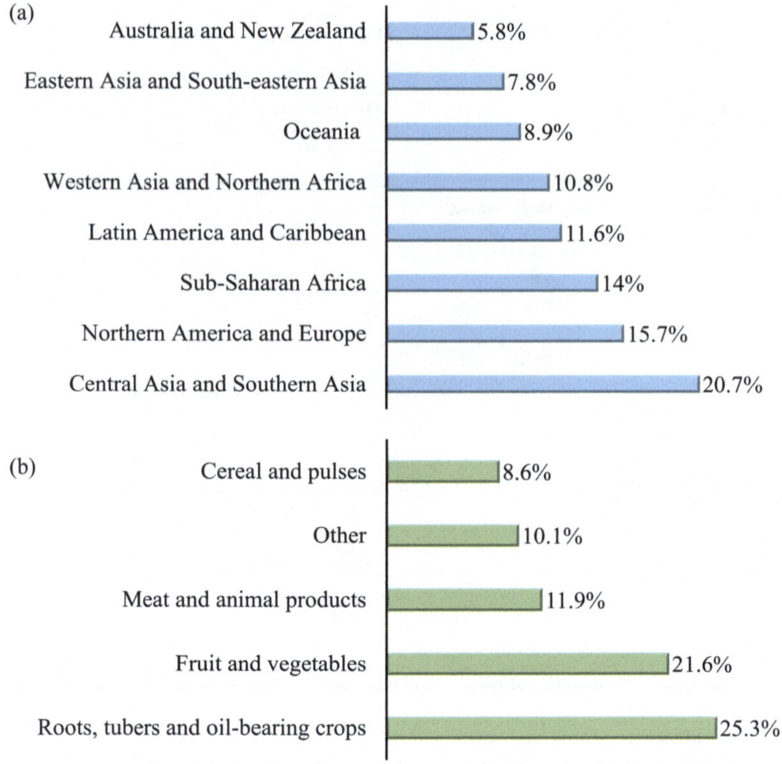

Fig. 10.3 Share of global food loss in 2016 by (**a**) region and (**b**) commodity from the farm to, but excluding, retail stage

If, on the one hand, there is scarcity and a lack of access to nutritionally healthy food, on the other, we find the problem of waste generation and mismanagement inherent in food production (Lai et al. 2017). It is estimated that about 1.3 billion tons of food intended for human consumption is wasted each year (Laso et al. 2021) and that on a global scale the nutritional cost resulting from these losses would feed approximately 940 million people around the world (Abbade 2020).

In addition to these drivers, the world is also currently facing the COVID-19 pandemic, with high potential to exacerbate worldwide malnutrition (Kurtz et al. 2021), which has increased the urgency to solve the problem of food loss and waste, as food systems have been trying to respond to sudden changes in demand, scarcity of jobs, and lower available incomes (Lipinski 2020). The pandemic has led to higher waste generation in the consumer stage mainly owing to the over-buying trend and improper storage of high quantities of foods, as well as a disruption in food supply chains, due to road closures which generated an accumulation of products, resulting in increased levels of food loss and waste (Boyacı-Gündüz et al. 2021). Aldaco et al. (2020) reported that COVID-19 has had a low impact on the overall food loss and waste generation, however it resulted in a 12% higher generation of food waste at the household level in Spain.

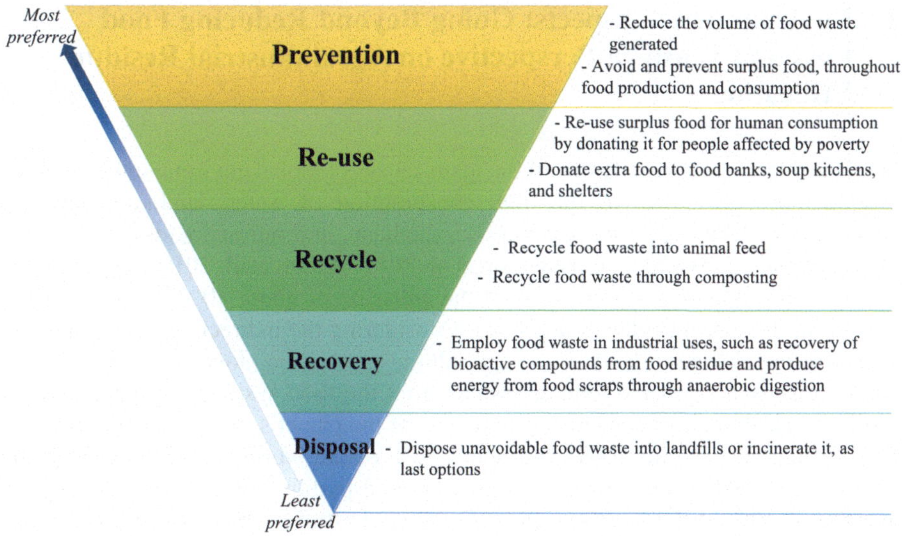

Fig. 10.4 Food waste hierarchy

To overcome the surplus generation of edible food and consequently the production of food waste, a food waste hierarchy is proposed to reduce and recover food excess, by separating and managing it. Reducing and recovering surplus food may result in both economic and environmental benefits (EPA 2012). The food recovery hierarchy establishes a hierarchical order between the different ways to reduce food waste, as follows: first, prevention, by reducing surplus at the source; second, recovery, by reusing for human consumption; and third, recycling, by feeding animals, creating energy or compost (Mourad 2016). From an economic point of view, preventing surplus food may save money since disposal costs would fall; a decrease in costs related to sewage and electricity treatment; decrease in purchase costs, since only what is needed is purchased; increased tax deductions for food donations to charities; and increased revenue from the sale of compost made of food scraps (EPA 2012). Figure 10.4 shows the food waste hierarchy.

Traditionally, agricultural and food waste management included incineration, landfills, composting and application as animal feed. Currently, most of the food residues end up in landfills, which leads to environmental and health concerns, e.g., emission of GHG, bacterial contaminations, and infectious diseases (Ng et al. 2020). However, to achieve more sustainable processes, and decrease costs while increasing the effectiveness of processes, industries are looking for innovative strategies (Kavitha et al. 2020), such as an approach for biofuel and biofertilizer production, besides energy recovery from food waste using ultra-fast hydrolysis and co-digestion process, making possible zero-food waste disposal (Ma and Liu 2019).

3 Environmental Aspects: Going Beyond Reducing Food Loss and Waste, a Perspective on Agroindustrial Residue Utilization

Food production is currently one of the main causes of global environmental change. Degradation occurs due to the high consumption of water, land, and energy, generation of GHG (e.g., carbon dioxide, methane, and nitrous oxide), and a loss of biodiversity (FAO 2013; Gustavsson et al. 2011; Willett et al. 2019). The impacts of the food production system include the emission of about 30–35% of all GHG emitted in the world (Foley et al. 2011). Considering the indirect emissions associated with the change in land cover, agricultural production alone is responsible for approximately 86% of the total emissions from the food system (Vermeulen et al. 2012). Agriculture also consumes about 70% of the planet's freshwater (Foley et al. 2011; Khan and Hanjra 2009) and about 85% of the world's water consumption (Pfister et al. 2011). Furthermore, cultivation and pasture areas already occupy an average of 40% of the land's surface (Foley et al. 2005). These data reinforce the need to transform the global food system to make it healthier and more environmentally sustainable, mainly due to the world population projections that point to 10 billion people in 2050 (Willett et al. 2019).

An issue that must be highlighted is the loss of food along the supply chain. In addition to being expressive, reaching one-third of all food produced for human consumption, especially fruit and vegetables (Gustavsson et al. 2011), it also competes for limited natural resources (Lemaire and Limbourg 2019). In this sense, food loss has a double impact on the environment: the excessive production of food and the management of food losses. In short, food loss is related to the waste of all resources used along the food chain, from agricultural production to consumption, including energy, human labor, and natural resources (Corrado et al. 2017). Therefore, it is necessary to include these aspects to decipher the real environmental impact of food loss (FAO 2014a, 2013).

It is estimated that the global loss of food and waste emits 4.4 $GtCO_2eq$ annually, considering changes in land use, corresponding to approximately 8% of the total global emissions of GHG (FAO 2015b). In addition, it occupies 0.9 million hectares of soil and consumes 306 km^3 of water (FAO 2014b).

Currently, several definitions of food loss and food waste are proposed. The FAO definition (2011) is widely accepted, used, and was adopted in this chapter, as stated above. However, waste resulting from inedible parts of foods or their processing as well as the loss and waste of edible parts of food along the food supply chain are also considered in these other definitions. Thus, in this work, we denominate as "residues" the processing waste residues, whether they are edible (e.g., fruit peel) or inedible (e.g., fruit seed), as well as the naturally inedible parts (e.g., some fruit leaves), based on Teigiserova et al. (2019b).

The food processing industry generates approximately 140 billion tons of residues annually. These residues are, in most cases, underutilized, even though they are rich in nutrients and bioactive compounds that could be used as bioproducts and/or raw material for products with higher added value (Hang 2004; Zuin and Ramin 2018).

Actions to reduce food loss and waste, as well as food residues, are necessary and encouraged, to contribute to a more sustainable and resilient food system (Lemaire and Limbourg 2019; Willett et al. 2019) and guarantee food security (IPCC 2019), as previously discussed. However, as it is difficult to eradicate residues in food production (Lemaire and Limbourg 2019; IPCC 2019), it is necessary to outline ways of using such residues that result in fewer environmental impacts than, for example, landfills, which are one of the main destinations of food residues (Ng et al. 2020; Papargyropoulou et al. 2014).

Currently, with the valorization of green industrial processes, the use of agro-industrial plant residues has been identified as an important strategy to develop sustainable products with greater added value for the chemical, pharmaceutical, and food industries (Freitas et al. 2021), given the rich and heterogeneous chemical composition of these residues (Fidelis et al. 2019; Jiménez-Moreno et al. 2020; Majerska et al. 2019) and their low cost (Zuin and Ramin 2018). This is possible because food residues present a variable chemical composition on the basis of their generation and origin (Kavitha et al. 2020). They stand out for their high nutritional value due to the presence of satisfactory amounts of proteins, lipids, starch, micronutrients, bioactive compounds, and dietary fibers (Banerjee et al. 2017; Faustino et al. 2019), providing numerous alternatives for the use by the food industry, which may improve the nutritional characteristics of food (Sancho et al. 2015). In addition, the use of this rich resource, mostly fruit and vegetables, could directly benefit local communities, through the formulation of new foods and the strengthening of existing ones (Lai et al. 2017). Many studies are using treated agro-industrial residues as ingredients for food products of greater nutritional value (Table 10.2).

Furthermore, these sources are considered good raw-materials since they can be bioconverted into high-value bioproducts, e.g., biofuel, bioactive compounds, enzymes, fine chemicals, and biodegradable plastics (Ng et al. 2020; Sharma et al. 2021). Figure 10.5 presents the origin of food residues along the food supply chain, highlighting the main vegetable sources and types, as well as potential applications. Protein and lipid-rich food residues are suitable for animal feed, whereas cellulosic-rich food residues are suitable for cattle feed (Kavitha et al. 2020). Some residues may not be appropriate for the elaboration of technological and biotechnological products owing to their complexity, uncontrolled spoilage, and lack of traceability. In these cases, methods such as biogas production and composting can be applied (Poovazhahi and Thakur 2020).

Residues and by-products from the industrial processing of fruit and vegetables are exploited for animal feed (Papargyropoulou et al. 2014; Velarde et al. 2020), biofuel manufacture (Abdelhady et al. 2020; Suhartini et al. 2020), recovery of bioactive compounds (Jiménez-Moreno et al. 2020; Rahimi and Mikani 2019;

Table 10.2 Studies that used agro-industrial residues to improve the nutritional value of food products between 2020 and 2021

Agro-industrial residue	Compounds	Application	Reference
Cocoa	Polyphenols	Chocolate/cocoa drinks	Manuela et al. (2020)
Beet leaves extract	Phenolics and antioxidants	Fruit and vegetable smoothie	Fernandez et al. (2020)
Peppermint hydrodistillation by-products	Phenolics	Ice creams	Berktas and Cam (2020)
Apple pomace	Dietary fibers and phenolics	Yogurt and yogurt drinks	Wang et al. (2020)
Banana peel	Phenolics	Yogurt	Kabir et al. (2021)
Apple peels and grape seeds	Phenolics and flavonoids	Yogurt	Brahmi et al. (2021)
Carrot waste extract	β carotene	Yogurt	Šeregelj et al. (2021)
Apple pomace flour	Polyphenols	Yogurt	Jovanović et al. (2020)
Mango and potato peel flour	Dietary fibers and polyphenols	Yogurt	Pérez-Chabela et al. (2021)
Grape pomace powder	Polyphenols	"Primosale" cheese	Gaglio et al. (2021)
Pomegranate peel	Phenolics	Cookies	Kaderides et al. (2020)
Apple peel	Dietary fibers and polyphenols	Bread and wheat cookies	Nakov et al. (2020)
Blackcurrant by-product	Dietary fibers and antioxidants	Chocolate cookies	Gagneten et al. (2021)
Prickly pear peel flour	Dietary fibers and phenolics	Cookies	Bouazizi et al. (2020)
Rice bran	Phenolics and antioxidants	Cookies	Christ-Ribeiro et al. (2021)
Cashew waste	Dietary fibers	Cookies and flour	de Araújo et al. (2021)
Guava peels and cashew bagasse	Protein	Cereal bars	Muniz et al. (2020)
Coffee silverskin	Dietary fibers, protein and polyphenols	Extruded cereal-based ready-to-eat food product	Beltrán-Medina et al. (2020)
Unripe papaya powder	Dietary fibers and polyphenols	Pancake	Joymak et al. (2021)
Flour and coconut residue	Dietary fibers	Pasta	Sykut-Domańska et al. (2020)

(continued)

Table 10.2 (continued)

Agro-industrial residue	Compounds	Application	Reference
Citrus reticulata (Kinnow) pomace	Dietary fibers and antioxidants	Pasta	Singla et al. (2020)
Grape pomace	Phenolics	Pasta	Tolve et al. (2020)
Onion skin powder	Dietary fibers, phenolics and flavonoids	Pasta	Michalak-Majewska et al. (2020)
Grape and olive pomace	Dietary fibers and phenolics	Pasta	Balli et al. (2021)
Whey and grape peels powders	Dietary fibers	Pasta	Ungureanu-Iuga et al. (2020)
Olive pomace	Polyphenols	Bread and pasta	Cedola et al. (2020)
Wheat bran	Protein and essential amino acids	Bread and pasta	Alzuwaid et al. (2021)
Onion powder and onion peel extract	Dietary fibers and protein	Bread	Masood et al. (2020)
Broad bean hull	Dietary fibers	Bread	Ni et al. (2020)
Prickly pear Peel flour	Betalains and flavonoids	Bread	Parafati et al. (2020)
Whole green banana flour	Dietary fibers and resistant starch	Bread	Khoozani et al. (2020)
Milk permeate, psyllium husk and apple by-products	Galactooligosaccharides (GOS) and lactic acid bacteria, desirable hydrocolloids and phenolic compounds	Nutraceutical chewing candy	Zokaityte et al. (2021)
Sweet potato peel flour	Minerals, carbohydrate and dietary fiber	Beef hamburger	Marconato et al. (2020)
Pumpkin peel flour	Minerals, carbohydrate and dietary fiber	Beef hamburger	Hartmann et al. (2020)

Sánchez-Camargo et al. 2019) that can be applied as sustainable and potentially functional ingredients (Fidelis et al. 2020), enzyme production (Costa et al. 2017; Mojumdar and Deka 2019), biofertilizers (Du et al. 2018; Lim and Matu 2015), raw material for application in microalgae cultivation (de Medeiros et al. 2020; Koutra et al. 2018), pigment production (Arikan et al. 2020; Aruldass et al. 2016; Lopes and Ligabue-Braun 2021), of biosurfactants (Jadhav et al. 2019; Rivera et al. 2019) and nanoparticles (Huang et al. 2020; Shruthy and Preetha 2019), in addition to the extensive development of active and biodegradable packaging for food (Luchese et al. 2019; Vedove et al. 2021).

There are several studies in the literature that highlight the potential of extracting bioactive compounds from agro-industrial residues. Vodnar et al. (2017) identified bioactive compounds in some plant residues using different treatments and, in some

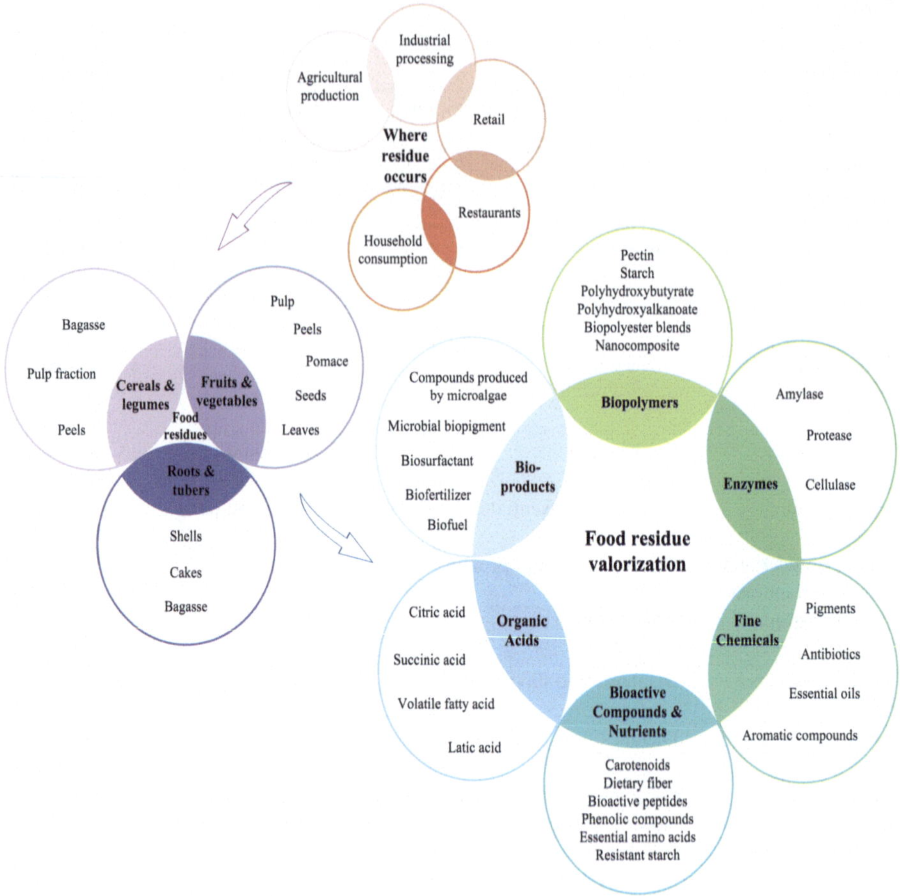

Fig. 10.5 Food residues: Sources, types and opportunities for their valorization

of them, observed antioxidant, antimutagenic, and antimicrobial activities. Velarde et al. (2020) evaluated the content of phenolic compounds and antimicrobial potential of hydroalcoholic extracts from three plant residues – avocado, guava, and cherry plum leaves – and suggested their application in additives for animal feed. Sánchez-Camargo et al. (2019) recovered carotenoids from mango peel and observed their action against lipid oxidation in sunflower oil, acting as a natural antioxidant. Rahimi and Mikani (2019) recovered lycopene from tomato residues employing an edible solvent (sunflower oil) in a green process to generate a safe and high-quality extract. Fidelis et al. (2020) observed the effect of lyophilized camu-camu seed extract on the antioxidant and sensory characteristics of yogurt. The extract showed high phenolic content, in addition to high antioxidant activity. It was also able to inhibit the cell proliferation of two cancer cell lines and the in vitro activity of some enzymes of clinical importance. The addition of the extract to yogurt in different concentrations proved to be adequate from a sensory point of view and was able to increase the antioxidant and chemical reducing properties in yogurts.

Other application examples involve the production of enzymes. Mojumdar and Deka (2019) compared some agro-industrial residues as substrates to produce alpha-amylase through solid-state fermentation using *Bacillus amyloliquefaciens*. The substrates of wheat bran and potato skins, both alone and in combination, showed higher enzymatic production when compared to rice bran. Moreover, the purest and most active amylases were those obtained with wheat bran substrate alone and combined with potato skins.

Another possibility is the production of biofertilizers from vegetable residues, which was explored by Lim and Matu (2015) using solid-state fermentation. Biofertilizers made from watermelon, papaya, and banana residues proved to be suitable for plant treatment; watermelon resulted in the greatest weight and the average length from mustard plant samples.

The use of agro-industrial residues as a means to produce biosurfactants is also extensively explored. Jadhav et al. (2019) produced a sophorolipid based on sunflower oil acid by submerged fermentation using *Starmerella bombicola*, which showed better emulsification and wetting properties and less foaming than the polysorbate 20 synthetic surfactant evaluated. Paraszkiewicz et al. (2018) identified and characterized lipopeptide biosurfactants produced by two strains of *Bacillus subtilis* grown in various media prepared from plant residues, including extract from apple and carrot peel. Studies like these reinforce the potential for the development of biosurfactants using residues as a fermentative medium, since they are widely generated and low in cost, and can therefore contribute to better managing these residues.

In addition, the residues can also serve as an alternative culture medium for the development of microalgae. De Medeiros et al. (2020) evaluated the use of agro-industrial residues from fruits and vegetables as an alternative means for the cultivation of three freshwater microalgae, resulting in adequate cell growth, absence of chemical and microbiological contaminants, in addition to better antioxidant activity and mono and polyunsaturated fatty acid levels in comparison with the conventional synthetic medium. The use of this type of lower-cost raw material can improve the cost-benefit ratio of microalgae cultivation and contribute to the production of compounds of interest to the food industry as well as other products with ecologically correct applications (Koutra et al. 2018).

Some biopigments have already been produced by microorganisms using agro-industrial residues as raw material. Arikan et al. (2020) used by-products by processing apple, pomegranate, black carrot, and red beet as a substrate for the filamentous fungus *Aspergillus carbonarius* to produce pigments. The authors observed the great potential of pomegranate pulps in pigment production, especially yellow, suggesting that *A. carbonarius* can produce melanin. Aruldass et al. (2016) optimized the production of yellowish-orange pigment by cultivating *Chryseobacterium artocarpi* CECT 8497 in a pineapple residue liquid medium. In this work, the researchers were able to produce a bacterial pigment, on a pilot scale, using a cheaper and more economical medium than the usual nutrient broth. These works highlight the potential of using vegetable residues as an alternative growth medium to produce biopigments.

Several studies have evaluated the incorporation of agro-industrial residues, such as blueberry peel and powdered jaboticaba (Luchese et al. 2019) and grape skins (Vedove et al. 2021), in biodegradable starch-based films, as a potential colorimetric indicator for application as smart food packaging. In these studies, the potential of the anthocyanins in the residues to change the color of the films in response to changes in the pH of the medium was observed. Agro-industrial residues are also used for the elaboration of nanoparticles. Shruthy and Preetha (2019) developed cellulose nanoparticles prepared from potato skins through alkaline treatment, bleaching, and acid hydrolysis, and incorporated them in an ecological film based on biodegradable polyvinyl alcohol added to fennel seed oil. The nanoparticles developed provided greater tensile strength to the films, in addition to good transparency, thermal stability, and biodegradability. Huang et al. (2020) developed modified cellulose nanofibrils from cassava residue as a reinforcing agent for starch-based films, resulting in a biodegradable film with hydrophobic characteristics and good mechanical and barrier properties.

Another interesting possibility is the use of fruit that would otherwise be discarded due to difficulties in their commercialization, as raw materials for the elaboration of biodegradable films, thus contributing to reducing losses (Matheus et al. 2021, 2020b). For example, persimmon, which has high rates of waste in Brazil (Matheus et al. 2020a), showed great potential regarding the development of biodegradable films used as packaging for minimally processed vegetables (Matheus et al. 2021).

Finally, despite several studies addressing some options for the use of agro-industrial residue, it is necessary to evaluate cases on an individual basis to understand the real environmental impacts of alternative uses, to guarantee that they cause less degradation to the ecosystems and are inserted in the context of the circular economy and sustainable development.

4 Circular Economy: A Holistic Approach

It is already evident that food loss is a global problem (Corrado et al. 2017; Laso et al. 2021) and still generates large amounts of agro-industrial residues. Food loss, food waste and food residues are often poorly managed, culminating in negative impacts on different sectors of human life and the environment (Papargyropoulou et al. 2014). Thus, it is essential to search for solutions that consider socioeconomic, nutritional, and environmental aspects. In this sense, one possible approach is a circular economy, which can contribute to human activities that favor the reduction of waste generation by transforming such waste into new resources to be exploited, making it possible to align economic development with the promotion of environmental quality and social equity under the spectrum of human rights (Velenturf and Purnell 2021). From this perspective, the residues generated in the food industry can go from being merely waste to becoming renewable resources with potential applications in different sectors.

Fig. 10.6 The UN's 17 Sustainable Development Goals highlighting those that may be favored when using food residues (2, 7, 8, 11, 12, 14, and 15)

In this context, the SDGs stand out as a strategic, global project of sustainable development linked to the circular economy (Schroeder et al. 2019). Their main purposes are to eradicate poverty, reduce inequality and minimize climate change, preserving oceans and forests (UN 2015). The use of agro-industrial residues could contribute directly or indirectly to some SDG targets (Fig. 10.6) since this strategy integrates environmental, socioeconomic, and nutritional issues.

Looking at it in greater detail, using food residues can minimize the environmental impact, thus contributing to goals 11, 12, 14, and 15, more specifically, subitems 11.4, 11.6, 14.1, and 15.5. More environmentally conscious management allocates the residues to non-polluting purposes, in order to assist in substituting non-renewable resources for the production of energy, bioplastics, and chemicals (Freitas et al. 2021). Therefore, harnessing food waste and food residues can contribute to the protection of the world's natural heritage and a reduction in the environmental impact of cities (11.4 and 11.6), a lower release of waste into oceans and coastal regions (14.1), and protection of biodiversity as a result of less environmental degradation (15.5) (UN 2015).

The use of waste and residues is directly related to goal 12, since it addresses the need to change current consumption and production patterns to a more sustainable model. In general, it is expected: to use and manage natural and renewable resources more efficiently, in addition to adopting actions towards the gradual transition to the circular economy model (12.1 and 12.2); halve world food waste per capita (12.3); reduce the generation of waste based on different strategies (prevention, reduction, recycling, and reuse) (12.5); and to properly manage chemical products and residues, avoiding contamination of different ecosystems and impacts on the environment and human health (12.4) (UN 2015).

Given the wide range of compounds, versatility, and potential of agro-industrial plant residues (Jiménez-Moreno et al. 2020), their by-products can be extensively explored to contribute indirectly to the different SDG subitems. For example, the development of nutritionally richer ingredients and/or food products may favor a healthier diet (Majerska et al. 2019), especially for poor and vulnerable people, thus relating to subitem 2.2 which aims to end all forms of malnutrition. If part of the waste is destined to produce biodegradable food packaging (de Moura et al. 2017), capable of replacing conventional plastic packaging of petrochemical origin, environmental pollution, especially in the oceans, can be minimized. In addition, taking advantage of liquid residues from food processing can contribute to such residues not being disposed of in rivers, which leads to a series of ecological imbalances. One example is the industrial effluent from cassava processing that can be used to produce biogas and biosurfactants or as a growth medium for microorganisms (Zevallos et al. 2018). These applications can contribute to the prevention and reduction of marine and coastal pollution (14.1). If the use of waste is destined to produce fuel and renewable energy (Suhartini et al. 2020), thus relating to subitem 7.2, as it aims to increase the global percentage of renewable energy. Finally, the increase in the efficiency of global resources in both consumption and production, will contribute to subitem 8.4, making it possible to combine economic growth and sustainability from the circular economy.

Despite the potential to apply agro-industrial residues as more sustainable alternatives, it is essential to carry out scientific research that analyzes specific cases of exploitation from the point of view of the circular economy, assessing whether the environmental, economic, and social impacts are indeed positive. Velenturf and Purnell (2021) carried out an extensive analysis of the circular economy, demonstrating that, in many cases, theory and practice have been contradictory, either not being truly in line with what is proposed or even negatively impacting the environment and society. They stressed that sustainable development can be achieved with the support of circular economy precepts, but that it is necessary to take a more holistic approach in understanding these concepts, so that it is not reduced to closed-loop recycling with short-term economic benefits (Velenturf and Purnell 2021).

It is also important to guarantee that these strategies do not stimulate the generation of residues, contradicting one of the most important modern-day needs: to prevent food loss and the generation of waste (Laso et al. 2021; Teigiserova et al. 2019a). Therefore, coordinated actions between all the actors involved (society, governments, and industries) must be strengthened to create a culture of circular economy and sustainable development, guaranteeing economic growth, environmental quality, well-being, and social equity for current and future generations (Velenturf and Purnell 2021).

5 Futures Perspectives

The use of agro-industrial residue can increase economic gains, reduce production costs and increase the value of residues, increase the nutritional value of food, in addition to strengthening sustainable practices inserted in the circular economy. These changes are in line with reducing our anthropogenic footprint on an already extensively affected planet (Nazzaro et al. 2018; Willett et al. 2019).

Due to the diverse and interesting chemical composition, agro-industrial plant residues can be better utilized from a circular economy perspective, in which bioconversion can generate different bioproducts with potential socioeconomic benefits coupled with lower environmental impact (Fidelis et al. 2019, 2021; Kavitha et al. 2020; Ng et al. 2020).

The need for changes in the food sector and waste management is increasingly urgent. The agreed deadlines for achieving the SDG targets are approaching and there are still many adjustments to be made. Furthermore, with the global crisis generated by the COVID-19 pandemic, it will be necessary to devise strategies for a worldwide better recovery. Efforts must be made to mitigate negative impacts and prevent them from lasting for many years. It is essential to make the food system more resilient, reducing food losses and waste generation, as well as environmental, socioeconomic, and nutritional setbacks.

Food waste management strategies must consider the numerous challenges involved to achieve optimal performance in the environmental, health, economic and social dimensions. As food waste is heterogeneous, knowing its composition is one of the determining factors for the most appropriate treatment approach, as well as the necessary infrastructure to be developed in each region in order to properly proceed with the collection, segregation, and processing of different food waste (Arumdani et al. 2021; Thakur et al. 2020; Singh 2020).

There are several processes for treating food waste, such as biological, physical, and chemical methods. Currently, the most common waste destination is sanitary landfills, representing the destination of 80% of solid waste discarded (Kaushik and Sharma 2020), followed by incineration, which causes environmental consequences (Thakur et al. 2020). Arumdani et al. (2021) observed that among the five largest countries that generate waste in Southeast Asia, only Indonesia and Thailand classify and recycle municipal solid waste, which is 64% organic or food waste.

Further advances must also be made towards efficiently reducing food loss by improving data on the quantification, causes, and destination of food waste (Hartikainen et al. 2020). In this sense, Vandana and Anshika (2020) point out that integrated and technological systems can contribute to better management of the entire food supply chain, reducing waste and sustainably supporting the population. Finally, expanding the techno-scientific use in the stages of post-harvest, transport, storage, and food processing can contribute to increasing the shelf life of food, preserving and/or increasing its sensory and nutritional quality. In the future, more actions related to sustainable development, linked to the circular economy, are expected to be taken. Therefore, initiatives for the reuse of food waste are of great importance and interest and must be encouraged by all social actors.

Acknowledgments The authors are grateful to UNIRIO, UERJ, FAPERJ (Research Support Foundation of the State of Rio de Janeiro, Project number: E-26/010/002738/2019), CAPES (Coordination for the Improvement of Higher Education Personnel Program), and CNPq (National Council for Scientific and Technological Development) for their financial support.

References

Abbade EB (2020) Estimating the nutritional loss and the feeding potential derived from food losses worldwide. World Dev 134:105038. https://doi.org/10.1016/j.worlddev.2020.105038

Abdelhady HH, Elazab HA, Ewais EM, Saber M, El-Deab MS (2020) Efficient catalytic production of biodiesel using nano-sized sugar beet agro-industrial waste. Fuel 261:116481. https://doi.org/10.1016/j.fuel.2019.116481

Aldaco R, Hoehn D, Laso J, Margallo M, Ruiz-Salmón J, Cristobal J, Kahhat R, Villanueva-Rey P, Bala A, Batlle-Bayer L, Fullana-i-Palmer P, Irabien A, Vazquez-Rowe I (2020) Food waste management during the COVID-19 outbreak: a holistic climate, economic and nutritional approach. Sci Total Environ 742:140524. https://doi.org/10.1016/j.scitotenv.2020.140524

Alzuwaid NT, Pleming D, Fellows CM, Sissons M (2021) Fortification of durum wheat spaghetti and common wheat bread with wheat bran protein concentrate-impacts on nutrition and technological properties. Food Chem 334:127497. https://doi.org/10.1016/j.foodchem.2020.127497

Arikan EB, Canli O, Caro Y, Dufossé L, Dizge N (2020) Production of bio-based pigments from food processing industry by-products (apple, pomegranate, black carrot, red beet pulps) using aspergillus carbonarius. J Fungi 6:1–18. https://doi.org/10.3390/jof6040240

Aruldass CA, Aziz A, Venil CK, Khasim AR, Ahmad WA (2016) Utilization of agro-industrial waste for the production of yellowish-orange pigment from Chryseobacterium artocarpi CECT 8497. Int Biodeterior Biodegrad 113:342–349. https://doi.org/10.1016/j.ibiod.2016.01.024

Arumdani IS, Puspita AS, Budihardjo MA (2021) MSW handling of top 5 leading waste-producing countries in Southeast Asia. IOP Conf Ser: Earth Environ Sci 896:012003. https://doi.org/10.1088/1755-1315/896/1/012003

Arun KB, Madhavan A, Sindhu R, Binod P, Pandey A, Reshmy R, Sirohi R (2020) Remodeling agro-industrial and food wastes into value-added bioactives and biopolymers. Ind Crop Prod 154:112621. ISSN 0926-6690 https://doi.org/10.1016/j.indcrop.2020.112621. https://www.sciencedirect.com/science/article/pii/S0926669020305379

Balli D, Cecchi L, Innocenti M, Bellumori M, Mulinacci N (2021) Food by-products valorisation: grape pomace and olive pomace (pâté) as sources of phenolic compounds and fiber for enrichment of tagliatelle pasta. Food Chem 355:129642. https://doi.org/10.1016/j.foodchem.2021.129642

Banerjee J, Singh R, Vijayaraghavan R, MacFarlane D, Patti AF, Arora A (2017) Bioactives from fruit processing wastes: green approaches to valuable chemicals. Food Chem 225:10–22. https://doi.org/10.1016/j.foodchem.2016.12.093

Beltrán-Medina EA, Guatemala-Morales GM, Padilla-Camberos E, Corona-González RI, Mondragón-Cortez PM, Arriola-Guevara E (2020) Evaluation of the use of a coffiee industry by-product in a cereal-based extruded food product. Foods 9:1–15. https://doi.org/10.3390/foods9081008

Berktas S, Cam M (2020) Peppermint leaves hydrodistillation by-products: bioactive properties and incorporation into ice cream formulations. J Food Sci Technol 58:4282–4293. https://doi.org/10.1007/s13197-020-04903-7

Bouazizi S, Montevecchi G, Antonelli A, Hamdi M (2020) Effects of prickly pear (Opuntia ficus-indica L.) peel flour as an innovative ingredient in biscuits formulation. LWT – Food Sci Technol 124:109155. https://doi.org/10.1016/j.lwt.2020.109155

Boyacı-Gündüz CP, Ibrahim SA, Wei OC, Galanakis CM (2021) Transformation of the food sector: security and resilience during the COVID-19 pandemic. Foods 10:497. https://doi.org/10.3390/foods10030497

Brahmi F, Merchiche F, Mokhtari S, Smail L, Guemghar-Haddadi H, Yalaoui-Guellal D, Achat S, Fahmi Elsebai M, Madani K, Boulekbache L (2021) Optimization of some extraction parameters of phenolic content from apple peels and grape seeds and enrichment of yogurt by their powders: a comparative study. J Food Process Preserv 45(2):e15126. https://doi.org/10.1111/jfpp.15126

Brito TBN, Ferreira MSL, Fai AEC (2020a) Utilization of agricultural by-products: bioactive properties and technological applications. Food Rev Int 38:1–25. https://doi.org/10.1080/87559129.2020.1804930

Brito TBN, Pereira APA, Pastore GM, Moreira RFA, Ferreira MSL, Fai AEC (2020b) Chemical composition and physicochemical characterization for cabbage and pineapple by-products flour valorization. LWT – Food Sci Technol 124:109028. https://doi.org/10.1016/j.lwt.2020.109028

Brito TBN, Lima LRS, Santos MCB, Moreira RFA, Cameron LC, Fai AEC, Ferreira MSL (2021) Antimicrobial, antioxidant, volatile and phenolic profiles of cabbage-stalk and pineapple-crown flour revealed by GC-MS and UPLC-MSE. Food Chem 339:127882. https://doi.org/10.1016/j.foodchem.2020.127882

Brown AGM, Esposito LE, Fisher RA, Nicastro HL, Tabor DC, Walker JR (2019) Food insecurity and obesity: research gaps, opportunities, and challenges. Transl Behav Med 9:980–987. https://doi.org/10.1093/tbm/ibz117

Buzby JC, Wells HF, Hyman J (2014) The estimated amount, value, and calories of postharvest food losses at the retail and consumer levels in the United States. EIB-121, USDA, Economic Research Service. Retrieved from https://www.ers.usda.gov/webdocs/publications/43833/43680_eib121.pdf

Cedola A, Cardinali A, D'Antuono I, Conte A, Del Nobile MA (2020) Cereal foods fortified with by-products from the olive oil industry. Food Biosci 33:100490. https://doi.org/10.1016/j.fbio.2019.100490

Chen C, Chaudhary A, Mathys A (2020) Nutritional and environmental losses embedded in global food waste. Resour Conserv Recycl 160:104912. https://doi.org/10.1016/j.resconrec.2020.104912

Christ-Ribeiro A, Chiattoni LM, Fagundes Mafaldo CR, Badiale-Furlong E, de Souza-Soares LA (2021) Fermented rice-bran by Saccharomyces cerevisiae: nutritious ingredient in the formulation of gluten-free cookies. Food Biosci 40:100859. https://doi.org/10.1016/j.fbio.2020.100859

Corrado S, Ardente F, Sala S, Saouter E (2017) Modelling of food loss within life cycle assessment : from current practice towards a systematisation. J Clean Prod 140:847–859. https://doi.org/10.1016/j.jclepro.2016.06.050

Costa TM, Hermann KL, Garcia-Roman M, De Valle RCSC, Tavares LBB (2017) Lipase production by aspergillus Niger grown in different agro-industrial wastes by solid-state fermentation. Brazilian J Chem Eng 34:419–427. https://doi.org/10.1590/0104-6632.20170342s20150477

de Araújo LBA, da Luz JRD, Batista D, Bezerra A d C, Ramalho HMM, Soares AKM d C, Maciel MAM (2021) Physicochemical, microbiological and sensorial characteristics of flour and prepared cookies from the reject cashew (Anacardium occidentale L.) neglected. Res Soc Dev 10:e38410313417. https://doi.org/10.33448/rsd-v10i3.13417

de Brito Nogueira TB, da Silva TPM, de Araújo LD, de Andrade CJ, de Andrade LM, Ferreira MSL, Fai AEC (2020) Fruits and vegetable-processing waste: a case study in two markets at Rio de Janeiro, RJ, Brazil. Environ Sci Pollut Res 27:18530–18540. https://doi.org/10.1007/s11356-020-08244-y

de Medeiros VPB, Pimentel TC, Varandas RCR, dos Santos SA, de Souza Pedrosa GT, da Costa Sassi CF, da Conceição MM, Magnani M (2020) Exploiting the use of agro-industrial residues from fruit and vegetables as alternative microalgae culture medium. Food Res Int 137:109722. https://doi.org/10.1016/j.foodres.2020.109722

de Moura IG, de Sá AV, Abreu ASLM, Machado AVA (2017) Bioplastics from agro-wastes for food packaging applications, food packaging. Elsevier Inc, pp 223–263. https://doi.org/10.1016/b978-0-12-804302-8.00007-8
Du C, Abdullah JJ, Greetham D, Fu D, Yu M, Ren L, Li S, Lu D (2018) Valorization of food waste into biofertiliser and its field application. J Clean Prod 187:273–284. https://doi.org/10.1016/j.jclepro.2018.03.211
EPA (2012) Putting surplus food to good use. Washington. Retrieved from https://archive.epa.gov/wastes/conserve/tools/rogo/web/pdf/food-guide.pdf
FAO (2011) Global food losses and food wastes: extent, causes and prevention. Rome, FAO. Retrieved from http://www.fao.org/3/i2697e/i2697e.pdf
FAO (2013) Food wastage footprint: impacts on natural resources. Rome, FAO. Retrieved from http://www.fao.org/3/i3347e/i3347e.pdf
FAO (2014a) Food wastage footprint: fool cost-accounting. Rome, FAO. Retrieved from http://www.fao.org/3/i3991e/i3991e.pdf
FAO (2014b) Mitigation of food wastage – societal costs and benefits. Rome, FAO. Retrieved from http://www.fao.org/3/i3989e/i3989e.pdf
FAO (2015a) Global initiative on food loss and waste reduction. Rome, FAO. Retrieved from http://www.fao.org/3/i4068e/i4068e.pdf
FAO (2015b) Food wastage footprint & climate change. Rome, FAO. Retrieved from http://www.fao.org/3/bb144e/bb144e.pdf
FAO (2019) The state food and agriculture. Moving forward on food loss and waste reduction. Rome, FAO. Retrieved from http://www.fao.org/3/ca6030en/ca6030en.pdf
FAO, IFAD, UNICEF, WDP, WHO (2019) The state of food security and nutrition in the world. Safeguarding against economic slowdowns and downturns. Rome, FAO. Retrieved from http://www.fao.org/3/ca5162en/ca5162en.pdf
FAO, IFAD, UNICEF, WFP, WHO (2021) The state of food security and nutrition in the world. Transforming food systems for food security, improved nutrition and affordable healthy diets for all. Rome, FAO. Retrieved from https://doi.org/10.4060/cb4474en
Faustino M, Veiga M, Sousa P, Costa EM, Silva S, Pintado M (2019) Agro-food byproducts as a new source of natural food additives. Molecules 24(6):1056. https://doi.org/10.3390/molecules24061056
Fernandez MV, Bengardino M, Jagus RJ, Agüero MV (2020) Enrichment and preservation of a vegetable smoothie with an antioxidant and antimicrobial extract obtained from beet by-products. LWT – Food Sci Technol 117:108622. https://doi.org/10.1016/j.lwt.2019.108622
Fidelis M, De Moura C, Kabbas T, Pap N, Mattila P, Mäkinen S, Putnik P, Kovačević DB, Tian Y, Yang B, Granato D (2019) Fruit seeds as sources of bioactive compounds: sustainable production of high value-added ingredients from by-products within circular economy. Molecules 24(21):3854. https://doi.org/10.3390/molecules24213854
Fidelis M, de Oliveira SM, Sousa Santos J, Bragueto Escher G, Silva Rocha R, Gomes Cruz A, Araújo Vieira do Carmo M, Azevedo L, Kaneshima T, Oh WY, Shahidi F, Granato D (2020) From byproduct to a functional ingredient: Camu-camu (Myrciaria dubia) seed extract as an antioxidant agent in a yogurt model. J Dairy Sci 103(2):1131–1140. https://doi.org/10.3168/jds.2019-17173
Foley JA, DeFries R, Asner GP, Barford C, Bonan G, Carpenter SR, Chapin FS, Coe MT, Daily GC, Gibbs HK, Helkowski JH, Holloway T, Howard EA, Kucharik CJ, Monfreda C, Patz JA, Prentice IC, Ramankutty N, Snyder PK (2005) Global consequences of land use. Science 309:570–574. https://doi.org/10.1126/science.1111772
Foley JA, Ramankutty N, Brauman KA, Cassidy ES, Gerber JS, Johnston M, Mueller ND, O'Connell C, Ray DK, West PC, Balzer C, Bennett EM, Carpenter SR, Hill J, Monfreda C, Polasky S, Rockström J, Sheehan J, Siebert S, Tilman D, Zaks DPM (2011) Solutions for a cultivated planet. Nature 478:337–342. https://doi.org/10.1038/nature10452

Freitas LC, Barbosa JR, da Costa ALC, Bezerra FWF, Pinto RHH, de Carvalho Junior RN (2021) From waste to sustainable industry: how can agro-industrial wastes help in the development of new products? Resour Conserv Recycl 169:105466. https://doi.org/10.1016/j.resconrec.2021.105466

Gaglio R, Barbaccia P, Barbera M, Restivo I, Attanzio A, Maniaci G, Di Grigoli A, Francesca N, Tesoriere L, Bonanno A, Moschetti G, Settanni L (2021) The use of winery by-products to enhance the functional aspects of the fresh ovine "primosale" cheese. Foods 10(2):461. https://doi.org/10.3390/foods10020461

Gagneten M, Archaina DA, Salas MP, Leiva GE, Salvatori DM, Schebor C (2021) Gluten-free cookies added with fibre and bioactive compounds from blackcurrant residue. Int J Food Sci Technol 56(4):1734–1740. https://doi.org/10.1111/ijfs.14798

Gustavsson J, Cederberg C, Sonesson U, Otterdijk RV, Meybeck A (2011) Global food losses and food waste – extent, causes and prevention, food loss and food waste: causes and solutions. Food and agriculture Organization of the United Nations (FAO), Rome. https://doi.org/10.4337/9781788975391

Hang YD (2004) Management and utilization of food processing wastes. J Food Sci 69(3):104–107. https://doi.org/10.1111/j.1365-2621.2004.tb13341.x

Hartikainen H, Riipi I, Katajajuuri JM, Silvennoinen K (2020) From measurement to management: food waste in the finnish food chain, food waste management. Springer, pp 415–439. https://doi.org/10.1007/978-3-030-20561-4_16

Hartmann GL, Marconato AM, Romeiro Santos MM, Do Amaral LA, Dos Santos EF, Novello D (2020) Addition of pumpkin Peel flour affect physicochemical and sensory characteristics of bovine burger. Int J Res – Granthaalayah 8(2):254–263. https://doi.org/10.29121/granthaalayah.v8.i2.2020.216

Huang L, Zhao H, Yi T, Qi M, Xu H, Mo Q, Huang C, Wang S, Liu Y (2020) Reparation and properties of cassava residue cellulose nanofibril/cassava starch composite films. Nano 10(4):755. https://doi.org/10.3390/nano10040755

IPCC (2019) Climate Change and Land: an IPCC special report on climate change, desertification, land degradation, sustainable land management, food security, and greenhouse gas fluxes in terrestrial ecosystems [in: Shukla PR, Skea J, Buendia EC, Masson-Delmotte V, Pörtner H-O, Roberts DC, Zhai P, Slade R, Connors S, Diemen RV, Ferrat M, Haughey E, Luz S, Neogi S, Pathak M, Petzold J, Pereira JP, Vyas P, Huntley E, Kissick K, Belkacemi M, Malley J (Eds.)] pp. 437–550. Retrieved from https://www.ipcc.ch/site/assets/uploads/2019/11/SRCCL-Full-Report-Compiled-191128.pdf

Jadhav JV, Pratap AP, Kale SB (2019) Evaluation of sunflower oil refinery waste as feedstock for production of sophorolipid. Process Biochem 78:15–24. https://doi.org/10.1016/j.procbio.2019.01.015

Jiménez-Moreno N, Esparza I, Bimbela F, Gandía LM, Ancín-Azpilicueta C (2020) Valorization of selected fruit and vegetable wastes as bioactive compounds: opportunities and challenges. Crit Rev Environ Sci Technol 50(20):2061–2108. https://doi.org/10.1080/10643389.2019.1694819

Jovanović M, Petrović M, Miočinović J, Zlatanović S, Petronijević JL, Mitić-Ćulafić D, Gorjanović S (2020) Bioactivity and sensory properties of probiotic yogurt fortified with apple pomace flour. Foods 9(6):763. https://doi.org/10.3390/foods9060763

Joymak W, Ngamukote S, Chantarasinlapin P, Adisakwattana S (2021) Unripe papaya by-product: from food wastes to functional ingredients in pancakes. Foods 10(3):615. https://doi.org/10.3390/foods10030615

Kabir MR, Hasan MM, Islam MR, Haque AR, Hasan SMK (2021) Formulation of yogurt with banana peel extracts to enhance storability and bioactive properties. J Food Process Preserv 45(3):e15191. https://doi.org/10.1111/jfpp.15191

Kaderides K, Mourtzinos I, Goula AM (2020) Stability of pomegranate peel polyphenols encapsulated in orange juice industry by-product and their incorporation in cookies. Food Chem 310:125849. https://doi.org/10.1016/j.foodchem.2019.125849

Kaushik M, Sharma D (2020) Characterization and treatment of waste from food processing industries, sustainable food waste management. Springer Nature Singapore Pte Ltd, pp 41–58. https://doi.org/10.1007/978-981-15-8967-6_3

Kavitha S, Kannah RY, Kumar G, Gunasekaran M, Banu JR (2020) Introduction: sources and characterization of food waste and food industry wastes. In: Banu JR, Kumar G, Gunasekaran M, Kavitha S (eds) Food waste to valuable resources, Applications and Management. Academic, pp 1–13. https://doi.org/10.1016/B978-0-12-818353-3.00001-8

Khan S, Hanjra MA (2009) Footprints of water and energy inputs in food production – global perspectives. Food Policy 34(2):130–140. https://doi.org/10.1016/j.foodpol.2008.09.001

Khoozani AA, Kebede B, Birch J, El-Din Ahmed Bekhit A (2020) The effect of bread fortification with whole green banana flour on its physicochemical, nutritional and in vitro digestibility. Foods 9(2):152. https://doi.org/10.3390/foods9020152

Koutra E, Economou CN, Tsafrakidou P, Kornaros M (2018) Bio-based products from microalgae cultivated in Digestates. Trends Biotechnol 36(8):819–833. https://doi.org/10.1016/j.tibtech.2018.02.015

Kurtz A, Grant K, Marano R, Arrieta A, Grant K Jr, Feaster W, Steele C, Ehwerhemuepha L (2021) Long-term effects of malnutrition on severity of COVID-19. Sci Rep 11:14974. https://doi.org/10.1038/s41598-021-94138-z

Lai WT, Khong NMH, Lim SS, Hee YY, Sim BI, Lau KY, Lai OM (2017) A review: modified agricultural by-products for the development and fortification of food products and nutraceuticals. Trends Food Sci Technol 59:148–160. https://doi.org/10.1016/j.tifs.2016.11.014

Laso J, Campos C, Fernández-Ríos A, Hoehn D, Del Río A, Ruiz-Salmón I, Cristobal J, Quiñones A, Amo-Setién FJ, Ortego MDC, Tezanos S, Abajas R, Bala A, Fullana-I-Palmer P, Puig R, Margallo M, Aldaco R, Abejón R (2021) Looking for answers to food loss and waste management in Spain from a holistic nutritional and economic approach. Sustainability 13(1): 125. https://doi.org/10.3390/su13010125

Lemaire A, Limbourg S (2019) How can food loss and waste management achieve sustainable development goals? J Clean Prod 234:1221–1234. https://doi.org/10.1016/j.jclepro.2019.06.226

Lim SF, Matu SU (2015) Utilization of agro-wastes to produce biofertilizer. Int J Energy Environ Eng 6:31–35. https://doi.org/10.1007/s40095-014-0147-8

Lipinski B (2020) SDG Target 12.3 on food loss and waste: 2020 progress report. Retrieved from https://champions123.org/sites/default/files/2020-09/champions-12-3-2020-progress-report.pdf

Lipinski B, Hanson C, Lomax J, Kitinoja L, Waite R, Searchinger T (2013) Reducing food loss and waste. World Resour Inst 1–40. Retrieved from https://files.wri.org/d8/s3fs-public/reducing_food_loss_and_waste.pdf

Lopes FC, Ligabue-Braun R (2021) Agro-industrial residues: eco-friendly and inexpensive substrates for microbial pigments production. Front Sustain Food Syst 5:589414. https://doi.org/10.3389/fsufs.2021.589414

Luchese CL, Pavoni JMF, Spada JC, Tessaro IC (2019) Influence of blueberry and Jaboticaba agroindustrial residue particle size on color change of corn starch based films submitted to different pH values solutions. J Renew Mater 7(3):235–243. https://doi.org/10.32604/jrm.2019.00033

Ma Y, Liu Y (2019) Turning food waste to energy and resources towards a great environmental and economic sustainability: an innovative integrated biological approach. Biotechnol Adv 37(7): 107414. https://doi.org/10.1016/j.biotechadv.2019.06.013

Majerska J, Michalska A, Figiel A (2019) A review of new directions in managing fruit and vegetable processing by-products. Trends Food Sci Technol 88:207–219. https://doi.org/10.1016/j.tifs.2019.03.021

Manuela P, Drakula S, Cravotto G, Verpoorte R, Hruškar M, Radojčić Redovniković I, Radošević K (2020) Biological activity and sensory evaluation of cocoa by-products NADES extracts used in food fortification. Innov Food Sci Emerg Technol 66:102514. https://doi.org/10.1016/j.ifset.2020.102514

Marconato AM, Hartmann GL, Santos MMR, do Amaral LA, de Souza GHO, dos Santos EF, Novello D (2020) Sweet potato peel flour in hamburger: effect on physicochemical, technological and sensorial characteristics. Brazilian J Food Technol 23:e2019115. https://doi.org/10.1590/1981-6723.11519

Masood S, Rehman A, Bashir S, Imran M, Khalil P, Khursheed T, Iftikhar F, Jaffar H, Farooq S, Rizwan B, Javaid N (2020) Proximate and sensory analysis of wheat bread supplemented with onion powder and onion Peel extract. Biosci Res 17(4):4071–4078

Matheus JRV, de Andrade CJ, Miyahira RF, AEC F (2020a) Persimmon (Diospyros Kaki L.): chemical properties, bioactive compounds and potential use in the development of new products–a review. Food Rev Int 38:384–401. https://doi.org/10.1080/87559129.2020.1733597

Matheus JRV, Miyahira RF, Fai AEC (2020b) Biodegradable films based on fruit puree: a brief review. Crit Rev Food Sci Nutr 61:2090–2097. https://doi.org/10.1080/10408398.2020.1772715

Matheus JRV, de Assis RM, Correia TR, da Costa Marques MR, Leite MCAM, Pelissari FM, Miyahira RF, Fai AEC (2021) Biodegradable and edible film based on persimmon (Diospyros kaki L.) used as a lid for minimally processed vegetables packaging. Food Bioprocess Technol 14:765–779. https://doi.org/10.1007/s11947-021-02595-1

Michalak-Majewska M, Teterycz D, Muszyński S, Radzki W, Sykut-Domańska E (2020) Influence of onion skin powder on nutritional and quality attributes of wheat pasta. PLoS One 15(1): e0227942. https://doi.org/10.1371/journal.pone.0227942

Mojumdar A, Deka J (2019) Recycling agro-industrial waste to produce amylase and characterizing amylase–gold nanoparticle composite. Int J Recycl Org Waste Agric 8:263–269. https://doi.org/10.1007/s40093-019-00298-4

Morales ME, Berkowitz SA (2016) The relationship between food insecurity, dietary patterns, and obesity. Curr Nutr Rep 5:54–60. https://doi.org/10.1007/s13668-016-0153-y

Mourad M (2016) Recycling, recovering and preventing "food waste": competing solutions for food systems sustainability in the United States and France. J Clean Prod 126:461–477. https://doi.org/10.1016/j.jclepro.2016.03.084

Muniz CES, Santiago ÂM, Gusmão TAS, Oliveira HML, Conrado L d S, de Gusmão RP (2020) Solid-state fermentation for single-cell protein enrichment of guava and cashew by-products and inclusion on cereal bars. Biocatal Agric Biotechnol 25:101576. https://doi.org/10.1016/j.bcab.2020.101576

Nakov G, Brandolini A, Hidalgo A, Ivanova N, Jukić M, Komlenić DK, Lukinac J (2020) Influence of apple peel powder addition on the physico-chemical characteristics and nutritional quality of bread wheat cookies. Food Sci Technol Int 26(7):574–582. https://doi.org/10.1177/1082013220917282

Nazzaro F, Fratianni F, Ombram MN, D'Acierno A, Coppola R (2018) Recovery of biomolecules of high benefit from food waste. Curr Opin Food Sci 22:43–54. https://doi.org/10.1016/j.cofs.2018.01.012

Ng HS, Kee PE, Yim HS, Chen PT, Wei YH, Chi-Wei Lan J (2020) Recent advances on the sustainable approaches for conversion and reutilization of food wastes to valuable bioproducts. Bioresour Technol 302:122889. https://doi.org/10.1016/j.biortech.2020.122889

Ni Q, Ranawana V, Hayes HE, Hayward NJ, Stead D, Raikos V (2020) Addition of broad bean hull to wheat flour for the development of high-fiber bread: effects on physical and nutritional properties. Foods 9(9):1192. https://doi.org/10.3390/foods9091192

Otles S, Despoudi S, Bucatariu C, Canan K (2015) Food waste management, valorization, and sustainability in the food industry. In: Galanakis CM (ed) Food waste recovery: processing technologies and industrial techniques. Academic Press, pp 3–23. https://doi.org/10.1016/B978-0-12-800351-0.00001-8

Papargyropoulou E, Lozano R, Steinberger JK, Wright N, Ujang ZB (2014) The food waste hierarchy as a framework for the management of food surplus and food waste. J Clean Prod 76:106–115. https://doi.org/10.1016/j.jclepro.2014.04.020

Parafati L, Restuccia C, Palmeri R, Fallico B, Arena E (2020) Characterization of prickly pear peel flour as a bioactive and functional ingredient in bread preparation. Foods 9(9):1189. https://doi.org/10.3390/foods9091189

Paraszkiewicz K, Bernat P, Kuśmierska A, Chojniak J, Płaza G (2018) Structural identification of lipopeptide biosurfactants produced by Bacillus subtilis strains grown on the media obtained from renewable natural resources. J Environ Manag 209:65–70. https://doi.org/10.1016/j.jenvman.2017.12.033

Pérez-Chabela ML, Cebollón-Juárez A, Bosquez-Molina E, Totosaus A (2021) Mango peel flour and potato peel flour as bioactive ingredients in the formulation of functional yogurt. Food Sci Technol 42. https://doi.org/10.1590/fst.38220

Pfister S, Bayer P, Koehler A, Hellweg S (2011) Projected water consumption in future global agriculture: scenarios and related impacts. Sci Total Environ 409(20):4206–4216. https://doi.org/10.1016/j.scitotenv.2011.07.019

Poovazhahi A, Thakur M (2020) Environmental standards & regulations for waste management in food industries. In: Thakur M, Modi VK, Khedkar R, Singh K (eds) Sustainable food waste management. Springer, Singapore, pp 21–39. https://doi.org/10.1007/978-981-15-8967-6

Rahimi S, Mikani M (2019) Lycopene green ultrasound-assisted extraction using edible oil accompany with response surface methodology (RSM) optimization performance: application in tomato processing wastes. Microchem J 146:1033–1042. https://doi.org/10.1016/j.microc.2019.02.039

Ritchie H, Reay DS, Higgins P (2018) Beyond calories: a holistic assessment of the global food system. Front Sustain Food Syst 2(57):1–12. https://doi.org/10.3389/fsufs.2018.00057

Rivera ÁD, Urbina MÁM, López VEL (2019) Advances on research in the use of agro-industrial waste in biosurfactant production. World J Microbiol Biotechnol 35(155):1–18. https://doi.org/10.1007/s11274-019-2729-3

Roodhuyzen DMA, Luning PA, Fogliano V, Steenbekkers LPA (2017) Putting together the puzzle of consumer food waste: towards an integral perspective. Trends Food Sci Technol 68:37–50. https://doi.org/10.1016/j.tifs.2017.07.009

Sánchez-Camargo A d P, Gutiérrez LF, Vargas SM, Martinez-Correa HA, Parada-Alfonso F, Narváez-Cuenca CE (2019) Valorisation of mango peel: proximate composition, supercritical fluid extraction of carotenoids, and application as an antioxidant additive for an edible oil. J Supercrit Fluids 152:104574. https://doi.org/10.1016/j.supflu.2019.104574

Sancho SDO, Raquel A, Nilson A, Dantas DS, Magalhães TA, Lopes GS, Rodrigues S, Maria J, André F, Fernandes N, Goretti M, Silva DV (2015) Characterization of the industrial residues of seven fruits and prospection of their potential application as food supplements. J Chem 2015: 264284–264288. https://doi.org/10.1155/2015/264284

Schroeder P, Anggraeni K, Weber U (2019) The relevance of circular economy practices to the sustainable development goals. J Ind Ecol 23(1):77–95. https://doi.org/10.1111/jiec.12732

Šeregelj V, Pezo L, Šovljanski O, Lević S, Nedović V, Markov S, Tomić A, Čanadanović-Brunet J, Vulić J, Šaponjac VT, Ćetković G (2021) New concept of fortified yogurt formulation with encapsulated carrot waste extract. LWT – Food Sci Technol 138:110732. https://doi.org/10.1016/j.lwt.2020.110732

Sharma P, Gaur VK, Sirohi R, Varjani S, Hyoun Kim S, Wong JWC (2021) Sustainable processing of food waste for production of bio-based products for circular bioeconomy. Bioresour Technol 325:124684. https://doi.org/10.1016/j.biortech.2021.124684

Shruthy R, Preetha R (2019) Cellulose nanoparticles from agro-industrial waste for the development of active packaging. Appl Surf Sci 484:1274–1281. https://doi.org/10.1016/j.apsusc.2019.04.003

Singh K (2020) Sustainable food waste management: a review. In: Thakur M, Modi VK, Khedkar R, Singh K (eds) Sustainable food waste management. Springer, Singapore, pp 3–19. https://doi.org/10.1007/978-981-15-8967-6_1

Singla G, Panesar PS, Sangwan RS, Krishania M (2020) Enzymatic processing of Citrus reticulata (Kinnow) pomace using naringinase and its valorization through preparation of nutritionally enriched pasta. J Food Sci Technol 58:3853–3860. https://doi.org/10.1007/s13197-020-04846-z

Suhartini S, Nurika I, Paul R, Melville L (2020) Estimation of biogas production and the emission savings from anaerobic digestion of fruit-based agro-industrial waste and agricultural crops residues. Bioenergy Res 14:844–859. https://doi.org/10.1007/s12155-020-10209-5

Sykut-Domańska E, Zarzycki P, Sobota A, Teterycz D, Wirkijowska A, Blicharz-Kania A, Andrejko D, Mazurkiewicz J (2020) The potential use of by-products from coconut industry for production of pasta. J Food Process Preserv 44(7):e14490. https://doi.org/10.1111/jfpp.14490

Teigiserova DA, Hamelin L, Thomsen M (2019a) Review of high-value food waste and food residues biorefineries with focus on unavoidable wastes from processing. Resour Conserv Recycl 149:413–426. https://doi.org/10.1016/j.resconrec.2019.05.003

Teigiserova DA, Hamelin L, Thomsen M (2019b) Towards transparent valorization of food surplus, waste and loss: clarifying definitions, food waste hierarchy, and role in the circular economy. Sci Total Environ 706:136033. https://doi.org/10.1016/j.scitotenv.2019.136033

Tester JM, Rosas LG, Leung CW (2020) Food insecurity and pediatric obesity: a double whammy in the era of COVID-19. Curr Obes Rep 9:442–450. https://doi.org/10.1007/s13679-020-00413-x

Thakur D, Ramawat N, Shyam V (2020) Agricultural waste produce: utilization and management, sustainable food waste management. Springer Nature Singapore Pte Ltd, pp 227–240. https://doi.org/10.1007/978-981-15-8967-6_13

Tolve R, Pasini G, Vignale F, Favati F, Simonato B (2020) Effect of grape pomace addition on the technological, sensory, and nutritional properties of durum wheat pasta. Foods 9(3):354. https://doi.org/10.3390/foods9030354

Torres-León C, Ramírez-Guzman N, Londoño-Hernandez L, Martinez-Medina GA, Díaz-Herrera R, Navarro-Macias V, Alvarez-Pérez OB, Picazo B, Villarreal-Vázquez M, Ascacio-Valdes J, Aguilar CN (2018) Food waste and byproducts: an opportunity to minimize malnutrition and hunger in developing countries. Front Sustain Food Syst 2:52. https://doi.org/10.3389/fsufs.2018.00052

UN (2015) Transforming our world: the 2030 agenda for sustainable development. United Nations. Retrieved from https://sdgs.un.org/2030agenda

Ungureanu-Iuga M, Dimian M, Mironeasa S (2020) Development and quality evaluation of gluten-free pasta with grape peels and whey powders. LWT – Food Sci Technol 130:109714. https://doi.org/10.1016/j.lwt.2020.109714

Vandana SC, Anshika SS (2020) Digital knowledge ecosystem: a new weapon to achieve sustainable food waste management, sustainable food waste management. Springer Nature Singapore Pte Ltd, pp 403–413. https://doi.org/10.1007/978-981-15-8967-6_22

Vedove TMARD, Maniglia BC, Tadini CC (2021) Production of sustainable smart packaging based on cassava starch and anthocyanin by an extrusion process. J Food Eng 289:110274. https://doi.org/10.1016/j.jfoodeng.2020.110274

Velarde EDA, Martínez DLP, Salem AZM, García PGM, Berasain MDM (2020) Antioxidant and antimicrobial capacity of three agroindustrial residues as animal feeds. Agrofor Syst 94:1393–1402. https://doi.org/10.1007/s10457-018-00343-7

Velenturf APM, Purnell P (2021) Principles for a sustainable circular economy. Sustain Prod Consum 27:1437–1457. https://doi.org/10.1016/j.spc.2021.02.018

Vermeulen SJ, Campbell BM, Ingram JSI (2012) Climate change and food systems. Annu Rev Environ Resour 37:195–222. https://doi.org/10.1146/annurev-environ-020411-130608

Vodnar DC, Călinoiu LF, Dulf FV, Ştefănescu BE, Crişan G, Socaciu C (2017) Identification of the bioactive compounds and antioxidant, antimutagenic and antimicrobial activities of thermally processed agro-industrial waste. Food Chem 231:131–140. https://doi.org/10.1016/j.foodchem.2017.03.131

Wakefield A, Axon S (2020) "I'm a bit of a waster": identifying the enablers of, and barriers to, sustainable food waste practices. J Clean Prod 275:122803. https://doi.org/10.1016/j.jclepro.2020.122803

Wang X, Kristo E, LaPointe G (2020) Adding apple pomace as a functional ingredient in stirred-type yogurt and yogurt drinks. Food Hydrocoll 100:105453. https://doi.org/10.1016/j.foodhyd.2019.105453

Watson M, Browne A, Evans D, Foden M, Hoolohan C, Sharp L (2020) Challenges and opportunities for re-framing resource use policy with practice theories: the change points approach. Glob Environ Chang 62:102072. https://doi.org/10.1016/j.gloenvcha.2020.102072

Willett W, Rockström J, Loken B, Springmann M, Lang T, Vermeulen S, Garnett T, Tilman D, DeClerck F, Wood A, Jonell M, Clark M, Gordon LJ, Fanzo J, Hawkes C, Zurayk R, Rivera JA, De Vries W, Majele Sibanda L, Afshin A, Chaudhary A, Herrero M, Agustina R, Branca F, Lartey A, Fan S, Crona B, Fox E, Bignet V, Troell M, Lindahl T, Singh S, Cornell SE, Srinath Reddy K, Narain S, Nishtar S, Murray CJL (2019) Food in the Anthropocene: the EAT–lancet commission on healthy diets from sustainable food systems. Lancet 393:447–492. https://doi.org/10.1016/S0140-6736(18)31788-4

Zevallos DMP, Querol MP, Ambrogi BG (2018) Cassava wastewater as a natural pesticide: current knowledge and challenges for broader utilisation. Ann Appl Biol 173(3):191–201. https://doi.org/10.1111/aab.12464

Zokaityte E, Siriakovaite K, Starkute V, Zavistanaviciute P, Lele V, Mozuriene E, Klupsaite D, Viskelis P, Ruibys R, Guiné RPF, Bartkiene E (2021) Characteristics of nutraceutical chewing candy formulations based on fermented Milk permeate, psyllium husk, and apple by-products. Foods 10(4):777. https://doi.org/10.3390/foods10040777

Zuin VG, Ramin LZ (2018) Green and sustainable separation of natural products from agro-industrial waste: challenges, potentialities, and perspectives on emerging approaches. Top Curr Chem 376(3):1–54. https://doi.org/10.1007/s41061-017-0182-z

Chapter 11
Food Safety and Quality Testing: Recent Areas of Focus and Research Perspectives

Singam Suranjoy Singh and K. V. Ragavan

1 Introduction

Globalization of trade and post-harvest processing enables continuous supply and access to almost any food material across the globe most of the time. It has supported overcoming food insecurity and hunger at many locations, especially where resources for food production are lacking. On the other hand, it has led to the introduction of food hazards, entirely new to a particular location due to various activities involved in food production (Nardi et al. 2020). Advances in food processing methods are responsible for reduced loss of agricultural produce and in enhancing the shelf-life of food products without compromising their safety and quality attributes. To an extent, it is accountable for achieving food security in some parts of the world (Augustin et al. 2016). However, certain processing conditions, additives, quality of raw materials, and their combinations lead to the formation of processing contaminants, which pose a risk of health hazard to humans (Ragavan et al. 2016). Estimation of the contaminants in food matrices is essential to ascertain the safety of processed food. A wide range of agrochemicals has helped increase food production to feed the human population. However, it has also led to a serious food safety issue due to agrochemical residues in the food matrix beyond permitted levels (Carvalho 2017; Medina-Pastor and Triacchini 2020; Thakur and Ragavan 2013). In the case of industrialized animal farming, extensive use of antibiotics and

S. S. Singh
School of Engineering, University of Guelph, Guelph, ON, Canada
e-mail: singam@uoguelph.ca

K. V. Ragavan (✉)
Agro-Processing and Technology Division, CSIR-National Institute for Interdisciplinary Science and Technology, Thiruvananthapuram, India

Academy of Scientific and Innovative Research (AcSIR), Ghaziabad, India
e-mail: kvragavan@niist.res.in

H. U. Hebbar et al. (eds.), *Engineering Aspects of Food Quality and Safety*, Food Engineering Series, https://doi.org/10.1007/978-3-031-30683-9_11

growth promoters results in their residues in animal products such as milk, meat, and meat products. It is held responsible for the rising concerns of antibiotic-resistant pathogens impacting human health (Boeckel et al. 2019; Moore 2019).

One of the most neglected food safety issues is related to the food-borne parasites, especially protozoans and helminths transmitted through contaminated majorly through pork, vegetables, dairy products, freshwater fish, and crustaceans (FAO 2021; Koutsoumanis et al. 2018). The global food supply chain increases the risk of introducing these food-borne parasites to new environments (Robertson et al. 2014). It causes around 90 million infections and is responsible for 52,000 deaths worldwide every year. However, regulations or testing protocols for these food-borne parasites are not in place (Chalmers et al. 2020; Torgerson et al. 2015). Due to the quantum of impact, regulatory agencies such as FAO/WHO and EFSA have come up with committees and recommendations to bridge the gap (Codex Alimentarius 2016; FAO 2021; Koutsoumanis et al. 2018).

In recent years, food fraud in terms of adulteration and food authentication is increasingly reported. The consumer is at a loss for receiving a less valuable product and, in some cases, a low-quality and unsafe product. Gaps in food regulations, economic status, and lack of traceability are reasoned for food fraud (Danezis et al. 2016; Manning 2016). Addressing food fraud requires the most advanced analytical techniques and continuous communication of raw material sourcing and processing (Danezis et al. 2016; van Ruth et al. 2017). Many groups are actively working to bring in practical solutions to prevent and identify food fraud.

The current generation of consumers is more aware of their food in terms of nutrition, source, quality, and safety attributes. Even though the awareness in middle and low-income countries is lower than in high-income countries, it might gradually increase due to the access to information through the internet and smartphones (Hoffmann et al. 2019; Huang et al. 2018a). It is worth noting that a section of consumers is willing to pay a premium price to ensure food safety (Alimi and Workneh 2016). Advancements in camera optics, wireless data transfer, and processing capacity in the smartphone are equipping it into mobile-based testing platforms. It renders the consumers test the food they consume for safety and quality parameters (Purohit et al. 2020).

Food safety issues discussed so far highlights the need for rapid and easy-to-use food safety and quality testing methods. Global food regulations are helping to an extent to overcome the food hazards in the supply chain and recommend a set of safe practices to produce safe food. However, it has mandated the necessity to have a robust food testing infrastructure across the globe to ensure the safety and quality of food obtained by agricultural practices, trade, and processing. It is a critical challenge to developing and poor economies to create and maintain the infrastructure. Advancement in electronics and material science has driven sensors and biosensors research to develop novel and straightforward food testing methods. Developed methods have the role of fulfilling the need of the consumer to test their food tested with accuracy. Overall, the development of food testing devices requires a highly multidisciplinary approach, with inputs from basic science and engineering topics. The link between advancement in food processing and food analysis is discussed concerning food quality and safety in the following sections.

2 Advances in Food Processing

Food processing is advancing with novel technologies to bring in required sensory attributes, food structures and ensure food safety. On the other hand, outbreaks of food-borne illness due to pathogens including *Salmonella, Staphylococcal* strains are reported every day, especially in ready-to-eat (RTE) foods, fresh and fresh-cut foods, and fish, meat, dairy products, and seafood. Flawed implementation of Hazard Analysis & Critical Control Points (HACCP) in processing plants coupled with Food fraud and Food adulteration for economic gain are commonly reported (Manning et al. 2019; Tibola et al. 2018). Intense collaborative efforts among the industry, regulatory, and research stakeholders to strengthen strategies and identify the best tools to ensure food safety is effectively implemented across the food supply chain (Castro-Ibáñez et al. 2017; WHO 2018; Zeaki et al. 2019). Food safety stakeholders are well aware that there is no "silver bullet" technology that can fully eliminate pathogens/contaminants from the food supply chain.

Nevertheless, substantial progress has been made in recent years, both in terms of enhancing existing prevention tools and developing novel technologies for microbial inactivation and detection of food contaminants. Hybrid techniques such as "Hurdle technologies" incorporate multiple processing operations to inactivate pathogens in foods (Chen et al. 2012). Along with the processing operations, researchers persistently report a range of food analytical techniques for food quality and safety. It includes chemical, biological, and nanomaterial-based sensors including bacterial adenosine triphosphate (ATP) based bioluminescence sensors and nucleic acid-based methods like polymerase chain reaction (PCR), etc., (Böhme et al. 2019; Cesewski and Johnson 2020; Parate et al. 2020; Zhang et al. 2017). Biosensors and chemical sensors are potential techniques to ensure food quality and safety assurance in the global food supply chains, especially in detecting pathogens or determining quality attributes such as shelf-life (Cesewski and Johnson 2020). Similarly, biosensors based on imaging and spectroscopic methods are onsite monitoring and screening of food products and raw materials for quality and safety attributes (Rady and Adedeji 2018). Free radicals and DNA are the most desirable targets for biosensor-based food analytical methods (Law et al. 2015; Poltronieri et al. 2014).

Mass spectroscopy techniques are powerful tools to detect adulterated components and detect inferior meat (presence of substantial pathogenic microorganisms and poor quality), called 'zombie meat' in China, which poses significant health risks. Huang et al. (2016) developed two-dimensional gel electrophoresis coupled with mass spectrometry (2DE-MS)-based proteomics system for detecting meat type and its quality. They identified 450 protein spots in the meat exudates, along with 22 proteins. Among them, myofibrillar protein and myoglobin, were chosen as markers to distinguish between freeze-thawed and fresh pork.

Global milk production is expected to grow at a rate of 1.6% per year, reaching 997 Mt. by 2029, outpacing major agricultural commodities (OCED/FAO 2020). Dairy products, second only to green leafy vegetables in terms of adulteration, account for 14% of all food-borne illnesses (Painter et al. 2013). Adulterants and the rate of adulteration in milk and milk products are reported higher than earlier with the notorious Chinese milk scandal containing melamine to artificially inflate the protein content of dairy products. Chronic melamine exposure can lead to nephropathy and various other health problems, and the detection of this adulterant is crucial for food safety. Some of the recently reported methods for rapid detection of melamine include silver nanoparticles (Daniel et al. 2017), magnetite nanoparticle-based immunochromatographic strip (Huang et al. 2018), smartphone-based optical sensor containing fluorescent gold nanoparticles and carbon quantum dots nanocomposites (Hu et al. 2019). Raman chemical imaging (RCI), a novel technique that combines Raman spectroscopy (signals from vibrational modes of a molecule) and digital imaging capabilities, has the advantage of accurate detection of adulterants/contaminants and their distribution in a food matrix. RCI with NIR chemical imaging is reported to increase the accuracy of melamine detection in skim milk powder (Betz et al. 2012).

2.1 Novel Interventions

Cold plasma is an emerging non-thermal food processing technique applied to decontaminate vegetables, fruits, dairy, and animal products from pathogenic and spoilage microbes. Plasma is commonly generated through the application of high potential difference, high voltage alternating current (AC), direct current (DC), radio frequency (RF), or microwave (MW) across a non-conducting dielectric fluid/gas. Reactive oxygen species (ROS) and reactive nitrogen species (RNS) are the two effective primary species responsible for antimicrobial action. (Tappi et al. 2014). UV light and reactive chemical species generated by the cold plasma ionization process are the primary mechanisms of action for decontamination. Cold plasma inactivates pathogens by three main pathways (Niemira 2012):

(i) Interaction between free radicals, charged particles or reactive species with microbial cell membranes
(ii) UV radiation damages cell membranes and internal cellular components
(iii) UV radiation has the potential to break DNA strands.

Cherry tomatoes subjected to dielectric barrier discharge (DBD) plasma at 80 kV for 5 min effectively reduced the microbial load (*E. coli, S. typhimurium*, and *L. monocytogenes*) by 3.5, 3.8, and 4.2 log CFU, respectively (Ziuzina et al. 2014), while in fresh strawberries a 2 log reduction in aerobic mesophilic bacteria, mold, and yeast population was reported (Misra et al. 2015). Degradation of carotenoids was found to be responsible for color loss in food products after plasma treatment (Bagheri and Abbaszadeh 2020; Misra et al. 2015).

Pulsed electric field (PEF), a non-thermal technology that has been commercialized since 2005, has proven to be an effective microbial inactivation tool for liquid foods as well as wastewater treatment. PEF uses intense electric pulses to break down the cell membranes of vegetative bacteria, molds, and yeasts. Besides the inactivation of microbes, PEF treatment has shown to be successful in certain in-package microbial decontamination, allergen reduction, and shelf-life extension of certain foods while maintaining their nutritional and quality attributes (Alirezalu et al. 2020). Pasteurization of foods such as milk, soups, juices, yoghurt, meat, and liquid eggs has been successfully tested using PEF technology (Bhat et al. 2019). However, it is restricted to foods that don't have any air bubbles and have a low electrical conductivity. A recent study on the application of PEF in pasteurization of liquid whole egg or liquid egg white to inactivate *L. monocytogenes, S. typhimurium,* and *S. enteritidis*. (Bricher and Keener 2007). From the above discussions, it is evident that PEF is a powerful processing technique to ensure the safety of high-risk food products prone to rapid pathogen/microbial contamination.

Other promising intervention technologies for food preservation and inactivation of spoilage and pathogens include, irradiation or ionization radiation treatment (Albert et al. 2021), UV for surface sterilization (Bintsis et al. 2000), and high-pressure processing (HPP) (Rajashri et al. 2020; Rastogi 2013; Zhang et al. 2019). At present, irradiation as a pasteurization method for fresh produce is being debated by industries and regulatory organizations such as the USDA. Irradiation of ready-to-eat (RTE) meat products and juices has recently received regulatory approvals. However, low-level ionizing radiation to inactivate microorganisms, yeasts, spores, molds, naturally occurring chemical toxins, and parasites in spices are followed for several years. Ionizing radiation was used in a study conducted by Food Safety Intervention Technologies Research Unit, USA, to reduce potential carcinogens (furan and acrylamide) in foods. Furan and acrylamide in water were entirely destroyed by low-dose ionizing radiation (2–3.5 kGy), and the levels of furan in RTE meats were substantially decreased from 25% to 40%, whereas a minor effect on the inactivation of acrylamide in potato chips and oil were observed (Fan and Mastovska 2006).

The effect of food processing methods on the physiology and behaviour of microorganisms in foods, such as homeostasis, stress reactions, and metabolic fatigue, has recently been studied, leading to the development of the novel concept of multi-target food preservation techniques (Leistner 2000; Peleg 2020). Studies involving combined intervention technologies demonstrate the significance of using a multiple hurdle strategy (Singh and Shalini 2016). Hurdle technology comprising antimicrobial agents, thermal processing, advanced non-thermal processing methods and antimicrobial packaging is expected to play a major role in retaining nutrients and ensuring food safety.

3 Advancement in Conventional Analytical Methods

Food authenticity testing is a primary criterion for food and food products becoming more common due to global food legislation. From quality and authenticity standpoints, product analysis in the food and beverage industry is critical to ensure that the products have the appropriate nutritional levels, contain all the required constituents, are what they claim to be (to avoid food fraud), and adhere to international and domestic standards. Its goal is to classify foods based on their chemical composition, nutritional value, sensory perception, traceability, protection, and consistency. Among different food categories, milk and milk products, oils and fats, fish and seafood, meat and meat products, fruit juice, alcoholic drinks, coffee and tea, sweeteners (including honey), spices, cereals, and pulses, were reported to have the highest numbers of adulteration incidence (Hong et al. 2017). Conventional food quality analytical techniques have lower precision, efficiency, time-consuming at quantifying or predicting food fraud activities. In the last two decades, some of the most widely used techniques for detection of food fraud and adulteration are: (i) chromatographic techniques: gas chromatography (GC), high-performance liquid chromatography (HPLC), thin-layer chromatography (TLC), (ii) mass spectrometry (MS) methods: gas chromatography-mass spectrometry (GC-MS), liquid chromatography-mass spectrometry (LC-MS), (iii) spectroscopic methods: Fourier transform infrared (FTIR), nuclear magnetic resonance (NMR), Raman, mid-infrared (MIR), near-infrared (NIR), and (iv) electrophoretic techniques: polymerase chain reaction (PCR) and random amplified polymorphic DNA (RAPD) (Fig. 11.1). Among all techniques, mass spectrometry (MS) constituted the most

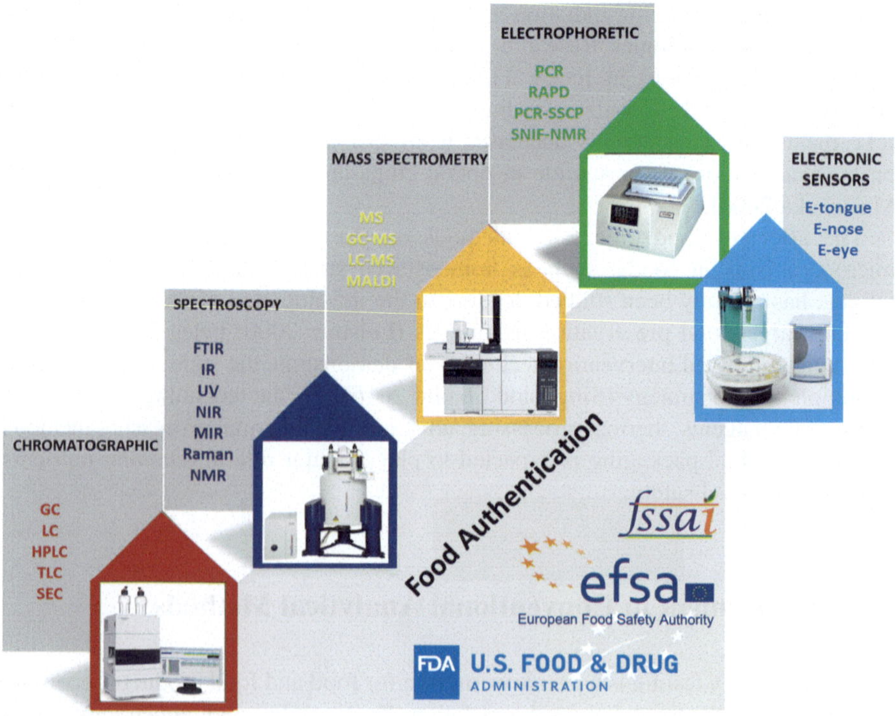

Fig. 11.1 Different analytical techniques followed by several regulatory organizations for the authentication of food samples

extensively and frequently used technique for cereals, spices, grains, and pulses. PCR techniques are most commonly employed in samples where DNA/RNA has to be tested, and it includes animal products such as meat and meat products and fish and seafood. Liquid chromatography (LC) and HPLC are regularly used for analyzing sweeteners, alcoholic beverages, fruits, and fruit juices. Owing to the chemical complexity of food products and high market demand for food quality and safety, high-resolution chromatographic techniques, including gas chromatography (GC) or liquid chromatography (LC) coupled with mass spectrometry (MS), have been identified as important food authentication methods.

Apple juice is among the most widely consumed juices in the world. For detecting the addition of low-cost commercial sugar syrups (beet and cane syrup) to pure apple juices and similar products, an advanced technique for determining ^{2}H and ^{13}C isotope ratios using gas chromatography-isotope ratio mass spectrometry (GC-IRMS) has been developed (Kelly et al. 2003). This technique can precisely detect added sugars such as inverted cane sugar, glucose, and fructose in authentic apple juices by confirming the variation in sugar contents in the juices. On the other hand, DNA-based methods, such as real-time PCR, species-specific PCR, and multiplex PCR, are undoubtedly the most common techniques used to assess the authenticity of meats and meat products. The Food Safety Authority of Ireland released a press report on January 15, 2013, announcing the application of real-time PCR to detect horse and donkey DNA in ground beef items such as sausage, burgers, and meatballs (Chisholm et al. 2005; Meira et al. 2017; O'Mahony 2013; Walker et al. 2013). Another example of meat fraud is murine meat as a replacement for mutton meat, frequently reported in China (Fang and Zhang 2016). The TaqMan@ real-time PCR method was used to detect the adulteration in which the study suggested a limit of detection of fewer than 1 picograms (pg) of DNA per reaction and 0.1% murine contamination in the adulterated meat. DNA and mass spectrometry methods are reported to be frequently employed methods for detecting food fraud. Along with these methods, it is very crucial to share the validated information among the concerned stakeholders for better traceability and monitoring (Huck et al. 2016; Ulberth 2020).

Due to the production of olive oils with unique regional and varietal features (protected designation of origin-PDO) and customer demand for high quality, authentication and quality control of olive oil are of primary importance. NMR spectroscopy and stable isotope analysis can reveal a pool of information on chemical composition and the chemical structure of oil metabolites (Dais and Hatzakis 2013). Stable isotope ratios can determine isotopes whose relative abundance is influenced by isotope fractionation in nature. Different elements in olive oil, including $^{13}C/^{12}C$, $^{18}O/^{16}O$, enabled Italian oils to be differentiated as per their geographical origin and between PDOs from the same area in some cases (Camin et al. 2010).

Consumers are increasingly looking for foods that are safe and nutritious and have a high organoleptic quality. Generally, acceptable sensory analysis findings needed a well-trained panel of human sensory analyzers. Even if the panellists are well trained, there is still a requirement to standardize the sensory interpretation, which is highly subjective. Instrumental food quality testing using perception sensors rather than human panel testing has recently gained popularity. An innovative cross-perception multi-sensors data fusion method has been proposed that mimics multiple human perceptions (Ouyang et al. 2014). Data were collected from rice wine samples using three sensors: an electronic tongue, eye, and nose. Principal components analysis (PCA) and multiple linear regression (MLR) were used to establish three cross-perception parameters: color, scent, and taste, used as inputs to models. Furthermore, a team of scientists from the UK has designed the first-ever 3D printed synthetic soft biomimetic surface that duplicates the wettability, elasticity, and topography of a real human tongue (Andablo-Reyes et al. 2020) (Fig. 11.2). The biomimetic tongues allow researchers to test newly developed products and speed up new development processes without expensive and time-consuming preliminary human testing. Oral tribological research with this advanced tongue-like surface can set the standard for understanding fundamental oral lubrication pathways, allowing basic mechanobiological questions to be addressed. At the same time, experimental and computational insights from this study can be extended to the biomimicry of other biological surfaces in the future to match the desired biophysical performance requirement.

Different food safety and control authentication techniques generate a humongous volume of data. Since a large volume of information needs to be analyzed, chemometrics and bioinformatics tools are essential for food authentication studies. Food science and technology have recently embraced novel and promising multivariate statistical methods such as chemometrics designed for analytical chemistry. Chemometric tools enable optimal applications of analytical techniques (chromatography, mass spectrometry, spectroscopy, PCR, calorimetry, wet chemistry, etc.) by extracting and interpreting valuable data from large and complex data sets. It also helps identify patterns in the data and develop calibration models (Capuano et al. 2014). However, the chemometric methods, such as principal component analysis (PCA), factorial discriminate analysis (FDA), quadratic discriminant analysis (QDA), and partial least squares discriminant analysis (PLS-DA), have certain shortcomings in delivering efficient and robust prediction models. Especially in cases involving large datasets presented in different formats (databases, images, texts, sounds, and video), which can be solved by incorporating statistical learning theories including artificial neural network (ANN), support vector machine (SVMs), probabilistic neural networks (PNN) (Kamal and Karoui 2015; Medina et al. 2019).

The development of databases containing standardized and comprehensive information regarding the origin of foods, such as geographical origin, species and subspecies, processing methods, and so on, will be critical to food authentication. Simultaneously, the availability of reference samples and well-identified databases is essential for predictive models that can relate an unknown sample to a known product. Although various regulatory organizations have compiled databases of food fraud and adulteration, there is a greater need for a pool of databases that can be used to classify these unknown samples. Some of the governmental/non-governmental platforms and databases that provide food authentication-related information are DOOR (records origin and registration of traditional specialties

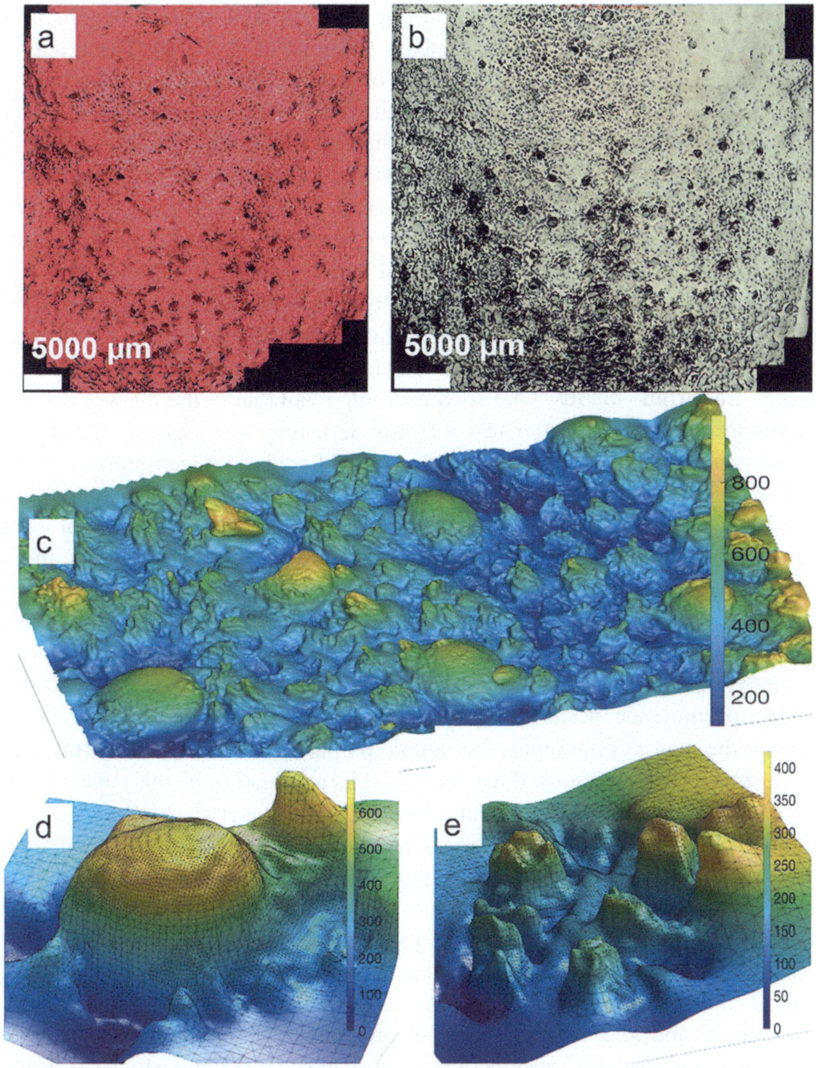

Fig. 11.2 3D printed biomimetic tongue surface to quantify oral friction and function as tribometer. Tongue impression on (**a**) polyvinyl siloxane (**b**) alginate masking materials imaged through negative 3D optical scans. (**c**) Tongue impression on polydimethylsiloxane obtained through positive 3D optical scan. (**d**) Fungiform papillae and (**e**) Filiform papillae on the surface of masking material reconstructed using screened Poisson surface. (Andablo-Reyes et al. 2020) (CC-BY)

guaranteed, protected designation of origin (PDOs), and protected geographical indication (PGIs), and Food Fraud Database (Medina et al. 2019). These databases are made available to the public through online websites such as Rapid Alerts System for Food and Feed (RASFF), which the public can access for recent incidents and any problems documented previously, thereby helping to identify food frauds.

The conventional analytical model involves sending samples to a laboratory and receiving results in several days or weeks. Today, food industries greatly benefit from several rapid testing techniques that can be performed at the point of use and provide a real-time result. The next expected paradigm shift will be towards testing kits/methods that customers can perform on supermarket shelves or at home. The use of small immunoassay test kits (similar to home pregnancy test kits) connected to smartphones to upload results into public databases is a specific area of interest for verifying "free-from" claims. One such example available in the market is AlerTox (https://glutentox.com) allergen test kits for detecting common allergens in soy, milk, and peanut. Although the R&D of advanced analytical techniques is going at a fast pace, most of the food authentication testing will continue to be performed in specialized laboratories due to the need for purchase and disposal of specific reagents, the intrinsic capital cost of equipment, stricter food regulations, the need for expert interpretation and so on.

It is evident from the literature that there has been considerable improvement in sample preparation, process automation, operating cost, and efficiency of conventional analytical techniques. Among them, supercritical fluid-based chromatography is gaining prominence because it requires less solvent (modifier and co-solvents, CO_2 being the major component) for sample preparation. When it is combined with mass spectroscopy techniques, it has increased its application in analytical chemistry due to its rapid detection and greener approach (Pilařová et al. 2019).

Magnetic extractants are increasingly used for sample preparative steps in food analysis for being economical in operation compared to other sample preparation steps. It is driven by the advancement in the synthesis of magnetic nanomaterials with fascinating properties such as superparamagnetism at room temperature. Magnetic nanomaterials are conjugated with selective adsorbents towards the molecule of interest and separated from the matrix with the combination of complementary shape, charge, and size. Novel chemical and physical functionalization methods boost magnetic extractants' application in food analysis (Li and Shi 2020; Ragavan and Rastogi 2017).

4 Advancement in Recognition Elements and Nanomaterials

From the beginning of twenty-first century, there has been a significant advancement in the synthesis and characterization of nanomaterials and related materials. It is mainly due to the new chemical routes for the synthesis of nanomaterials and progress in the instrumentation techniques to study their structure and properties.

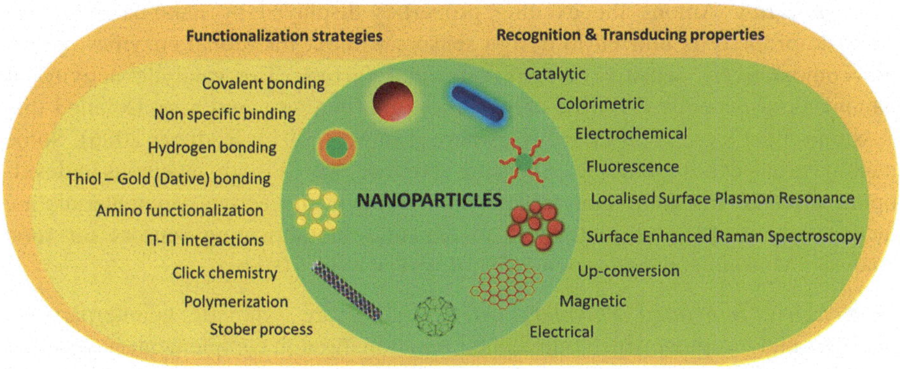

Fig. 11.3 Major types of nanomaterials utilized for the development of novel food sensors and detection methods. Left wing lists different physical and chemical routes followed for functionalization of nanomaterials with other sensor components. Right wing lists the role of these nanomaterials in the developed sensors (1). Permission obtained

As a result, numerous novel applications were reported in almost every possible field. Nanomaterials contributed significantly to food analysis and testing, which is evident from the quantum of research publications reported in the past two decades. Specific property and role of different nanomaterials in terms of their functionality for developing sensors and biosensors for food analysis are listed as a table elsewhere (Ragavan and Neethirajan 2019). Nanomaterial combination with different types of novel recognition elements has resulted in new detection methods with better analytical attributes (Fig. 11.3) (Ragavan et al. 2013a; Sharma et al. 2015). Recognition elements commonly used to integrate with nanomaterials for food analysis are briefly discussed below.

Antibodies Antibodies are the most reliable bio-recognition elements used to produce immunobased assays and kits (Ayyar et al. 2019; Yakes et al. 2016). They offer the advantage of conjugation with a wide range of molecules, including proteins, dyes, and nanoparticles, to develop food detection methods (Abhijith et al. 2013; Selvakumar et al. 2013).

Aptamers Aptamers are single-stranded nucleic acids such as DNA, RNA, or peptide sequences having a strong affinity to bind with different molecules with better affinity than antibodies. As a result, they are increasingly being used as a recognition element in the development of sensors. It also offers excellent functionalization with nanomaterials that improved the sensor attributes (Ragavan et al. 2013b; Sharma et al. 2015).

Enzymes Enzymes exhibit high specificity towards their target molecules through complementary structure and bond formation. The above property was the driving force in fabricating biosensors during the initial stages of biosensor development. It also offers good compatibility in functionalization with other sensor elements such as optical dyes, nanomaterials, and so on (Asal et al. 2018; Vaidya and Annapure 2019).

Nanomaterials Among the exciting properties displayed by nanomaterials, the catalytic property is utilized to design sensors without traditional enzymes, which are comparatively sensitive to environmental factors. The catalytic activity of nanomaterial depends upon the composition, morphology, and size. Detailed discussion on the topic is discussed elsewhere (Lin et al. 2014; Roduner 2006). Some nanomaterials exhibit multiple enzymatic activities, which is advantageous in developing a sensor for multi-analyte detection. Catalytically active nanomaterials are increasingly used as an alternative to enzymes in fabricating sensors for food analysis (Mustafa and Andreescu 2020; Ragavan et al. 2018b).

Molecularly Imprinted Polymers (MIPs) MIPs are synthetic counterparts of enzymes and receptors which mimic their function through complementary structure and chemical interaction brought about by polymers. The inherent nature of MIPs exhibits better electrical properties than enzymes and finds application in the fabrication of electrochemical sensors for food analysis with superlative analytical performance. Simultaneously, they are cost-effective and relatively stable compared to enzymes and other biological recognition elements, which require specific buffers for optimal activity. MIPs being selective towards a particular molecule find application in sample preparation for various analytical methods (Ashley et al. 2017; BelBruno 2019; Rhouati et al. 2019).

Receptors Receptors are highly selective and sensitive biomolecules that are part of signal transduction in living organisms. Purified receptors are used as bio-recognition elements in fabricating sensors for primarily volatile organic compounds. Disadvantages include difficulties in integration with other sensor components, limited receptors, high cost of purification, and less stability compared to its counterparts (Bohbot and Vernick 2020; Wu et al. 2014).

Whole Cells Instead of cellular components such as nucleic acids, enzymes, receptors, peptides, etc., whole bacterial or mammalian cells are used for sensing and screening different types of food components, including toxins and contaminants. Optical and electrochemical detection platforms based on whole cells are reported with the advantage of being stable and low cost. However, it requires specific conditions to culture them, and the present application is limited to very few analytes (Ye et al. 2019; Yu et al. 2017).

In the following section, important nanomaterials and their role in food testing are discussed below (Table 11.1).

4.1 Graphene and Related Materials

Graphene is an atom-thick two-dimensional carbon material known for fascinating optical, mechanical, electrical, and magnetic properties. It offers to tune the properties of graphene through changes in its composition and structure, resulting in graphene-related materials such as graphene oxide, graphene quantum dots, reduced

Table 11.1 Functional role of nanomaterials in various biosensor platforms along with their strengths and weakness

Nano-biosensor platform	Function of nanomaterial	Strengths	Weakness
Chemiluminescence (CL)	Catalyst in oxidation reaction, overcomes the need of enzymes	Highly sensitive (pM – fM), better signal to noise ratio	Not selective, quantum yields of the CL reaction is low
Aptasensor	Acts as an optical transducer (molar extinction coefficient is higher than chemical dyes)	Sensitive (sub nM), doesn't require instrument for interpretation,	Stability of aptamers is a concern, coloured compounds in the sample may interfere with results
Enzyme sensor		Fairly sensitive (μM – nM)	Most of the enzymes are not selective and presence of inhibitors reduces the efficiency of enzymes
Immunosensor		Sensitive (sub nM)	Dissociation constants of polyclonal antibodies are low
Electrochemical	Electrode modifier improves the surface area, electron conductivity of electrodes, sensitivity of the sensor and biocompatible	Highly accurate and sensitive analysis	Sample preparation is required
Fluorescence	Stable to photo bleaching and emission can be tuned	Highly sensitive (nM – fM)	Some of the dyes are highly toxic and prone to photo-bleaching
Microfluidics	Improves the analytical performance of microfluidic chip and device	Miniaturization, multiplex detection, high throughput analysis and less sample and reagent requirement	In nascent stage of growth, understanding the basic properties of fluids is required for better application
Quartz crystal microbalance (QCM)	Improves the sensitivity of the sensor	Possibility of miniaturization	Sensitivity of small molecules are low
Surface plasmon resonance (SPR)	Act as plasmonic substrate	Highly accurate analysis and overcomes the sample preparation steps, label free analysis	Advanced and sophisticated instruments are required
Surface enhanced Raman spectroscopy (SERS)	Nanoparticles enhanced the Raman signals by 10^{10}–10^{11}	Highly accurate and sensitive technique.	Advanced and sophisticated instruments are required

graphene oxide, graphene aerogels, and graphene-based composites. It is one of the best available materials to serve as an anchor for nanoparticles in the formation of composites while preserving nanomaterial properties (Ragavan and Rastogi 2016, 2017). Graphene acts as a quencher in its two-dimensional morphology and as an emitter of fluorescence signals while in spherical format (graphene quantum dots and carbon quantum dots), which finds huge applications in sensors as an alternative to semiconductor quantum dots (Pan et al. 2020; Ragavan and Neethirajan 2019; Zheng and Wu 2017). Carbon allotropes such as carbon nanotubes (CNTs) and fullerene are known for their electrical properties especially electrical conductivity, which makes them the material of choice for the fabrication of electrochemical sensors (Merkoçi et al. 2005; Taouri et al. 2021). Similar to graphene, other 2D materials such as MoS_2 and other transitional metal dichalcogenides, transitional metal carbides, nitrides hexagonal boron nitrides, carbonitrides, metal oxides, and metal-organic frameworks bring in interesting optical and electronic properties for designing sensors and detection methods meant for food analysis (Boroujerdi et al. 2020; Shavanova et al. 2016).

4.2 Metal and Metal Oxide-Based Nanomaterials

Transition metal and metal oxide nanoparticles such as gold, silver, zinc, copper, Iron, Titanium, etc., are commonly used nanomaterials. Among them, gold, silver, and their alloy in the form of nanoclusters, nanoparticles, nanorods, and other morphologies exhibit the best optical properties compared to other nanomaterials and dyes in terms of extinction coefficient (Ragavan et al. 2013b). Unique catalytic, distance-, and size-dependent optical properties are widely employed in developing colorimetric, fluorescence, chemiluminescence, surface plasmon resonance, and surface-enhanced Raman spectroscopy-based sensors for food analysis (Chen et al. 2018). Recently, gold and silver nanoclusters are being investigated for their optical properties and found applications in detecting food analytes (Hu et al. 2020; Li et al. 2019). Iron oxide nanoparticles exhibit superparamagnetism at room temperature, along with catalytic properties. It is the material of choice for sample preparation in food analysis, and it is conjugated with other nanomaterials to bring in multiple roles (Cao et al. 2012; Li and Shi 2020; Ragavan and Rastogi 2017). In many cases, they help overcome the matrix effect by separating the molecule of interest in relatively simple steps due to its magnetic property. Titanium-based nanomaterials and their hybrids find specific applications in the development of electrochemical sensors for their semiconductor-like electrical properties (Romero-Arcos et al. 2016; Shetti et al. 2019).

4.3 Miscellaneous Nanomaterials

Silica-Based Nanomaterials Silica and its nanomaterials for their biocompatibility and less toxicity offer an excellent platform for conjugation of biomolecules. It finds major applications for in-vivo applications. However, their role in sensors for food analysis can't be ignored.

Nanodiamonds Nanodiamonds are an allotrope of carbon, known for their mechanical and electrical properties, to fabricate quartz crystal microbalance-based sensors (Yao and Xue 2015). Doping of nanodiamonds with boron overcomes electrode surface fouling and improves the sensor attributes (Jiang et al. 2021).

Cerium-Based Nanomaterials These nanomaterials are known for their catalytic and electrochemical properties, find application in the development of electrochemical sensors for food analysis (Esmaeili et al. 2019; Yang et al. 2017). The catalytic property of cerium nanoparticles towards different antioxidants in food products resulted in products with different hues, which was utilized to demonstrate cerium nanoparticle-based optical sensors (Sharpe et al. 2013).

Palladium-Based Nanomaterials They are known for their attractive multi catalytic properties used to fabricate paper-based color sensors (Fig. 11.4) (Ragavan et al. 2018b). Palladium nanoparticles are reported to be compatible for conjugation with aptamers for designing fluorescence sensors to detect tetracycline in milk (Ahmed et al. 2021).

Semiconductor QDs They are highly fluorescent nanomaterials with very high quantum yield and also offer the advantage of tuning their emission through composition and size. They are the nanomaterial of choice as optical tags in optical sensors and fluorescence resonance energy transfer (FRET) based sensors (Chern et al. 2019; Freeman et al. 2013; Pedrero et al. 2017).

Upconversion Nanoparticles A unique class of nanomaterials known for their optical property of absorbing low energy electrons and emitting them in high energy or shorter wavelength. These nanoparticles have a larger stokes shift than conventional fluorescent dyes and semiconductor quantum dots. Sensors for detecting antibiotic residues are designed using these upconversion nanomaterials (Peltomaa et al. 2021; Wen et al. 2018).

5 Mobile/Smart Phone-Based Sensors/Biosensors

Consumer access to smartphones across the globe is increasing, and the day is not far when almost every human has a smartphone. It is perceived as a digital companion and an extension of the user (Carolus et al. 2019; Harkin and Kuss 2021). Smartphones are increasingly replacing conventional imaging infrastructure for measuring various parameters, including contact angle (Chen et al. 2018). Similarly,

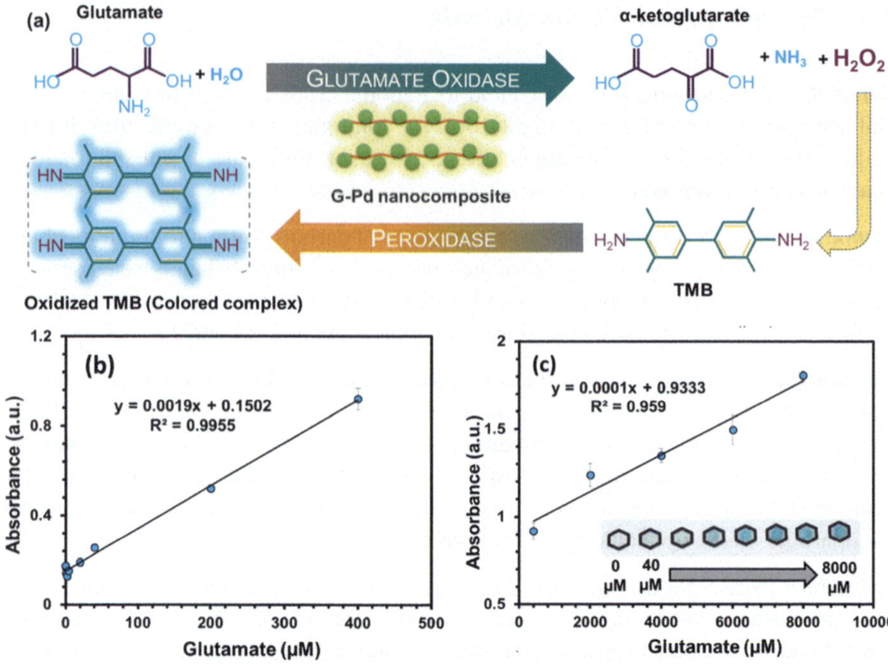

Fig. 11.4 (**a**) Working principle of graphene-palladium nanocomposite containing paper based colorimetric paper sensor. Graphene-palladium nanocomposite exhibits dual catalytic activity for the detection of glutamate in water samples. (**b–c**) Response of the developed colometric paper sensor towards glutamate (2). (Permission obtained)

interest in using smartphones for point of care and onsite analysis is mainly due to the following advantages. In terms of hardware design, they are compact and easy to operate; also, it provides access to location data and communicates necessary information to concerned stakeholders (Fig. 11.5). Smartphones are mainly integrated with most of the assay types for detecting a wide range of food compounds in solid and liquid matrices (Fig. 11.6) (Lu et al. 2019; Nelis et al. 2020). Smartphone-based sensors are reported to have applications in different types of food matrix with.

5.1 *Electrochemical Sensors*

Portable electrochemical sensors are developed with the help of a miniaturized potentiostat for generating the required electrical signals. Output from the electrodes can be processed through smartphones which overcome the use of bulky computers. An exclusive app for the quantification of electrochemical signals into a qualitative or quantitative measurement is necessary for the setup. Various types of food

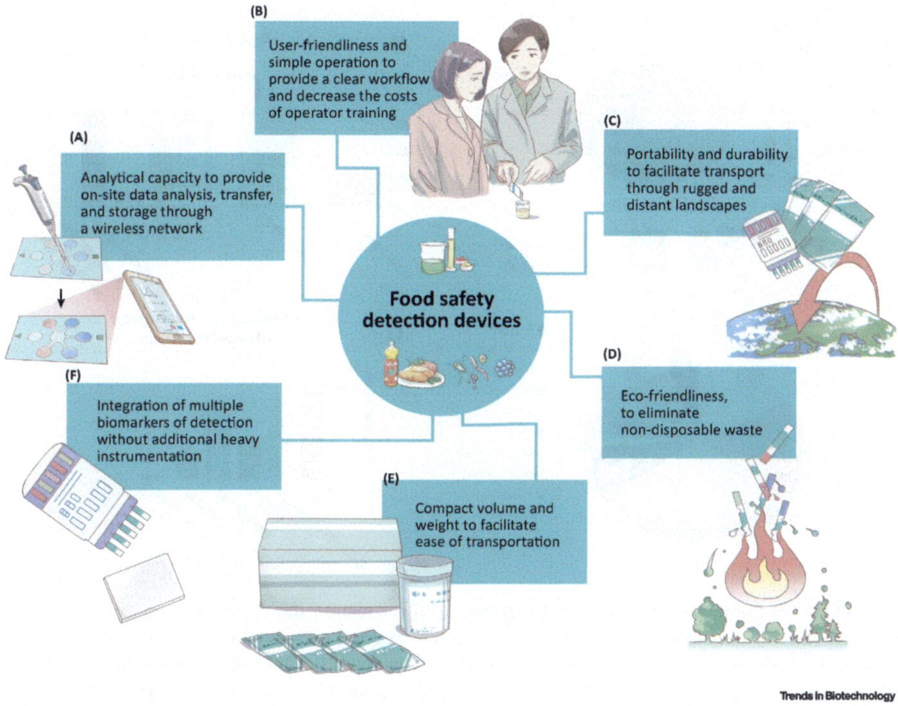

Fig. 11.5 Important features of smartphone based food safety detection devices compared to conventional analytical methods (3). (Permission obtained)

analytes are reported to be detected through smartphone-based electrochemical sensors (Nelis et al. 2020; Seo et al. 2019; Sivakumar and Lee 2021).

5.2 *Optical Sensors and Microscopy*

Compared to the smartphone-based electrochemical sensors, optical sensor integration requires the smartphone's optical features and data processing facilities. For the data processing, unlike electrochemical sensors, certain commonly available apps can be used, however transducing the optical signals to an analyte concentration requires specific programs or apps for the smartphone (Fig. 11.7) (Nelis et al. 2020; Seo et al. 2019; Sivakumar and Lee 2021). Smartphone LEDs specifications vary widely from manufacturer to model, however, recent smartphones have advanced LED features such as provision for warm and cold colour temperature, better intensity and illumination. In colorimetric, fluorescence, luminescence, and spectroscopic methods, specific filters, excitation sources, and an external setup are necessary. It includes UV based LEDs in the sensor kit (mobile accessory) to excite

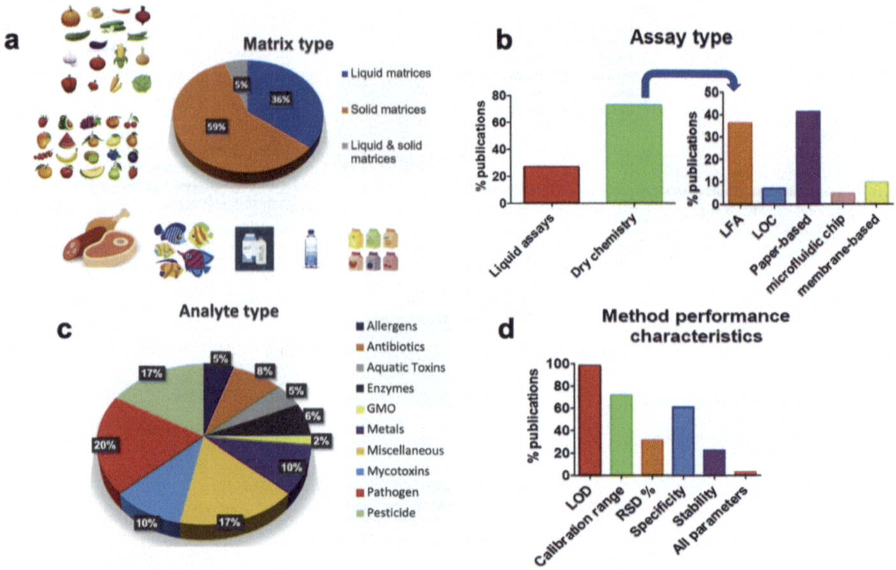

Fig. 11.6 (**a**) Summary of type of food matrix used for testing the developed smartphone based sensors. (**b**) Dry chemistry techniques are more often reported compared to liquid based sensing or testing methods. (**c**) Food analytes, toxins, contaminants against which the sensors are developed. (**d**) Analytical performance indicators analysed in the reported literature (4). (CC BY 4.0)

fluorophores in the assays (Rateni et al. 2017). Information related to the spectrum, colour temperature, intensity and other relevant information are seldom collected and presented in the literature. Overall, smartphones might serve as an excellent platform to test the quality and safety of food meant for consumption (traceability) as well.

6 Miscellaneous

Numerous types of sensors/biosensors for testing food quality and safety have been reported in the literature, and some of the promising technology/types are discussed below.

6.1 *Microfluidics*

Microfluidic devices are miniature devices designed through lithography methods predominantly composed of silica-containing compounds that are increasingly used for food testing. In a typical microfluidic device, microcapillary paths/pores are designed to move fluid by inherent capillary forces overcoming the need for an

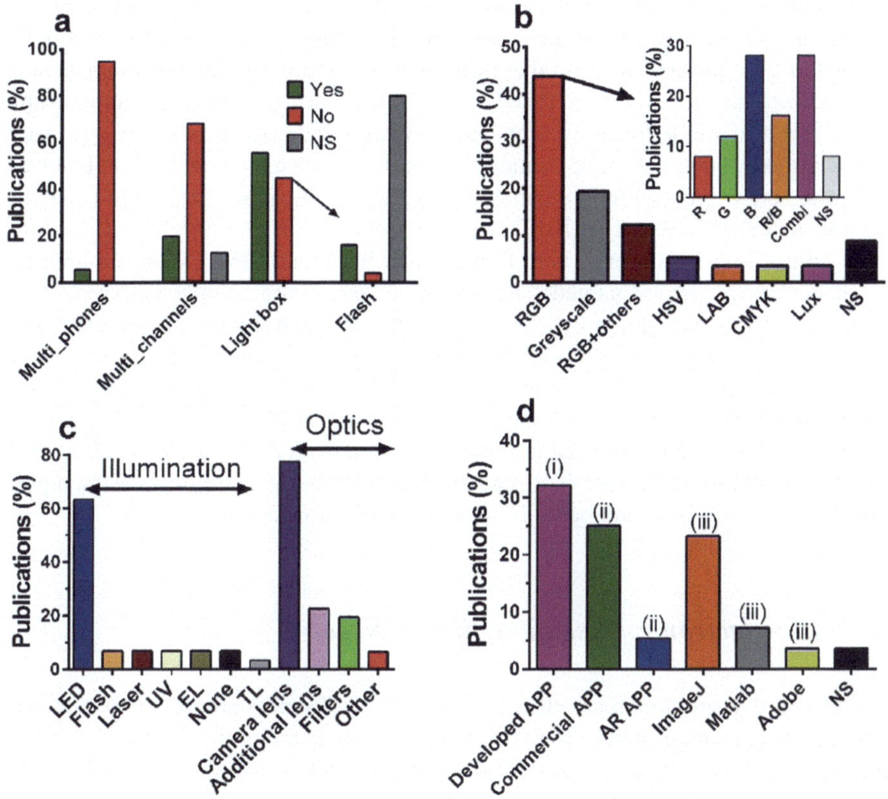

Fig. 11.7 Summary of smartphone based sensor publications (**a**): comparison of developed method with different phones, channels, light boxes and flashes to evaluate the performance. (**b**): Different type of color modes/space followed in the publications. (**c**): Illumination and optics used in the developed sensors. (**d**): Apps used in the reported literature, in which custom made and commercial apps used predominantly (4). (CC BY 4.0)

external force or pump. They are familiarly known as "lab on a chip" and "μpad" devices, which are further classified into microelectromechanical systems (MEMS) and micro total analysis systems (μTAS). As the name suggests, these devices require less volume of sample, can integrate multiple analysis into a single device, and importantly offers onsite analysis without the requirement of sample transportation. It also offers other advantages such as automation and high-throughput screening at an economical cost compared to conventional analytical methods to estimate food quality and safety parameters (Romao et al. 2017; Weng and Neethirajan 2017; Wu et al. 2017). Lab on a chip are the most promising devices with the potential for onsite detection/analysis. It combines multiple operations to be carried out in a small chip with approximately having an area of 20–30 cm^2.

Paper-based microfluidic devices are known as 'lab on paper' in which cellulose-based paper is used as a substrate rather than silica-based molecules. Micro 2D and 3D patterns are fabricated to aid the movement of hydrophilic fluid through absorption and capillary action. Paper-based microfluidic devices are practical for one-time usage, low cost, biodegradable compared to silica-based microfluidic devices. Apart from cellulose, other cellulose-containing materials such as lignocellulose, bamboo, and cotton are explored for their suitability to develop sensors to detect food analytes (Malon et al. 2017; Wu et al. 2017).

In an exciting invention, Manu Prakash and his group have fabricated a paper-based centrifuge device named "paperfuge" to overcome sample preparation for onsite detection and diagnosis. It overcomes the need for using an expensive centrifuge for sample preparation at low resource environments for analysis. It is basically a microfluidic device containing a capillary tube utilizing human power to operate it akin to a "whirligig" capable of achieving 125,000 rpm or 30,000 g. Even though it has been demonstrated to separate plasma from the blood to detect malaria parasites (Bhamla et al. 2017). It can be generalized for handling food samples, which might overcome cumbersome sample preparation steps.

6.2 *Bio-electronic Nose or Artificial Nose*

Visual, hearing, and touch are classified into physical senses among the five human senses, whereas odour and taste are chemical senses. Among them, vision, hearing, and touch are standardized to a great extent through colour, sound, and texture estimation. In the case of taste, trained sensory panellist's responses are universally accepted and followed. However, quantification of olfactory senses through panellists is contradictory, raising the need for a device-based evaluation. Bioelectronic nose or artificial nose or electronic nose (Fig. 11.8) is a complex device fabricated to sense and reconstitute odor as perceived by the human olfactory system (Fitzgerald et al. 2017; Gancarz et al. 2017). Conventionally, volatile organic compounds responsible for odour perception are primarily quantified through gas chromatography coupled with mass spectroscopy (GC-MS). However, in the case of electronic nose, detection/sensing mechanism is classified into three broad groups,

(i) Electric – change in electrical response (mainly conductivity) in the presence of the analyte. The design includes field-effect transistors coupled with conducting polymers.
(ii) Gravimetric – change in frequency due to the binding of analyte – Piezoelectric crystals conjugated to receptor binds with the analyte, which leads to change in mass on crystal, dampening the resonant frequency.
(iii) Optical – Change in signal (fluorescence/chemiluminescence) intensity or shift in absorption or emission of the dyes – interaction between the dye and analyte results in the distinguishable optical signal, quantitatively or qualitatively correlated to the analyte concentration. In recent times, bar-coded resins with

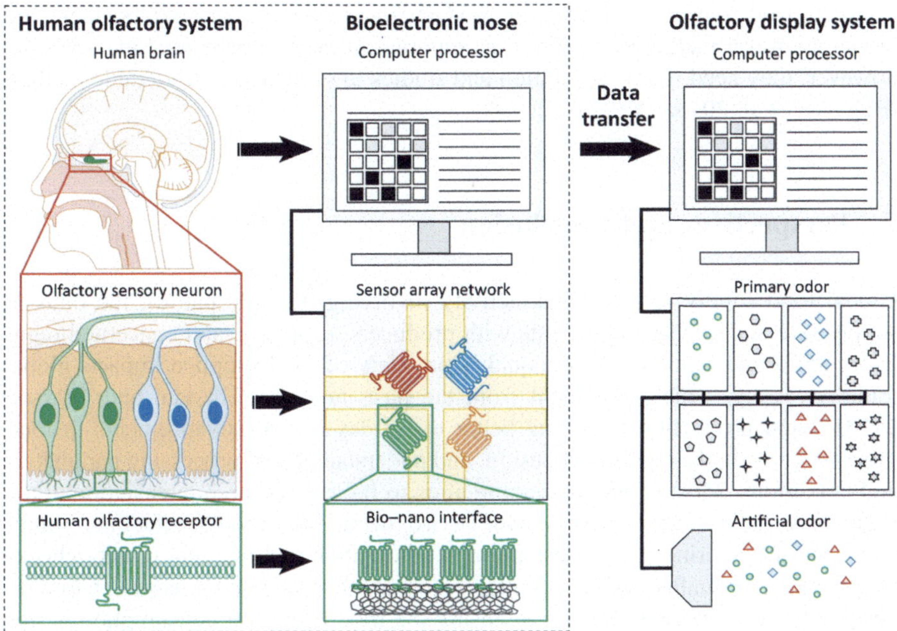

Fig. 11.8 (**a**) Comparison of human olfactory mechanism with working principle of bioelectronics nose. Receptor of human olfactory system are mimicked by fabricating a delicate composite of nanomaterial with biomolecules (bio-nano interface). Signals from the interface is transferred and processed with the aid of computers which is reconstituted by combining basic odours as perceived by the human interface in a standard format (5). (Permission obtained)

unique Raman or infra-red spectra (fingerprint) are used for rapid detection of volatiles (Fitzgerald et al. 2017; Sanaeifar et al. 2017; Son et al. 2017). More studies concerning calibration and validation are required to further improve the real-time application of these attractive electronic nose based devices.

DNA metabarcoding helps in the identification of species in a food matrix, which includes microbial, plant or animal origin proposed initially by researchers at the University of Guelph (Hebert et al. 2003). It involves extracting DNA from the food samples, its amplification, sequencing, bioinformatics analysis and species identification to figure out the DNA of different species present in the sample. It is better than DNA barcoding because it identifies multi-species DNA rather than a single species in a single reaction by including multi-species DNA data. Hence, it finds application in food authenticity to validate the ingredients and their source as claimed in the products, to identify the pathogenic/spoilage microbes in contaminated foods and other food quality and safety aspects (Bruno et al. 2019; Grützke et al. 2019).

Apart from the above discussed food testing methods and sensors, there are numerous novel methods and materials reported in the literature for food analysis; however, they need more exploration and studies to grow as a considerable method (Ragavan et al. 2018a; Ye et al. 2019).

7 Perspectives and Conclusion

Global food supply is a complex activity involving multiple parties, the onus to supply safe and quality goods rests with producers, suppliers, and concerned regulatory agencies. Any lapse in the quality or safety of food supplied impacts global trade adversely, which is evident from the past incidents (Hussain and Dawson 2013). Food production is getting better every day due to advancements in food processing methods, equipment design, and understanding of processing parameters and conditions. Most of the processing aims to be less resource-intensive with the least waste generation to comply with environmental and food regulations. Simultaneously, monitoring the process through automation and frequent quality checks results in better quality and safe products sustainably compared to earlier. Comprehensive food testing is necessary to ensure the food meant for consumption is free from hazards and quality as per marketing claims. However, it is affected by one or more following means in some instances, including food fraud, ineffective food laws and regulations, fraudulent science, and miscommunication (Neethirajan et al. 2018). In recent times, following food products and food analytes derived from new and alternative sources (GMOs, insect proteins, amnesic shellfish poisoning, and food processing contaminants) require more attention from the researchers and regulatory agencies to develop food testing methods and regulations, respectively. It is crucial to develop regulations and testing methods for most of the above products to gain consumer acceptability.

Smartphone-based methods offer multiple advantages for onsite food testing. It is relatively new in the market and is constantly evolving, so the developed methods might have a shorter life span than conventional analytical methods. Another important aspect is the plethora of detection strategies has to be communicated in a globally standardized format (Global standards). An additional requirement of sensor accessories/attachments might be a hindrance to its adoption by consumers. Also, the integration of these sensor attachments with different smartphones for hardware and software is a challenge. However, lack of communication and information can be solved with mobile/smartphone-based sensors for real-time monitoring and data sharing. Similarly, microfluidic-based devices, bioelectronics nose, and nanomaterial-based food testing methods are gaining prominence owing to their features. It is expected that they might have an even greater role to play in the coming days. Food matrix being a complex one, interferes with the outcome of the testing method in multiple ways leading to less useful information from the tests. Considerable improvement to overcome matrix effects in food samples is necessary to render these techniques adopted everywhere. Research towards the development of innovative and straightforward steps for sample preparation without sophisticated instruments and solvents might help to an extent.

It is worth noting that, even though several thousand research articles are published every year with novel detection methods, they seldom move to industry for product development and marketing, evident from very few food testing devices in the market (Luong et al. 2008). Despite these advancements, reported sensors are in the nascent stage in real-life applications, which requires rigorous on-site testing and validation. Moreover, the food matrix is one of the most complexes and challenging due to multiphasic in nature with a plethora of compounds with a wide range of functional properties (Aguilera 2019). Hence, most detection methods are conveniently tested in simple matrices such as water (Nelis et al. 2020). Due to the above-said drawbacks, currently, there is no single method which can be used to detect most of the target compounds related to food quality, food fraud, food adulteration, or food safety. Analytical parameters of the developed method or the food testing device including data analysis need to be rigorously evaluated comprehensively to render them reliable and get approval from regulatory agencies (Wu et al. 2020). Emphasis on validation and calibration of the developed methods by the researchers is one of the solutions to come up with robust food testing methods/devices. Such recognition for food testing devices might increase the usage among consumers and expand the market share (Nelis et al. 2020). It is expected that developed sensors and biosensors to act as reliable food testing devices for the initial screening (a mode for early warning system) of samples for various parameters, which will certainly reduce the burden of overall cost spent for food analysis at the industry and marketplace. From a consumer, it is necessary to have a reliable qualitative food testing device to ensure the food /food product is safe and quality is the same as marketing.

Acknowledgments Singam acknowledges the International Doctoral Tuition Scholarship (IDTS) for his Ph.D. work at the University of Guelph, Canada.

References

Abhijith KS, Ragavan KV, Thakur MS (2013) Gold nanoparticles enhanced chemiluminescence – a novel approach for sensitive determination of aflatoxin-B1. Anal Methods 5:4838–4845. https://doi.org/10.1039/C3AY40694F

Aguilera JM (2019) The food matrix: implications in processing, nutrition and health. Crit Rev Food Sci Nutr 59:3612–3629. https://doi.org/10.1080/10408398.2018.1502743

Ahmed SR, Kumar S, Ortega GA, Srinivasan S, Rajabzadeh AR (2021) Target specific aptamer-induced self-assembly of fluorescent graphene quantum dots on palladium nanoparticles for sensitive detection of tetracycline in raw milk. Food Chem 346:128893. https://doi.org/10.1016/j.foodchem.2020.128893

Albert T, Braun PG, Saffaf J, Wiacek C (2021) Physical methods for the decontamination of meat surfaces. Curr Clin Micro Rpt 8:9–20. https://doi.org/10.1007/s40588-021-00156-w

Alimi BA, Workneh TS (2016) Consumer awareness and willingness to pay for safety of street foods in developing countries: a review. Int J Consum Stud 40:242–248. https://doi.org/10.1111/ijcs.12248

Alirezalu K, Munekata PES, Parniakov O, Barba FJ, Witt J, Toepfl S, Wiktor A, Lorenzo JM (2020) Pulsed electric field and mild heating for milk processing: a review on recent advances. J Sci Food Agric 100:16–24. https://doi.org/10.1002/jsfa.9942

Andablo-Reyes E, Bryant M, Neville A, Hyde P, Sarkar R, Francis M, Sarkar A (2020) 3D biomimetic tongue-emulating surfaces for Tribological applications. ACS Appl Mater Interfaces 12:49371–49385. https://doi.org/10.1021/acsami.0c12925

Asal M, Özen Ö, Şahinler M, Polatoğlu İ (2018) Recent developments in enzyme, DNA and Immuno-based biosensors. Sensors 18:1924. https://doi.org/10.3390/s18061924

Ashley J, Shahbazi M-A, Kant K, Chidambara VA, Wolff A, Bang DD, Sun Y (2017) Molecularly imprinted polymers for sample preparation and biosensing in food analysis: Progress and perspectives. Biosens Bioelectron 91:606–615. https://doi.org/10.1016/j.bios.2017.01.018

Augustin MA, Riley M, Stockmann R, Bennett L, Kahl A, Lockett T, Osmond M, Sanguansri P, Stonehouse W, Zajac I, Cobiac L (2016) Role of food processing in food and nutrition security. Trends Food Sci Technol 56:115–125. https://doi.org/10.1016/j.tifs.2016.08.005

Ayyar BV, Arora S, O'Kennedy R (2019) Chapter 2: Production and use of antibodies. In: Rapid antibody-based technologies in food analysis, pp 6–31. https://doi.org/10.1039/9781788016322-00006

Bagheri H, Abbaszadeh S (2020) Effect of cold plasma on quality retention of fresh-cut produce. J Food Qual 2020:e8866369. https://doi.org/10.1155/2020/8866369

BelBruno JJ (2019) Molecularly imprinted polymers. Chem Rev 119:94–119. https://doi.org/10.1021/acs.chemrev.8b00171

Betz JF, Cheng Y, Rubloff GW (2012) Direct SERS detection of contaminants in a complex mixture: rapid, single step screening for melamine in liquid infant formula. Analyst 137:826–828. https://doi.org/10.1039/C2AN15846A

Bhamla MS, Benson B, Chai C, Katsikis G, Johri A, Prakash M (2017) Hand-powered ultralow-cost paper centrifuge. Nat Biomed Eng 1:1–7. https://doi.org/10.1038/s41551-016-0009

Bhat ZF, Morton JD, Mason SL, Bekhit AE-DA (2019) Current and future prospects for the use of pulsed electric field in the meat industry. Crit Rev Food Sci Nutr 59:1660–1674. https://doi.org/10.1080/10408398.2018.1425825

Bintsis T, Litopoulou-Tzanetaki E, Robinson RK (2000) Existing and potential applications of ultraviolet light in the food industry – a critical review. J Sci Food Agric 80:637–645. https://doi.org/10.1002/(SICI)1097-0010(20000501)80:6<637::AID-JSFA603>3.0.CO;2-1

Boeckel TPV, Pires J, Silvester R, Zhao C, Song J, Criscuolo NG, Gilbert M, Bonhoeffer S, Laxminarayan R (2019) Global trends in antimicrobial resistance in animals in low- and middle-income countries. Science 365. https://doi.org/10.1126/science.aaw1944

Bohbot JD, Vernick S (2020) The emergence of insect odorant receptor-based biosensors. Biosensors 10:26. https://doi.org/10.3390/bios10030026

Böhme K, Calo-Mata P, Barros-Velázquez J, Ortea I (2019) Review of recent DNA-based methods for Main food-authentication topics. J Agric Food Chem 67:3854–3864. https://doi.org/10.1021/acs.jafc.8b07016

Boroujerdi R, Abdelkader A, Paul R (2020) State of the art in alcohol sensing with 2D materials. Nano-Micro Lett 12:33. https://doi.org/10.1007/s40820-019-0363-0

Bricher J, Keener L, (2007) Innovations in technology promising food safety technologies. Is there a silver lining when there is no silver bullet? Foodsafety magazine

Bruno A, Sandionigi A, Agostinetto G, Bernabovi L, Frigerio J, Casiraghi M, Labra M (2019) Food tracking perspective: DNA Metabarcoding to identify plant composition in complex and processed food products. Genes 10:248. https://doi.org/10.3390/genes10030248

Camin F, Larcher R, Perini M, Bontempo L, Bertoldi D, Gagliano G, Nicolini G, Versini G (2010) Characterisation of authentic Italian extra-virgin olive oils by stable isotope ratios of C, O and H and mineral composition. Food Chem Food Authenticity Traceability 118:901–909. https://doi.org/10.1016/j.foodchem.2008.04.059

Cao M, Li Z, Wang J, Ge W, Yue T, Li R, Colvin VL, Yu WW (2012) Food related applications of magnetic iron oxide nanoparticles: enzyme immobilization, protein purification, and food analysis. Trends Food Sci Technol 27:47–56. https://doi.org/10.1016/j.tifs.2012.04.003

Capuano E, Rademaker J, van den Bijgaart H, van Ruth SM (2014) Verification of fresh grass feeding, pasture grazing and organic farming by FTIR spectroscopy analysis of bovine milk. Food Res Int Authenticity Typicality Traceability Intrinsic Quality Food Prod 60:59–65. https://doi.org/10.1016/j.foodres.2013.12.024

Carolus A, Binder JF, Muench R, Schmidt C, Schneider F, Buglass SL (2019) Smartphones as digital companions: characterizing the relationship between users and their phones. New Media Soc 21:914–938. https://doi.org/10.1177/1461444818817074

Carvalho FP (2017) Pesticides, environment, and food safety. Food Energy Secur 6:48–60. https://doi.org/10.1002/fes3.108

Castro-Ibáñez I, Gil MI, Allende A (2017) Ready-to-eat vegetables: current problems and potential solutions to reduce microbial risk in the production chain. LWT – Food Sci Technol Fruit Veg Proc 2016 – Towards Sustain 85:284–292. https://doi.org/10.1016/j.lwt.2016.11.073

Cesewski E, Johnson BN (2020) Electrochemical biosensors for pathogen detection. Biosens Bioelectron 159:112214. https://doi.org/10.1016/j.bios.2020.112214

Chalmers RM, Robertson LJ, Dorny P, Jordan S, Kärssin A, Katzer F, La Carbona S, Lalle M, Lassen B, Mladineo I, Rozycki M, Bilska-Zajac E, Schares G, Mayer-Scholl A, Trevisan C, Tysnes K, Vasilev S, Klotz C (2020) Parasite detection in food: current status and future needs for validation. Trends Food Sci Technol 99:337–350. https://doi.org/10.1016/j.tifs.2020.03.011

Chen H, Muros-Cobos JL, Amirfazli A (2018) Contact angle measurement with a smartphone. Rev Sci Instrum 89:035117. https://doi.org/10.1063/1.5022370

Chen JH, Ren Y, Seow J, Liu T, Bang WS, Yuk HG (2012) Intervention Technologies for Ensuring Microbiological Safety of meat: current and future trends. Compr Rev Food Sci Food Saf 11: 119–132. https://doi.org/10.1111/j.1541-4337.2011.00177.x

Chern M, Kays JC, Bhuckory S, Dennis AM (2019) Sensing with photoluminescent semiconductor quantum dots. Methods Appl Fluoresc 7:012005. https://doi.org/10.1088/2050-6120/aaf6f8

Chisholm J, Conyers C, Booth C, Lawley W, Hird H (2005) The detection of horse and donkey using real-time PCR. Meat Sci 70:727–732. https://doi.org/10.1016/j.meatsci.2005.03.009

Codex Alimentarius (2016) Guidelines on the application of general principles of food hygiene to the control of foodborne parasites. CAC/GL 88-2016

Dais P, Hatzakis E (2013) Quality assessment and authentication of virgin olive oil by NMR spectroscopy: a critical review. Anal Chim Acta 765:1–27. https://doi.org/10.1016/j.aca.2012.12.003

Danezis GP, Tsagkaris AS, Brusic V, Georgiou CA (2016) Food authentication: state of the art and prospects. Curr Opin Food Sci Innov Food Sci Foodomics Technol 10:22–31. https://doi.org/10.1016/j.cofs.2016.07.003

Daniel SCGK, Nirupa Julius LA, Gorthi SS (2017) Instantaneous detection of melamine by interference biosynthesis of silver nanoparticles. Sensors Actuators B Chem 238:641–650. https://doi.org/10.1016/j.snb.2016.07.112

Esmaeili C, Norouzi P, Zar MS, Eskandari M, Faridbod F, Ganjali MR (2019) A FFT Square wave voltammetry sensing method for highly sensitive detection of Phytic acid using a cerium oxide nanoparticles decorated graphene oxide. J Electrochem Soc 166:B1630–B1636. https://doi.org/10.1149/2.1191914jes

Fan X, Mastovska K (2006) Effectiveness of ionizing radiation in reducing furan and acrylamide levels in foods. J Agric Food Chem 54:8266–8270. https://doi.org/10.1021/jf061151+

Fang X, Zhang C (2016) Detection of adulterated murine components in meat products by TaqMan© real-time PCR. Food Chem 192:485–490. https://doi.org/10.1016/j.foodchem.2015.07.020

FAO (2021) Parasites in foods - an invisible threat, food safety and quality. Food and Agriculture Organization of the United Nations, Bangkok

Fitzgerald JE, Bui ETH, Simon NM, Fenniri H (2017) Artificial nose technology: status and prospects in diagnostics. Trends Biotechnol 35:33–42. https://doi.org/10.1016/j.tibtech.2016.08.005

Freeman R, Girsh J, Willner I (2013) Nucleic acid/quantum dots (QDs) hybrid Systems for Optical and Photoelectrochemical Sensing. ACS Appl Mater Interfaces 5:2815–2834. https://doi.org/10.1021/am303189h

Gancarz M, Wawrzyniak J, Gawrysiak-Witulska M, Wiącek D, Nawrocka A, Tadla M, Rusinek R (2017) Application of electronic nose with MOS sensors to prediction of rapeseed quality. Measurement 103:227–234. https://doi.org/10.1016/j.measurement.2017.02.042

Grützke J, Malorny B, Hammerl JA, Busch A, Tausch SH, Tomaso H, Deneke C (2019) Fishing in the soup – pathogen detection in food safety using Metabarcoding and metagenomic sequencing. Front Microbiol 10:1–15

Harkin LJ, Kuss D (2021) "My smartphone is an extension of myself": a holistic qualitative exploration of the impact of using a smartphone. Psychol Pop Media 10:28–38. https://doi.org/10.1037/ppm0000278

Hebert PDN, Cywinska A, Ball SL, de Waard JR (2003) Biological identifications through DNA barcodes. Proceedings of the Royal Society of London. Ser B Biol Sci 270:313–321. https://doi.org/10.1098/rspb.2002.2218

Hoffmann V, Moser C, Saak A (2019) Food safety in low and middle-income countries: the evidence through an economic lens. World Dev 123:104611. https://doi.org/10.1016/j.worlddev.2019.104611

Hong E, Lee SY, Jeong JY, Park JM, Kim BH, Kwon K, Chun HS (2017) Modern analytical methods for the detection of food fraud and adulteration by food category. J Sci Food Agric 97: 3877–3896. https://doi.org/10.1002/jsfa.8364

Hu X, Shi J, Shi Y, Zou X, Arslan M, Zhang W, Huang X, Li Z, Xu Y (2019) Use of a smartphone for visual detection of melamine in milk based on au@carbon quantum dots nanocomposites. Food Chem 272:58–65. https://doi.org/10.1016/j.foodchem.2018.08.021

Hu Y, Li Y, Liao Y, Jiang X, Cheng Z (2020) Poly(sodium-p-styrenesulfonate)-enhanced fluorescent silver nanoclusters for the assay of two food flavors and silicic acid. Food Chem 318: 126502. https://doi.org/10.1016/j.foodchem.2020.126502

Huang F, Li Y, Wu J, Dong J, Wang Y (2016) Identification of repeatedly frozen meat based on near-infrared spectroscopy combined with self-organizing competitive neural networks. Int J Food Prop 19:1007–1015. https://doi.org/10.1080/10942912.2014.968789

Huang W-C, Wu K-H, Hung H-C, Wang J-C, Chang S-C (2018) Magnetic nanoparticle-based lateral flow Immunochromatographic strip as a reporter for rapid detection of melamine. J Nanosci Nanotechnol 18:7190–7196. https://doi.org/10.1166/jnn.2018.16020

Huck CW, Pezzei CK, Huck-Pezzei VA (2016) An industry perspective of food fraud. Curr Opin Food Sci Innov Food Sci Foodomics Technol 10:32–37. https://doi.org/10.1016/j.cofs.2016.07.004

Hussain MA, Dawson CO (2013) Economic impact of food safety outbreaks on food businesses. Foods 2:585–589. https://doi.org/10.3390/foods2040585

Jiang L, Santiago I, Foord J (2021) A comparative study of fouling-free nanodiamond and nanocarbon electrochemical sensors for sensitive bisphenol a detection. Carbon 174:390–395. https://doi.org/10.1016/j.carbon.2020.11.073

Kamal M, Karoui R (2015) Analytical methods coupled with chemometric tools for determining the authenticity and detecting the adulteration of dairy products: a review. Trends Food Sci Technol 46:27–48. https://doi.org/10.1016/j.tifs.2015.07.007

Kelly SD, Rhodes C, Lofthouse JH, Anderson D, Burwood CE, Dennis MJ, Brereton P (2003) Detection of sugar syrups in apple juice by δ2H‰ and δ13C‰ analysis of hexamethylenetetramine prepared from fructose. J Agric Food Chem 51:1801–1806. https://doi.org/10.1021/jf021044p

Koutsoumanis K, Allende A, Alvarez-Ordóñez A, Bolton D, Bover-Cid S, Chemaly M, Davies R, Cesare AD, Herman L, Hilbert F, Lindqvist R, Nauta M, Peixe L, Ru G, Simmons M,

Skandamis P, Suffredini E, Cacciò S, Chalmers R, Deplazes P, Devleesschauwer B, Innes E, Romig T, van der Giessen J, Hempen M, der Stede YV, Robertson L (2018) Public health risks associated with food-borne parasites. EFSA J 16:e05495. https://doi.org/10.2903/j.efsa.2018.5495

Law JW-F, Ab Mutalib N-S, Chan K-G, Lee L-H (2015) Rapid methods for the detection of foodborne bacterial pathogens: principles, applications, advantages and limitations. Front Microbiol 5. https://doi.org/10.3389/fmicb.2014.00770

Leistner L (2000) Basic aspects of food preservation by hurdle technology. Int J Food Microbiol 55: 181–186. https://doi.org/10.1016/S0168-1605(00)00161-6

Li H, Jin R, Kong D, Zhao X, Liu F, Yan X, Lin Y, Lu G (2019) Switchable fluorescence immunoassay using gold nanoclusters anchored cobalt oxyhydroxide composite for sensitive detection of imidacloprid. Sensors Actuators B Chem 283:207–214. https://doi.org/10.1016/j.snb.2018.12.026

Li W-K, Shi Y-P (2020) Recent advances of magnetic extractants in food analysis. TrAC Trends Anal Chem 129:115951. https://doi.org/10.1016/j.trac.2020.115951

Lin Y, Ren J, Qu X (2014) Catalytically active nanomaterials: a promising candidate for artificial enzymes. Acc Chem Res 47:1097–1105. https://doi.org/10.1021/ar400250z

Lu Y, Shi Z, Liu Q (2019) Smartphone-based biosensors for portable food evaluation. Curr Opin Food Sci Cannabis Oil Extrac Purification, Utilization – Innov Food Sci Omics Approach Food Analy Authen 28:74–81. https://doi.org/10.1016/j.cofs.2019.09.003

Luong JHT, Male KB, Glennon JD (2008) Biosensor technology: technology push versus market pull. Biotechnol Adv 26:492–500. https://doi.org/10.1016/j.biotechadv.2008.05.007

Malon RSP, Heng LY, Córcoles EP (2017) Recent developments in microfluidic paper-, cloth-, and thread-based electrochemical devices for analytical chemistry. Rev Anal Chem 36. https://doi.org/10.1515/revac-2016-0018

Manning L (2016) Food fraud: policy and food chain. Curr Opin Food Sci Innov Food Sci Foodomics Technol 10:16–21. https://doi.org/10.1016/j.cofs.2016.07.001

Manning L, Luning PA, Wallace CA (2019) The evolution and cultural framing of food safety management systems—where from and where next? Compr Rev Food Sci Food Saf 18:1770–1792. https://doi.org/10.1111/1541-4337.12484

Medina S, Perestrelo R, Silva P, Pereira JAM, Câmara JS (2019) Current trends and recent advances on food authenticity technologies and chemometric approaches. Trends Food Sci Technol 85: 163–176. https://doi.org/10.1016/j.tifs.2019.01.017

Medina-Pastor P, Triacchini G (2020) The 2018 European Union report on pesticide residues in food. EFSA J 18:e06057. https://doi.org/10.2903/j.efsa.2020.6057

Meira L, Costa J, Villa C, Ramos F, Oliveira MBPP, Mafra I (2017) EvaGreen real-time PCR to determine horse meat adulteration in processed foods. LWT 75:408–416. https://doi.org/10.1016/j.lwt.2016.08.061

Merkoçi A, Pumera M, Llopis X, Pérez B, del Valle M, Alegret S (2005) New materials for electrochemical sensing VI: carbon nanotubes. TrAC Trends Anal Chem 24:826–838. https://doi.org/10.1016/j.trac.2005.03.019

Misra NN, Pankaj SK, Frias JM, Keener KM, Cullen PJ (2015) The effects of nonthermal plasma on chemical quality of strawberries. Postharvest Biol Technol 110:197–202. https://doi.org/10.1016/j.postharvbio.2015.08.023

Moore CE (2019) Changes in antibiotic resistance in animals. Science 365:1251–1252. https://doi.org/10.1126/science.aay9652

Mustafa F, Andreescu S (2020) Nanotechnology-based approaches for food sensing and packaging applications. RSC Adv 10:19309–19336. https://doi.org/10.1039/D0RA01084G

Nardi VAM, Auler DP, Teixeira R (2020) Food safety in global supply chains: a literature review. J Food Sci 85:883–891. https://doi.org/10.1111/1750-3841.14999

Neethirajan S, Ragavan V, Weng X, Chand R (2018) Biosensors for sustainable food engineering: challenges and perspectives. Biosensors 8:23. https://doi.org/10.3390/bios8010023

Nelis JLD, Tsagkaris AS, Dillon MJ, Hajslova J, Elliott CT (2020) Smartphone-based optical assays in the food safety field. TrAC Trends Anal Chem 129:115934. https://doi.org/10.1016/j.trac.2020.115934

Niemira BA (2012) Cold plasma decontamination of foods. Annu Rev Food Sci Technol 3:125–142. https://doi.org/10.1146/annurev-food-022811-101132

OCED/FAO 2020. OECD agriculture statistics

O'Mahony PJ (2013) Finding horse meat in beef products—a global problem. QJM Int J Med 106:595–597. https://doi.org/10.1093/qjmed/hct087

Ouyang Q, Zhao J, Chen Q (2014) Instrumental intelligent test of food sensory quality as mimic of human panel test combining multiple cross-perception sensors and data fusion. Anal Chim Acta 841:68–76. https://doi.org/10.1016/j.aca.2014.06.001

Painter JA, Hoekstra RM, Ayers T, Tauxe RV, Braden CR, Angulo FJ, Griffin PM (2013) Attribution of foodborne illnesses, hospitalizations, and deaths to food commodities by using outbreak data, United States, 1998–2008. Emerg Infect Dis 19:407–415. https://doi.org/10.3201/eid1903.111866

Pan M, Xie X, Liu K, Yang J, Hong L, Wang S (2020) Fluorescent carbon quantum dots—synthesis, functionalization and sensing application in food analysis. Nano 10:930. https://doi.org/10.3390/nano10050930

Parate K, Pola CC, Rangnekar SV, Mendivelso-Perez DL, Smith EA, Hersam MC, Gomes CL, Claussen JC (2020) Aerosol-jet-printed graphene electrochemical histamine sensors for food safety monitoring, 034002. 2D Mater 7. https://doi.org/10.1088/2053-1583/ab8919

Pedrero M, Campuzano S, Pingarrón JM (2017) Quantum dots as components of electrochemical sensing platforms for the detection of environmental and food pollutants: a review. J AOAC Int 100:950–961. https://doi.org/10.5740/jaoacint.17-0169

Peleg M (2020) The hurdle technology metaphor revisited. Food Eng Rev 12:309–320. https://doi.org/10.1007/s12393-020-09218-z

Peltomaa R, Benito-Peña E, Gorris HH, Moreno-Bondi MC (2021) Biosensing based on upconversion nanoparticles for food quality and safety applications. Analyst 146:13–32. https://doi.org/10.1039/D0AN01883J

Pilařová V, Plachká K, Khalikova MA, Svec F, Nováková L (2019) Recent developments in supercritical fluid chromatography – mass spectrometry: is it a viable option for analysis of complex samples? TrAC Trends Anal Chem 112:212–225. https://doi.org/10.1016/j.trac.2018.12.023

Poltronieri P, Mezzolla V, Primiceri E, Maruccio G (2014) Biosensors for the detection of food pathogens. Foods 3:511–526. https://doi.org/10.3390/foods3030511

Purohit B, Kumar A, Mahato K, Chandra P (2020) Smartphone-assisted personalized diagnostic devices and wearable sensors. Curr Opin Biomed Eng Future Biomed Eng Bioeng Organoids Tissue Develop Biomaterial Biosensors 13:42–50. https://doi.org/10.1016/j.cobme.2019.08.015

Rady A, Adedeji A (2018) Assessing different processed meats for adulterants using visible-near-infrared spectroscopy. Meat Sci 136:59–67. https://doi.org/10.1016/j.meatsci.2017.10.014

Ragavan KV, Ahmed SR, Weng X, Neethirajan S (2018a) Chitosan as a peroxidase mimic: paper based sensor for the detection of hydrogen peroxide. Sensors Actuators B Chem 272:8–13. https://doi.org/10.1016/j.snb.2018.05.142

Ragavan KV, Egan P, Neethirajan S (2018b) Multi mimetic graphene palladium nanocomposite based colorimetric paper sensor for the detection of neurotransmitters. Sensors Actuators B Chem 273:1385–1394. https://doi.org/10.1016/j.snb.2018.07.048

Ragavan KV, Neethirajan S (2019) Chapter 7 – Nanoparticles as biosensors for food quality and safety assessment. In: López Rubio A, Fabra Rovira MJ, Martínez Sanz M, Gómez-Mascaraque LG (eds) Nanomaterials for food applications, Micro and Nano technologies. Elsevier, pp 147–202. https://doi.org/10.1016/B978-0-12-814130-4.00007-5

Ragavan KV, Rastogi N, Srivastava A (2016) Industrial food processing contaminants. In: Debasis B, Anand S, Stohs S (eds) Food toxicology. CRC Press, Boca Raton, FL, pp 395–432. https://doi.org/10.1201/9781315371443-22

Ragavan KV, Rastogi NK (2017) β-Cyclodextrin capped graphene-magnetite nanocomposite for selective adsorption of bisphenol-a. Carbohydr Polym 168:129–137. https://doi.org/10.1016/j.carbpol.2017.03.045

Ragavan KV, Rastogi NK (2016) Graphene–copper oxide nanocomposite with intrinsic peroxidase activity for enhancement of chemiluminescence signals and its application for detection of bisphenol-a. Sensors Actuators B Chem 229:570–580. https://doi.org/10.1016/j.snb.2016.02.017

Ragavan KV, Rastogi NK, Thakur MS (2013a) Sensors and biosensors for analysis of bisphenol-a. TrAC Trend Anal Chem Modern Food Anal Foodomics 52:248–260. https://doi.org/10.1016/j.trac.2013.09.006

Ragavan KV, Selvakumar LS, Thakur MS (2013b) Functionalized aptamers as nano-bioprobes for ultrasensitive detection of bisphenol-a. Chem Commun 49:5960–5962. https://doi.org/10.1039/C3CC42002G

Rajashri K, Rastogi NK, Negi PS (2020) Non- thermal processing of tender coconut water – a review. Food Rev Intl 38:1–22. https://doi.org/10.1080/87559129.2020.1847142

Rastogi NK (2013) Recent developments in high pressure processing of foods, SpringerBriefs in food, health, and nutrition. Springer, US. https://doi.org/10.1007/978-1-4614-7055-7

Rateni G, Dario P, Cavallo F (2017) Smartphone-based food diagnostic technologies: a review. Sensors 17:1453. https://doi.org/10.3390/s17061453

Rhouati A, Bakas I, Marty JL (2019) MIPs and aptamers as artificial receptors in advanced separation techniques. In: handbook of smart materials in analytical chemistry. Wiley, pp 825–857. https://doi.org/10.1002/9781119422587.ch26

Robertson LJ, Sprong H, Ortega YR, van der Giessen JWB, Fayer R (2014) Impacts of globalisation on foodborne parasites. Trends Parasitol 30:37–52. https://doi.org/10.1016/j.pt.2013.09.005

Roduner E (2006) Size matters: why nanomaterials are different. Chem Soc Rev 35:583–592. https://doi.org/10.1039/B502142C

Romao VC, Martins SAM, Germano J, Cardoso FA, Cardoso S, Freitas PP (2017) Lab-on-Chip devices: gaining ground losing size. ACS Nano 11:10659–10664. https://doi.org/10.1021/acsnano.7b06703

Romero-Arcos M, Garnica-Romo MG, Martinez-Flores HE, Vázquez-Marrufo G, Ramírez-Bon R, González-Hernández J, Barbosa-Cánovas GV (2016) Enzyme immobilization by Amperometric biosensors with TiO2 nanoparticles used to detect phenol compounds. Food Eng Rev 8:235–250. https://doi.org/10.1007/s12393-015-9129-8

Sanaeifar A, ZakiDizaji H, Jafari A, de la Guardia M (2017) Early detection of contamination and defect in foodstuffs by electronic nose: a review. TrAC Trends Anal Chem 97:257–271. https://doi.org/10.1016/j.trac.2017.09.014

Selvakumar LS, Ragavan KV, Abhijith KS, Thakur MS (2013) Immunodipstick based gold nanosensor for vitamin B12 in fruit and energy drinks. Anal Methods 5:1806–1810. https://doi.org/10.1039/C3AY26320G

Seo SE, Tabei F, Park SJ, Askarian B, Kim KH, Moallem G, Chong JW, Kwon OS (2019) Smartphone with optical, physical, and electrochemical nanobiosensors. J Ind Eng Chem 77: 1–11. https://doi.org/10.1016/j.jiec.2019.04.037

Sharma R, Ragavan KV, Thakur MS, Raghavarao KSMS (2015) Recent advances in nanoparticle based aptasensors for food contaminants. Biosens Bioelectron 74:612–627. https://doi.org/10.1016/j.bios.2015.07.017

Sharpe E, Frasco T, Andreescu D, Andreescu S (2013) Portable ceria nanoparticle -based assay for rapid detection of food antioxidants (NanoCerac). Analyst 138:249–262. https://doi.org/10.1039/C2AN36205H

Shavanova K, Bakakina Y, Burkova I, Shtepliuk I, Viter R, Ubelis A, Beni V, Starodub N, Yakimova R, Khranovskyy V (2016) Application of 2D non-graphene materials and 2D oxide nanostructures for biosensing technology. Sensors 16:223. https://doi.org/10.3390/s16020223

Shetti NP, Bukkitgar SD, Reddy KR, Reddy CV, Aminabhavi TM (2019) Nanostructured titanium oxide hybrids-based electrochemical biosensors for healthcare applications. Colloids Surf B: Biointerfaces 178:385–394. https://doi.org/10.1016/j.colsurfb.2019.03.013

Singh S, Shalini R (2016) Effect of hurdle Technology in Food Preservation: a review. Crit Rev Food Sci Nutr 56:641–649. https://doi.org/10.1080/10408398.2012.761594

Sivakumar R, Lee NY (2021) Recent progress in smartphone-based techniques for food safety and the detection of heavy metal ions in environmental water. Chemosphere 275:130096. https://doi.org/10.1016/j.chemosphere.2021.130096

Son M, Lee JY, Ko HJ, Park TH (2017) Bioelectronic nose: an emerging tool for odor standardization. Trends Biotechnol 35:301–307. https://doi.org/10.1016/j.tibtech.2016.12.007

Taouri L, Bourouina M, Bourouina-Bacha S, Hauchard D (2021) Fullerene-MWCNT nanostructured-based electrochemical sensor for the detection of Vanillin as food additive J Food Compos Anal 103811. https://doi.org/10.1016/j.jfca.2021.103811

Tappi S, Berardinelli A, Ragni L, Dalla Rosa M, Guarnieri A, Rocculi P (2014) Atmospheric gas plasma treatment of fresh-cut apples. Innovative Food Sci Emerg Technol 21:114–122. https://doi.org/10.1016/j.ifset.2013.09.012

Thakur MS, Ragavan KV (2013) Biosensors in food processing. J Food Sci Technol 50:625–641. https://doi.org/10.1007/s13197-012-0783-z

Tibola CS, da Silva SA, Dossa AA, Patrício DI (2018) Economically motivated food fraud and adulteration in Brazil: incidents and alternatives to minimize occurrence. J Food Sci 83:2028–2038. https://doi.org/10.1111/1750-3841.14279

Torgerson PR, Devleesschauwer B, Praet N, Speybroeck N, Willingham AL, Kasuga F, Rokni MB, Zhou X-N, Fèvre EM, Sripa B, Gargouri N, Fürst T, Budke CM, Carabin H, Kirk MD, Angulo FJ, Havelaar A, de Silva N (2015) World Health Organization estimates of the global and regional disease burden of 11 foodborne parasitic diseases, 2010: a data synthesis. PLoS Med 12:e1001920. https://doi.org/10.1371/journal.pmed.1001920

Ulberth F (2020) Tools to combat food fraud – a gap analysis. Food Chem 330:127044. https://doi.org/10.1016/j.foodchem.2020.127044

Vaidya AM, Annapure US (2019) Chapter 38 – Enzymes in biosensors for food quality assessment. In: Kuddus M (ed) Enzymes in food biotechnology. Academic Press, pp 659–674. https://doi.org/10.1016/B978-0-12-813280-7.00038-4

van Ruth SM, Huisman W, Luning PA (2017) Food fraud vulnerability and its key factors. Trends Food Sci Technol 67:70–75. https://doi.org/10.1016/j.tifs.2017.06.017

Walker MJ, Burns M, Burns DT (2013) Horse meat in beef products- species substitution 2013. J Assoc Publ Analysts 41:67–106

Wen S, Zhou J, Zheng K, Bednarkiewicz A, Liu X, Jin D (2018) Advances in highly doped upconversion nanoparticles. Nat Commun 9:2415. https://doi.org/10.1038/s41467-018-04813-5

Weng X, Neethirajan S (2017) Ensuring food safety: quality monitoring using microfluidics. Trends Food Sci Technol 65:10–22. https://doi.org/10.1016/j.tifs.2017.04.015

WHO (2018) Salmonella (non-typhoidal) [WWW Document]. URL https://www.who.int/news-room/fact-sheets/detail/salmonella-(non-typhoidal) (accessed 4.18.21)

Wu C, Du L, Zou L, Zhao L, Huang L, Wang P (2014) Recent advances in taste cell- and receptor-based biosensors. Sensors Actuators B Chem 201:75–85. https://doi.org/10.1016/j.snb.2014.04.021

Wu H-L, Wang T, Yu R-Q (2020) Recent advances in chemical multi-way calibration with second-order or higher-order advantages: multilinear models, algorithms, related issues and applications. TrAC Trends Anal Chem 130:115954. https://doi.org/10.1016/j.trac.2020.115954

Wu MY-C, Hsu M-Y, Chen S-J, Hwang D-K, Yen T-H, Cheng C-M (2017) Point-of-care detection devices for food safety monitoring: proactive disease prevention. Trends Biotechnol 35:288–300. https://doi.org/10.1016/j.tibtech.2016.12.005

Yakes BJ, Buijs J, Elliott CT, Campbell K (2016) Surface plasmon resonance biosensing: approaches for screening and characterising antibodies for food diagnostics. Talanta 156–157: 55–63. https://doi.org/10.1016/j.talanta.2016.05.008

Yang X, Ouyang Y, Wu F, Hu Y, Ji Y, Wu Z (2017) Size controllable preparation of gold nanoparticles loading on graphene sheets@cerium oxide nanocomposites modified gold electrode for nonenzymatic hydrogen peroxide detection. Sensors Actuators B Chem 238:40–47. https://doi.org/10.1016/j.snb.2016.07.016

Yao Y, Xue Y (2015) Impedance analysis of quartz crystal microbalance humidity sensors based on nanodiamond/graphene oxide nanocomposite film. Sensors Actuators B Chem 211:52–58. https://doi.org/10.1016/j.snb.2014.12.134

Ye Y, Guo H, Sun X (2019) Recent progress on cell-based biosensors for analysis of food safety and quality control. Biosens Bioelectron 126:389–404. https://doi.org/10.1016/j.bios.2018.10.039

Yu Y-Y, Wang J-X, Si R-W, Yang Y, Zhang C-L, Yong Y-C (2017) Sensitive amperometric detection of riboflavin with a whole-cell electrochemical sensor. Anal Chim Acta 985:148–154. https://doi.org/10.1016/j.aca.2017.06.053

Zeaki N, Johler S, Skandamis PN, Schelin J (2019) The role of regulatory mechanisms and environmental parameters in staphylococcal food poisoning and resulting challenges to risk assessment. Front Microbiol 10. https://doi.org/10.3389/fmicb.2019.01307

Zhang Z, Wang C, Zhang L, Meng Q, Zhang Y, Sun F, Xu Y (2017) Fast detection of Escherichia coli in food using nanoprobe and ATP bioluminescence technology. Anal Methods 9:5378–5387. https://doi.org/10.1039/C7AY01607G

Zhang Z-H, Wang L-H, Zeng X-A, Han Z, Brennan CS (2019) Non-thermal technologies and its current and future application in the food industry: a review. Int J Food Sci Technol 54:1–13. https://doi.org/10.1111/ijfs.13903

Zheng P, Wu N (2017) Fluorescence and sensing applications of graphene oxide and graphene quantum dots: a review. Chem Asian J 12:2343–2353. https://doi.org/10.1002/asia.201700814

Ziuzina D, Patil S, Cullen PJ, Keener KM, Bourke P (2014) Atmospheric cold plasma inactivation of Escherichia coli, salmonella enterica serovar typhimurium and listeria monocytogenes inoculated on fresh produce. Food Microbiol 42:109–116. https://doi.org/10.1016/j.fm.2014.02.007

Chapter 12
Spectroscopy Based In-Line Monitoring and Control of Food Quality and Safety

Praveena Bhatt, Sarma Mutturi, and M. S. Thakur

1 Introduction

The food industry involves complex interrelated processes that must be monitored and controlled to ensure consistent high quality and undisputable safety of the product being manufactured. Traditionally, most of the quality and safety parameters are monitored off-line, which are time-consuming, require skilled labour, involve several intermediary steps and are subjected to misinterpretation. With the industry 4.0 and smart manufacturing movement, the food industry today has huge opportunities to upgrade its processes to align itself to the latest industrial revolution (Udugama et al. 2020; Yadav et al. 2022). This implies that with advancement in technology, the food industry could adopt "in-line" and "on-line" systems to monitor the performance of its processes, rapidly identify defects or faults if any, check for quality and ensure safety of the product, practically "real-time" (Gargalo et al. 2020). This also assumes importance in the background of increasing need and demand of consumers for safe, hygienic, properly labelled food as well as stricter laws and regulatory requirements for safe and high-quality product (Hassoun et al. 2020).

The technical definitions of these monitoring and control systems adopted by the industry in food quality and safety is described below (Claβen et al. 2017):

P. Bhatt (✉) · S. Mutturi
Microbiology & Fermentation Technology Department, CSIR-Central Food Technological Research Institute, Mysore, India

Academy of Scientific and Innovative Research (AcSIR), Ghaziabad, UP, India
e-mail: praveena@cftri.res.in

M. S. Thakur
Department of Chemistry, Centre for Nanomaterials and MEMS, NITTE Meenakshi Institute of Technology, Bangalore, India

H. U. Hebbar et al. (eds.), *Engineering Aspects of Food Quality and Safety*, Food Engineering Series, https://doi.org/10.1007/978-3-031-30683-9_12

- **"In-line"** refers to a measurement that is made using a device or sensor which measures from the process line without sample removal or diversion. Measurements are made on a continuous basis "real-time".
- **"On-line"** refers to measurement where a sample from the process line is diverted by a bypass, immediately followed by its analysis. Measurements are made on a continuous basis "real-time".
- **"At-line"** refers to usage of a device near the process line which separates the material from the sampling point, followed by its conditioning such as filtration, separation, addition of reagents etc., and then analysis. Measurements are not made on a continuous basis and data generated depends on the frequency of analysis over pre-determined time intervals.
- **"Off-line"** refers to analysis of sample that is withdrawn from the process line and is analysed in a laboratory or centralized facility.

Figure 12.1 is a schematic diagram of monitoring systems used in process analysis. It may be noted; since the difference between "on-line" and "in-line" sensors is very narrow, the terms have often been used interchangeably in literature. In the chapter, we have included "in-line" as those technologies which have sensor probes, device

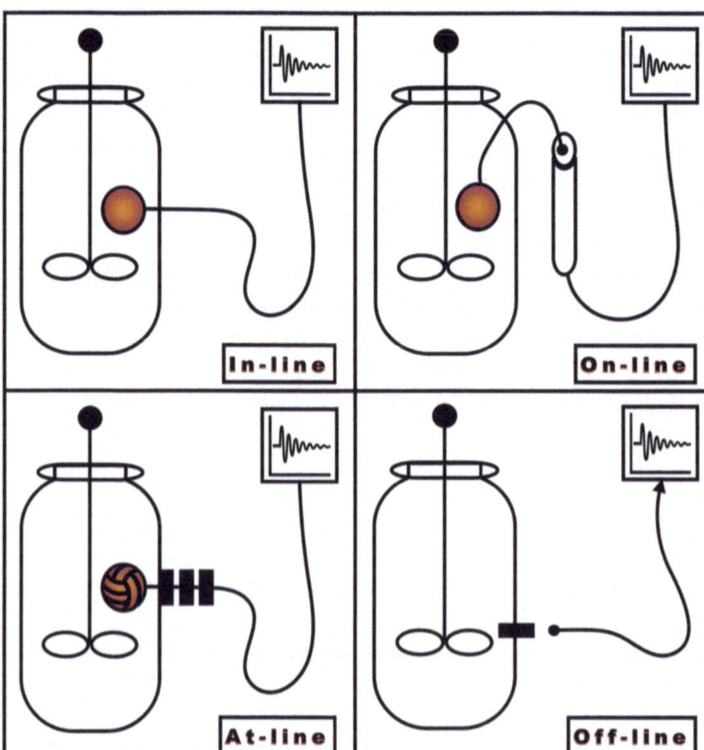

Fig. 12.1 Various modes of process analysis (Adapted from Gargalo et al. 2020)

or instrument that is integrated with the process operation as part of the process flow, with or without conditioning, in a continuous manner, without sample preparation (unlike atline or offline) and which give results of parameters without the requirement to stop or discontinue the process. Emphasis is therefore on non-invasive or non-destructive and rapid techniques that give the results "real time".

By convention, the most common sensors used in the industry for the purpose of process monitoring include those that measure physical parameters such as the temperature, pressure, level and flow sensors (flowmeters) (Patel and Doddamani 2019). In this chapter, our focus would be on sensors and devices which are used to monitor physical, chemical and biological parameters of food products in terms of their quality and safety rather than on physical and mechanical process parameters such as pressure, temperature, etc. Table 12.1 illustrates the different parameters that can be measured in an industrial set up as inline operations. These parameters are some examples and in no way exhaustive and there are many others which are being applied/explored for measurement by the food industry.

Table 12.1 Possible parameters that can be measured "inline" for food quality and safety determination

	Parameters that may be applied for inline operations		
	Physical	Chemical	Biological
Food quality	Colour (Kamruzzaman et al. 2016)	Glucose (Craven et al. 2014)	Biomass (Abu-Absi et al. 2014)
	Size (Tibayrenc et al. 2010)	Lactic acid (Mehdizadeh et al. 2015)	Enzymes (Moretto et al. 2011)
	Shape (Camisard et al. 2002)	Fatty acids (Iversen et al. 2014)	
	Texture (Bocker et al. 2007)	Fat (Osborne 2006)	
	Rheology (Ozbekova and Kulmyrzaev 2017)	Protein (Osborne 2006)	
	Firmness (Ozbekova and Kulmyrzaev 2017)	Moisture (Osborne 2006)	
	Elasticity	Volatiles (Li et al. 2013)	
	Freshness (Lohumi et al. 2015)	Nutritional value	
		Authentic labelling (Hassoun et al. 2019)	
Food safety		Allergens (Poms et al. 2004)	Cell density/load (Argyri et al. 2013)
		Pesticides	Pathogens
		Heavy metals	Faecal contamination (Park et al. 2011)
		Toxins (Tripathi and Mishra 2009)	
		Specific adulterants (Alamprese et al. 2013)	

1.1 Pre-requisites and Desirable Quality of In-Line Sensors

1.1.1 Requisites

- Suitable design for easy process integration
- Rapid and accurate measurement in real-time with fast data processing and analysis
- No requirement of sample preparation
- Automatic data acquisition capability
- Robust with ability to collect proper representative, reliable and reproducible data amidst challenging industrial environments such as temperature fluctuations, sample movement, sample inhomogeneity, presentation etc.
- Compact, with minimum requirement of space
- Ability to withstand harsh process environment, such as high or low temperatures, vibrations, dust, humidity, etc.
- Easy to maintain with features such as explosion-proof, waterproof, easy to clean etc.
- Should not cause any disruption to the production process.
- Sensors having any contact with food material must be food-safe, inert and not affected by any chemical or physical changes in the process

1.1.2 Desirable

- Enabled by remote control via fibre-optic probes or ethernet
- User friendly with no requirement of trained or skilled personnel during operation
- Cost effective with requirement of minimum or low investment by the industry
- Easily adaptable to harsh industrial set up

2 Role of In-Line Sensors for Monitoring Quality and Safety in Food Industry

Food samples need to be measured for their physical, chemical and nutritive aspects to produce quality and safe product in order to meet consumer satisfaction and also regulatory requirements. Any food product may have limited shelf life or low quality due to a number of factors including poor quality of raw material, low or poor process control, incorrect method of packing, transportation, handling, time-limited supply chains, type and method of storage, etc., (Alander et al. 2013). An early, easy and rapid analysis of products at different stages of its production is extremely desirable (and in many cases necessary) to ensure freshness, higher yield, safety, consistency etc., which ultimately dictates economic profit for an industry. Elimination of low-quality material, defective items, un-conforming products, contaminated products, possible hazardous operations, etc., at early stages, particularly with non-invasive techniques which do not require to disrupt the process, do not require sample preparation and yet have the ability to give fast and accurate results, have shown to add tremendous value to the food industry (Dietzsch et al. 2013). In this context, spectroscopic techniques are the most suitable for such "inline" monitoring and measurement. This chapter thus, focuses on spectroscopic techniques for inline monitoring and control of food operations in the industry.

3 Spectroscopy Based In-Line Sensors and Monitoring Systems

The most common and promising in-line sensors or devices for monitoring food processes is based on spectroscopy. The biggest advantage of using spectroscopic techniques is that, several chemical, physical, and biological species of interest, relevant to quality and safety of a product/process can be measured over a wide range of electromagnetic spectrum of light which ranges from near infra-red, mid-infra-red, visible and UV range, to low frequency radio waves and high frequency γ-rays (Abu-Absi et al. 2014). For any process, the characteristic emission and/or absorption spectra are observed for the sample or selected molecule or compounds in the sample, providing valuable and required information on the quality or safety parameter being analysed. Spectroscopy based techniques can be non-invasive, extremely rapid, reliable and non-laborious. It is precisely due to this reason, that several food industries have adopted spectroscopic analysis of their processes and products to monitor as well as control food quality and safety. This technique will continue to be the method of choice for PAT initiative for Industry 4.0 (Eifert et al. 2020). Advancement in instrumentation, computation and data analysis through machine learning has made spectroscopic techniques, the method of choice for inline monitoring of industrial processes.

Spectroscopic techniques are based on the interaction between matter and electromagnetic radiation. Atoms contain electrons that exist at discrete energy levels which correspond to their resonant vibrational frequencies. Electrons can absorb radiation get excited to higher energy state and emit energy as they come back to their ground state. The major spectroscopic techniques used in inline sensors include those that are based on reflectance, transmittance and interactance (Fig. 12.2). These modes depend on the position of the illumination source and the detector. When the illumination source and the detector are above the sample, and light reflected from the sample is captured, it is referred to as reflectance or diffused reflectance, when light source and detector are placed opposite to each other, the light that is transmitted through the sample is captured, the mode is called transmittance, when the illumination source and the detector are placed parallel to each other, it is referred as interactance (Hassoun et al. 2020).

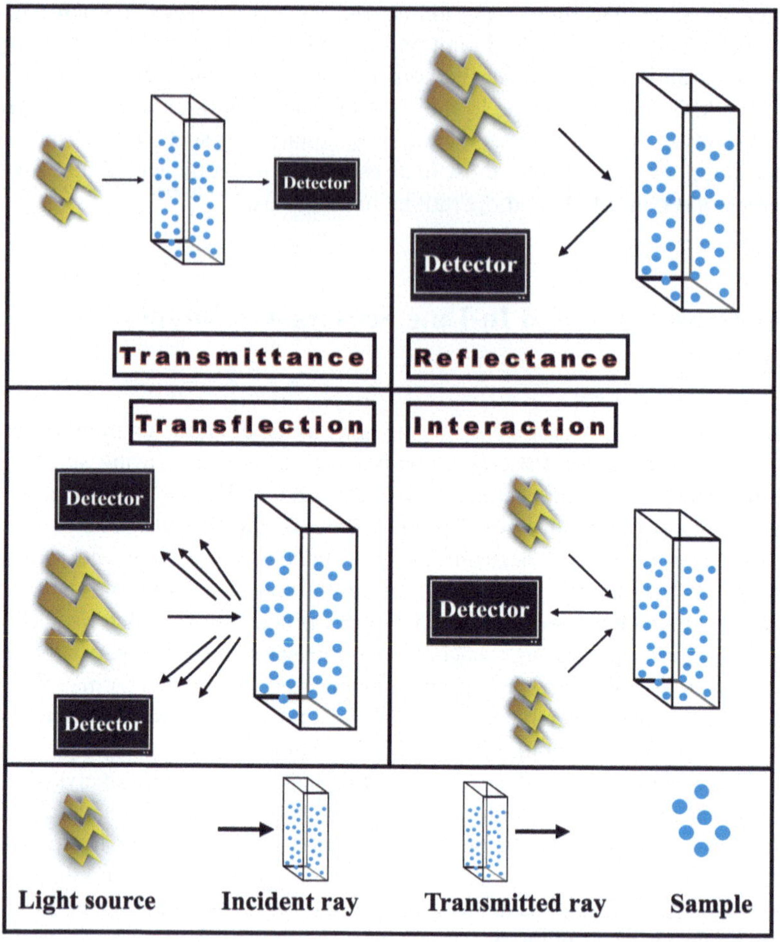

Fig. 12.2 The different modes of spectroscopic analysis based on position of illumination source and detector. (Adapted from Kang 2011)

All spectroscopic techniques follow the basic steps described below:

(a) Optimization of measurement conditions based on the complexity of the process, type of components being analysed, nature of sample matrix, suitability of type of technology and the instrument that needs to be used.
(b) Appropriate and accurate calibrations between the analyte of interest and the spectra collected from the system followed by generation and selection of an appropriate dataset that will effectively capture the variations and complexity of the system.
(c) Application of multivariate chemometric methods to develop models, transforming the measured spectra into useful information.

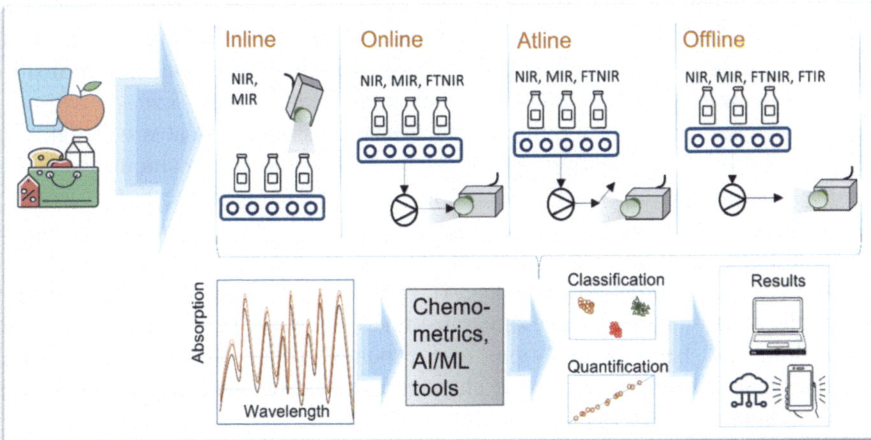

Fig. 12.3 Flowsheet for analysis of quality and safety of food matrix using IR- based tools in supply chain

(d) Validation of the developed models with conventional techniques to ensure reliable and repeatable measurement of the samples.

In the following sections of the chapter, we discuss the principles of four spectroscopic techniques which have been more widely used in the food industry for the evaluation of processes or products for process monitoring and control. They are: (a) Infrared Spectroscopy: which include near infra-red (NIRS) and mid-infra-red spectroscopy (MIRS) (b) Fluorescence Spectroscopy (FS) (c) Raman spectroscopy (RS) and (d) Dielectric spectroscopy (DS). Spectroscopic methods can provide both qualitative and quantitative identification of a chemical or biological species, since the wavelengths of the absorbed and/or emitted radiation are chemical specific while the intensity of the radiation depends on the concentration of the species. Figure 12.3 gives the flow diagram for the common vibrational techniques used for sample analysis.

The following section discusses, in brief, the principles of spectroscopic techniques used in inline sensors.

3.1 *Infrared Spectroscopy*

3.1.1 Near-Infrared Spectroscopy (NIRS)

Near infra-red spectroscopy (NIRS) is the most applied technique and has been extensively used in the food industry at various stages of food processing. It has been used in pre-harvest steps (for example, to evaluate quality of raw produce) to postharvest assessment of final processed product. The greatest advantage of NIR spectroscopy in inline industrial applications is their ability to provide non-invasive, rapid, and accurate results with no sample preparation, ease of instrumentation as well as multiple parameter measurement in a single scan.

NIRS is based on the absorption of electromagnetic radiation at wavelengths in the range 780–2500 nm. During food analysis, the vibrational transitions characterized by low energy values are reflected in the NIR region of the light spectrum (Herold et al. 2009). In NIRS, the determination of molecules or any chemical species in food is based on the chemical bonds of the organic constituents present in it. The most common bonds include C-H, N-H, O-H, and S-H. When light falls on the sample, electromagnetic waves are transmitted, and wave behaviour changes due to stretching and bending vibrations of the bonds. These observed changes are captured by spectroscopy to provide characteristic and detailed fingerprints of the samples (Huang et al. 2014).

The general procedure to develop a NIR based analysis of target of interest involves the following steps (Wang 2019)

(a) NIR spectra of samples is acquired and their chemical profile along with variances is analysed
(b) This is followed by chemical composition analysis using a standard detection method
(c) A prediction model is constructed, and unknown samples are analysed using chemometrics.

The NIR spectral data is represented as “reflection”, “transflection”, “transmission” or “interaction” (Huang et al. 2008). The most common measurement modes used in inline applications based on NIR are diffuse transmittance and diffuse reflectance. It is imperative that NIR spectra acquisition must be followed with data pre-processing or treatment because the spectral data is usually characterized by several overlaps and strong collinearity making interpretation difficult and resulting in high noise levels and baseline drifts when food samples are analyzed (Wang and Paliwal 2007). Spectral variations may also arise from light scattering in samples, temperature fluctuations, difference in particle size, density etc., which may be frequently encountered in industrial settings. Pre-treatment methods are applied in order to cause noise reduction, enhance resolution, introduce baseline correction, data centring, normalization etc. (Porep et al. 2015). Multiplicative scatter correction (MSC) and standard variate correction (SVC) are one of the most common NIR pre-processing treatments used to correct spectral data. The data acquired thus, is then subjected to statistical and mathematical analysis which is referred to as chemometrics. This may involve non-linear techniques or linear techniques which are applied to analytical information from the spectra (Herold et al. 2009). Chemometrics will be dealt in detail in the later part of the chapter. After chemometrics, calibration models are then developed using sample sets with known concentrations of target obtained with reference methods and then validating the model with sample sets other than the calibration set. It is noteworthy that NIR spectroscopy integrated as an “in line” system can provide not only a fingerprint of

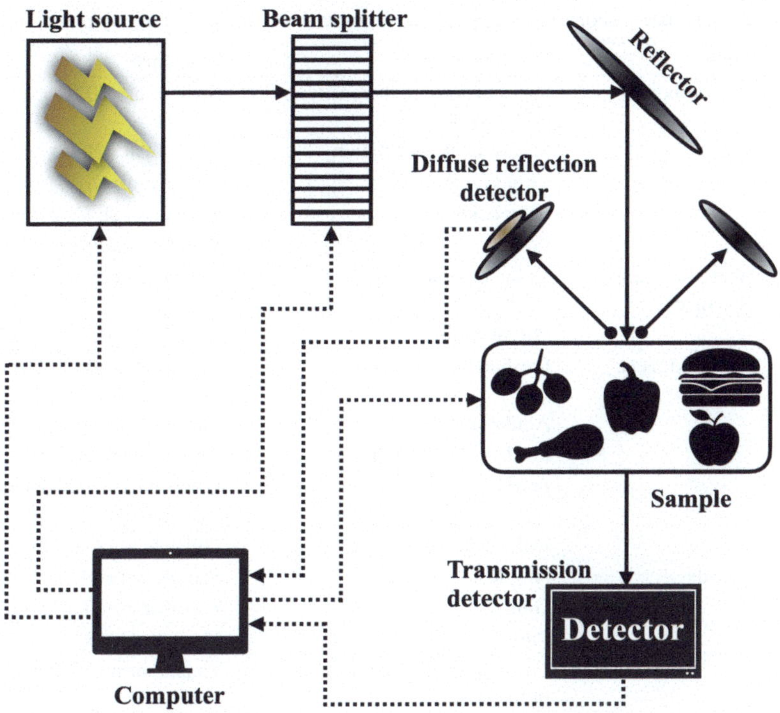

Fig. 12.4 Schematic of NIR Spectrometer (Adapted from Wang 2019)

the chemical composition but can also generate information on physical properties of the sample under investigation, which is generally not possible with several other techniques.

A typical NIR spectrometer consists of a radiation source, a monochromator, a photoelectric detector for the measurement of the intensity of detected light and conversion into electrical signals, and a computer integrated system for spectral data acquisition and processing (Herold et al. 2009). Figure 12.4 depicts the schematics of NIR spectrometer.

The earliest demonstration of NIR for practical application in inline measurement of various parameters relevant to agriculture and food products was done by Karl Norris in several of his works (William and Norris 2001). Over the years, it has been used in the non-destructive determination of many quality and safety parameters in several food types that include dairy, fruits and vegetables, fish and fish products, meat and products, oils, honey, cereal and cereal products, grains, seeds etc. Readers may refer to reviews by Lu et al. (2017) and Wang X. (2019) which give a comprehensive account of applications of NIR in food applications. Table 12.2 details some examples of use of NIR in food analysis for quality and safety assessment. Plate 12.1 below is an image of online NIR technology to measure addition of seasoning as the sample is moved across the conveyor belt. The process integrated device can also measure fat and moisture simultaneously.

Table 12.2 NIR spectroscopy-based application for monitoring quality and safety of food products

S. No	Sample	Determination	Reference
1.	Dairy products	Fat, moisture, bulk density in milk powder	Khan et al. (2021)
		Milk rennet coagulation	Strani et al. (2021)
		Cholesterol	Chitra et al. (2017)
		Fatty acids	Muncan et al. (2021a)
		Protein	Wang et al. (2019)
2.	Wine, beer and beverages	Volatile compounds	Genisheva et al. (2018)
		Polyphenols	Baca-Bocanegra et al. (2018)
3.	Meat and meat products	Freshness	Peyvasteh et al. (2020)
		Adulteration	Zheng et al. (2019)
		Fresh and thawed meat	Parastar et al. (2020)
		Fraud	P'erez-Marín and Garrido-Varo (2020)
		IMF,SFA, MUFA, PUFA	Pullanagari et al. 2015
4.	Cereals	Amylose	Sampaio et al. (2018)
		Gluten	Erkinbaev et al. (2017)
		Protein	Ye et al. (2018)
		Cooking quality, texture, pasting properties	Thanathornvarakul et al. (2016)
		Starch and protein	Izso et al. (2018)
5.	Seeds	Coffee bean quality	Santos et al. (2012)

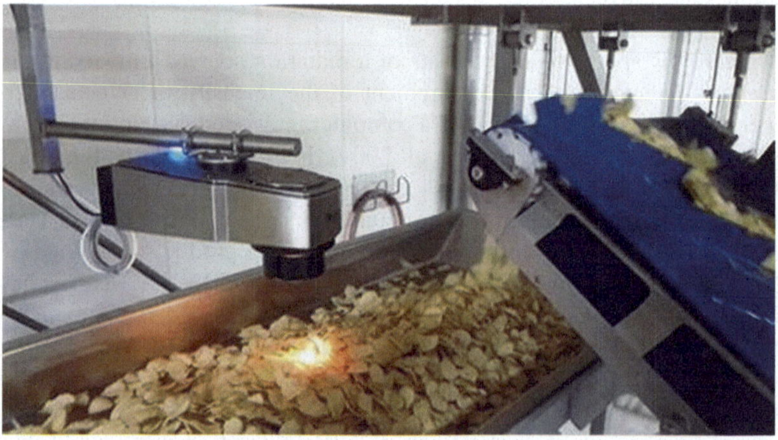

Plate 12.1 In-line NIR instrument based on diode array technology for continuous measurement of seasoning addition, moisture and fat content in snack foods; Photograph courtesy; Perkin Elmer. (https://www.perkinelmer.com/uk/libraries/app_measuring_seasoning_addition_in_snackfoods_da7440)

3.1.2 Mid-Infra-Red Spectroscopy

MIR spectroscopy is also an IR based vibrational spectroscopic technique that uses a beam of light through the sample and measures transmission and absorption of the light in the mid-infra-red region (2.5 to 25 μm). Transmission, transflection, and attenuated total reflectance (ATR) are the three main sampling methods of MIR spectroscopy. Like NIR, MIR spectroscopy recognizes organic and inorganic chemicals based on their unique absorption frequencies characteristic of their structure. Each chemical bond of a molecule has a unique vibrational energy, which indicates that each compound has a unique fingerprint which can be used to determine its structure.

The MIR technique has been successfully applied to assess the quality and the safety of food products such as adulteration of meat (Alamprese et al. 2013), protein content and protein genetic variants in milk (Bonfatti et al. 2016), sugar analysis of fruits and vegetables (Clark et al., 2018), gelatinization in cereals, adulteration of oil (Upadhyay et al. 2018), etc. However, NIRS still continues to be the preferred technique in the industry in spite of high sensitivity and chemical specificity of MIRS. This is because of the high cost of spectrophotometers and in several cases requirement of additional factors like liquid nitrogen for detector cooling or requirement of nitrogen gas atmosphere during measurements.

Readers may refer to Su and Sun (2019) to get a comprehensive review of application of MIR in liquid foods. In recent times, Fourier Transform spectroscopy (FT-MIR) and attenuated transmission reflection spectroscopy (ATR-MIR) have emerged as a promising tool in "inline" monitoring systems for the industry. They are discussed in brief below:

3.1.3 Fourier Transform Infra-Red Spectroscopy (FTIR)

There are two types of IR instruments that find application in sensing, they are: dispersive and Fourier transform (FT).

FT-IR uses an interferometer to measure all the frequencies simultaneously. The interferogram is then subjected to Fourier-transformation (a mathematical expression) where data is transformed into a spectrum. FT-IR spectroscopy is generally integrated with MIR than NIR because it works best at longer wavelengths and the chemical information derived is more specific (Abu-Absi et al. 2014). FT-IR instruments have several distinct advantages over the dispersive type such as higher throughput and accuracy. FT instruments enhance sensitivity, permit higher energy throughput, and dramatically increase the speed of spectral acquisition (Su and Sun 2019).

3.1.4 Attenuated Transmission Reflection Spectroscopy

Attenuated transmission reflectance or (ATR) works on the principle of measuring the changes that occur in a totally internally reflected infra-red beam when it comes in contact with the sample. The beam is directed to an optical crystal which has a high refractive index and is in contact with the sample. The internal reflectance creates an evanescent wave which is altered or attenuated in the regions of the IR spectrum where the sample absorbs energy. ATRS which is also generally in the MIR region, successfully overcomes the limitations of sample preparation and spectral reproducibility which are commonly encountered problems in spectroscopy.

3.1.5 Limitations of IRS

The main limitation attributed to the use of IR spectroscopy is its inability to analyse chemicals present in trace levels in samples because of the weak absorption by the target in comparison to other constituents. It is generally accepted, that it can be used to detect only those samples whose counts are more than 0.1% mass ratio. The other challenge of IRS is; the sample data is greatly influenced by other chemical constituents in the sample. IRS is also heavily dependent on statistical and mathematical tools specially to analyse variance in the chemical profile. It is generally also considered to be less precise without a sample separation process (Wang 2019). The use of IRS is also limited because of variations that arise in data as a result of the complexity of the samples in question, for example, varietal differences of plant based raw material, variations arising due to movement of samples hindering precise capture of spectra, environmental variations affecting the sample etc. Moreover, not all constituents in food are IR active and so cannot be detected. Since most food commodities contain water, the influence of water on the IR spectra is of major concern.

However, in spite of the disadvantages, IR spectroscopy continues to be the most popular technique for inline sensing.

3.2 *Dielectric Spectroscopy*

Dielectric spectroscopy (DS) also called impedance spectroscopy or electrochemical impedance spectroscopy, involves the study of a sample which has been subjected to an electric field of fixed or changing frequency. Microwave dielectric spectroscopy (MDS) has been widely popularized as a potential tool for inline monitoring systems. MDS is based on the rotation of molecules and their functional groups in the presence of an electromagnetic field in the frequency range of 0.3–300 GHz, which is then used to differentiate and fingerprint chemical composition in foods for safety and quality aspects. Literature however, seems to be limited to lab-scale alone, in application of DS for inline monitoring of desired parameter.

3.3 Florescence Spectroscopy

Florescence spectroscopy is based on the emission of radiation by molecules upon absorption of light. Molecules generally occupy the lowest vibrational level of the ground electronic state. On absorption of light, they are elevated to produce excited states. Energy, which is absorbed as discrete quanta, results in a series of distinct absorption bands. Having absorbed energy and reached one of the higher vibrational levels of an excited state, the molecule rapidly loses its excess of vibrational energy by collision and falls to the lowest vibrational level of the excited state. When the molecule returns to the vibrational levels of the ground state, it emits its energy in the form of fluorescence. Jablonski diagram depicts the fluorescence and phosphorescence emission of light as a result of electronic states of molecules and transitions between them (Fig. 12.5).

Time-integrated laser induced fluorescence spectroscopy is a sensitive technique which can be effectively used for the inline monitoring and detection of particularly surface related quality defects. In principle, the surface analysis can be divided into two important application areas: (a) analysis of functional coatings, (b) analysis of food surface. The advancement in technology which has enabled effective capture of even a single photon, offers detection of contaminants or substance of interest present even at extremely low quantities in a sample with high sensitivity by FS. This is clearly an advantage over spectroscopic techniques like NIRS and MIRS discussed earlier. However, since capture of spectral intensity distribution of fluorescence does not necessarily result in good resolution, a time-integrated approach is included in the procedure to observe the decay times of fluorescence signals in the selected wavelength range. Additionally, after excitation, the time decay of the fluorescence radiation is registered at different and appropriately

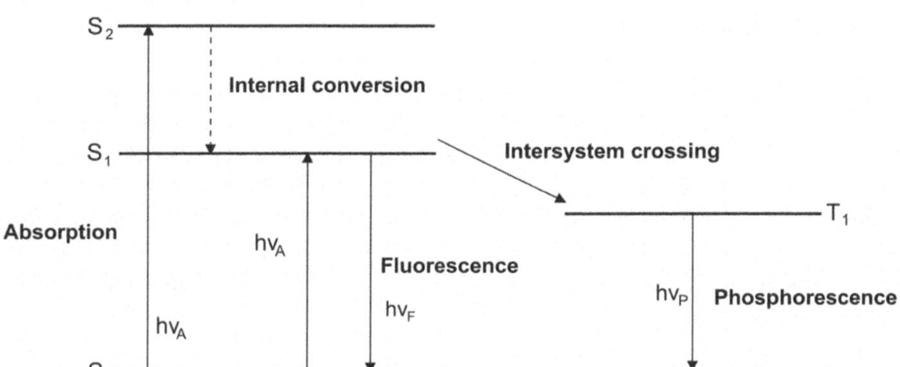

Fig. 12.5 Jablonski diagram depicting the fluorescence and phosphorescence emission due to electronic states of molecules and transition between them (Reproduced from Nawrocka and Lamorska 2013)

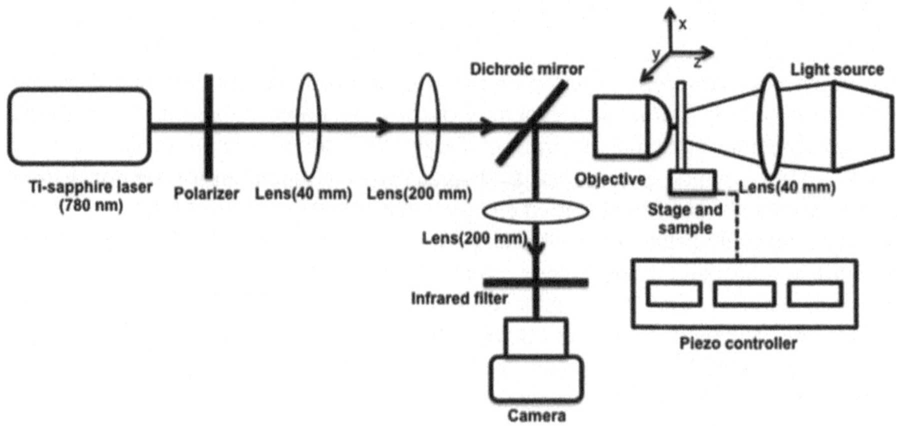

Fig. 12.6 Flow diagram of a fluorescence spectrometer (Reprinted from Kunwar et al. 2014)

positioned windows in the process operation to integrate the data. This separates the wanted signal from the background (Fig. 12.6). Fibre optic probes are used in the system which increase the depth of sample penetration and can also be cleaned easily using compressed air, gas flushing or by ultrasound techniques.

The fluorescence sensitivity is increased manifold by use of photomultiplier tube which amplifies the single-fluorescence light events. Moreover, since each single-fluorescence measurement takes place in the nanosecond timescale, the results of surface measurements on moving parts or sheets does not depend on the process speed, which is an essential requirement of inline monitoring especially in the food processing industry.

3.3.1 Limitations of FS

The major limitation of FS especially in the food industry, is the noise and interference that occur due to the complex food matrix. Moreover, most compounds have broad absorption spectra which may make it difficult to identify individual species. Although a very sensitive technique which can be applied to get useful data, fluorescence measurements are sometimes not consistent over a period of time. They also require amplification devices like the photomultiplier tube and multiple measurements at different time and locations in a system to get reliable and accurate data on the sample.

3.4 Raman Spectroscopy

Raman spectroscopy (RS) is based on Raman scattering which is inelastic scattering of radiation that produces a vibrational spectrum of sample molecules. Unlike, IR spectroscopy, Raman spectra are not subjected to large interference from polar molecules such as water which makes them superior to other vibrational spectroscopies in monitoring of liquid food samples. RS was not widely investigated as a tool for monitoring and detection of organic compounds until 1980s. With advancement in technology such as development of silicon-based detector arrays, stable and high-power laser diodes, low noise, high-resolution spectrometers etc., RS promises to be an important tool in the detection of analytes/parameters in the food industry (Collette and Williams 2002; Li et al. 2014). In addition, improved hardware and advent of advanced Raman techniques such as surface enhanced Raman spectroscopy (SERS), the limit of detection has dramatically improved, thus making RS a suitable technique for inline sensing and in monitoring of chemical and biological contaminants.

Typically, a Raman spectrometer measures the "Raman shift", which is a plot between the Raman signal intensity and shift in frequency of the Raman signal, relative to the excitation source (Fig. 12.7). In recent years, RS has been studied as a potential substitute of NIRS for application especially in high moisture foods.

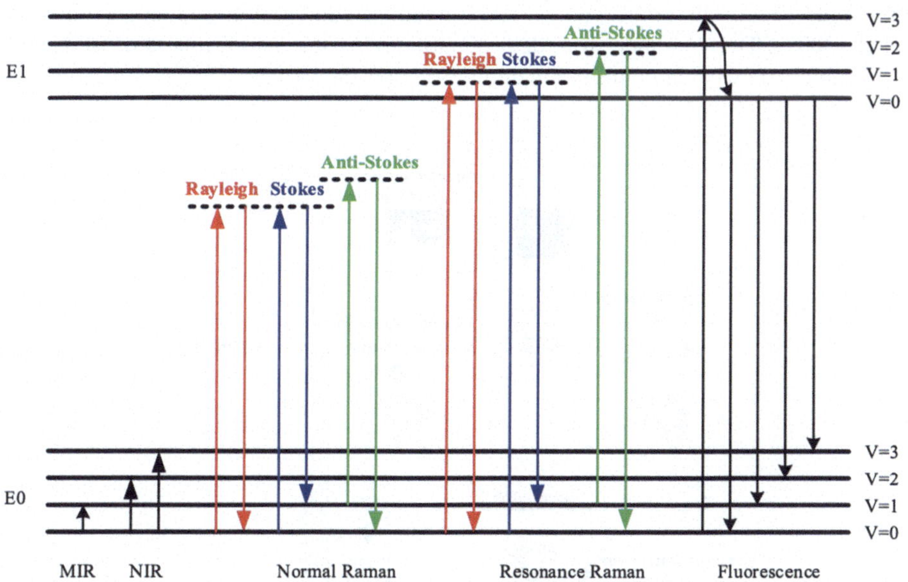

Fig. 12.7 Energy diagram for Raman scattering and fluorescence (Reproduced from Li et al. 2014)

3.5 *Hyperspectral Imaging in Spectroscopy: An Advance Technique for Inline Monitoring*

Hyperspectral imaging (HIS) is an advanced technique which is combined with optical spectroscopy to generate a two-dimensional image of an object or sample. Essentially, in HSI, each pixel of the image contains spectral information, which is added as a third dimension of values to the two-dimensional spatial image (Vo-dinh 2004). Hyperspectral data could combine absorption, fluorescence, or reflectance spectrum data for each image pixel (Lu and Fei 2014). Generally, as a thumb rule, HSI data is spectrally sampled at more than 20 equally distributed wavelengths. Spectroscopic chemical imaging such as HIS, not only increase the mass of material sampled, but also provide spatial distribution of spectral information, and have several advantages over color imaging such as RGB (red-green-blue) or spectroscopy alone. Figures 12.8 and 12.9 gives the flow diagram for HIS of sample on a conveyor belt. Many literature reports have recognized the role and application of hyperspectral imaging in food quality and safety which include detection of defects (Nagata et al. 2006), quality parameters of a sample (Qiao et al. 2007), microbial contamination (Yao et al. 2013), etc. Readers may refer to review papers by Feng and Sun (2012), Zhang et al. (2012), Zhang et al. 2017, ElMasry et al. (2012), Kamruzzaman et al. (2015), that cover the principles as well as application of hyperspectral imaging in several aspects of food quality and safety. HIS has been combined with NIR, MIR and Raman to derive valuable information about quality of a sample or aspects related to its contamination, adulteration and safety (Gowen et al., 2007). These include from determination of texture and firmness of fruits and vegetables to faecal contamination and defects on the surface.

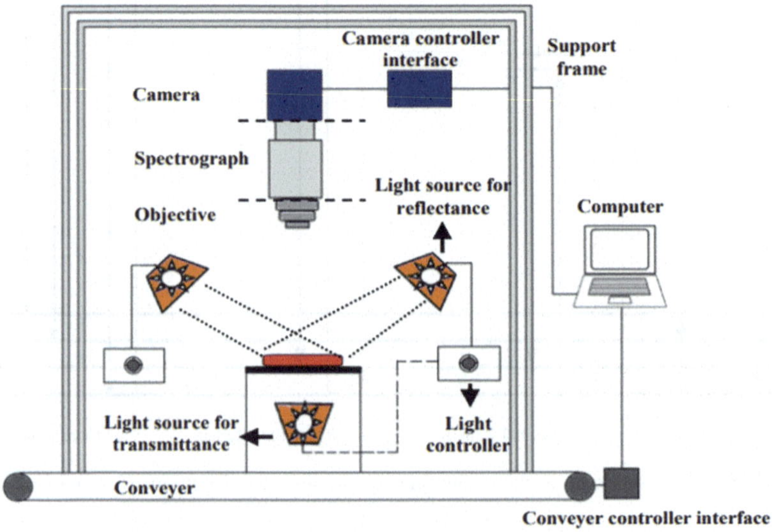

Fig. 12.8 Schematic diagram for hyperspectral imaging inline system (Reprinted from Huang et al. 2014)

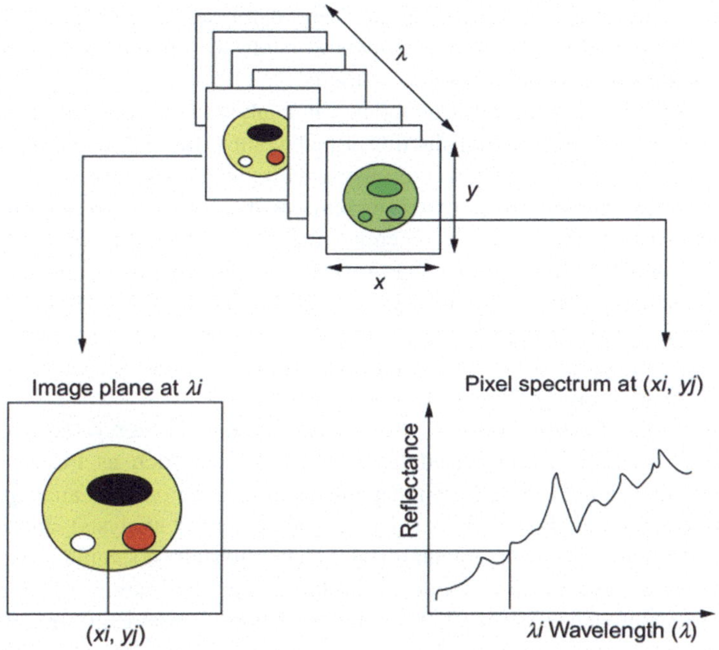

Fig. 12.9 Hyperspectral imaging hypercube showing the relationship between spectral and spatial dimensions (Reprinted from Wang 2019)

3.6 Soft Sensors

Soft sensors are advanced process monitoring systems, which use algorithms to assess measurements in an on-line manner to generate information. Spectroscopic sensors have sometimes also been described as 'soft sensors' due to the fact that spectroscopic data is modelled using software. There are two types of soft sensors– (a) data driven sensors and (b) model-driven sensors. Each one of them have their own advantages and drawbacks. Data driven sensors are based on conclusions that are derived from data which do not require previous process knowledge while model driven sensors are based on first principle approaches which can generally be extrapolated to new process conditions.

4 Chemometrics in Spectroscopic Analysis

As mentioned earlier, chemometrics is primarily applied to extract information from multivariate spectral data derived from NIR, FT-IR, Raman, and UV-VIS spectroscopy. Chemometrics can be defined as the use of mathematical and statistical methods to analyse data generated from a sample. The goal of chemometric analysis or multivariate data analysis is to either classify, calibrate or sometimes carry both these analyses using the multivariable data sets. Application of chemometrics provides qualitative and/or quantitative models to study the analyte, which otherwise cannot be understood using univariate methods.

In case of simple sensing reaction having an analyte and a receptor, a univariate regression for calibrating the transduction signal against the analyte concentration is used (e.g., the widely used glucose biosensing with glucose oxidase as bioreceptor). However, in real samples, the analyte is a component amongst different components (sometimes structurally similar), wherein mathematical multivariate regression might offer simpler solution rather than investing in the chemistry and physics for selective biosensing (Martynko and Kirsanov 2020). Traditional chemometrics and machine learning advancements are paving the way for newer detection, analysis and diagnosis in the field of sensors particularly inline monitoring and control to make the sensing more intelligent (Cui et al. 2020).

In case of classification problem, the multivariate data is segregated into known groups using specific unsupervised learning algorithms such as PCA (principal component analysis), SIMCA (self-independent modelling of class analogy), LDA (linear discriminant analysis), HCA (hierarchical clustering analysis), and PLS-DA (partial least squares-discriminant analysis). Later an untrained sample is used to test the established classification model. Similarly, the calibration of data using chemometric approach requires additionally the response variable information (measured using other analytical procedures) for training the data set to generate a linear or non-linear calibration models, using the supervised learning procedure. Here, PLS (partial least squares) is widely used as a linear calibration algorithm. For detailed understanding of PLS, Brereton (2000) and Wold et al. (2001) provides comprehensive information. With the explosion of machine learning tools in the last couple of decades, chemometrics received leverage in terms of newer algorithms for both classification and segregation problems. Tools such as kNN clustering (k-nearest neighbour), SVM (support vector machine), NB (Naïve Bayes classifiers), ensemble classifiers etc. are being used for classification problems, whereas SVR (support vector regression), ANN (artificial neural networks), CART (classification and regression tree), RF (random forest), MCR-ALS (multivariate curve resolution-alternating least squares) etc. are being used for regression analysis (Cui et al. 2020). One of the advantages these regression algorithms hold is the handling of non-linearity in data (Rentería-Gutiérrez et al. 2014). Deep learning methods such as CNN (convolutional neural network) and RNN (recurrent neural network) are also been used for robust analysis having non-analyte signal, and ability to handle non-linear models (Thrift and Ragan 2019; Uddin et al. 2020).

In general, chemometrics facilitates the data processing and thereby extracts the critical information for analysis as a tool. Merits and benefits of chemometrics and machine learning tools for biosensing applications are detailed comprehensively in the Cui et al. (2020). These include: (i) categorizing of signal data, (ii) anomaly detection and signal correction, (iii) improvement in signal to noise ratio, (iv) ability to conduct pattern recognition and object identification, and (v) rapid and sensitive detection by lowering the time of sensing. Traditional chemometrics can also be used as standalone soft sensor without requirement of any (bio) receptor. In such procedure, multivariate data arising from spectral analysis such as UV-Vis, NIR, FT-IR, GC, HPLC, DSC, LC-MS, IR-MS etc. are used for carrying chemometric estimations. Examples of such applications are many in food authentication and adulterant detection. Oliveira et al. (2019), Callao and Ruisánchez (2018), and Granato et al. (2018) provides comprehensive details on chemometrics- based food authentication and detection of adulterants.

Application of chemometrics- based sensing in different fields such as environmental monitoring, water quality assessment, food and beverages analysis, biological and medical chemistry are covered in Martynko and Kirsanov (2020). Some of the selected applications of chemometrics/machine learning in sensing is provided in Table 12.3.

Chemometrics- based data analysis, not only facilitates the sensory analysis, but also can make the process more robust, sensitive, and importantly cheaper. Batch to batch variability in the mass production of sensors can be lowered significantly if chemometrics is employed in the analytical method. Moreover, optimizing the analytical conditions for real samples is a tedious job, and chemometrics can provide a great advantage to overcome such variability in the samples. Also, with advancements in the newer tools, the chemometrics has significantly improved the analysis of spectroscopic sensing devices.

5 Selection of Spectroscopic Technique in Food Sensing

There are several factors that govern the choice of the spectroscopic technique/s in the sensing of different parameters in food when applied as an inline system. For example, NIRS is the method of choice in the determination of several physical and chemical entities in solid or dry food matrix but is usually not applied for liquid or high moisture content samples. This is because, NIR measures the absorbance of the vibrational modes of a sample and therefore large water absorbance prohibits proper measurement of the target of choice in high water content foods. In drier samples, these large absorbance bands make NIRS suitable for moisture detection. Foods with high moisture can be measured using Raman or ATR. Similarly, samples containing complex structures such as botanicals and organic materials with impurities, are preferably measured using NIRS or MIRS since RS and FS suffer from fluorescence of molecules other than the target of interest. It is also a challenge to capture Raman peaks when fluorescence is observed in the same excitation range which tends to scatter in all directions. Time resolved fluorescence is a sensitive technique but the requirement of multiple measurements at different time points and locations, restricts its use in several food applications. FS however, continues to be the suitable choice in animal and plant cell cultivation and for fermentation-based processes. Since Raman emission peaks are sharp and distinct, molecules in a mixture can be easily identified without the need for a library database for all of the sample or mixture of the sample unlike NIRS. This makes RS highly appropriate for qualitative analysis while NIRS is best suited for quantitative analysis.

Table 12.3 Chemometric application in spectroscopy-based determination of quality and safety parameters

Sl. No.	Analyte	Transduction principle	Chemometric tool	Details	Reference
1	BOD (biological oxygen demand) of waste water	Amperometry	PLS	Clark-type electrode using immobilized microorganism	Raud and Kikas (2013)
2	BOD, COD, TOC	Amperometric	PCA	Horseradish peroxidase and pure platinum were observed to be critical among eight sensors	Tønning et al. (2005)
3	Acetic acid, malonic acid, lysine, and ammonia	Colorimetry	PCA, LDA, PLS-DA, RPART, SIMCA and SVM	Colorometric sensors using different colored dyes were used to classify organic acids	Kangas et al. (2018)
4	Chlorpyruphos-oxon and malaoxon in milk	Amperometry	ANN	Flow-injection system (via AChE inhibition)	Mishra et al. (2015)
5	Captan in apple samples	Voltammetric	PCA	AChE inhibition	Nesakumar et al. (2015)
6	Glucose and polyphenols	Cyclic voltammetry	PCA PLS	Glucose oxidase or tyrosinase assisted biosensors	Medina-Plaza et al. (2015)
7	Rice syrup in honey samples	Cyclic voltammetry	PCA, LDA MLR	Electrochemical sensor	Cai et al. (2013)
8	*Escherichia coli* and *salmonella typhimurium*	Pulse voltammetry	PLS	Electrochemical sensor	Berrettoni et al. (2004)
9	Alcohol %, glucose + fructose, acetic acid, and lactic acid	UV-VIS spectral data	PLS	For monitoring the cider fermentation quality	Villar et al. (2017)
10	Green tea recognition	Fluorometry	PLS-DA	Fluorescent "turn-off" sensors based on double quantum dots	Hu et al. (2018)
11	Discrimination of monofloral honey	Colorimetric	PLS-DA SVM	Nanomaterial based colorimetric sensor array	Chaharlangi et al. (2020)
12	Discrimination of wine	Voltammetry	PCA, LDA HCA	Sensor array based on modified screen-printed carbon electrodes (SPCE)	Geană et al. (2020)

(continued)

Table 12.3 (continued)

Sl. No.	Analyte	Transduction principle	Chemometric tool	Details	Reference
13	Cumin quality	Electronic nose (olfactory)	LDA, U-PLS-DA, PARAFAC-LDA	Gas sensors	Ghasemi-Varnamkhasti et al. (2018)

It is extremely important that in-line spectroscopic instruments/sensors are able to measure required parameters of quality and safety in a dynamic environment, which is hallmark of any food industry. This may include variable environmental factors like pH and temperature, continuous movement of samples (e.g conveyor belts), continuous mixing (eg. fermentation processes), non-homogenous sample, in-flow of ingredients etc. This essentially means that any spectrometer should be designed and constructed keeping in mind the changing environment and the affect it may have on the overall result derived from the measurement. A lot of spectroscopic choices also depend on the extent of sensitivity required, type of sample, requirement and purpose of measurement, instrumentation required and ease of their integration into the process etc. As a result of cheaper instrumentation cost, robustness, availability and upgradation of computational models, NIRS has been widely used in inline monitoring.

6 In Line Monitoring of Food Quality and Safety Using Spectroscopy

The section below describes the application of spectroscopic inline sensors/devices used in various industries based on the food types. The purpose of the measurement is to contribute and describe the quality and safety of the food product. It is also necessary to mention here that spectroscopic techniques in general, have been used predominantly for qualitative analysis of samples. Majority of the publications, report the results as coefficient of correlation between developed method and conventional method for the parameter analysed by the spectroscopic method.

6.1 Dairy Products

Milk and its products are one of the most consumed food products across the world for their nutritional value. They have been evaluated for their quality parameters such as fat, protein, lactose and moisture content “inline” using NIR spectroscopy for more than 30 years in some countries (Osborne 2006). The measurement of these parameters in milk which relate to its quality, also helps to decide the further processing it requires to make products. One may refer to reviews by Kunes et al. (2021) and Pereira et al. (2020) which cover spectroscopic techniques for quality and safety assessment of milk and its products in good detail.

The protein content in milk powder-based products was analysed in a study using NIRS and values were compared to the conventional Dumas method. Results indicated that the maximum bias between the NIR method and Dumas was 3% and the developed spectroscopic method was capable of predicting the protein content (± 2%) which was present in samples in the range of 22–90% (Ingle et al. 2016). Cholesterol content was predicted using FT-NIRS coupled with partial least square (PLS) regression model in diary powders, which was claimed by the authors to have applicability as an inline monitoring tool during downstream processing of milk. Their results indicated reliable data with good comparability with the conventional HPLC method. The PLS model applied to the data was found to be satisfactory with the best performance indicators with r2 = 0.9998 and root mean square error of cross validation (RMSECV) of 1.05 mg cholesterol/100 g).

Ozbekova and Kulmyrzaev (2017), in order to predict rheological characteristics namely yield stress and flow stress, as well as chemical composition of Tilsit cheeses at melting temperatures between 20 to 70 °C used fluorescence spectroscopy. Principal component analysis (PCA) and PLSR chemometric tools were applied to the fluorescence emission and excitation spectra obtained from tryptophan residues (305–480 nm; ex: 290 nm) and Vitamin A (340–620 nm, ex: 320 nm) present in the linear viscoelastic region. PLSR predicted the yield stress and flow stress with an $R^2 = 0.90$ from the vitamin A emission and excitation spectra, while predicted values with tryptophan residues had a regression co-efficient of $R^2 = 0.8$. Other parameters such as melting temperatures, moisture, protein, and fat contents could also be predicted from the vitamin A emission spectra with $R^2 = 0.98$.

In an attempt to conduct sensory evaluation of Cheddar cheese using fluorescence spectroscopy, Chiba et al. (2019) used the PLS chemometric analysis for cheese body measurements. A higher coefficient of determination was obtained for calibration ($R^2 = 0.80$) and the predicted values were comparable to those obtained by conventional methods (Chiba et al. 2019).

Parameters such as nutritional composition (Comin et al. 2008), fatty acid composition (Ferrand-Calmels et al. 2014), and milk coagulation properties (Toffanin et al. 2015) using spectroscopic techniques have been investigated. Apart from routine quality measurements like fat, moisture and protein, several studies have focused on minerals, volatile compounds, firmness, ripening time, as well as sensory attributes of milk products like cheese, yogurt, buttermilk, etc. (Bonfatti et al. 2016; Arango and Castillo 2018; Muncan et al. 2021a; Loudiyi et al. 2017). The latest trend in the field of inline sensors, apart from traditional measurements, is to link spectral data of milk and its composition to genomic and molecular data of cattle to improve dairy cattle breeding programs and relate animal health and wellness to this data (De Marchi et al. 2018; Tiplady et al. 2020).

German dairy cooperative "Berchtesgaden", has adopted Foss analytics based on NIR for measuring key quality parameters in their butter and cheese production process (Mills 2015). The industry claims to have improved its yield, saved on costly raw material, and improved product quality using the inline NIR system which measures fat, protein, lactose, sucrose, total dry mass, fat-free dry mass, density and acids. They report that the inline system enabled maintaining the organization's high standards and has resulted in high brand value with satisfied customers and end users.

A very recent study investigated the use of RS and chemometrics for the determination of eight mineral elements in infant formula (Zhao et al. 2020). The authors concluded from their study that RS equipped with a non-contact fiber-optic probe had the potential for inline quantification of mineral content of infant formulas during manufacturing. The PLSR model developed using all samples for calibration, achieved a predicted mineral content in samples with R^2CV values of 0.51–0.95 and RMSECVs of 0.13–2.96 ppm. Validation of the method was also carried out with R^2CV values between 0.93–0.97 for minerals tested (prediction of Ca, Mg, K, Na, Fe, and Zn). ICP-AES was used as a reference method for the determination of the mineral content. This study assumes importance in the background of quick and easy detection of proper labelling in the commercial formulas for mineral content which can truly reflect their nutritional value.

Further, milk being a highly valued and consumed natural product, is often adulterated with cheap and unsafe chemicals like melamine, starch, citrate, sucrose, urea, cheaper sources of proteins etc., which also require early detection. Adulteration of melamine in milk and milk products has been investigated using spectroscopic techniques (Liang et al. 2021). It may be noted that many of these studies and reports are publications and still need to be demonstrated in industrial settings. The purpose of including publications is to appraise the readers of the latest research work in the area.

6.2 *Meat and Their Products*

Meat and their products are important source of dietary components such as proteins, polyunsaturated fatty acids, vitamins, and minerals. However, they are highly perishable food commodities and their quality declines rapidly during storage due to enzymatic autolysis, microbial growth and oxidation (Kondjoyan et al. 2018). Several intrinsic and extrinsic factors in meat make them easily susceptible to both chemical and microbial spoilage. Since, they are considered reasonably expensive products, monitoring their quality and composition during industrial operations has a direct bearing on the final product and ultimately affects consumer satisfaction, safety and also profit margins.

Meat quality indicators like pH, colour, water-holding capacity, tenderness, intramuscular fat, protein and moisture content, adulteration with other types of animal tissues, collagen, etc., have been the main focus of research and development of inline spectroscopic sensors. Categorising and grading meat, detecting frozen-thawed from fresh meat, and discriminating feeding regimes, have been implemented in several industries. Among them, NIR spectroscopy has been applied for inline monitoring of water, fat and protein in meat, since a very long time. The first report of NIRS application "in-line" in an industrial setting was reported by Isaksson et al. (1996). A diffuse NIR instrument was set at the outlet of the meat grinder to determine key quality parameters of ground beef on a conveyor belt. A multiple linear regression (MLR) used as the calibration model determined the fat, moisture, and protein contents in ground beef. Kamruzzaman et al. (2016) successfully used hyperspectral imaging as an online monitoring system to determine colour of red meat, an extremely important quality attribute that governs purchase decisions of consumers. For ease in industrial application, a set of feature wavelength for red meat color (L*, a*, b) was selected. Multiple linear regression models developed were able to predict L* ($R^2 = 0.97$), a* ($R^2 = 0.84$) and b* ($R^2 = 0.82$) with a root mean square error (RMSE) of 1.72, 1.73, and 1.35, respectively, indicating potential to be used for rapid assessment of meat color. The work of Robert et al. (2021) demonstrated the ability of RS to rapidly discriminate intact beef, venison and lamb meat and highlights the applicability of the technique in meat sorting and authentication. The authors used three chemometric techniques in combination with RS to discriminate the meat samples. The linear and non-linear support vector machine (SVM) model used by the authors could achieve sensitivities between 87 and 90% respectively with specificity above 88% in the validation set.

A plethora of information is available on NIRS for quality determination in meat and meat products as published literature and several reviews on the same are also available (Porep et al. 2015; Dixit et al. 2017; Preito et al. 2017). The use of MIR for evaluation of meat and its products for quality and safety has been sufficiently covered by Su and Sun (2019). Wang et al. (2018) in their review have covered in detail spectroscopic techniques for determination of fresh red meat quality, safety and classification.

In real-world scenario, several industries have already established dedicated instruments and have factory-set calibration systems (mostly NIR) that determine the protein, fat and moisture contents of meat and meat products like cooked meat, cooked ham, pepperoni, liver sausage, etc. (Osborne 2006). The other applications of spectral techniques are limited largely to publications and need to see the light of the day in "inline" settings of the industry.

6.3 Cereals and Cereal Products

Grading of grains is an important parameter not only to ensure quality of product which would be derived after processing but also for economic gains during export. Since the value and price of grains is fixed based on their quality, any minor variation has a direct consequence on revenue to the exporter. The quality of grains is determined by its protein and starch content as well as hardness. Canada, Australia, USA and Europe have implemented NIRS for protein estimation of grains like wheat and barley as early as 1960s (Osborne 2006). It has also been used to predict the optimum fertilizer requirements of cereal crops by analysing the nitrogen content and total carbohydrates in plant tissue samples. It is indeed interesting to note that spectroscopic inline monitoring of wheat quality by means of analysing its protein content, has led to huge cost savings in countries like Canada, where the technique has become routine in its wheat segregation programmes.

In the context of developing nations like India, application of IoT and real time monitoring is paramount since loss of grains have huge economic implications. A review of real time monitoring and control of grain quality during transportation, purchase and storage is provided by Hema et al. 2020. Readers may refer to Tian et al. (2020) who have provided a review of sensor technologies including NIR, MIR, Raman and FS in monitoring of wheat quality.

In an attempt to study the rheological and baking properties of wheat flour. Ahmad et al. (2016a), used FS and applied PLSR. The amount of protein, wet gluten and sedimentation coefficient was determined in the wheat flour samples and results indicated linear regression (R2) values of 0.90, 0.92 and 0.81 for the three determinants, respectively. The same authors in another study (Ahmad et al. 2016b) observed that, nutritional values of different commercially available wheat flours could also be reasonably well predicted by FS using the local weighted regression (LWR) model. When the wheat flours were evaluated for the energy values ($R^2 = 0.96$), protein $R^2 = 0.93$), fat ($R^2 = 0.99$), moisture ($R^2 = 0.99$), carbohydrates ($R^2 = 0.98$), sucrose ($R^2 = 0.99$), salt ($R^2 = 0.89$) and saturated fatty acids ($R^2 = 0.99$), was obtained indicating its capability to make accurate "inline" measurements.

Protein, gluten, moisture, and starch was estimated using NIRS, to grade the quality as well as rheological property of wheat samples. The spectral region between 1000 to 2500 nm was found to be the most suitable for determination of protein, gluten and starch while 680 to 2500 nm could determine the moisture content in samples, using the PLS model. Good coefficient of prediction (R^2p) values between 0.94–0.98 and acceptable standard error of prediction (SEP) between 4.82–9.79 were obtained for the samples (Ibrahim 2018).

Buhler, the world leader in cereal processing, has established online sensor systems integrated with various processing steps in several of its facilities. These include among others, protein and moisture determination in incoming wheat to select the right silo, adjust the mill to specific ash content, add gluten powder to increase protein levels, or blend different flours for a perfect product (https://www.buhlergroup.com).

NIRS has been used to monitor batter mixing and physicochemical changes of dough with respect to consistency variation and gluten network (Kaddour et al. (2008). Dough mixing characteristics have also been monitored inline by Wesley et al. (1998) using NIRS. In comparison to other spectroscopic techniques, NIRS seems to be the method of choice for monitoring several chemical and physical parameters of cereal and cereal products.

6.4 Fermentation Based Processes

Spectroscopic techniques have been used to monitor analyte concentrations in microbial fermentations mostly as at-line measurements, where a sample is removed from the reactor and measured on a spectrometer situated close to the process (Guo et al. 2012; Liang et al. 2013). However, although this reduces the time as compared to off-line assessment, it still requires removal of sample for the determination. Several authors have attempted in-line spectroscopic techniques for various microbial parameters in bioprocess monitoring and have relatively been successful (Lee et al. 2004, Petersen et al. 2010; Bogomolov et al. 2015). Alves-Rausch et al. (2014) were able to demonstrate the use of NIRS where *Bacillus* fermentation was monitored at an industrial scale in bioreactors (50 L), under harsh industrial conditions. They used a BioPAT® Spectro NIR sensor, with clean in place (CIP) and sterilization in place (SIP) capabilities to detect variations and classify media. Additionally, spore counts, acetoin, dry mass, and sugar concentrations could be determined using multiparametric, multivariate analysis for fast, sensitive, non-destructive and robust measurements (Alves-Rausch et al. 2014).

It is highly desirable that sensors employed for bioreactor monitoring must be capable of measuring even low concentrations of various nutrients or metabolites without interference from the complex, multiphasic matrix which is inherent in a fermentation process (Lourenço et al. 2012, Abu-Absi et al. 2014, Sivakesava et al. 2001a, Sivakesava et al. 2001b). For this reason, Bonk et al. 2011, used two *in-situ* online-methods namely in situ microscopy (ISM) and 2D fluorescence spectroscopy to monitor the cell density as well as the glucose, lactate and glutamate concentration during cultivation of CHO-K1 cells. It was demonstrated that ISM could monitor cell density with the same accuracy as that of conventional technique (Neubauer counting chamber) and fluorescence spectroscopy was equally capable of monitoring the selected metabolites with good accuracy and repeatability. Such on-line techniques in bioprocess monitoring could be extremely helpful in reducing the risk of contamination especially during cultivation of sensitive cells, by avoiding frequent sampling which is the drawback of offline measurements (Bonk et al. 2011).

The last two decades have seen a rise in number of commercial particle-monitoring sensors which have been extensively applied in inline monitoring especially in bioprocesses like fermentation (Muncan et al. 2021b). Some examples include probe-based sensors like SOPAT, Mettler Toledo Particle View, flow-cell based sensors Sympatec, ParticleTech, etc. (Gargalo 2020).

Raman spectroscopy was used along with chemometric procedures for in-line monitoring of glucose fermentation by *Saccharomyces* sp. The use of multivariate control charts enabled easy and rapid detection of any fault in the process line without requirement of sample preparation and was based only on the spectra of the system (Avila et al. 2012).

NIR spectroscopy was used inline along with electronic nose (EN) in a fed btch cultivation process to monitor and control a cholera-toxin producing *Vibrio cholerae*. Biomass, glucose and acetate production was monitored based on spectral identification and prediction models were built. The PLSR model could generate high correlation to reference data with appreciable R^2p for biomass (0.20 g l^{-1}), glucose (0.26 g l^{-1}) and acetate (0.28 g l^{-1}). The authors built a trajectory representation of the fed batch cultivation using the NIR and EN data using the PCA. Any bacterial contamination could be easily detected with a change or deviation in the normal trajectory. This in situ monitoring with NIRS was claimed to be robust with an SEP of 0.020 g l^{-1} for determination of the cholera toxin. The acetate formation by the bacteria could also be controlled efficiently using the data for biomass concentration (Navrátil et al. 2005).

6.5 *Wine, Brewing and Distilleries*

The wine, brewing and distillery industry has been using inline sensors for monitoring the original gravity and alcohol content in the samples. In the brewing industry, the sensors are generally NIR based and data is collected online from flow-through cells. Transmittance or transflectance cells are used and it is established that the standard errors for the evaluation are less than 0.2%. NIR has also been used to monitor fruit quality and determine the alcohol content of wine and dedicated filter instruments for wine analysis are commercially available.

Although performed ex-situ, Grassi et al. (2014), demonstrated that (FTIR-ATR) spectroscopy could be a good technique to assess sugars and ethanol concentration for the inline monitoring of beer fermentation. Multivariate curve resolution-alternating least squares (MCR-ALS) models developed by them could successfully predict the fermentation progression with a 99.9% of explained variance, 3.5% lack of fit, and standard deviation of the residuals lower than 0.023. The FT-IR and MCR-ALS models could describe spectral changes of the main components of wort namely the sugars and ethanol concentration.

Trivellin et al. 2018, developed a completely different strategy based on fluorescence behavior of metal/porphyrin complex to measure oxygen levels at different time points during fermentation. The system was based on use of an optical fiber probe to measure luminescent lifetime variation of the complex which decayed in the presence of oxygen. Dynamic modelling techniques were used to predict the nutrient evolution in space and time at defined measuring points for the purpose of process monitoring and control. The experimental validation was done at an actual Italian winery.

6.6 *Fruits and Vegetables*

Quality assessment of fruit and vegetables involves evaluation of its appropriate maturity, structure, texture, chemical composition, and the absence of defects like bruises, browning, microbial growth, insect damage etc. They have also been evaluated for total soluble solids content as an indicator of sweetness, total acidity as an indicator of sourness, total dry matter as an indication of maturity, moisture content as an index of juiciness, lycopene content for nutraceutical value, overall texture including for firmness and toughness. The conventional method of measurement of the internal quality in most food industries still happens to be offline, destructive analysis. However, spectroscopic techniques (MIR, NIR, RS, FS) and others like x-ray imaging, and nuclear magnetic resonance spectroscopy (NMR), have been explored and some have also been adopted in industries (Irudayaraj and Reh 2008). An overview of spectroscopic, multispectral imaging and hyperspectral imaging techniques for quality attributes, measurement and variety discrimination of fruit and vegetable species is presented by Wang et al. (2015) and Sirisomboon (2018) and may be referred for further reading.

6.7 Freshness of Food Products

In many sectors of the food industry, sensory assessment (which include colour, texture, taste, odour, appearance etc.) of food product has been traditionally used to evaluate the freshness of a product such as fruits, vegetables, meat and fish etc. The sensory assessment is done by a panel of trained experts who score the quality of the product by established protocols. For example, "quality index method" is used to assess the quality of the fish (Hassoun et al. 2019). The fish processing units evaluate and monitor the quality and safety of fish by methods such as pH, ATP, total volatile basic nitrogen, trimethylamine, microbial plate count technique, etc. These methods are time consuming and labour intensive and moreover not "real-time". The traditional methods of sensory assessment are slowly being replaced by instrumental sensory methods that mimic the human system. Referred to as the "biomimetic" sensors, instruments such as "e-nose" that mimic olfactory system, "e-tongue" that mimic the gustatory system and "e-eye" that mimic the visual system are now being used to evaluate several parameters reflecting the quality and safety of food products. However, these systems are not covered in the chapter. Readers may refer to reviews by Jiang et al. (2018) and Tan and Xu (2021) for further reading on applications of these techniques in quality and freshness monitoring.

Vis-NIRS reflectance spectroscopy was used for authenticating fresh and frozen-thawed swordfish by Fasolato et al. (2012). Authors integrated Visible and NIR to draw predictions and found that the results were better than (accuracy >96.7%) when the data was taken individually. VIS/NIRS was found be a useful tool to differentiate fresh and frozen-thawed fish in another study with an overall classification rate in the range of 80% and 91% (Ottavian et al. 2013).

A very recent review by Franceschelli et al. (2021), covers a wide range of sensor technologies for monitoring fish freshness and quality. Article by Sarkar et al. (2019) may also be referred which compares the advantages of polarization reflectance spectroscopy over other non-invasive, rapid and real time tools like NIR, hyperspectral imaging and machine vision for monitoring the freshness of fruits.

6.8 Authenticity and Adulteration

It is a consumer right, that the food that is purchased should be in compliance with its label description, whether in respect to nutritional composition, allergen declaration, geographical origin, method of production, etc. Additionally, adulteration and food fraud also create health hazard, apart from economic losses in case of trade (Ropodi et al. 2016). Inline sensors can help manufacturers evaluate the raw material or ingredient which may have been sourced from different regions. For example, fruit distillates are affected by botanical origin as well as the region and climatic conditions in which the fruits are grown (geographical origin). Especially with respect to the quality of the distillate, the chemical constituents dictate the uniqueness of the product which is highly dependent on the raw material specific to the region and have been traditionally conserved. In this direction, it is very important to not only ensure high-quality but also detect false claims of assigning a region for product origin. In a recent study, Raman spectroscopy was used to differentiate distillates with respect to their trademark, geographical and botanical origin by Berghian-Grosan and MagDas (2020). Authors evaluated eight fruit distillates (apple, apricot, cherry, grape, pear, plum, quince, sour-cherry) containing between 40 to 80% alcohol by volume. The proposed approach had a model accuracy of 95.5% for trademark fingerprint while an accuracy of (90.9%) was achieved for the geographical discrimination of the fruit spirits (Berghian-Grosan and Magdas 2020). Inline application of this method was suggested by the authors to rule out adulteration and flavour masking of the product.

Adulteration in food samples is rampant especially in underdeveloped and developing countries due to several socio-economic reasons. Adulteration in high value products like olive oil with cheaper alternates like sunflower oil using applied VIS and NIR transflectance spectra has been studied (Downey et al. 2002). In another example, a modified real coded-GA coupled to PLS (RCGA-PLS) was developed which was found to be better in terms of sensitivity and fingerprinting of tartarzine in comparison to other chemometric tools such as PLS, GA-PLS, BiPLS and CARS-PLS for the detection of adulteration of tartrazine in tea (Amsaraj and Mutturi 2021). The detection range was found to be 0–30 mg/g using the FT-IR coupled system. Such studies can be extremely helpful especially for regulatory agencies to monitor adulteration particularly "on the field".

In a study by Downey and Kelly (2004), strawberry and raspberry purees were adulterated with cooked apples (10–75% w/w) and NIR transflectance measurement was used to predict the adulteration. The prediction of apple content was achieved in the 1100–1880 nm range for strawberry and 400–1880 nm range for raspberry after using PLS chemometric tool. The study concluded that the detection was possible when adulteration exceed 25% of raspberry and 20% of strawberry purees.

Su and Sun (2017), explored the application of spectral imaging for detection of adulteration of organic flour (Irish organic wheat flour; OWF) with cassava flour (CaF), common wheat flour (WF), and corn flour (CoF). OWF samples were adulterated with different percentages of other flours. RC-FMCIA-PLSR model was reported to be the best for the determination of adulteration with a coefficient of prediction (R^2P) of 0.97 and a root mean square error of prediction (RMSEP) of 0.036 for CoF adulteration in OWF, R^2P of 0.986 and RMSEP of 0.026 for OWF adulterated with CaF, and R^2P of 0.971 and RMSEP of 0.038 for OWF adulterated with WF. The applicable range for authentication of the admixtures in specific wheat flour was found to be 3–75% (w/w).

6.9 Microbial Safety and Hygiene

Microbial safety and hygiene are one of the most important parameters to be monitored in a food processing industry. Processing machines, equipments, conveyor belts, pipes, wash waters, packaging material, personnel, etc., are important sources of microbes. Visual detection, off-line plating and swabbing for ATP analysis are the commonly used techniques of microbial analysis. Generally, Cleaning-in-Place (CIP) and Sterilization-in-place (SIP) are available to clean and disinfect without disassembling or assembling any components in a process. Non-invasive techniques can be very useful to monitor as well as control microbial growth in any process. Readers may refer article by Lobete et al. (2015) to get useful insights on non-invasive techniques for microbial load analysis. Much of the literature on spectroscopic analysis of microbial safety and hygiene are publications, and ATRS, RS etc. have been widely used for total biomass analysis in fermentation experiments. Fluorescence microscopy coupled gel-cassette has also been reported for fermentation-based process especially to study function of inoculum level in cheese as the model matrix (Jeanson et al. 2011).

7 Limitations of Spectroscopy

Although used widely because of the several advantages they offer for "inline" monitoring, spectroscopic techniques suffer from certain drawbacks. Since many spectroscopic applications are based on reflectance mode, the presence of the source and the detector on the same probe results in low penetration depth of the radiation in the sample. In addition, low sampling mass due to restricted area of the probe also results in non-representative measurements, increase in standard deviation as well as underestimates the degree of homogeneity. Several attempts and interventions have been made to overcome these limitations specially to increase the sampling size. Many industries therefore have adopted inclusion of multiple probes at different locations, combine spectroscopy with other supporting techniques such as imaging, and automate the probes by mounting them on motorized translational stages to get repeatable data at different time points.

Apart from some technical drawbacks mentioned above, spectroscopic sensors also suffer from high cost of instrumentation and sometimes requirement of highly qualified personnel. Innovation in the field of spectroscopic sensors and their actual implementation in the food industry has also been slow (although publications are in explosion) due to lack of clear understanding and documentation of analytical systems actually used by the industries. Moreover, many industries do not prefer to disclose or publish their monitoring procedures mainly to avoid competition and regulatory attention.

8 Emerging Technologies for Inline Monitoring and Control of Food Quality and Safety

8.1 Biosensors

Biosensors are devices that convert biological signal into an electrical one. They have been extensively used in the field of medical diagnostics and less explored as inline probes for process monitoring although some attempts have been made in the past especially in the field of fermentation.

Tric et al. (2017), reported application of enzyme-based biosensor for continuous monitoring of glucose which was applied for animal cell culture optimization studies in bioreactors. The optical biosensor which enabled assessment of internal concentration of hydrogen peroxide; the by-product of the glucose oxidation reaction, also reported the turnover rate of the enzyme glucose oxidase as an important factor to be considered for the monitoring purpose. The sensor performance was validated using experimental data with conventional techniques and numerical simulations were derived for the process. More importantly, the sensor was easily sterilizable using beta and UV irradiation, demonstrating its application in real-life industrial processes. Automated online biosensors to detect microbes and their toxins have been developed for water monitoring (Shi et al. 2013; Ettenauer et al. 2015). A sulphur-oxidising bacteria was used for real time monitoring of heavy metals and other toxic chemicals in water (Hassan et al. 2019). Several researchers have reported detection of *E.coli* "on line" for monitoring of water quality with sensitivity as high as two colony forming units (Kellner et al. 2016).

Thakur and Ragavan (2013) have presented a comprehensive review on application of biosensors in food processing, including their potential role in inline monitoring of food quality and safety. The group has worked extensively in the development of biosensors for detection of multiple food contaminants like pesticides (Kumar et al. 2001; Gulla et al. 2002; Lisa et al. 2009), heavy metals (Ranjan et al. 2012), microbial toxins and pathogens (Vinayaka and Thakur 2011; Thakur et al. 2010), and adulterants like formaldehyde (Akshath et al. 2012; Akshath and Bhatt 2018). Apart from enzymes and antibodies, development of aptasensors for food relevant molecules such as antibiotics, myco- and algal toxins, etc. have also been investigated (Sharma et al. 2019; Mukherjee et al. 2017; Mukherjee et al. 2021). Many of these sensing platforms can further be fine-tuned for on line sensing with appropriate integration of other technologies to develop devices to monitor food contamination.

Despite a huge number of research studies on biosensors for the food industry, this technology has not been translated widely as successful commercial products for food diagnostics. There are multiple reasons which include factors like stability of the biosensing element, performance fluctuations with changing environment, sensor fouling, problems with reusability and regeneration issues, cost involved, etc. However, with advancement in science and technology, biosensors continue to hold the promise to be exploited in the agrifood industry especially for onsite and online monitoring of quality and safety.

8.2 Acoustic Sensors

Acoustic sensors use scattering and reflecting of sound waves when they interact with matter. These waves which travel through matter, cause oscillations without causing any alteration to the structure of the material. Passive acoustics introduce no external sound waves while active acoustic analysis refers to introduction of sound waves to a system and then monitor the changes caused. This sensor technology has wide applications in the food industry especially as a noninvasive technique. Some examples of application of acoustic technology include assessing the crispiness of product (Arimi et al. 2012), texture of fruit (Costa et al. 2011), firmness of fruit or vegetables (Jancso et al. 2001), discriminating between material for further processing (Elbatawi 2008), etc. Although investigated since 2001, especially for assessing sensorial aspects of a product, this technology has not been explored in a big way in "inline" sensing and carries the potential to be applied in food quality monitoring and control.

8.3 Magnetic Resonance Imaging

Magnetic resonance is referred to as the interaction which occurs between atomic particles and an applied external magnetic field. Resonance occurs due to absorption and emission of energy at specific frequencies which is in turn a function of individual atomic particles as well as strength of the applied magnetic field. When magnetic resonance is applied to develop images of an object or its internal structures, it is referred to as magnetic resonance imaging (MRI). MRI is obtained as a signal of spatial co-ordinates within a sample. MRI is also a non-destructive, non-invasive technique which has been used to assess the quality of products particularly fruit and vegetables and meat and meat products. Review by Hamed et al. (2018) may be referred to for literature on MRI based sensing platforms.

9 Future Perspectives

FDA in its guidance framework of 2004, has emphasized the use of monitoring and control approaches such as PAT tools, to improve and guarantee product quality in the "manufacturing" sector. The intent is to replace established product release and validation protocols which are presently being carried out by costly and time consuming laboratory analysis, to a more process-oriented real-time monitoring and control which ensures "Quality by Design" (QbD). PAT will have to be implemented in the food industry and technologies that are rapid, non-destructive and robust can play a very important role to increase profits for the food business as well as satisfy consumer demands with consistency and uniformity in the quality and safety of the product. In this direction, spectroscopy-based sensors will continue to be extremely useful tools for "inline" monitoring and determination of food quality and safety.

Spectroscopy based inline sensors offer several advantages over conventional methods of analysis. These include among others, minimal to negligible sample preparation, analysis of large composite varieties, non-destructive and non-invasive nature, less overall analysis time, less processing cost, environment friendly, easy to operate, capability to be coupled to cloud –based IoT devices, no specialized training for operation, etc. However, the main drawback in the use of spectroscopy-based sensors in the food industry for real-time monitoring has been to overcome challenges related to sensitivity of calibration, specificity, spectral changes accompanying varietal differences, climate and season variations, environmental variations influencing the sample, internal and external constituents impacting the determinants and most importantly high initial cost of instrumentation and its maintenance (although analysis per sample becomes cheaper when used for routine analysis in the long run). Many of these challenges have been addressed partially with advancement in computation, technological developments in optical sensors, companies venturing into mass production, use of AI and machine learning tools, exploration of other regions of the electromagnetic spectra, information-driven automation, metadata acquisition etc. The role of software, especially for sensor application in inline or online monitoring has also been extremely important in overcoming many of these challenges.

There is no doubt, that monitoring and control of processes and products using cloud-based services for traceable performance and safety verification will have to be integrated to enable huge profits as well as impart credibility to a food industry. In future, optimization of process steps using inline sensor tools will result in maximizing economic dividends. Apart from early detection of events, predictive maintenance that indicate immediate action will have great implications to the industry in forthcoming years. Sensors that are able to generate data on process know-how and detect hazards real-time is the need of the hour today.

Apart from advanced programming software, and data processing algorithms, it is also necessary that sensors developed in future should be field-deployable, compact and can be easily integrated into existing industrial processes. The future is therefore for smart, small and sensible sensors. Both the sensor and the software should be able to predict and present repeatable, reliable and robust measurements of variables that dictate process and product safety and quality with less or no requirement of trained personnel.

In conclusion, empowering food industry to transition to Industry 4.0 operations, is a win-win for all stakeholders, both the industry and the customers. Improved productivity, product quality and safety by introducing more advanced inline monitoring and control strategies is the way forward for the food manufacturing sector. At present, the progress on inline sensors are restricted mainly to publications or restricted only to certain parameters specific to a food product. The advantages offered by inline sensor systems are far greater than what appears on the surface and the food industry needs to implement and plan actionable strategies to reap its full benefits.

References

Abu-Absi NR, Martel RP, Lanza AM, Clements SJ, Borys MC, Li ZJ (2014) Application of spectroscopic methods for monitoring of bioprocesses and the implications for the manufacture of biologics. Pharm Bioprocess 2:267–284. https://doi.org/10.4155/pbp.14.24

Ahmad MH, Nache M, Waffenschmidt S, Hitzmann B (2016a) Characterization of farinographic kneading process for different types of wheat flours using fluorescence spectroscopy and chemometrics. Food Control 66:44–52

Ahmad MH, Nache M, Waffenschmidt S, Hitzmann B (2016b) A fluorescence spectroscopic approach to predict analytical, rheological and baking parameters of wheat flours using chemometrics. J Food Eng 182:65–71

Akshath US, Vinayaka AC, Thakur MS (2012) Quantum dots as nano plug-in for efficient NADH resonance energy routing. Biosens Bioelectron. https://doi.org/10.1016/j.bios.2012.05.003

Akshath US, Bhatt P (2018) Supramolecular nanosniffer for ultrasensitive detection of formaldehyde. Biosens Bioelectron 100:201–207

Alander JT, Bochko V, Martinkauppi B, Saranwong S, Mantere T (2013) A review on optical nondestructive visual and near-infrared methods for food quality and safety. Int J Speleol 2013(341402):1e36–1e36

Alamprese C, Casale M, Sinelli N, Lanteri S, Casiraghi E (2013) Detection of minced beef adulteration with Turkey meat by UV-Vis, NIR and MIR spectroscopy. LWT – Food Sci Technol 53:225–232

Alves-Rausch J, Bienert R, Grimm C, Bergmaier D (2014) Real time in-line monitoring of large scale bacillus fermentations with near-infrared spectroscopy. J Biotechnol 189:120–128. https://doi.org/10.1016/j.jbiotec.2014.09.004

Amsaraj R, Mutturi S (2021) Real-coded GA coupled PLS for rapid detection and quantification of tartrazine in tea using FT-IR spectroscopy. LWT 139:110583

Arango O, Castillo M (2018) A method for the inline measurement of milk gel firmness using an optical sensor. J Dairy Sci 101:3910–3917. https://doi.org/10.3168/jds.2017-13595

Argyri AA, Jarvis RM, Wedge D, Xu Y, Panagou EZ, Goodacre R et al (2013) A comparision of Raman and FT-IR spectroscopy for the prediction of meat spoilage. Food Control 29:461e470

Arimi JM, Duggan E, O'sullivan M, Lyng JG, O'riordan ED (2012) Crispiness of a microwave-expanded imitation cheese: mechanical, acoustic and sensory evaluation. J Food Eng 108:403–409

Avila TC, Poppi RJ, Lunardi I, Tizei PAG, Pereira GAG (2012) Raman spectroscopy and chemometrics for on line control of glucose fermentation by Saccharomyces cerevisiae. AiCHE, Biotechnol Prog. https://doi.org/10.1002/btpr.1615

Baca-Bocanegra B, Nogales-Bueno J, Hernández-Hierro JM, Heredia FJ (2018) Evaluation of extractable polyphenols released to wine from cooperage byproduct by near infrared hyperspectral imaging. Food Chem 244:206–212. https://doi.org/10.1016/j.foodchem.2017.10.027

Berghian-Grosan C, Magdas DA (2020) Application of Raman spectroscopy and machine learning algorithms for fruit distillates discrimination. Sci Rep 10:1–9. https://doi.org/10.1038/s41598-020-78159-8

Berrettoni M, Carpani I, Corradini N, Conti P, Fumarola G, Legnani G, Lanteri S, Marassi R, Tonelli D (2004) Coupling chemometrics and electrochemical-based sensor for detection of bacterial population. Anal Chim Acta 509:95–101. https://doi.org/10.1016/j.aca.2003.12.025

Bogomolov A, Heßling M, Wenzel U, Princz S, Hellmuth T, Bernal MJB, Sakharova T, Usenov I, Artyushenko V, Meyer H (2015) Development and testing of mid-infrared sensors for in-line process monitoring in biotechnology. Sensors Actuators B Chem 221:1601–1610. https://doi.org/10.1016/j.snb.2015.07.118

Bocker U, Ofstad R, Bertram HC, Sockalingum GD, Manfait M, Egelandsdal B et al (2007) Revealing covariabce structures in Fourier transform infrared and Raman microspectroscopy spectra: a study on pork muscle fiber tissue subjected to different processing parameters. Appl Spectrosc 61(10):1032e1039

Bonfatti V, Degano L, Menegoz A, Carnier P (2016) Short communication: mid-infrared spectroscopy prediction of fine milk composition and technological properties in Italian Simmental. J Dairy Sci 99:8216–8221. https://doi.org/10.3168/jds.2016-10953

Bonk S, Sandor M, Rüdinger F, Tscheschke B, Prediger A, Babitzky A, Solle D, Beutel S, Scheper T (2011) In-situmicroscopy and 2D fluorescence spectroscopy as online methods for monitoring CHO cells during cultivation. BMC Proc 5:2–4. https://doi.org/10.1186/1753-6561-5-s8-p76

Brereton RG (2000) Introduction to multivariate calibration in analytical chemistry. Analyst 125: 2125–2154. https://doi.org/10.1039/b003805i

Cai J, Wu X, Yuan L, Han E, Zhou L, Zhou A (2013) Determination of Chinese Angelica honey adulterated with rice syrup by an electrochemical sensor and chemometrics. Anal Methods 5: 2324–2328. https://doi.org/10.1039/c3ay00041a

Callao MP, Ruisánchez I (2018) An overview of multivariate qualitative methods for food fraud detection. Food Control 86:283–293. https://doi.org/10.1016/j.foodcont.2017.11.034

Camisard V, Brienne JP, Baussart H, Hammann J, Suhr H (2002) Inline characterization of cell concentration and cell volume in agitated bioreactors using in situ microscopy: application to volume variation induced by osmotic stress. Biotechnol Bioeng. https://doi.org/10.1002/bit.10178

Chaharlangi M, Tashkhourian J, Bordbar MM, et al (2020) A paper-based colorimetric sensor array for discrimination of monofloral European honeys based on gold nanoparticles and chemometrics data analysis. Spectrochim Acta Part A Mol Biomol Spectrosc 119076. https://doi.org/10.1016/j.saa.2020.119076

Chiba A., Kokawa M, Tsuta M, Todoriki S (2019) Predicting sensory evaluation indices of Cheddar cheese texture by fluorescence fingerprint measurement coupled with an optical fibre. Intl Dairy J 91:129–136

Chitra J, Ghosh M, Mishra HN (2017) Rapid quantification of cholesterol in dairy powders using Fourier transform near infrared spectroscopy and chemometrics. Food Control 78:342–349

Collette TW, Williams TL (2002) The role of Raman spectroscopy in the analytical chemistry of potable water. J Environ Monit 4:27–34. https://doi.org/10.1039/b107274a

Comin A, Cassandro M, Chessa S, Ojala M, Dal Zotto R, De Marchi M, Carnier P, Gallo L, Pagnacco G, Bittante G (2008) Effects of composite β- and κ-casein genotypes on milk coagulation, quality, and yield traits in Italian Holstein cows. J Dairy Sci 91:4022–4027. https://doi.org/10.3168/jds.2007-0546

Claßen J, Aupert F, Reardon KF, Solle D, Scheper T (2017) Spectroscopic sensors for in-line bioprocess monitoring in research and pharmaceutical industrial application. Anal Bioanal Chem 409(3):651–666. https://doi.org/10.1007/s00216-016-0068-x

Clark CJ, Shaw ML, Wright KM, McCallum JA (2018) Quantification of free sugars, fructan, pungency and sweetness indices in onion populations by FT-MIR spectroscopy. J Sci Food Agric 98:5525–5533.

Costa F, Cappellin L, Longhi S, Guerra W, Magnago P, Porro D, Soukoulis C, Salvi S, Velasco R, Biasioli F et al (2011) Assessment of apple (Malus × domestica Borkh.) fruit texture by a combined acoustic-mechanical profiling strategy. J Postharvest Biol Technol 61:21–28

Craven S, Whelan J, Glennon B (2014) Glucose concentration control of a fed-batch mammalian cell bioprocess using a nonlinear model predictive controller. J Process Control 24:344–357. https://doi.org/10.1016/j.jprocont.2014.02.007

Cui F, Yue Y, Zhang Y, Zhang Z, Zhou HS (2020) Advancing biosensors with machine learning. ACS Sensors 5:3346–3364. https://doi.org/10.1021/acssensors.0c01424

De Marchi M, Penasa M, Zidi A, Manuelian CL (2018) Invited review: use of infrared technologies for the assessment of dairy products—applications and perspectives. J Dairy Sci 101:10589–10604. https://doi.org/10.3168/jds.2018-15202

Dietzsch C, Spadiut O, Herwig C (2013) On-line multiple component analysis for effecient quantitative bioprocess development. J Biotechnol 163(4):362–370. https://doi.org/10.1016/j.jbiotec.2012.03.010

Dixit Y, Casado-Gavalda MP, Cama-Moncunill R, Cama-Moncunill X, Markiewicz-Keszycka M, Cullen PJ, Sullivan C (2017) Developments and challenges in online NIR spectroscopy for meat processing. Compr Rev Food Sci Food Saf 16:1172–1187. https://doi.org/10.1111/1541-4337.12295

Downey G, Kelly JD (2004) Detection and quantification of apple adulteration in diluted and sulfited strawberry and raspberry purees using visible and near-infrared spectroscopy. J Agric Food Chem 52:204–209

Downey G, McIntyre P, Davies AN (2002) Detecting and quantifying sunflower oil adulteration in extra virgin olive oils from the Eastern Mediterranean by visible and near-infrared spectroscopy. J Agric Food Chem 50:5520–5525. https://doi.org/10.1021/jf0257188

Eifert T, Eisen K, Maiwald M, Herwig C (2020) Current and future requirements to industrial analytical infrastructure—part 2: smart sensors. Anal Bioanal Chem 412(9):2037–2045. https://doi.org/10.1007/s00216-020-02421-1

Elbatawi IE (2008) An acoustic impact method to detect hollow heart of potato tubers. J Biosyst Eng 100:206–213

ElMasry G, Kamruzzaman M, Sun DW, Allen P (2012) Principles and applications of hyperspectral imaging in quality evaluation of agro-food products: a review. Crit Rev Food Sci Nutr 52:999–1023. https://doi.org/10.1080/10408398.2010.543495

Erkinbaev C, Henderson K, Paliwal J (2017) Discrimination of gluten-free oats from contaminants using near infrared hyperspectral imaging technique. Food Control 80:197–203. https://doi.org/10.1016/j.foodcont.2017.04.036

Ettenauer J, Zuser K, Kellner K, Posnicek T, Brandl M (2015) Development of an automated biosensor for rapid detection and quantification of E. coli in water. Procedia Eng 120:376–379. https://doi.org/10.1016/j.proeng.2015.08.643

Fasolato L, Balzan S, Riovanto R, Berzaghi P, Mirisola M, Ferlito JC, Serva L, Benozzo F, Passera R, Tepedino V, Novelli E (2012) Comparison of visible and near-infrared reflectance spectroscopy to authenticate fresh and frozen-thawed Swordfish (Xiphias gladius L). J Aquat Food Prod Technol 21(5):493–507

Feng YZ, Sun DW (2012) Application of hyperspectral imaging in food safety inspection and control: A review. Crit Rev Food Sci Nutr 52:1039–1058. https://doi.org/10.1080/10408398.2011.651542

Ferrand-Calmels M, Palhière I, Brochard M, Leray O, Astruc JM, Aurel MR, Barbey S, Bouvier F, Brunschwig P, Caillat H, Douguet M, Faucon-Lahalle F, Gelé M, Thomas G, Trommenschlager

JM, Larroque H (2014) Prediction of fatty acid profiles in cow, ewe, and goat milk by mid-infrared spectrometry. J Dairy Sci 97:17–35. https://doi.org/10.3168/jds.2013-6648

Franceschelli L, Berardinelli A, Dabbou S, Ragni L, Tartagni M (2021) Sensing technology for fish freshness and safety: a review. Sensors (Switzerland) 21:1–24. https://doi.org/10.3390/s21041373

Gargalo CL, Udigam I, Pontius K, Lopez PC, Nielsen RF, Hasanzadeh A, Mansouri SS, Bayer C, Junicke H, Gernaey KV (2020) Towards smart biomanufacturing: A perspective on recent developments in industrial measurement and monitoring technologies for bio-based production processes. J Indus Microbiol Biotechnol 47:947–964. https://doi.org/10.1007/s10295-020-02308-1

Geană EI, Ciucure CT, Artem V, Apetrei C (2020) Wine varietal discrimination and classification using a voltammetric sensor array based on modified screen-printed electrodes in conjunction with chemometric analysis. Microchem J 159:105451. https://doi.org/10.1016/j.microc.2020.105451

Genisheva Z, Quintelas C, Mesquita DP, Ferreira EC, Oliveira JM, Amaral AL (2018) New PLS analysis approach to wine volatile compounds characterization by near infrared spectroscopy (NIR). Food Chem 246:172–178. https://doi.org/10.1016/j.foodchem.2017.11.015

Ghasemi-Varnamkhasti M, Amiri ZS, Tohidi M, Dowlati M, Mohtasebi SS, Silva AC, Fernandes DDS, Araujo MCU (2018) Differentiation of cumin seeds using a metal-oxide based gas sensor array in tandem with chemometric tools. Talanta 176:221–226. https://doi.org/10.1016/j.talanta.2017.08.024

Gonzalez-Navarro FF, Stilianova-Stoytcheva M, Renteria-Gutierrez L, Belanche-Muñoz LA, Flores-Rios BL, Ibarra-Esquer JE (2016) Glucose oxidase biosensor modeling and predictors optimization by machine learning methods. Sensors (Switzerland) 16:1–13. https://doi.org/10.3390/s16111483

Gowen AA, O'Donnell CP, Cullen PJ, Downey G, Frias JM (2007) Hyperspectral imaging – an emerging process analytical tool for food quality and safety control. Trends Food Sci Technol 18:590–598. https://doi.org/10.1016/j.tifs.2007.06.001

Granato D, Putnik P, Kovačević DB, Santos JS, Calado V, Rocha RS, Da Cruz AG, Jarvis B, Rodionova OY, Pomerantsev A (2018) Trends in Chemometrics: food authentication, microbiology, and effects of processing. Compr Rev Food Sci Food Saf 17:663–677. https://doi.org/10.1111/1541-4337.12341

Grassi S, Amigo JM, Lyndgaard CB, Foschino R, Casiraghi E (2014) Assessment of the sugars and ethanol development in beer fermentation with FT-IR and multivariate curve resolution models. Food Res Int 62:602–608. https://doi.org/10.1016/j.foodres.2014.03.058

Gulla KC, Gouda MD, Thakur MS, Karanth NG (2002) Reactivation of immobilized acetyl cholinesterase in an amperometric biosensor for organophosphorus pesticide. Biochim Biophys Acta 1597:133–139

Guo WL, Du YP, Zhou YC, Yang S, Lu JH, Zhao HY, Wang Y, Teng LR (2012) At-line monitoring of key parameters of nisin fermentation by near infrared spectroscopy, chemometric modeling and model improvement. World J Microbiol Biotechnol 28:993–1002. https://doi.org/10.1007/s11274-011-0897-x

Hassoun A, Aït-Kaddour A, Sahar A, Cozzolino D (2020) Monitoring thermal treatments applied to meat using traditional methods and spectroscopic techniques: a review of advances over the last decade. Food Bioproc Tech. https://doi.org/10.1007/s11947-020-02510-0

Ibrahim A (2018) Monitoring the quality attributes of different wheat varieties by infrared technologies. Agic Engg Intl: CIGR J 20:201–210

Ingle PD, Christian R, Purohit P, Zarraga V, Handley E, Freel K, Abdo S (2016) Determination of protein content by NIR spectroscopy in protein powder mix products. J AOAC Int 99(2): 360–363. https://doi.org/10.5740/jaoacint.15-0115

Isaksson T, Nilsen BN, Tøgersen G, Hammond RP, Hildrum KI (1996) On-line, proximate analysis of ground beef directly at a meat grinder outlet. Meat Sci 43:245–253. https://doi.org/10.1016/S0309-1740(96)00016-2

Izso E, Bartalne-Berceli M, Gergely S (2018) Monitoring of heat-treated wheat milling fractions by near infrared spectroscopic method. Qual Assur Saf Crop Foods 10:93–101

Hamed E, Hadi E, Salajegheh A, Barghi H (2018) Use of magnetic energy resonance in food quality control: a review. J Biomed Phys Eng. https://doi.org/10.22086/jbpe.v0i0.628

Hassan SHA, Gurung A, Kang W-C, Shin BS, Rahimnejad M, Jeon BH, Kim JR, Oh SE (2019) Real-time monitoring of water quality of stream water using sulfur-oxidizing bacteria as bio-indicator. Chemosphere 223:58–63. https://doi.org/10.1016/j.chemosphere.2019.01.089

Hassoun A, Sahar A, Lakhal L, Aït-Kaddour A (2019) Fluorescence spectroscopy as a rapid and non-destructive method for monitoring quality and authenticity of fish and meat products: impact of different preservation conditions. LWT 103:279–292. https://doi.org/10.1016/j.lwt.2019.01.021

Hema LK, Velmurugan S, Sunil DN, Thariq Aziz S, Thirunavkarasu S (2020) IOT based real-time control and monitoring system for food grain procurement and storage. IOP Conf Ser Mater Sci Eng 993:012079. https://doi.org/10.1088/1757-899X/993/1/012079

Herold B, Kawano S, Sumpf B, Tillmann P, Walsh KB (2009) Chapter 3. Vis/NIR spectroscopy. In: Zude M (ed) Optical monitoring of fresh and processed Agricultural crops. CRC Press, Boca Raton, USA, pp 141–249

Hu O, Xu L, Fu H, Yang T, Fan Y, Lan W, Tang H, Wu Y, Ma L, Wu D, Wang Y, Xiao Z, She Y (2018) "Turn-off" fluorescent sensor based on double quantum dots coupled with chemometrics for highly sensitive and specific recognition of 53 famous green teas. Anal Chim Acta 1008:103–110. https://doi.org/10.1016/j.aca.2017.12.042

Huang H, Liu L, Ngadi MO (2014) Recent developments in hyperspectral imaging for assessment of food quality and safety. Sensors (Switzerland) 14:7248–7276. https://doi.org/10.3390/s140407248

Huang H, Yu H, Xu H, Ying Y (2008) Near infrared spectroscopy for on/in-line monitoring of quality in foods and beverages: a review. J Food Eng 87:303–313. https://doi.org/10.1016/j.jfoodeng.2007.12.022

Irudayaraj J, Reh C (eds) (2008) Nondestructive testing of food quality. Blackwell Publishing, Ames, IA

Isaksson T, Nilsen BN, Tøgersen G, Hammond RP, Hildrum KI (1996) On-line, proximate analysis of ground beef directly at a meat grinder outlet. Meat Sci 43:245–253. https://doi.org/10.1016/S0309-1740(96)00016-2

Iversen JA, Berg RW, Ahring BK (2014) Quantitative monitoring of yeast fermentation using Raman spectroscopy. Anal Bioanal Chem 406:4911–4919. https://doi.org/10.1007/s00216-014-7897-2

Jancso PT, Clijmans L, Nicolaei BM, Baerdemaeker JD (2001) Investigation of the effect of shape on the acoustic response of "conference" pears by finite element modeling. J Postharvest Biol Technol 23:1–12

Jeanson S, Chadoeuf J, Maddec MN, Aly S, Floury J, Brocklehurst TF, Lortal S (2011) Spatial distribution of bacterial colonies in a model cheese. Appl Environ Microbiol 77:1493–1500

Jiang H, Zhang M, Bhandari B, Adhikari B (2018) Application of electronic tongue for fresh foods quality evaluation: A review. Food Rev Int 34:746–769. https://doi.org/10.1080/87559129.2018.1424184

Kaddour A, Barron C, Robert P, Cuq B (2008) Physico-chemical description of bread dough mixing using two-dimensional near-infrared correlation spectroscopy and moving-window two-dimensional correlation spectroscopy. J Cereal Sci 48:10–19. https://doi.org/10.1016/j.jcs.2007.07.008

Kamruzzaman M, Makino Y, Oshita S (2015) Non-invasive analytical technology for the detection of contamination, adulteration, and authenticity of meat, poultry, and fish: A review. Anal Chim Acta 853:19–29. https://doi.org/10.1016/j.aca.2014.08.043

Kamruzzaman M, Makino Y, Oshita S (2016) Online monitoring of red meat color using hyperspectral imaging. Meat Sci 116:110–117. https://doi.org/10.1016/j.meatsci.2016.02.004

Kang S (2011) Chpt 5 – NIR spectroscopy for chemical composition and internal quality in foods. In: Cho Y-J (ed) Emerging technologies for food quality and food safety analysis. CRC Press, Taylor and Francis, LLC, pp 113–138

Kangas MJ, Wilson CL, Burks RM, et al (2018) An Improved Comparison of Chemometric Analyses for the Identification of Acids and Bases With Colorimetric Sensor Arrays. 10:36–55. https://doi.org/10.5539/ijc.v10n2p36

Kellner K, Ettenauer J, Zuser K, Posnicek T, Brandl M (2016) An automated, robotic biosensor for the electrochemical detection of *E Coli*. Water Procedia Eng 168:594–597. https://doi.org/10.1016/j.proeng.2016.11.222

Khan A, Munir MT, Yu W, Young BR (2021) Near-infrared spectroscopy and data analysis for predicting milk powder quality attributes. Int J Dairy Technol 74:235–245

Kondjoyan A, Portanguen S, Duchène C, Mirade PS, Gandemer G (2018) Predicting the loss of vitamins B3 (niacin) and B6 (pyridoxamine) in beef during cooking. J Food Eng 238 (June):44–53. https://doi.org/10.1016/j.jfoodeng.2018.06.008

Kunes R, Bartos P, Iwasaka GK, Lang A, Hankovec T, Smutny L, Cerny P, Poborska A, Smetana P, Kriz P, Kernerova N (2021) In-line technologies for the analysis of important milk parameters during the milking process: A review. Agric 11:1–17. https://doi.org/10.3390/agriculture11030239

Kumar M, Thakur M, Senthuran A et al (2001) An automated flow injection analysis system for on-line monitoring of glucose and L-lactate during lactic acid fermentation in a recycle bioreactor. World J Microbiol Biotechnol 17:23–29. https://doi.org/10.1023/A:1016699701903

Lee HLT, Boccazzi P, Gorret N, Ram RJ, Sinskey AJ (2004) In situ bioprocess monitoring of *Escherichia coli* bioreactions using Raman spectroscopy. Vib Spectrosc 35:131–137. https://doi.org/10.1016/j.vibspec.2003.12.015

Li YQ, Kong DX, Wu H (2013) Analysis and evaluation of essential oil components of cinnamon barks using GC–MS and FTIR spectroscopy. Ind Crop Prod 2013(41):269–278. https://doi.org/10.1016/j.indcrop.2012.04.056

Li Z, Deen MJ, Kumar S, Selvaganapathy PR (2014) Raman spectroscopy for in-line water quality monitoring- instrumentation and potential. Sensors (Switzerland) 14:17275–17303. https://doi.org/10.3390/s140917275

Lisa M, Chouhan RS, Vinayaka AC, Manonmani HK, Thakur MS (2009) Gold nanoparticles based dipstick immunoassay for the rapid detection of dichlorodiphenyltrichloroethane: an organochlorine pesticide. Biosens Bioelectron 25(1):224–227

Liang J, Zhang D, Guo X, Xu Q, Xie X, Zhang C, Bai G, Xiao X, Chen N (2013) At-line near-infrared spectroscopy for monitoring concentrations in temperature-triggered glutamate fermentation. Bioprocess Biosyst Eng 36:1879–1887. https://doi.org/10.1007/s00449-013-0962-y

Liang W, Zhu Z, Yang B, Zhu X, Guo W (2021) Detecting melamine-adulterated raw milk by using near-infrared transmission spectroscopy. J Food Process Eng 44. https://doi.org/10.1111/jfpe.13685

Liu YJ, André S, Saint Cristau L, Lagresle S, Hannas Z, Calvosa É, Devos O, Duponchel L (2017) Multivariate statistical process control (MSPC) using Raman spectroscopy for in-line culture cell monitoring considering time-varying batches synchronized with correlation optimized warping (COW). Anal Chim Acta 952:9–17. https://doi.org/10.1016/j.aca.2016.11.064

Lobete MM, Fernandez EN, Van Impe JFM (2015) Recent trends in non-invasive in situ techniques to monitor bacterial colonies in solid (model) food. Front Microbiol 6:1–9. https://doi.org/10.3389/fmicb.2015.00148

Lohumi S, Lee S, Lee H, Cho B-K, (2015) A review of vibrational spectroscopic techniques for the detection of food authencity and adulteration. Trend Food Sci Technol 46:85–98

Loudiyi M, Temiz H-T, Sahar A, Ahmad MH, Boukria O, Hassoun A, Aït-Kaddour A (2017) Spectroscopic techniques for monitoring changes in the quality of milk and other dairy products during processing and storage. Crit Rev Food Sci Nutr 62:3063–3087

Lourenço ND, Lopes JA, Almeida CF, Sarraguça MC, Pinheiro HM (2012) Bioreactor monitoring with spectroscopy and chemometrics: a review. Anal Bioanal Chem 404:1211–1237. https://doi.org/10.1007/s00216-012-6073-9

Lu G, Fei B (2014) Medical hyperspectral imaging: a review. J Biomed Opt 19:010901. https://doi.org/10.1117/1.jbo.19.1.010901

Martynko E, Kirsanov D (2020) Application of Chemometrics in biosensing: a brief review. Biosensors 10. https://doi.org/10.3390/bios10080100

Masateru N, Tallada JG, Taiichi K (1967) Bruise detection using NIR Hyperspectral imaging for strawberry (fragaria ananassa Duch). Angew Chemie Int Ed 6(11):951–952, 44:133–142

McGovern AC, Broadhurst D, Taylor J, Kaderbhai N, Winson MK, Small DA, Rowland JJ, Kell DB, Goodacre R (2002) Monitoring of complex industrial bioprocesses for metabolite concentrations using modern spectroscopies and machine learning: application to gibberellic acid production. Biotechnol Bioeng 78:527–538. https://doi.org/10.1002/bit.10226

Medina-Plaza C, García-Hernández C, de Saja JA, Fernández-Escudero JA, Barajas E, Medrano G, García-Cabezón C, Martin-Pedrosa F, Rodriguez-Mendez ML (2015) The advantages of disposable screen-printed biosensors in a bioelectronic tongue for the analysis of grapes. LWT – Food Sci Technol 62:940–947. https://doi.org/10.1016/j.lwt.2015.02.027

Mehdizadeh H, Lauri D, Karry KM, Moshgbar M, ProcopioMelino R, Drapeau D (2015) Generic Raman-based calibration models enabling real-time monitoring of cell culture bioreactors. Biotechnol Prog 31:1004–1013. https://doi.org/10.1002/btpr.2079

Mills R (2015, Dec 9) Dairy testing with NIR, FOSS. www.fossanalytics.com/en/news-articles/dairy/dairy-process-analysis-with-nir

Mishra RK, Alonso GA, Istamboulie G, Bhand S, Marty JL (2015) Automated flow based biosensor for quantification of binary organophosphates mixture in milk using artificial neural network. Sensors Actuators B Chem 208:228–237. https://doi.org/10.1016/j.snb.2014.11.011

Miyamoto K, Kitano Y (1995) Non-destructive determination of sugar content in Satsuma mandarin fruit by near infrared transmittance spectroscopy. J Near Infrared Spectrosc 3:227–237. https://doi.org/10.1255/jnirs.73

Moretto J, Smelko JP, Cuellar M, Berry B, Doane A, Ryll T et al (2011) Process Raman spectroscopy for in-line CHO cell culture monitoring. Am Pharm Rev 14:18–25

Mukherjee M, Bhatt P, Manonmani HK (2017) Fluorescent competitive aptasensor for detection of aflatoxin B1. J Mol Recognit 30:e2650. (1–6)

Mukherjee M, Veerabhadraiah S, Bettadaiah BK, Thakur MS, Bhatt P (2021) DNA aptamer selection and detection of marine biotoxin 20 Methylspirolide G. Food Chem 363:130332

Muncan J, Kovacs Z, Tsenkova R (2021a) Near infrared aquaphotomics study on common dietary fatty acids in cow's liquid, thawed milk. Food Control 122:107805

Muncan J, Tei K, Tsenkova R (2021b) Real-time monitoring of yogurt fermentation process by aquaphotomics near-infrared spectroscopy. Sensors 21:1–18. https://doi.org/10.3390/s21010177

Nagata M, Tallada JG, Kobayashi T (2006) Bruise detection using NIR hyperspectral imaging for strawberry (Fragaria x ananassa duch.). Environ Control Biol 44:133

Navrátil M, Norberg A, Lembrén L, Mandenius CF (2005) On-line multi-analyzer monitoring of biomass, glucose and acetate for growth rate control of a vibrio cholerae fed-batch cultivation. J Biotech 115:67–79

Nawrocka A, Lamorska J (2013) Advances in agrophysical research. Adv Agrophysical Res. https://doi.org/10.5772/3341

Nesakumar N, Sethuraman S, Krishnan UM, Rayappan JBB (2015) Cyclic voltammetric acetylcholinesterase biosensor for the detection of captan in apple samples with the aid of chemometrics. Anal Bioanal Chem 407:4863–4868. https://doi.org/10.1007/s00216-015-8687-1

Oliveira MM, Cruz-Tirado JP, Barbin DF (2019) Nontargeted analytical methods as a powerful tool for the authentication of spices and herbs: a review. Compr Rev Food Sci Food Saf 18:670–689. https://doi.org/10.1111/1541-4337.12436

Osborne BG (2006) Near-infra red spectroscopy in food analysis, encyclopedia of analytical chemistry. Wiley. https://doi.org/10.1002/9780470027318.a1018
Ottavian M, Fasolato L, Facco P, Barolo M (2013) Foodstuff authentication from spectral data: toward a species-independent discrimination between fresh and frozen-thawed fish samples. J Food Eng 19(4):765–775
Ozbekova Z, Kulmyrzaev A (2017) Fluorescence spectroscopy as a non-destructive method to predict rheological characteristics of Tilsit cheese. J Food Eng 210:42–49. https://doi.org/10.1016/j.jfoodeng.2017.04.023
Parastar H, van Kollenburg G, Weesepoel Y, van den Doel A, Buydens L, Jansen J (2020) Integration of handheld NIR and machine learning to "Measure & Monitor" chicken meat authenticity. Food Control 112(2020):107149. https://doi.org/10.1016/j.foodcont.2020.107149
Park B, Yoon S-C, Windham W, Lawrence K, Kim M, Chao K (2011) Line-scan hyperspectral imaging for real-time in-line poultry fecal detection. Sens & Instrumen Food Qual 5:25–32
Patel P, Doddamani A (2019) Role of sensor in the food processing industries. Intl Arch App Sci Technol 10:10–18
Pell RJ et al (1998) Chemometrics: a practical guide. Wiley, United Kingdom
Peng Y, Zhang J, Wang W, Li Y, Wu J, Huang H, Gao X, Jiang W (2011) Potential prediction of the microbial spoilage of beef using spatially resolved hyperspectral scattering profiles. J Food Eng 102:163–169
Pereira, EV, Fernandes, DDS, de Araújo, MCU, Diniz, PHGD, and Maciel, MIS (2020) Simultaneous determination of goat milk adulteration with cow milk and their fat and protein contents using NIR spectroscopy and PLS algorithms, LWT 127:109427
P'erez-Marín D, Garrido-Varo A (2020) NIR sensors for the in-situ assessment of iberian ham. Ref Modul Food Sci., Elsevier. https://doi.org/10.1016/B978-0-08-100596-5.22860-6
Petersen N, Ödman P, Cervera Padrell AE, Stocks S, Lantz AE, Gernaey KV (2010) In situ near infrared spectroscopy for analyte-specific monitoring of glucose and ammonium in *Streptomyces coelicolor* fermentations. Biotechnol Prog 26:263–271. https://doi.org/10.1002/btpr.288
Peyvasteh M, Popov A, Bykov A, Meglinski I (2020) Meat freshness revealed by visible to near-infrared spectroscopy and principal component analysis. J Phys Commun 4:1–11. https://doi.org/10.1088/2399-6528/abb322
Picard A, Daniel I, Montagnac G, Oger P (2007) In situ monitoring by quantitative Raman spectroscopy of alcoholic fermentation by Saccharomyces cerevisiae under high pressure. Extremophiles 11:445–452. https://doi.org/10.1007/s00792-006-0054-x
Poms RE, Klein CL, Anklam E (2004) Methods for allergen analysis in food: a review. Food Addit Contam 21(1):1–31
Porep JU, Kammerer DR, Carle R (2015) Online application of near infrared (NIR) spectroscopy in food production. Trends Food Sci Technol 46:211–230
Preito N, Roehe R, Lavin P, Batten G, Andres S (2017) Application of near infrared reflectance spectroscopy to predict meat and meat products quality: a review. Meat Sci 83:175–186
Pullanagari RR, Yule IJ, Agnew M (2015) On-line prediction of lamb fatty acid composition by visible near infrared spectroscopy. Meat Sci 100:156–163
Kunwar P, Hassinen J, Bautista G, Rus RHA, Toivonen J (2014) Direct laser writing of photostable fluorescent silver nanoclusters in polymer films. ACS Nano 11:11165–11171
Qiao J, Ngadi MO, Wang N, Gariépy C, Prasher SO (2007) Pork quality and marbling level assessment using a hyperspectral imaging system. J Food Eng 83:10–16. https://doi.org/10.1016/j.jfoodeng.2007.02.038
Ranjan R, Rastogi NK, Thakur MS (2012) Development of immobilized biophotonic beads consisting of *Photobacterium leigonathi* for the detection of heavy metals and pesticide. J Hazard Mater. https://doi.org/10.1016/j.jhazmat.2012.04.076
Raud M, Kikas T (2013) Bioelectronic tongue and multivariate analysis: a next step in BOD measurements. Water Res 47:2555–2562. https://doi.org/10.1016/j.watres.2013.02.026
Rekha K, Gouda MD, Thakur MS, Karanth NG (2000) Ascorbate oxidase based amperometric biosensor for organophosphorous pesticide monitoring. Biosens Bioelectron 15:499–502

Rentería-Gutiérrez L, González-Navarro FF, Stilianova-Stoytcheva M et al (2014) Glucose oxidase biosensor modeling by machine learning methods. Lect Notes Comput Sci (including Subser Lect Notes Artif Intell Lect Notes Bioinformatics) 8857:464–473. https://doi.org/10.1007/978-3-319-13650-9_40

Robert C, Fraser-Miller SJ, Jessep WT, Bain WE, Hicks TM, Ward JF, Craigie CR, Loeffen M, Gordon KC (2021) Rapid discrimination of intact beef, venison and lamb meat using Raman spectroscopy. Food Chem 343:128441. https://doi.org/10.1016/j.foodchem.2020.128441

Ropodi AI, Panagou EZ, Nychas GJE (2016) Data mining derived from food analyses using non-invasive/non-destructive analytical techniques; determination of food authenticity, quality & safety in tandem with computer science disciplines. Trends Food Sci Technol 50:11–25. https://doi.org/10.1016/j.tifs.2016.01.011

Sampaio PS, Soares A, Castanho A, Almeida AS, Oliveira J, Brites C (2018) Optimization of rice amylose determination by NIR-spectroscopy using PLS chemometrics algorithms. Food Chem 242:196–204. https://doi.org/10.1016/j.foodchem.2017.09.058

Santos JR, Saragucca MC, António O.S.S. Rangel AOSS, Lopes JA (2012) Evaluation of green coffee beans quality using near infrared spectroscopy: A quantitative approach, Food Chem, 135, 1828–35

Sarkar M, Gupta N, Assad M (2019) Monitoring of fruit freshness using phase information in polarization reflectance spectroscopy. Appl Opt 58:6396–6407

Sharma R, Akshath US, Bhatt P, RaghavaRao KSMS (2019) Fluorescent aptaswitch for detection of chloramphenicol- quantification enabled by immobilization of aptamer. Sensors Actuators B Chem 290:110–117

Shi HC, Song BD, Long F, Zhou XH, He M, Lv Q, Yang HY (2013) Automated online optical biosensing system for continuous real-time determination of microcystin-LR with high sensitivity and specificity: early warning for cyanotoxin risk in drinking water sources. Environ Sci Technol 47:4434–4441. https://doi.org/10.1021/es305196f

Sirisomboon P (2018) NIR spectroscopy for quality evaluation of fruits and vegetables. Mater Today Proc 5:22481–22486. https://doi.org/10.1016/j.matpr.2018.06.619

Sivakesava S, Irudayaraj J, Ali D (2001a) Simultaneous determination of multiple components in lactic acid fermentation using FT-MIR, NIR, and FT-Raman spectroscopic techniques. Process Biochem 37:371–378. https://doi.org/10.1016/S0032-9592(01)00223-0

Sivakesava S, Irudayaraj J, Demirci A (2001b) Monitoring a bioprocess for ethanol production using FT-MIR and FT-Raman spectroscopy. J Ind Microbiol Biotechnol 26:185–190. https://doi.org/10.1038/sj.jim.7000124

Strani L, Grassi S, De Juan A (2021) Effect of physicochemical factors and use of milk powder on milk rennet-coagulation: process understanding by near infrared spectroscopy and chemometrics. Food Contl 119:107494

Su W-H, Sun D-W (2017) Evaluation of spectral imaging for inspection of adulterants in terms of common wheat flour, cassava flour and corn flour in organic avatar wheat (Triticum spp.) flour. J Food Eng 200:59–69

Su WH, Sun DW (2019) Mid-infrared (MIR) spectroscopy for quality analysis of liquid foods. Food Eng Rev 11:142–158. https://doi.org/10.1007/s12393-019-09191-2

Tan J, Xu J (2021) Applications of electronic nose (e-nose) and electronic tongue (e-tongue) in food quality-related properties determination: a review. Artif Intell Agric 4:104–115. https://doi.org/10.1016/j.aiia.2020.06.003

Thakur MS, Chouhan RS, Vinayaka AC (2010) Biosensors for pesticides and food borne pathogens. In: Mutlu M (ed) Biosensors in food processing, safety and quality control. CRC Press Taylor & Francis Group, Florida, pp 147–192

Thakur MS, Ragavan (2013) Biosensors in food processing. J Food Sci Technol 50:625–641. https://doi.org/10.1007/s13197-012-0783-z

Tian Y-G, Zhang Z-N, Tian S-Q (2020) Non-destructive testing for wheat quality with sensor technology based on big data. J Anal Methods Chem 2020:8851509. https://doi.org/10.1155/2020/8851509

Tibayrenc P, Preziosi-Belloy L, Roger JM, Ghommidh C (2010) Assessing yeast viability from cell size measurements? J Biotechnol 149(1–2):74–80. https://doi.org/10.1016/j.jbiotec.2010.06.019

Tiplady KM, Lopdell TJ, Jogn MD, Garrick DJ (2020) The evolving role of Fourier-transform mid-infrared spectroscopy in genetic improvement of dairy cattle. J Animal Sci Biotechnol 11:39. https://doi.org/10.1186/s40104-020-00445-2

Thanathornvarakul N, Anuntagool J, Tananuwong K (2016) Aging of low and high amylose rice at elevated temperature: mechanism and predictive modelling. J Cereal Sci 70:155–163

Thrift WJ, Ragan R (2019) Quantification of Analyte concentration in the single molecule regime using convolutional neural networks. Anal Chem 91:13337–13342. https://doi.org/10.1021/acs.analchem.9b03599

Toffanin V, De Marchi M, Lopez-Villalobos N, Cassandro M (2015) Effectiveness of mid-infrared spectroscopy for prediction of the contents of calcium and phosphorus, and titratable acidity of milk and their relationship with milk quality and coagulation properties. Int Dairy J 41:68–73. https://doi.org/10.1016/j.idairyj.2014.10.002

Tønning E, Sapelnikova S, Christensen J, Carlsson C, Winther-Nielsen M, Dock E, Solna R, Skladal P, Nørgaard L, Ruzgas T, Emnéus J (2005) Chemometric exploration of an amperometric biosensor array for fast determination of wastewater quality. Biosens Bioelectron 21:608–617. https://doi.org/10.1016/j.bios.2004.12.023

Tric M, Lederle M, Neuner L, Dolgowjasow I, Wiedemann P, Wölfl S, Werner T (2017) Optical biosensor optimized for continuous in-line glucose monitoring in animal cell culture. Anal Bioanal Chem 409:5711–5721. https://doi.org/10.1007/s00216-017-0511-7

Tripathi S, Mishra HN (2009) A rapid FT-NIR method for estimation of aflatoxin B1 in red chili powder. Food Control 20:840–846. https://doi.org/10.1016/j.foodcont.2008.11.003

Trivellin N, Barbisan D, Badacco D, Pastore P, Menegesso G, Meneghini M et al (2018) Study and development of fluorescence based sensor system for monitoring oxygen in wine production: the WOW project. Sensors 18:1130. https://doi.org/10.3390/s18041130

Uddin MZ, Hassan MM, Alsanad A, Savaglio C (2020) A body sensor data fusion and deep recurrent neural network-based behavior recognition approach for robust healthcare. Inf Fusion 55:105–115. https://doi.org/10.1016/j.inffus.2019.08.004

Udugama IA, Gargalo CL, Yamashita Y, Taube MA, Palazoglu A, Young BR, Gernaey KV, Kulahci M, Bayer C (2020) The role of big data in industrial (bio)chemical process operations. Ind Eng Chem Rese 59(34):15283–11529

Upadhyay N, Jaiswal P, Jha SN (2018) Application of attenuated total reflectance Fourier transform infrared spectroscopy (ATR– FTIR) in MIR range coupled with chemometrics for detection of pig body fat in pure ghee (heat clarified milk fat). J Mol Struct 1153:275–281

Villar A, Vadillo J, Santos JI, Gorritxategi E, Mabe J, Arnaiz A, Fernández LA (2017) Cider fermentation process monitoring by Vis-NIR sensor system and chemometrics. Food Chem 221:100–106. https://doi.org/10.1016/j.foodchem.2016.10.045

Vinayaka AC, Thakur MS (2011) Photoabsorption and resonance energy transfer phenomenon in CdTe-protein bioconjugates: an insight into QD-biomolecular interactions. Bioconjug Chem 22:968–975

Vo-Dinh T (2004) A hyperspectral imaging system for in vivo optical diagnostics. IEEE Engg Med Biol Mag 23:40–49. https://doi.org/10.1109/MEMB.2004.1360407

Wang H, Peng J, Xie C, Bao Y, He Y (2015) Fruit quality evaluation using spectroscopy technology: a review. Sensors (Switzerland) 15:11889–11927. https://doi.org/10.3390/s150511889

Wang L, Sun DW, Pu H, Cheng JH (2017) Quality analysis, classification, and authentication of liquid foods by near-infrared spectroscopy: A review of recent research developments. Crit Rev Food Sci Nutr 57:1524–1538. https://doi.org/10.1080/10408398.2015.1115954

Wang W, Paliwal J (2007) Near-infrared spectroscopy and imaging in food quality and safety. Sens & Instrumen Food Qual 1:193–207. https://doi.org/10.1007/s11694-007-9022-0

Wang W, Peng Y, Sun H, Zheng X, Wei W (2018) Spectral detection techniques for non-destructively monitoring the quality, safety, and classification of fresh red meat. Food Anal Methods 11:2707–2730. https://doi.org/10.1007/s12161-018-1256-4

Wang X (2019) Near infra-red spectroscopy for food quality evaluation. In: Zhong J, Wang X (eds) Evaluation technologies for food quality. Woodhead Publishing, pp 105–118. https://doi.org/10.1016/C2017-0-01187-4

Wang Y, Guo W, Zhu X, Liu Q (2019) Effect of homogenization on detection of milk protein content based on NIR diffuse. Intl J Food Sci Technol 54:387–395

Wesley IJ, Larsen N, Osborne BG, Skerritt JH (1998) Non-invasive monitoring of dough mixing by near infrared spectroscopy. J Cereal Sci 27:61–69. https://doi.org/10.1006/jcrs.1997.0151

William PC, Norris KS (2001) Near-infrared Technology in the Agricultural and Food Industries, 2nd edn. AACC, St. Paul, 312 p

Wold S, Sjöström M, Eriksson L (2001) PLS-regression: a basic tool of chemometrics. Chemom Intell Lab Syst 58:109–130. https://doi.org/10.1016/S0169-7439(01)00155-1

Yadav VS, Singh AR, Raut RD, Mangla SR, Luthra S, Kumar A (2022) Exploring the application of industry 4.0 technologies in the agricultural food supply chain: a systematic literature review. Comp Ind Engg 169:108304

Yao H, Hruska Z, Kincaid R, Brown RL, Bhatnagar D, Cleveland TE (2013) Detecting maize inoculated with toxigenic and atoxigenic fungal strains with fluorescence hyperspectral imagery. Biosyst Eng 115:125–135. https://doi.org/10.1016/j.biosystemseng.2013.03.006

Ye D, Sun L, Zou B, Zhang Q, Tan W, Che W (2018) Non-destructive prediction of protein content in wheat using NIRS. Spectrochim Acta – Part A Mol Biomol Spectrosc 189:463–472. https://doi.org/10.1016/j.saa.2017.08.055

Zhang C, Jiang H, Liu F, He Y (2017) Application of near-infrared hyperspectral imaging with variable selection methods to determine and visualize caffeine content of coffee beans. Food Bioprocess Technol 10:213–221. https://doi.org/10.1007/s11947-016-1809-8

Zhang R, Ying Y, Rao X, Li J (2012) Quality and safety assessment of food and agricultural products by hyperspectral fluorescence imaging. J Sci Food Agric 92:2397–2408. https://doi.org/10.1002/jsfa.5702

Zhao M, Shaikh S, Kang R, Markiewicz-Keszycka M (2020) Investigation of Raman spectroscopy (with fiber optic probe) and Chemometric data analysis for the determination of mineral content in aqueous infant formula. Foods 9. https://doi.org/10.3390/foods9080968

Zheng X, Li Y, Wei W, Peng Y (2019) Detection of adulteration with duck meat in minced lamb meat by using visible near-infrared hyperspectral imaging. Meat Sci 149:55–62. https://doi.org/10.1016/j.meatsci.2018.11.005

Chapter 13
Recent Trends in Nano Biosensors for Food Testing

Feby Luckose, Mamata Bangera, Nidhi S. Shetty, and Akshath Uchangi Satyaprasad

1 Introduction

The research and development of biosensors designed for the detection of food contaminants have gained significant interest in the past decade. The application of biosensors in the detection of food adulterants, microbial load, toxins, process control, packaging etc. is a multi and interdisciplinary field and one of the most active areas of research in analytical chemistry. In the past decade, considerable efforts were directed towards developing biosensors for the detection of food contaminants, adulterants, microbial load, and toxins, and more recently to aid process control and intelligent packaging of food. Employing biosensors in the food industry for quality control typically eliminates the requirement of sample preparation and purification which are indispensable practices for classical analytical methods as well as the need for skilled manpower and infrastructure. However, there are several parameters like selectivity, sensitivity, specificity, accuracy and precision that need to be extensively investigated to develop an efficient biosensing method. Parameters like reusability, operational and storage stability, portability and ease of use also play a crucial role in the development of robust biosensors for food quality analysis. Due to the complex nature of food, designing a commercial biosensor for specific food application poses significant challenges (Kissinger 2005). The success of biosensors in process monitoring is dependent on the reusability and stability of the platform. For instance, during online monitoring of sugar processing, the

F. Luckose · M. Bangera · N. S. Shetty · A. U. Satyaprasad (✉)
Nitte (Deemed to be University), Nitte University Centre for Science Education and Research, Department of Bio and Nanotechnology, Paneer Campus, Deralakatte, Mangalore, Karnataka, India
e-mail: akshath.uchangi@nitte.edu.in

H. U. Hebbar et al. (eds.), *Engineering Aspects of Food Quality and Safety*, Food Engineering Series, https://doi.org/10.1007/978-3-031-30683-9_13

biosensors should be invariably stable at different pH, temperature and fermentation environments (Ronkainen et al. 2010). Nevertheless, single-use biosensors for rapid detection of microbes/toxins/contaminates etc. have significant applications at user-end and quality monitoring agencies.

Nanobiosensors have played a crucial role in developing a new generation of food biosensors. The biosensors design consists of a transducing element, nanomaterial, biocompatible ligands on the surface, a bioreceptor and a detector. The choice of nanomaterial is crucial in developing an efficient biosensor. For sensitive detection using colourimetric or fluorometric outputs, gold nanoparticles (AuNPs) and quantum dots are preferred due to their unique opto-physical properties. Whereas, to develop an electrochemical biosensor gold/carbon-based materials are of choice due to their conductive properties. Nevertheless, food biosensors are still in infancy as compared to that biosensors in healthcare monitoring. There are many reasons for this and one of which is the matrix of food that spoils the nanomaterial during the oxidation/reduction process. The matrix itself is versatile in food considering the wide range of biomolecules. The present chapter introduces the reader to a few basic concepts of biosensors and gives insights into the recent trends in biosensing of food contaminants. Two main areas wherein considerable research to develop biosensors have taken place are: (a) quality control and detection of contaminants and (b) online monitoring of analytes like sugars, flavours, vitamins sweeteners and other products. This chapter explores the advancements in classical nano-bio sensors in the past decade and advantages/drawbacks of these systems and various technical obstacles/challenges that need to be addressed for their successful commercialization (Table 13.1).

2 Food Biosensors: Pre-requisites

To develop a stable and effective food biosensor, some of the prerequisites are indispensable. It should be noted that the food samples have a complex matrix and biosensors use samples directly without any pre-treatment. Some major factors to consider for the development of food biosensors are given below:

Selectivity Food consists of many components that are structurally related (small molecules) and difficult to detect in the matrix. Selectivity of the sensor towards the specific analyte is extremely important that decides the fate of the biosensor towards commercialization.

Sensitivity The detection sensitivity is crucial for molecules with low levels, like contaminates or process control etc. Sensitivity is the key issue in food biosensors mainly due to low levels and the matrix.

The Linearity of Response This feature provides insights on sensor performance and stability of recognition elements. This is also important in the detection of molecules over a concentration range.

Table 13.1 Major broad areas for biosensor development in food analysis

Food Composition		
Health labelling	Compositional labelling	Aesthetic labelling
Cholesterol Triglycerides Fatty acids Polyphenols Vitamins	Glucose Sucrose Lactose Organic acids	Flavour compounds (wine, tea) Polyphenols
Process control		
Fermentation process indicators		Freshness and storage indicators
Glucose Galactose Sorbitol Fructose Lactose Sucrose		Formaldehyde Lactic acid Malic acid Acetic acid Ethanol Methanol
Food safety		
Contaminants		Adulterants
Pesticides & Herbicides Fertilizers Antibiotics Pathogens Toxins Heavy metals		Starch Urea Formaldehyde Melamine Hydrogen peroxide Soda Industrial dyes Microplastics

Reproducibility of Signal Response The most important feature of any assay is its reproducibility in complex real-time samples.

Quick Response Time and Recovery Time Biosensors should be able to reduce the time taken by classical methods thereby enhancing the efficiency of the process itself. For reusability, the time required for recovery should be less.

Stability and Operating Life The biorecognition element partly decides the stability of the biosensing system. Stability is key for the accurate and long-term performance of the biosensors and also important in providing a decent operational life.

3 Classification of Biosensors

In general, biosensors consist of (a) a bio-recognition element that detects analyte (b) a platform where the reaction takes place (c) A transducing element that converts bio-signals to electrical (d) high-end user-interface software to convert the electrical signal to a digital one.

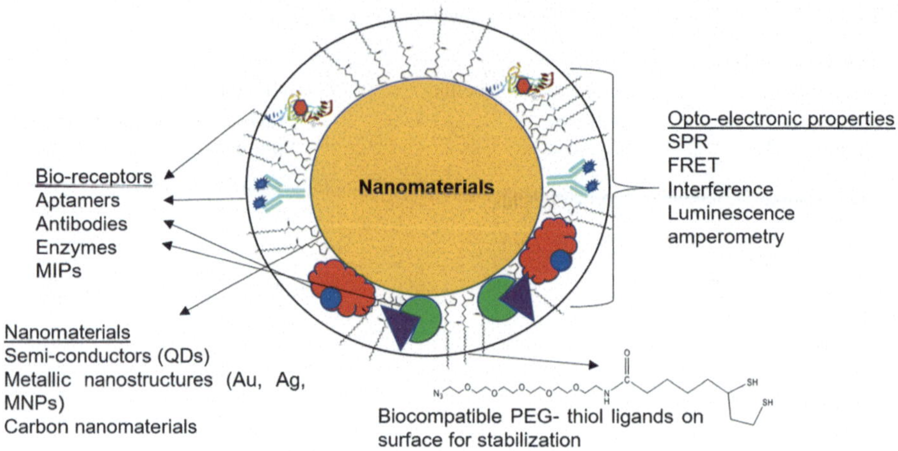

Fig. 13.1 Nanobiosensors: the core nanomaterial, biorecognition elements and optoelectronic properties used in biosensing

Based on the components that make up the nano-biosensor, they are divided as (Fig. 13.1):

(i) *Based on transduction element*s: electrochemical, optical, piezoelectric, and thermal/calorimetric biosensors (Kissinger, 2005).
(ii) *Based on biorecognition element*s: enzyme biosensors, immuno sensors, DNA sensors, etc.

4 Nanobiosensors Based on Transduction Element

4.1 Enzyme Based Electrochemical Nanobiosensors

Electrochemical biosensors are designed with a wide range of biorecognition elements like enzymes/antibodies/aptamers etc. for the development of commercial biosensors. Electrochemical biosensors have several advantages, (1) high selectivity-owing to their biological reaction (2) highly integrated-no need for sample pre-treatment or use of reagents thereby reducing cost (3) can be reduced to smaller size (4) biomolecule immobilized on a surface of a material can be reused several times and require less sample size (5) time required for output is very less (6) ease of integration in different materials-end user application is wide. A general method of the enzyme-based electrochemical biosensor is given in Fig. 13.2. It should be noted that most of the electrochemical biosensors in the market are enzyme biosensors owing to their specificity and selectivity.

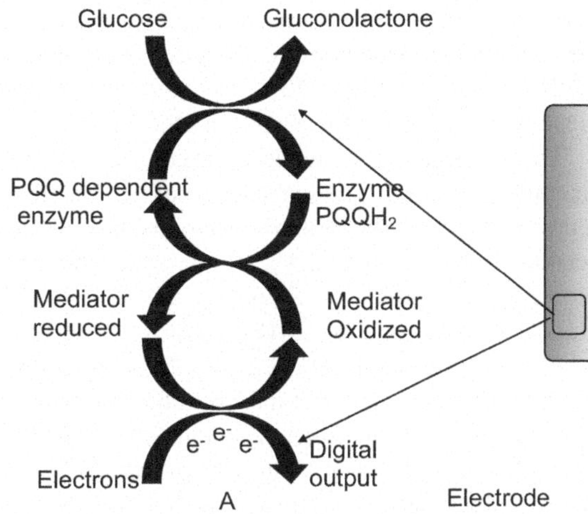

Fig. 13.2 Scheme of an electrochemical enzyme-biosensor. (Glucose biosensors)

Recent studies on electrochemical biosensors for the detection of food contaminants are focused on employing advanced surface chemistry methods. The reaction which is being monitored in an electrochemical biosensor typically generates a measurable current (amperometry), a measurable potential or charge accumulation (potentiometry) or alters the conductive properties of the medium between electrodes (conductometry) (Wang 2008). The three types of electrochemical biosensors based on the above output are: (a) Voltammetric/amperometric biosensor (b) Potentiometric biosensor (c) Conductometric biosensor.

4.1.1 Voltammetry/Amperometry

In general, the amperometric techniques work on applying a potential to a working electrode with a reference electrode and then measuring the current (Sawan et al. 2020). Voltammetry based detection of food contaminants is routinely used in food biosensors design and development. A recent study focused on the detection of heavy metals using 3-D printed metal electrodes. The design was sensitive and specific for lead and cadmium detection and performed well in comparison with glossy-carbon electrode-based methods (Lee et al. 2017). A nanofibrillated cellulose-based sensor was also developed for the detection of various heavy metals like Cd(II), Cu(II), Pb(II) and Hg(II) with appreciable sensitivity (Zinoubi et al. 2017).

In recent years, due to the ease of use and development of electrodes, the application of voltammetry in the detection of various food contaminants has significantly increased. An e-tongue based sensor for monitoring honey syrup contamination was developed by Sobrino-Gregorio et al. (2018). The voltammetry methods are routinely used in the detection of glucose (Amatatongchai et al. 2017), cholesterol (Jayanthi & Suja (2020), polyphenols (Alpar et al. 2018), dyes (Nayak & Shetti (2016) and many more (Ghasemi-Varnamkhasti et al. 2018). The voltammetry methods are versatile and have gained significant interest due to advancements in electrode design and fabrication methods.

The amperometric method, in contrast to voltammetry, differs in absence of scanning potential. The current/signal generated from electrochemical oxidation or reduction is directly monitored against a reference electrode. Amperometric biosensors have dominated the food biosensors research with the development of biosensors for glucose (Guzsvány et al. 2019), lactic acid (Vargas et al. 2016), cholesterol (Cevik et al. 2018), pesticides (Nasir et al. 2017; Song et al. 2019) and others (Artigues et al. 2017). It should be noted that in amperometric biosensors the oxidation/reduction potential is characteristic of enzymatic reaction/antigen-antibody interactions. Recent advancements in electrode fabrication and immobilization had led to the development of many biosensors for the detection of antigens (Ruiz-Valdepeñas Montiel et al. 2016), and toxins (Li et al. 2016), small molecules (Yu et al. 2017) etc. which are extremely important. In general, amperometric detection is simple, low-cost and easily designable which makes them an attractive choice for enzymes/affinity-based biosensors. To date, amperometric based biosensors rule the market both in clinical/food biosensors considering their advantages (Ibadullaeva et al. 2019). Nevertheless, these systems also pose many drawbacks in food samples due to the complex matrix effect leading to non-specific oxidation/reduction. Also, a pre-calibrated reference electrode is very much essential for field application which makes the design challenging.

4.1.2 Potentiometry

The commonly used potentiometric sensors are ion-selective probes and pH electrodes. In potentiometry, the potential of the electrochemical cells is measured. The advancements in the fabrication of pH-sensitive electrodes have led to the development of novel sensors for many contaminants. The reason for potentiometric biosensor attraction is its easy design and fabrication. Recent trends in potentiometric biosensors include the detection of toxins (Stepurska et al. 2015), microbial load (Shaibani et al. 2018), and trace metals (Ayenimo & Adeloju (2015)) etc. For developing a successful biosensor using this method, bio-recognition elements like enzymes, antibodies, and aptamers are commonly used. Nevertheless, the drawbacks of these systems are their poor performance and relatively high cost of fabrication. There are many recent review articles focused on food biosensors based on various transducing elements (Zeng et al. 2016; Neethirajan et al. 2018; Riu and Giussani 2020; Zhou & Tang (2020). We recommend the readers go through some of these for more information.

4.2 Optical Nanobiosensors

Optical biosensors exploit light absorption, fluorescence, luminescence, reflectance, Raman scattering, refractive index and other such optical techniques for measuring a chemical or biological species. These sensors have been used for the detection of toxins, microorganisms (bacteria and viruses), spores, proteins, and other small molecules from the air, water, environment (soil), food, and clinical specimens (Mendonca and Bhunia 2015). A hand-held highly sensitive fibre optic milk fat sensor that uses U-bent plastic optical fibre (POF) probes based on the refractive index (RI) of milk (an inherent physicochemical property of milk, which is significantly influenced by the milk fat content) has also been reported (Gowri et al. 2019).

Given the advanced determination of the pathogen and to decrease the risks of human diseases caused by *Salmonella*-antigens microorganism was studied (Viter et al. 2017). Photoluminescence of TiO_2 nanoparticles was applied for the detection of *Salmonella typhimurium*. The optical immunosensor sensitivity towards *Salmonella*-Ag ranged from 10^3 to 10^5 cells/mL. Improvement is needed for more rapid, accurate, and multiplex sensing of pathogens. There are challenges in identifying the nanomaterial-based optical biosensors.

4.2.1 Evanescent Wave Fibre Optic

When light propagates through a fibre optic based on total internal reflection (TIR), a thin electromagnetic field (the “evanescent wave”) generated decays exponentially with the distance from the interface with a typical penetration depth of up to several hundred nanometers. Fibre-optic evanescent wave (FOEW) sensors are promising in pollutant detection and evaluation of water quality because of their high resistance to corrosion, smart structure, anti-electromagnetic interference, and low cost (Jiao et al. 2020). An optical enzymatic biosensor for rapid and point-of-use detection of antibiotics in food and water has been developed (Nag et al. 2021). Enzymatic hydrolysis of β-lactams, on the electroactive polyaniline nanofibers, altered the polymeric backbone of the nanofibers, from emeraldine base form to emeraldine salt, which was measured as an increase in evanescent wave absorbance at 435 nm. The sensor developed and tested has a quick solution for the measurement of β-lactam residues in food and the environment. Based on the FE computation, a fibre evanescent wave spectroscopic (FEWS) sensor consisting of a 40 cm Ge–Te–Se fibre, coupled with Fourier transform infrared spectrometer and liquid nitrogen cooled mercury–cadmium–tellurium (MCT) detector was developed (Jiang and Jha 2015). The fabricated fibre sensor was used for the analysis of tocopherol (vitamin E), ascorbic acid (vitamin C), and fresh orange and lemon juice. A dual-colour fluorescence resonance energy transfer (FRET) based apt sensor is described for simultaneous determination of the mycotoxins aflatoxin M1 (AFM1) and ochratoxin A (OTA) (Song et al. 2018). A compact dual-colour evanescent wave all-fibre detection system with two lasers (635 nm; red; and 405 nm; purple) was used for the simultaneous collection of two-wavelength fluorescence signals. Graphene sensitization of glucose-imprinted polymer (G-IP)-coated optical fibre has been introduced as a new biosensor for evanescent wave trapping on the polymer optical fibre to detect low-level glucose (Azargoshasb et al. 2020). An aptamer-based evanescent wave fibre (EWF) sensor has been developed to quantify melamine in milk. A miniaturized optical sensor was developed for fluoride determination in tea samples to evaluate their specific risks of fluorosis for public health based on evanescent-wave interaction (Xiong et al. 2017). Fibre optic technique for the measurement of very low concentrations of curcumin also is reported (Smrithi et al. 2015). A novel aptamer-based biosensing strategy based on an evanescent wave all-fibre (EWA) platform was developed to detect Ochratoxin (Wang & Alocilja (2015). In a simple target capturing step using aptamer-functionalized magnetic microbeads, signal probes conjugated with streptavidin are released and further detected by an EWA biosensor via a facial desthiobiotin–streptavidin recognition.

4.2.2 SPR/SPR-Raman

Surface Plasmon Resonance (SPR) is an optical technique that allows label-free detection of biomolecular interaction. This phenomenon relies on the excitation of electron plasma (surface plasmon) of a thin metallic layer (gold or silver) on the surface of the waveguide. It is the basis of many standard tools that measure the adsorption of material onto the surface of planar metal nanoparticles. (Fang et al. 2003). The biorecognition element is immobilized on the metal surface and minute changes in the refractive index during the biorecognition event are captured and produced as an output signal. Low-fouling SPR biosensor for multi-step detection of foodborne bacterial pathogens in complex food samples which rapidly (< 80 mins) detected bacterial pathogens in foods as low as <50 CFU/mL was developed. SPR was used to detect βLG and Ara h1 by immobilizing the affinity-purified monoclonal antibodies on the biosensor chip to detect milk and peanut allergen (Wu et al. 2016). The interaction of bovine serum albumin with ascorbyl palmitate and ascorbyl stearate using SPR was studied to design sensors to detect food additives (Fathi et al. 2018). A nanoparticle integrated gold chip for Aflatoxin B1 (AFB1) detection using SPR apparatus was also developed. An aptamer-based SPR biosensor was developed to detect Aflatoxins (Wu et al. 2018). An SPR biosensor using an anti-OTA aptamer immobilized sensor chip was developed to measure ochratoxin A (OTA) in wine and peanut oil quantitatively through a straightforward direct binding assay (Zhu et al. 2015). Biomimetic nanoparticles based SPR biosensors for histamine detection in foods were also reported (Rahtuvanoğlu et al. 2020).

4.2.3 Interferometry

Bio-layer interferometry (BLI) is an optical, label-free technology for measuring biomolecular interactions by analysing the interference pattern of white light reflected from a layer of biomolecules that are immobilized on the surface of a sensor tip (bio-layers) (Lou et al. 2016). This technique gives real-time interactions along with monitoring of binding specificity, association and dissociation rates of molecules during the interaction. A fibre optic sensor (FOS) has been employed for food composition detection by using a Mach–Zehnder Interferometer (MZI) structure. The FOS has been developed by using a fusion arc splicing technique. The device was tested for water, 1 mol sucrose solution and oil with the refractive index of 1.333, 1.384, and 1.464, respectively. The best sensitivity achieved is 4.413 nm/RIU for an 8 cm length of sensor region (Razak et al. 2018). Interferograms of samples of oil placed under different heating conditions to establish the changes in their quality were obtained using a Mach-Zehnder Interferometer using a beam of light from a HeNe laser of 10 mW at 632.8 nm. Based on the results, the project highlights the importance of the quality of the oils used in the food industry and shows how interferometry can be a useful tool for this purpose.

4.2.4 Luminescence

Luminescence is the emission of light either from biological or chemical substances upon interaction with specific molecules. It has wide applications in developing optical biosensors. Luminescence is broadly classified into two types: (a) Bioluminescence (b) Chemiluminescence.

4.2.5 Bioluminescence

Bioluminescence is the emission of energy from a living organism in the form of visible light. Specific chemical interaction of ATP with enzyme luciferase produces photons in cells, which can be quantified using a luminometer. Bioluminescence sensors use luminescent microorganism or their related genes (Narsaiah et al. 2012). Luminescence is measured before and after exposure to target molecules like toxic compounds and the changes in photon emission are calculated. This system combines the advantages of biotoxicity testing and instrumental precision (Ranjan et al. 2012). The adenosine triphosphate (ATP) is the energy currency of all living microbes and can be used as a rapid indicator of microbial viability. An ATP bioluminescence-sensing assay was developed to detect microbial viability. A bioluminescent recombinant *Escherichia coli* strain was used with luciferase extracted from transformed bacteria. Bacterial counts from food samples were detected using a sensing assay and were compared with the Traditional Plate Counting method. Compared with the plate-counting method, the ATP bioluminescence-sensing assay is a rapid and efficient approach for detecting microbial viability.

4.3 *Piezoelectric Nanobiosensors*

Piezoelectric biosensors rely on the generation and transmission of acoustic waves through the crystal of a quartz disc. The wave frequency depends on crystal properties and the mass deposited onto the crystal surface, which allows for the detection of small mass changes on the crystal surface (Ronkainen et al. 2010; Rout and Chakraborty 2020). Volatile chemical compounds including food aromas from classes of aldehydes, alcohols, esters, hydrocarbons and ketones known to be detectable by piezoelectric biosensors (Mascini et al. 2017). Cofactors and prosthetic groups are known from biochemistry and can be simply utilized as recognition parts of a biosensor because of their specificity to enzymatic substrates. The electrode of a Quartz Crystal Microbalance was covered with tetraphenyl porphyrin and adsorption of zinc nitrate was recorded by this piezoelectric sensor (Yahia et al. 2016). An electrochemical quartz crystal microbalance (EQCM) based label-free immunosensor has been developed for the quantitative detection of aflatoxin B1 (AFB1) in groundnut (Chauhan et al. 2015). The Au coated quartz crystal (6 MHz) functionalized with a self-assembled monolayer of 4-amino thiophenol has been utilized to immobilize anti-aflatoxin antibodies. Gold nanoparticles were used as a nanostructuring agent on quartz crystal sensor chips to engineer *Staphylococcal* enterotoxin A piezoelectric biosensors with an amplified response (Haddada et al. 2018). Though the commercialization of piezoelectric biosensors has not been attempted yet, there is a lot of scope for developing technologies for the mass production of specific materials like nano-piezoelectric biosensors for their application in the field of food science.

4.4 *Thermometric Nanobiosensors*

Thermometric biosensors also known as calorimetric biosensors are based on biological reactions involving heat generation. The temperature difference between the substrate entering a reactor and the product is sensed and displayed as an output signal. Thermometric sensors are used for the determination of cholesterol, uric acid, penicillin, etc. (Wang 2008; Patel and Doddamani 2019). Recently, Xu et al. (2017) developed a thermometric biosensor device for rapid detection and quantification of Diazepam, a drug that is notoriously used for 'spiking' drinks and beverages in drug-facilitated sexual assault and robberies as well as in adulteration of herbal medicine and functional foods. The biosensor measures heat production from the enzyme label in a thermometric enzyme-linked immunosorbent assay (TELISA). The TELISA strategy is based on the competitive inhibition of the antigen-antibody reaction, which is reflected by thermal signals generated when enzyme-labeled analytes are catalytically degraded by the substrate. The device is capable of detecting diazepam at the lowest limit of 33.71 ng/ml with a linear range from 45.37 to 726.71 ng/ml and shows an excellent correlation with the HPLC measurements.

4.5 Nanoparticles as Enhancers of Transduction

In the past two decades, advancements in nanotechnology-based biosensing have gained significant interest. Due to the properties like conductivity, surface plasmon resonance (SPR), fluorescence resonance energy transfer (FRET), fluorescence quenching etc. the nanoparticles are used in biosensing (Sharma et al. 2015, Wang & Duncan (2017). Due to the unique opto-physical properties, the sensitivity of nanosensors has dramatically increased and can detect analytes at trace levels. The amalgamation of biological receptors with nanoparticles has led to the development of ultrasensitive yet specific biosensors which have revolutionized food biosensor research. Advancements in immobilization of enzymes/antibodies/aptamers, surface chemistry, fabrication etc. have a large impact on biosensing systems (Chen et al. 2017; Saha et al. 2017; Kim et al. 2016)). Among the various nanomaterials used zinc oxide, gold nanoparticles (GNPs), quantum dots, carbon nanotubes (CNTs) and graphene have received considerable interest in the development of sensitive biosensors.

4.5.1 Zinc Oxide

Due to the unique electronic properties like wide bandgap, piezo-electric, semiconducting and poly electric properties, ZnO nanoparticles are used extensively in optics, electronic devices and sensors. There are many studies wherein ZnO is used in electrochemical/amperometric biosensors for the detection of glucose (Naderi Asrami et al (2020), and lactate (Uzunoglu et al. 2016), methyl-glyoxal (Jayaprakasan et al. 2018) etc. Since ZnO exhibits high surface areas and is relatively non-toxic, stable and biocompatible, it has been widely used in food contaminants detection wherein enzymes are used as biorecognition elements (Beitollahi et al. 2020). Since in food samples, the complex matrix can give rise to non-specific response, the ability of ZnO to communicate electrons efficiently makes them a choice for electrochemical biosensors.

4.5.2 Gold Nanoparticles

Amongst all nanoparticles, gold nanoparticles (AuNPs) are a major transducer used in a wide variety of biosensors. The AuNPs offer an excellent biocompatible platform for stable immobilization of receptors due to their large-surface volume ratio. AuNPs are normally used as wires to connect the receptor with the electrode surface. Advancements in surface chemistry techniques for immobilization have enhanced the performance of biosensors (Sandhyarani 2019). Some recent trends include the detection of toxins (Li et al. 2020; Mazur et al. 2020), bacteria (Wang et al. 2015; Pissuwan et al. 2020), antibiotics (Peng et al. 2017; Ghanbari & Roushani (2018) etc. using electrochemical biosensors. Recent trends in using colorimetric platforms for the detection of food contaminants are seen. Several studies focus on contaminates that are very important to food safety like pesticides (Malarkodi et al. 2017; Xu et al. 2020) The immediate Yes or No tests can be of great benefit for the users and can be used as preliminary screening (Aldewachi et al. 2018). There are numerous studies on the use of AuNPs for the colorimetric detection of bacteria (Wachiralurpan et al. 2018; Shahbazi et al. 2018), toxins (Khan et al. 2015; Kong et al. 2016), antibiotics (Peng et al. 2016; Ramezani et al. 2016) and many more. The integration of the colorimetric output into a smart-phone based detection has made these tests very user-friendly (Zhong et al. 2019). There are many attempts to develop smart-phone assisted biosensors for the detection of bacteria (Zheng et al. 2019; Cheng et al. 2017), sugars (Nelis et al. 2020), antibiotics (Wu et al. 2019; Luo et al. 2021), etc. Optical monitoring of enzymes using gold nanoparticles has been the target of recent studies. Fluorescence quenching capability by AuNPs was exploited in the detection of formaldehyde, which involved NADH-mediated AuNP synthesis, responsible for the turn-on of fluorescence (Akshath and Bhatt 2016). The unique properties of AuNPs like surface-enhanced Raman scattering, size and aggregate-state dependent absorption, and SPR signal amplification has been used for diagnostic and biosensing applications. AuNPs with a dense layer of glycans is an efficient probe for multivalent lectin-glycan interactions. These glycoconjugates possess the ability to block binding targets of lectin, thereby inhibiting viral infection(Budhadev et al. 2020). Due to the unique opto-physical properties, AuNPs are the most sought nanoparticles to design both electrochemical and colorimetric detection systems for rapid and sensitive detection.

4.5.3 Quantum Dots

Quantum dots (QDs) possess unique optical properties arising from quantum confinement effects. The properties like high photostability, tunable emission, broad absorption and narrow emission properties make them attractive choices for biosensor systems (Reshma & Mohanan (2019). The spectral cross-talk has been extensively used for Forster resonance energy transfer (FRET) based biosensors for the detection of toxins, antibiotics (Song et al. 2015), bacteria (Mohamadi et al. 2017), pesticides (Karadurmus et al. 2021) and other food contaminants (Arora 2018). In our earlier studies, (Akshath and Bhatt 2018) we exploited locking and unlocking the interaction between QD-AuNP pair by forming nanoparticle hybrids. Quenching interaction between QD-GNP pair was unlocked by formaldehyde dependent dehydrogenase activity leading to QD fluorescence turn-on (Fig. 13.3). This phenomenon was applied for the successful detection of formaldehyde in fruit juice, wine and milk samples.

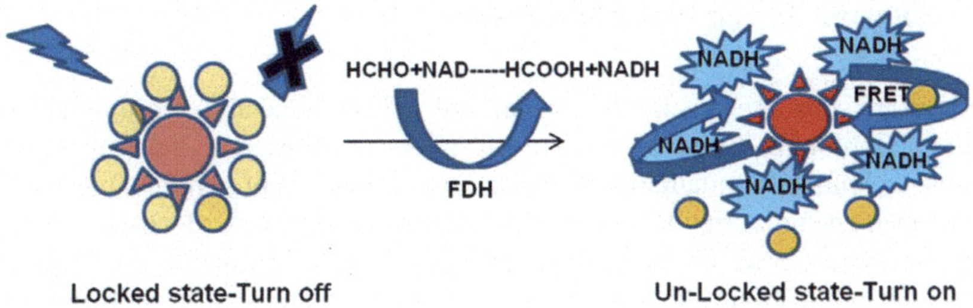

Fig. 13.3 Enzyme based FRET reaction. (Reproduced and reprinted with permission from Akshath & Bhatt (2018), copyright 2018 with permission)

4.5.4 Graphene

Graphene-based biosensors have achieved significant technological advancements in fabrication and miniaturization. Due to its unique optoelectrical properties and fast electron transfer, graphene is a choice for biosensors fabrication (Bobrinetskiy & Knezevic (2018). Recent studies are focused on the immobilization of antibodies (Hashemi et al. 2020), enzymes (Bolibok et al. 2017), aptamers (Ping et al. 2015) and many more biorecognition elements for the development of biosensors. The graphene derivatives like graphene oxide and graphene quantum dots are a leading nano-composite material extensively used in biosensors fabrication. Recent work is focused on the development of graphene-based biosensors for glucose, antibiotics (Tan et al. 2016), pesticides (Zhu et al. 2020) etc. The success of these biosensors relies on the effective immobilisation of biorecognition elements that can help to achieve stability and selectivity of biosensors. Nanocomposite based fabrication is popular due to its unique optoelectronic properties. The graphene-based materials have huge scope for the development of commercial biosensors due to fabrication advancements and can be a market leader soon.

4.5.5 Carbon Nano-Tubes (CNTs)

Carbon nanotubes have unique properties of high electro-catalysis and fast electron transfer. The layers of CNTs, multiwall and single wall (MWNT/SWNTs) have been extensively used in the past decade to fabricate nanoelectrode-based biosensors (Zaporotskova et al. 2016). Efficient receptor immobilization on surfaces and the development of miniaturized biosensors are being studied in recent times. There are many reports on the detection of various analytes including heavy metals (Jeromiyas et al. 2019), pesticides (Saraji et al. 2016), bacteria (Akbari et al. 2015) etc. The biocompatibility of CNTs is a huge benefit for the immobilization of various receptors on the surface. There are numerous reports and readers can go through some of these reviews for additional information on state-of-the-art biosensors (Rotariu et al. 2016).

4.6 Enzyme Based Nanobiosensors

Enzymes are substrate-specific biological molecules that catalyze specific chemical reactions. Exploiting this phenomenon, enzyme-based biosensors have emerged as a valuable technique for qualitative and quantitative analysis of several target analytes in food processing, agriculture, medicine, pharmaceuticals, food safety and monitoring etc. (Ispas et al. 2012; Kurbanoglu et al. 2020). The major reason for their widespread application is their use for on-site detection and significant benefits over classical analytical methods including high sensitivity, selectivity, no or less clean-up of samples etc. (Asal et al. 2018). Classical electrochemical enzyme biosensors are based on oxidoreductase enzyme bio-catalytic events coupled with amperometric detection mediated by an electron transfer mechanism (Algar et al. 2010, Costa-Fernández 2006). The most successful application of enzyme biosensors commercially has been the self-monitoring blood glucose biosensor (Accu Check, One Touch) (Ispas et al. 2012).

An NAD^+ dependent formaldehyde dehydrogenase enzyme was used to develop a biosensing platform for the ultrasensitive detection of formaldehyde (Akshath et al. 2012). The spectral cross-talk of NAD^+/NADH with QDs was used to monitor the target analyte levels. It was proposed that CdTe QD may undergo dipolar interaction with NADH as a result of broad spectral absorption due to multiple excitonic states resulting from quantum confinement effects. NAD^+ acts as an electron acceptor and gets reduced to NADH during the dehydrogenase reaction. Since NADH from the enzymatic reaction is a fluorescent compound, CdTe QD was used as a "plug-in" for the NAD^+ co-factor which can route the fluorescent energy from the formed NADH (Fig. 13.4). It was possible to detect formaldehyde in the range 1000 ng/mL-0.01 ng/mL with a limit of detection (LOD) of 0.01 ng/mL using the proposed method.

Phenolic compounds from industrial effluents contaminate water bodies and are environmental pollutants. In recent times, remarkable efforts have been directed to

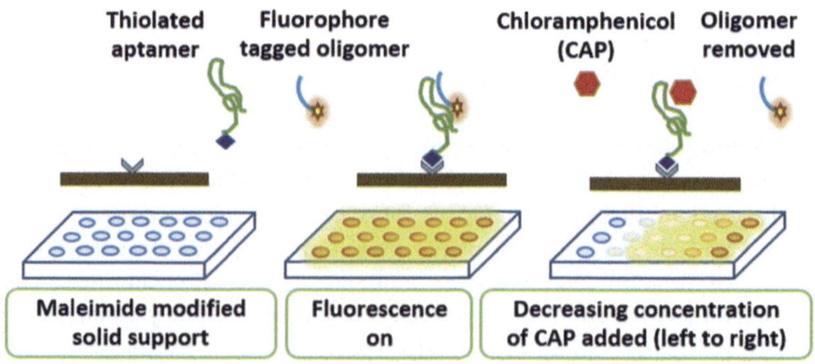

Fig. 13.4 Aptasensors for chloramphenicol detection. (Reproduced and reprinted with permission from Sharma et al. 2019, copyright 2019 with permission)

the development of polyphenol oxidase-based biosensors that can be applied to detect phenolic compounds in various biological and non-biological materials and can be standardized for food, environmental and medical industries (Campaña et al. 2019; Della Pelle and Compagnone 2018). Enzyme immobilization to avoid enzyme leakage and its stability are two major challenges that need to be addressed for this technology to be standardized for food, environmental and medical industries. The application of nanoparticles to preserve and/or improve enzyme activity needs to be investigated (Gul et al. 2017; Raymundo-Pereira et al. 2020).

Sucrose content is a good discriminant marker for the beverage quality as it is responsible for improving the aroma and overall flavour of the coffee. Stredansky et al. (2018) developed an amperometric multi enzymatic biosensor utilizing FAD-dependent glucose dehydrogenase (FAD-GDH) for selective quantification of sucrose in green coffee. FAD-GDH was co-immobilized with invertase and mutarotase on a thin-layer gold planar electrode using chitosan. The biosensor showed wide linearity (from 10 to 1200 μM), low detection limit (8.4 μM), and a fast response time (50 s) and showed good operational (3 days) and storage (1 year) stability.

Electrochemical tyrosinase biosensors for tyramine determination were developed by the immobilization of the enzyme in calcium phosphate material (brushite cement) followed by cross-linking with glutaraldehyde (Lopez et al. 2017). Tyramine is a toxic biogenic amine that is accumulated during the fermentation of cheese and is an indicator of food quality, as its levels gradually increase as a consequence of microbial contamination. Under optimal conditions (time of 15 min, a ratio of 1.0 and 50 μg of the brushite-enzyme mixture, 20 °C and pH 6,0), a linear range of 5.8×10^{-7} to 1.6×10^{-5}, sensitivity 1.50×10^{3} mA/ M.cm^{2}, detection limit, 4.85×10^{-8} M and a response time, 6 s were obtained.

The application of enzyme-based biosensors for pesticide detection has been explored. Acetylcholinesterase biosensors have emerged as simple, rapid and ultra-sensitive tools for toxicity detection of organophosphorus pesticides in the environment and food. Current research studies are directed toward the development of acetylcholinesterase biosensing systems with improved sensitivity, selectivity and analytical performance. In this regard, the utilization of nanomaterials could improve the efficiency of enzyme-based biosensors (Songa and Okonkwo 2016).

Compared with natural enzymes, enzymes offer higher efficiency and biocatalytic activity, prolonged life and high stability at lower cost and are now being studied for their applications in biosensing (Gao et al. 2007). Nanozyme-based biosensors can be used for the detection of ions, foodborne pathogens and microbial metabolites, and chemical contaminants such as pesticide residues veterinary drug residues and melamine in food. Future studies to enhance their catalysis ability can lead to more diverse applications of enzymes for biosensing in food and non-food materials (Wang and Gunasekaran 2020).

Enzyme based biosensors are reliable analytical devices and hold great potential for commercialization. The main challenge in developing enzyme-based biosensors is an immobilization platform which is crucial for the stability, shelf life, reproducibility and repeatability of the biosensors. The increased demand for biosensors as important tools in food processing has generated an increased need for research related to enzyme-based biosensors and in the years to come, the development of sophisticated and more efficient devices is foreseen (Kurbanoglu et al. 2020).

4.7 Immuno Nanobiosensors

Developing sensor technologies that are rapid and highly sensitive and selective for the detection of toxins is a high priority for food scientists and microbiologists. Immunosensors have been regarded as the gold-standard technique for biosensing in several fields including environmental monitoring and clinical diagnostics. An immunosensor relies on highly specific interactions of the two binding sites of an antibody with a particular molecule which is detected by the transducer (e.g., optical or electronic). Due to their unique ability to recognize and bind to other molecules (antigens) or specific structures, antibodies make an important component in biosensing (Ronkainen et al. 2010). Immunosensors are highly specific, require very small sample volumes, and have low detection limits. Moreover, they require little or no sample preparation, minimal use of chemicals, and are highly repeatable. These advantages give them an edge over immunological methods like ELISA (D'costa et al. 1986). There are three general formats of an immunosensor: (a) Competitive assay (b) Uncompetitive assay (c) Sandwich assay. Bu et al. (2019) developed a novel immunochromatographic assay for the determination of ochratoxin A. Microorganism loaded AuNP composites were synthesized by reacting Yeast/Lactobacillus with AuNPs via biosorption and subsequent bio-reduction. Antibodies loaded on the carriers, can be controlled for rapid and selective detection of ochratoxin A. The assay was successfully assessed for the detection of ochratoxin A in rice, corn, ginger and green bean samples exhibiting high sensitivity (with a low detection limit of 0.1 ng/mL). The microorganism loaded gold nanoparticle composites could be a potential substitute to construct immunosensors for biomolecules detection. Label-free immunosensor based on one-step electrodeposition of chitosan-gold nanoparticles biocompatible film on gold microelectrode for determination of aflatoxin B1 in maize was developed by Ma et al. (2016). The nano biocompatible film and assay format showed promising results with high sensitivity in lower concentration, low detection limit (0.06 ng/ml) and fast detection time (12 minutes). In another study published in the same year, a label-free electrochemical quartz crystal microbalance (EQCM) based immunosensor using a self-assembled monolayer of hexandithiol (HDT), cysteamine and 3D gold nanoparticles (AuNPs) was developed for aflatoxin B1 detection (Chauhan et al. 2016). Such an approach can be used to develop the sensor for other food mycotoxins such as ochratoxin (A and B), fumonisins etc. Solanki et al. (2017) fabricated an immunosensor based on bismuth oxide nanorods, electrophoretically deposited onto an indium-tin-oxide coated glass substrate. Anti-aflatoxin monoclonal antibodies and bovine serum albumin for aflatoxin B1 detection were immobilized on the sensor. Bismuth oxide nanoparticles maintain the biological activity of the antibody and improve the electron transfer between the analyte (aflatoxin B1) and immune-electrode surface resulting in enhanced sensitivity and selectivity for the toxin. A platform of Poly (3, 4-ethylene dioxythiophene) and graphene oxide composite loaded with spherical gold nanoparticles has been developed for rapid electrochemical detection of aflatoxin B1 (Sharma et al. 2018). The immunoassay was also applied for analysis of maize samples spiked with AFB1 and the sensitivity was found to be 11.81 μA ng/ml within the linear range 0.1 ng/ml – 1.8 ng/ml.

5 Receptor Based Nanobiosensors

Receptors are "small molecules" that are impregnated in the cellular membrane that specifically binds to their target analytes resulting in metabolic changes (Taylor et al. 1988). In biosensing, a change in conformation of the receptor upon binding to a target molecule is captured and translated into an optical/electrochemical signal (Wei et al. 2012). Receptors unlike antibodies are not specific for a single analyte but are common for a group of target analytes. Therefore, receptor-based biosensors may not be suitable for biosensor applications which require high specificity. Odour receptors are prime candidates as sensing elements for biosensors as they can distinguish diverse natural and synthetic volatile organic chemicals (VOCs) Bioelectronic odour receptor-based sensors generally utilize an odour receptor embedded in natural (nanovesicles, or artificial (nanodiscs) membranes (Barbosa et al. 2018; Bohbot and Vernick 2020). While most studies in the past have utilized mammalian odour receptors, a few recent studies explored the use of insect odour receptors as the sensing elements in a receptor-based biosensor. For instance, in a recent study, odorant receptors from the common fruit fly, *Drosophila melanogaster*, were recombinantly expressed, purified and integrated into nano-liposomes (100–200 nm) which were immobilized onto a gold surface, thus enabling the development of a novel sensor which could sensitively and selectively detect odorants down to femtomolar levels for food, agricultural and medical applications (Khadka et al. 2019).

6 Nucleic Acid-Based Nanobiosensors

Affinity biosensors based on RNA/DNA coupled with nanoparticles have been widely used for the detection of various analytes. Aptamers are short, structured, single-stranded sequences of DNA, RNA or peptides that can bind with high affinity, specificity and selectivity to a wide range of targets. Aptamers are generated by a selection process known as SELEX (Systematic Evolution of Ligands by

Exponential enrichment). The ability of aptamer to fold into three-dimensional structures facilitates the interaction with target molecules with high-affinity constants. Aptamer-based colorimetric detection for mercury ions by a simple, sensitive, and portable Hg^{2+} detection system on a smartphone was developed by Xiao et al. (2016). In this work, a smartphone equipped with a light meter app was used to detect based on the colorimetric readout of the aptamer nanosensor upon specific interaction between the aptamer and Hg^{2+} and used in the analysis of both tap water and Pearl River water samples. Recent work by Sharma et al. (2019) focussed on the detection of chloramphenicol using a molecular beacon-based fluorescent sensing method (Fig. 13.4). Fluorophore- and quencher-tagged oligonucleotides complementary to aptamer recognizing chloramphenicol were hybridized to construct an apt switch sensor complex. On analyte binding, the fluorescence was turned on, and the linear range of detection of chloramphenicol was observed to be from 10 pg/ml to 10^7 pg/ml of CAP and LOD in buffer was estimated to be 0.987 pg/ml.

7 Commercially Available Food Biosensors and Future Directions

Biosensors have found applications in almost all fields of engineering and science since the launch of the first commercial biosensor in 1987. Exponential growth in food biosensors research seen in the past two decades has led to the development of many biosensing methods, publications, patents and some products.

Unfortunately, the number of scientific communications published every year remains high, while the number of market-ready commercial products is abysmally small. The number of patent applications over the last decade on food biosensors has drastically come down. There may be many reasons for this (a) market potential for individual analytes amongst the general public may be less (b) The matrix effect in food samples makes them extremely hard to make specific biosensors (c) Non-specific interactions that block the sensor surface and lead to electrode fouling (d) the regulatory process involved in food biosensors development (e) not accepted by agencies for screening or commercial use.

Table 13.2 provides some of the commercial food sensors and the companies manufacturing them (Bahadır and Sezgintürk 2015; Kokkali and van Delft 2014; Valderrama et al. 2016; Adley 2014). Recent reviews provide an up-to-date information on these biosensors (Scognamiglio et al. 2014; Antonacci et al. 2016).

Biosensors are mainly focused on sugar detection, gas sensors, humidity detection pH sensors, ATP based microbial load sensors, etc. With emerging contaminants like pathogens, toxins, small molecules etc., there is a large scope to develop biosensors to detect a wide range of analytes. The biosensor should be robust, and highly stable considering the matrix in samples like milk, meat, fish, cereal, vegetable products etc. It is no surprise that the academic publications are not translated

Table 13.2 Commercially available food biosensors

Company	Products related	Link
DiagnoSwiss	Development and commercialization of biochip technologies for food, pharmaceuticals health care etc	http://www.diagnoswiss.com/
NeogenCorporation	Products are focused on the detection of pathogens in raw ingredients and finished food products. Test kits to detect foodborne bacteria, mycotoxins, allergens, genetic modifications, drug residues, plant diseases etc. are developed.	http://foodsafety.neogen.com/en/
Roche diagnostics AG	Biosensors used for analyzing blood samples	http://www.roche-diagnostics.ch/
IBA GmbH	Sucrose, glucose, alcohol etc	http://www.iba-lifesciences.com/product-shop.html
Hygiene	Freshness metre: Degradationproducts of ATP	http://www.hygiena.com/other-monitoring-systems-home.html
Biosentec	Lactate biosensors	http://www.biosentec.fr/en/products.html
Sensolytics	Glucose, lactate, sucrose, ethanol, glutamate	http://www.sensolytics.de/en/products/microelectrodes.html
Rapid biosensors	Microorganisms and toxic substances	http://www.rapidbiosensor.com/
Biosensors S.L.	Online monitoring of the concentration of total microorganisms in water samples in15 min	http://www.biosensores.com/EN/empresa.php
Texas instruments Inc.	Sensors for gas and pH determination in the food industry	http://www.ti.com/lsds/ti/applications/industrial/medical/overview.page
Integrated genetics	DNA probes for detection of microbial contamination	http://www.integratedgenetics.com/test-menu
Molecular devices corporation	Sensors and components for microbial determination	https://www.moleculardevices.com/
Nova biomedical	Glucose, L-lactate, L-glutamate, L-glutamine, alcohol, sucrose, methanol, ammonia	http://www.novabio.us/
Novas insignia technologies	Gas e modified atmosphere	http://insigniatechnologies.com/
Massachusetts institute of technology	Detection of E. coli O157:H7 in lettuce (Canary)	http://www.atrp.gatech.edu/pt16-3/16-3_p1.html
Universitat Autonoma de Barcelona in collaboration with CSIC	Detection of atrazine traces	http://www.uab.cat/PDF/PDF_1179901814052_en.pdf

(continued)

Table 13.2 (continued)

Company	Products related	Link
Research international	Proteins, toxins, virus, bacteria, spores, and fungi (simultaneous analysis)	http://www.resrchintl.com/
Yellow Springs instruments	Glucose, sucrose, lactose, L-lactate, galactose, L-glutamate, ethanol, H2O2, starch, glutamine, choline	https://www.ysi.com/
Biacore AB	Vitamins, chemical veterinary residues, and mycotoxins	https://www.biacore.com/lifesciences/index.html

to the market due to the above reasons. (http://www.freshplaza.com/article/127601/India-Formalin-detection-kit-found-unfit-to-test-fruits).

The food industries demand the development of sensitive and selective biosensors that can match up to classical detection methods. There is a need to have a focused approach and use advanced fabrication systems to develop food biosensors as in the case of glucometers.

References

Adley CC (2014) Past, present and future of sensors in food production. Foods. Aug 19;3(3): 491–510

Akbari E, Buntat Z, Afroozeh A, Zeinalinezhad A, Nikoukar A (2015) Escherichia coli bacteria detection by using a graphene-based biosensor. IET Nanobiotechnol 9(5):273–279

Akshath US, Bhatt P (2016) Gold nanoparticle synthesis coupled to fluorescence turn-on for sensitive detection of formaldehyde using formaldehyde dehydrogenase. RSC Adv 6(60): 54777–54784

Akshath US, Bhatt P (2018) Supramolecular nano-sniffers for ultrasensitive detection of formaldehyde. Biosens Bioelectron 100:201–207

Akshath US, Vinayaka AC, Thakur MS (2012) Quantum dots as nano plug-ins for efficient NADH resonance energy routing. Biosensors and Bioelectronics. Oct 1;38(1):411–5

Aldewachi H, Chalati T, Woodroofe MN, Bricklebank N, Sharrack B, Gardiner P (2018) Gold nanoparticle-based colourimetric biosensors. Nanoscale 10(1):18–33

Algar WR, Tavares AJ, Krull UJ (2010) Beyond labels: a review of the application of quantum dots as integrated components of assays, bioprobes, and biosensors utilizing optical transduction. Analytica chimica acta. Jul 12;673(1):1–25

Alpar N, Yardım Y, Şentürk Z (2018) Selective and simultaneous determination of total chlorogenic acids, vanillin and caffeine in foods and beverages by adsorptive stripping voltammetry using a cathodically pretreated boron-doped diamond electrode. Sensors Actuators B Chem 257:398–408

Antonacci A, Arduini F, Moscone Dinia D, Palleschi G, Scognamiglio V (2016) Commercially available (bio) sensors in the agrifood sector. Comprehensive Analytical Chemistry. 315–40

Amatatongchai M, Sroysee W, Chairam S, Nacapricha D (2017) Amperometric flow injection analysis of glucose using immobilized glucose oxidase on nano-composite carbon nanotubes-platinum nanoparticles carbon paste electrode. Talanta 166:420–427

Arora K (2018) Advances in nano-based biosensors for food and agriculture. In: Nanotechnology, food security and water treatment. Springer, Cham, pp 1–52

Artigues M, Abellà J, Colominas S (2017) Analytical parameters of an amperometric glucose biosensor for fast analysis in food samples. Sensors 17(11):2620

Asal M, Özen Ö, Şahinler M, Polatoğlu İ (2018) Recent developments in enzyme, DNA and immuno-based biosensors. Sensors 18(6):1924

Ayenimo JG, Adeloju SB (2015) Inhibitive potentiometric detection of trace metals with ultrathin polypyrrole glucose oxidase biosensor. Talanta 137:62–70

Azargoshasb T, Navid HA, Parvizi R, Heidari H (2020) Evanescent wave optical trapping and sensing on polymer optical fibres for ultra-trace detection of glucose. ACS Omega 5(35): 22046–22056

Bahadır EB, Sezgintürk MK (2015) Applications of commercial biosensors in clinical, food, environmental, and biothreat/biowarfare analyses. Analytical biochemistry. Jun 1;478:107–20

Barbosa AJ, Oliveira AR, Roque AC (2018) Protein-and peptide-based biosensors in artificial olfaction. Trends Biotechnol 36(12):1244–1258

Beitollahi H, Tajik S, Nejad FG, Safaei M (2020) Recent advances in ZnO nanostructure-based electrochemical sensors and biosensors. J Mater Chem B 8(27):5826–5844

Bobrinetskiy II, Knezevic NZ (2018) Graphene-based biosensors for on-site detection of contaminants in food. Anal Methods 10(42):5061–5070

Bohbot JD, Vernick S (2020) The emergence of insect odorant receptor-based biosensors. Biosensors 10(3):26

Bolibok P, Wiśniewski M, Roszek K, Terzyk AP (2017) Controlling enzymatic activity by immobilization on graphene oxide. Sci Natur 104(3–4):36

Bu T, Zhang M, Sun X, Tian Y, Bai F, Jia P et al (2019) Gold nanoparticles-functionalized microorganisms assisted the construction of immunobiosensor for sensitive detection of ochratoxin a in food samples. Sensors Actuators B Chem 299:126969

Budhadev D, Poole E, Nehlmeier I, Liu Y, Hooper J, Kalverda E et al (2020) Glycan-gold nanoparticles as multifunctional probes for multivalent lectin–carbohydrate binding: implications for blocking virus infection and nanoparticle assembly. J Am Chem Soc 142(42): 18022–18034

Campaña AL, Florez SL, Noguera MJ, Fuentes OP, Ruiz Puentes P, Cruz JC, Osma JF (2019) Enzyme-based electrochemical biosensors for microfluidic platforms to detect pharmaceutical residues in wastewater. Biosensors 9(1):41

Cevik E, Cerit A, Gazel N, Yildiz HB (2018) Construction of an amperometric cholesterol biosensor based on DTP (aryl) aniline conducting polymer-bound cholesterol oxidase. Electroanalysis 30(10):2445–2453

Chauhan R, Solanki PR, Singh J, Mukherjee I, Basu T, Malhotra BD (2015) A novel electrochemical piezoelectric label-free immunosensor for aflatoxin B1 detection in groundnut. Food Control 52:60–70

Chauhan R, Singh J, Solanki PR, Manaka T, Iwamoto M, Basu T, Malhotra BD (2016) Label-free piezoelectric immunosensor decorated with gold nanoparticles: kinetic analysis and biosensing application. Sensors Actuators B Chem 222:804–814

Chen M, Zeng G, Xu P, Lai C, Tang L (2017) How do enzymes meet'nanoparticles and nanomaterials? Trends Biochem Sci 42(11):914–930

Cheng N, Song Y, Zeinhom MM, Chang YC, Sheng L, Li H et al (2017) Nanozyme-mediated dual immunoassay integrated with a smartphone for use in simultaneous detection of pathogens. ACS Appl Mater Interfaces 9(46):40671–40680

Costa-Fernández JM, Pereiro R, Sanz-Medel A (2006) The use of luminescent quantum dots for optical sensing. TrAC Trends in Analytical Chemistry. Mar 1;25(3):207–18

Della Pelle F, Compagnone D (2018) Nanomaterial-based sensing and biosensing of phenolic compounds and related antioxidant capacity in food. Sensors 18(2):462

D'costa EJ, Higgins IJ, Turner AP (1986) Quinoprotein glucose dehydrogenase and its application in an amperometric glucose sensor. Biosensors. Jan 1;2(2):71–87

Fang A, Ng HT, Li SF (2003) A high-performance glucose biosensor based on monomolecular layer of glucose oxidase covalently immobilised on indium–tin oxide surface. Biosensors and bioelectronics. 19(1):43–9

Fathi F, Mohammadzadeh-Aghdash H, Sohrabi Y, Dehghan P, Dolatabadi JEN (2018) Kinetic and thermodynamic studies of bovine serum albumin interaction with ascorbyl palmitate and ascorbyl stearate food additives using surface plasmon resonance. Food Chem 246:228–232

Gao L, Zhuang J, Nie L, Zhang J, Zhang Y, Gu N et al (2007) The intrinsic peroxidase-like activity of ferromagnetic nanoparticles. Nat Nanotechnol 2(9):577–583

Ghanbari K, Roushani M (2018) A novel electrochemical aptasensor for highly sensitive and quantitative detection of the streptomycin antibiotic. Bioelectrochemistry 120:43–48

Ghasemi-Varnamkhasti M, Apetrei C, Lozano J, Anyogu A (2018) Potential use of electronic noses, electronic tongues and biosensors as multisensor systems for spoilage examination in foods. Trends Food Sci Technol 80:71–92

Gowri A, Rajamani AS, Ramakrishna B, Sai VVR (2019) U-bent plastic optical fibre probes as a refractive index-based fat sensor for milk quality monitoring. Opt Fiber Technol 47:15–20

Gul I, Ahmad MS, Naqvi SS, Hussain A, Wali R, Farooqi AA, Ahmed I (2017) Polyphenol oxidase (PPO) based biosensors for detection of phenolic compounds: a review. J Appl Biol Biotechnol 5(03):72–85

Guzsvány V, Anojčić J, Vajdle O, Radulović E, Madarász D, Kónya Z, Kalcher K (2019) Amperometric determination of glucose in white grape and tablets as ingredient by screen-printed electrode modified with glucose oxidase and composite of platinum and multiwalled carbon nanotubes. Food Anal Methods 12(2):570–580

Haddada MB, Salmain M, Boujday S (2018) Gold colloid-nanostructured surfaces for enhanced piezoelectric immunosensing of staphylococcal enterotoxin a. Sensors Actuators B Chem 255: 1604–1613

Hashemi P, Afkhami A, Baradaran B, Halabian R, Madrakian T, Arduini F et al (2020) Well-orientation strategy for direct immobilization of antibodies: development of the immunosensor using the boronic acid-modified magnetic graphene nanoribbons for ultrasensitive detection of lymphoma cancer cells. Anal Chem 92(16):11405–11412

Ibadullaeva SZ, Appazov NO, Tarahovsky YS, Zamyatina EA, Fomkina MG, Kim YA (2019) Amperometric multi-enzyme biosensors: development and application, a short review. Biophysics 64(5):696–707

Ispas CR, Crivat G, Andreescu S (2012) Recent developments in enzyme-based biosensors for biomedical analysis. Analytical letters. Jan 1;45(2–3):168–86

Jayanthi KG, Suja SK (2020) Cholesterol oxidase immobilized inulin based nanocomposite as the sensing material for cholesterol in biological and food samples. Enzym Microb Technol 140: 109631. https://doi.org/10.1016/j.enzmictec.2020.109631. Epub Jul 9. PMID: 32912691

Jayaprakasan A, Thangavel A, Bhat LR, Gumpu MB, Nesakumar N, Babu KJ et al (2018) Fabrication of an electrochemical biosensor with ZnO nanoflakes interface for methylglyoxal quantification in food samples. Food Sci Biotechnol 27(1):9–17

Jeromiyas N, Elaiyappillai E, Kumar AS, Huang ST, Mani V (2019) Bismuth nanoparticles decorated graphenated carbon nanotubes modified screen-printed electrode for mercury detection. J Taiwan Inst Chem Eng 95:466–474

Jiang X, Jha A (2015) Engineering of a Ge–Te–se glass fibre evanescent wave spectroscopic (FEWS) mid-IR chemical sensor for the analysis of food and pharmaceutical products. Sensors Actuators B Chem 206:159–169

Jiao L, Zhong N, Zhao X, Ma S, Fu X, Dong D (2020) Recent advances in fiber-optic evanescent wave sensors for monitoring organic and inorganic pollutants in water. TrAC Trends Anal Chem 127:115892

Karadurmus L, Kaya SI, Ozkan SA (2021) Recent advances of enzyme biosensors for pesticide detection in foods. J Food Meas Charact 15(5):1–14

Khadka R, Aydemir N, Carraher C, Hamiaux C, Colbert D, Cheema J et al (2019) An ultrasensitive electrochemical impedance-based biosensor using insect odorant receptors to detect odorants. Biosens Bioelectron 126:207–213

Khan SA, DeGrasse JA, Yakes BJ, Croley TR (2015) Rapid and sensitive detection of cholera toxin using gold nanoparticle-based simple colorimetric and dynamic light scattering assay. Anal Chim Acta 892:167–174

Kim YS, Raston NHA, Gu MB (2016) Aptamer-based nanobiosensors. Biosens Bioelectron 76:2–19

Kissinger PT (2005) Biosensors—a perspective. Biosensors and Bioelectronics. 20(12):2512–6

Kokkali V, van Delft W (2014) Overview of commercially available bioassays for assessing chemical toxicity in aqueous samples. TrAC Trends in Analytical Chemistry. Oct 1;61:133–55

Kong D, Liu L, Song S, Suryoprabowo S, Li A, Kuang H et al (2016) A gold nanoparticle-based semi-quantitative and quantitative ultrasensitive paper sensor for the detection of twenty mycotoxins. Nanoscale 8(9):5245–5253

Kurbanoglu S, Erkmen C, Uslu B (2020) Frontiers in electrochemical enzyme based biosensors for food and drug analysis. TrAC Trends Anal Chem 124:115809

Lee KY, Ambrosi A, Pumera M (2017) 3D-printed metal electrodes for heavy metals detection by anodic stripping voltammetry. Electroanalysis 29(11):2444–2453

Li M, Chen X, Guo Y, Zhang B, Tang F, Wu X (2016) Enhanced sensitivity and resolution for the analysis of paralytic shellfish poisoning toxins in water using capillary electrophoresis with amperometric detection and field-amplified sample injection. Electrophoresis 37(23–24): 3109–3117

Li Z, Li X, Jian M, Geleta GS, Wang Z (2020) Two-dimensional layered nanomaterial-based electrochemical biosensors for detecting microbial toxins. Toxins 12(1):20

Lopez MSP, Redondo-Gómez E, López-Ruiz B (2017) Electrochemical enzyme biosensors based on calcium phosphate materials for tyramine detection in food samples. Talanta 175:209–216

Lou X, Egli M, Yang X (2016) Determining functional aptamer-protein interaction by biolayer interferometry. Current Protocols in Nucleic Acid Chemistry. 67(1):7–25

Luo Z, Ali R, Khan MI, Li F, Qiu B (2021) Advances in gold nanoparticles-based colorimetric Aptasensors for the detection of antibiotics: an overview of the past decade. Nano 11(4):840

Ma H, Sun J, Zhang Y, Bian C, Xia S, Zhen T (2016) Label-free immunosensor based on one-step electrodeposition of chitosan-gold nanoparticles biocompatible film on au microelectrode for determination of aflatoxin B1 in maize. Biosens Bioelectron 80:222–229

Malarkodi C, Rajeshkumar S, Annadurai G (2017) Detection of environmentally hazardous pesticide in fruit and vegetable samples using gold nanoparticles. Food Control 80:11–18

Mascini M, Pizzoni D, Perez G, Chiarappa E, Di Natale C, Pittia P, Compagnone D (2017) Tailoring gas sensor arrays via the design of short peptides sequences as binding elements. Biosens Bioelectron 93:161–169

Mazur F, Tran H, Kuchel RP, Chandrawati R (2020) Rapid detection of Listeriolysin O toxin based on a nanoscale liposome–gold nanoparticle platform. ACS Appl Nano Mater 3(7):7270–7280

Mendonca M, Bhunia AK (2015) Fiber-optic sensors for high throughput screening of pathogens. In: High throughput screening for food safety assessment. Woodhead Publishing, pp 249–262

Mohamadi E, Moghaddasi M, Farahbakhsh A, Kazemi A (2017) A quantum-dot-based fluoroassay for detection of food-borne pathogens. J Photochem Photobiol B Biol 174:291–297

Naderi Asrami P, Aberoomand Azar P, Saber Tehrani M, Mozaffari SA (2020) Glucose oxidase/nano-ZnO/thin film deposit FTO as an innovative clinical transducer: a sensitive glucose biosensor. Front Chem 8:503

Nag P, Sadani K, Mohapatra S, Mukherji S, Mukherji S (2021) Evanescent wave optical fiber sensors using enzymatic hydrolysis on nanostructured polyaniline for detection of β-lactam antibiotics in food and environment. Anal Chem 93(4):2299–2308

Narsaiah K, Jha SN, Bhardwaj R, Sharma R, Kumar R (2012) Optical biosensors for food quality and safety assurance—a review. Journal of food science and technology. 49:383–406

Nasir MZM, Mayorga-Martinez CC, Sofer Z, Pumera M (2017) Two-dimensional 1T-phase transition metal dichalcogenides as nanocarriers to enhance and stabilize enzyme activity for electrochemical pesticide detection. ACS Nano 11(6):5774–5784

Nayak DS, Shetti NP (2016) A novel sensor for a food dye erythrosine at glucose modified electrode. Sensors Actuators B Chem 230:140–148

Neethirajan S, Ragavan V, Weng X, Chand R (2018) Biosensors for sustainable food engineering: challenges and perspectives. Biosensors 8(1):23

Nelis JLD, Tsagkaris AS, Dillon MJ, Hajslova J, Elliott CT (2020) Smartphone-based optical assays in the food safety field. TrAC Trends Anal Chem 129:115934

Patel P, Doddamani A (2019) Role of sensor in the food processing industries. Int Arch Appl Sci Technol 10:10–18

Peng J, Wang Y, Liu L, Kuang H, Li A, Xu C (2016) Multiplex lateral flow immunoassay for five antibiotics detection based on gold nanoparticle aggregations. RSC Adv 6(10):7798–7805

Peng J, Liu L, Xu L, Song S, Kuang H, Cui G, Xu C (2017) Gold nanoparticle-based paper sensor for ultrasensitive and multiple detection of 32 (fluoro) quinolones by one monoclonal antibody. Nano Res 10(1):108–120

Ping J, Zhou Y, Wu Y, Papper V, Boujday S, Marks RS, Steele TW (2015) Recent advances in aptasensors based on graphene and graphene-like nanomaterials. Biosens Bioelectron 64:373–385

Pissuwan D, Gazzana C, Mongkolsuk S, Cortie MB (2020) Single and multiple detections of foodborne pathogens by gold nanoparticle assays. Wiley Interdiscip Rev Nanomed Nanobiotechnol 12(1):e1584

Rahtuvanoğlu A, Akgönüllü S, Karacan S, Denizli A (2020) Biomimetic nanoparticles-based surface plasmon resonance biosensors for histamine detection in foods. ChemistrySelect 5(19):5683–5692

Ramezani M, Danesh NM, Lavaee P, Abnous K, Taghdisi SM (2016) A selective and sensitive fluorescent aptasensor for detection of kanamycin based on catalytic recycling activity of exonuclease III and gold nanoparticles. Sensors Actuators B Chem 222:1–7

Ranjan R, Rastogi NK, Thakur MS (2012) Development of immobilized biophotonic beads consisting of Photobacterium leiognathi for the detection of heavy metals and pesticide. Journal of hazardous materials. Jul 30;225:114–23

Raymundo-Pereira PA, Silva TA, Caetano FR, Riboviski L, Zapp E, Brondani D et al (2020) Polyphenol oxidase-based electrochemical biosensors: a review. Anal Chim Acta 1139:198–221

Razak HA, Sulaiman NH, Haroon H, Mohd Zain AS (2018) A fiber optic sensor based on Mach-Zehnder interferometer structure for food composition detection. Microw Opt Technol Lett 60(4):920–925

Reshma VG, Mohanan PV (2019) Quantum dots: applications and safety consequences. J Lumin 205:287–298

Riu J, Giussani B (2020) Electrochemical biosensors for the detection of pathogenic bacteria in food. TrAC Trends Anal Chem 126:115863

Ronkainen NJ, Halsall HB, Heineman WR (2010) Electrochemical biosensors. Chemical Society Reviews 39(5):1747–63

Rotariu L, Lagarde F, Jaffrezic-Renault N, Bala C (2016) Electrochemical biosensors for fast detection of food contaminants–trends and perspective. TrAC Trends Anal Chem 79:80–87

Rout P, Chakraborty C (2020) The commercial application of biosensors as an analytical device. Int J Curr Microbiol App Sci 9(5):864–880

Ruiz-Valdepeñas Montiel V, Torrente-Rodríguez RM, Campuzano S, Pellicanò A, Reviejo ÁJ, Cosio MS, Pingarrón JM (2016) Simultaneous determination of the main peanut allergens in foods using disposable amperometric magnetic beads-based immunosensing platforms. Chemosensors 4(3):11

Saha B, Songe P, Evers TH, Prins MW (2017) The influence of covalent immobilization conditions on antibody accessibility on nanoparticles. Analyst 142(22):4247–4256

Sandhyarani N (2019) Surface modification methods for electrochemical biosensors. In: Electrochemical biosensors. Elsevier, pp 45–75

Saraji M, Jafari MT, Mossaddegh M (2016) Carbon nanotubes@ silicon dioxide nanohybrids coating for solid-phase microextraction of organophosphorus pesticides followed by gas chromatography–corona discharge ion mobility spectrometric detection. J Chromatogr A 1429:30–39

Sawan S, Maalouf R, Errachid A, Jaffrezic-Renault N (2020) Metal and metal oxide nanoparticles in the voltammetric detection of heavy metals: a review. TrAC Trends Anal Chem 131:116014

Scognamiglio V, Arduini FA, Palleschi GI, Rea G (2014) Biosensing technology for sustainable food safety. TrAC Trends in Analytical Chemistry. Nov 1;62:1-0

Shahbazi R, Salouti M, Amini B, Jalilvand A, Naderlou E, Amini A, Shams A (2018) Highly selective and sensitive detection of *Staphylococcus aureus* with gold nanoparticle-based core-shell nano biosensor. Mol Cell Probes 41:8–13

Shaibani PM, Etayash H, Jiang K, Sohrabi A, Hassanpourfard M, Naicker S et al (2018) Portable nanofiber-light addressable potentiometric sensor for rapid Escherichia coli detection in orange juice. ACS Sensors 3(4):815–822

Sharma R, Ragavan KV, Abhijith KS, Thakur MS (2015) Synergistic catalysis by gold nanoparticles and metal ions for enhanced chemiluminescence. RSC Adv 5(40):31434–31438

Sharma A, Kumar A, Khan R (2018) A highly sensitive amperometric immunosensor probe based on gold nanoparticle functionalized poly (3, 4-ethylenedioxythiophene) doped with graphene oxide for efficient detection of aflatoxin B1. Synth Met 235:136–144

Sharma R, Akshath US, Bhatt P, Raghavarao KSMS (2019) Fluorescent aptaswitch for chloramphenicol detection–quantification enabled by immobilization of aptamer. Sensors Actuators B Chem 290:110–117

Smrithi V, Mathews CB, Radhakrishnan P (2015) Fiber optic evanescent wave sensor for the measurement of curcumin. In: In *2015 workshop on recent advances in photonics (WRAP)*. IEEE, pp 1–4

Sobrino-Gregorio L, Bataller R, Soto J, Escriche I (2018) Monitoring honey adulteration with sugar syrups using an automatic pulse voltammetric electronic tongue. Food Control 91:254–260

Solanki PR, Singh J, Rupavali B, Tiwari S, Malhotra BD (2017) Bismuth oxide nanorods based immunosensor for mycotoxin detection. Mater Sci Eng C 70:564–571

Song E, Yu M, Wang Y, Hu W, Cheng D, Swihart MT, Song Y (2015) Multi-color quantum dot-based fluorescence immunoassay array for simultaneous visual detection of multiple antibiotic residues in milk. Biosens Bioelectron 72:320–325

Song D, Yang R, Fang S, Liu Y, Long F (2018) A FRET-based dual-color evanescent wave optical fiber aptasensor for simultaneous fluorometric determination of aflatoxin M1 and ochratoxin a. Microchim Acta 185(11):1–10

Song D, Jiang X, Li Y, Lu X, Luan S, Wang Y et al (2019) Metal−organic frameworks-derived MnO2/Mn3O4 microcuboids with hierarchically ordered nanosheets and Ti3C2 MXene/au NPs composites for electrochemical pesticide detection. J Hazard Mater 373:367–376

Songa EA, Okonkwo JO (2016) Recent approaches to improving selectivity and sensitivity of enzyme-based biosensors for organophosphorus pesticides: a review. Talanta 155:289–304

Stepurska KV, Soldatkin OO, Arkhypova VM, Soldatkin AP, Lagarde F, Jaffrezic-Renault N, Dzyadevych SV (2015) Development of novel enzyme potentiometric biosensor based on pH-sensitive field-effect transistors for aflatoxin B1 analysis in real samples. Talanta 144: 1079–1084

Stredansky M, Redivo L, Magdolen P, Stredansky A, Navarini L (2018) Rapid sucrose monitoring in green coffee samples using multienzymatic biosensor. Food Chem 254:8–12

Tan B, Zhao H, Du L, Gan X, Quan X (2016) A versatile fluorescent biosensor based on target-responsive graphene oxide hydrogel for antibiotic detection. Biosens Bioelectron 83:267–273

Taylor RF, Marenchic IG, Cook EJ (1988) An acetylcholine receptor-based biosensor for the detection of cholinergic agents. Analytica Chimica Acta. Jan 1;213:131–8
Uzunoglu A, Ramirez I, Andreasen E, Stanciu LA (2016) Layer by layer construction of ascorbate interference-free amperometric lactate biosensors with lactate oxidase, ascorbate oxidase, and ceria nanoparticles. Microchim Acta 183(5):1667–1675
Vargas E, Ruiz MA, Campuzano S, de Rivera GG, López-Colino F, Reviejo AJ, Pingarrón JM (2016) Implementation of a new integrated d-lactic acid biosensor in a semiautomatic FIA system for the simultaneous determination of lactic acid enantiomers. Application to the analysis of beer samples. Talanta 152:147–154
Valderrama WB, Dudley EG, Doores S, Cutter CN (2016) Commercially available rapid methods for detection of selected food-borne pathogens. Critical reviews in food science and nutrition. 2Jul 3;56(9):1519–31
Viter R, Tereshchenko A, Smyntyna V, Ogorodniichuk J, Starodub N, Yakimova R et al (2017) Toward development of optical biosensors based on photoluminescence of TiO2 nanoparticles for the detection of salmonella. Sensors Actuators B Chem 252:95–102
Wachiralurpan S, Sriyapai T, Areekit S, Sriyapai P, Augkarawaritsawong S, Santiwatanakul S, Chansiri K (2018) Rapid colorimetric assay for detection of listeria monocytogenes in food samples using LAMP formation of DNA concatemers and gold nanoparticle-DNA probe complex. Front Chem 6:90
Wang J (2008) Electrochemical glucose biosensors. Chemical reviews. 108(2):814–25
Wang Y, Alocilja EC (2015) Gold nanoparticle-labeled biosensor for rapid and sensitive detection of bacterial pathogens. J Biol Eng 9(1):1–7
Wang Y, Duncan TV (2017) Nanoscale sensors for assuring the safety of food products. Curr Opin Biotechnol 44:74–86
Wang W, Gunasekaran S (2020) Nanozymes-based biosensors for food quality and safety. TrAC Trends Anal Chem 126:115841
Wang R, Xiang Y, Zhou X, Liu LH, Shi H (2015) A reusable aptamer-based evanescent wave all-fiber biosensor for highly sensitive detection of Ochratoxin a. Biosens Bioelectron 66:11–18
Wei X, Lin W, Ma N, Luo F, Lin Z, Guo L, Qiu B & Chen G (2012) Sensitive fluorescence biosensor for folate receptor based on terminal protection of small-molecule-linked DNA. Chem Comm 48(49):6184–6186
Wu X, Li Y, Liu B, Feng Y, He W, Liu Z, Huang H (2016) Two-site antibody immunoanalytical detection of food allergens by surface plasmon resonance. Food Anal Methods 9(3):582–588
Wu W, Zhu Z, Li B, Liu Z, Jia L, Zuo L, Luo SZ (2018) A direct determination of AFBs in vinegar by aptamer-based surface plasmon resonance biosensor. Toxicon 146:24–30
Wu YY, Liu BW, Huang P, Wu FY (2019) A novel colorimetric aptasensor for detection of chloramphenicol based on lanthanum ion–assisted gold nanoparticle aggregation and smartphone imaging. Anal Bioanal Chem 411(28):7511–7518
Xiao W, Xiao M, Fu Q, Yu S, Shen H, Bian H, Tang Y (2016) A portable smartphone readout device for the detection of mercury contamination based on an aptamer-assay nanosensor. Sensors 16(11):1871
Xiong Y, Wu J, Wang Q, Xu J, Fang S, Chen J, Duan M (2017) Optical sensor for fluoride determination in tea sample based on evanescent-wave interaction and fiber-optic integration. Talanta 174:372–379
Xu N, Bai J, Peng Y, Qie Z, Liu Z, Tang H et al (2017) Pretreatment-free detection of diazepam in beverages based on a thermometric biosensor. Sensors Actuators B Chem 241:504–512
Xu Y, Kutsanedzie FY, Hassan M, Zhu J, Ahmad W, Li H, Chen Q (2020) Mesoporous silica supported orderly-spaced gold nanoparticles SERS-based sensor for pesticides detection in food. Food Chem 315:126300
Yahia MB, Hsan LBH, Knani S, Yahia MB, Nasri H, Lamine AB (2016) Modeling of adsorption isotherms of zinc nitrate on a thin layer of porphyrin. J Mol Liq 222:576–585
Yu YY, Wang JX, Si RW, Yang Y, Zhang CL, Yong YC (2017) Sensitive amperometric detection of riboflavin with a whole-cell electrochemical sensor. Anal Chim Acta 985:148–154

Zaporotskova IV, Boroznina NP, Parkhomenko YN, Kozhitov LV (2016) Carbon nanotubes: sensor properties. A review. Mod Electro Mater 2(4):95–105

Zeng Y, Zhu Z, Du D, Lin Y (2016) Nanomaterial-based electrochemical biosensors for food safety. J Electroanal Chem 781:147–154

Zheng L, Cai G, Wang S, Liao M, Li Y, Lin J (2019) A microfluidic colorimetric biosensor for rapid detection of Escherichia coli O157: H7 using gold nanoparticle aggregation and smart phone imaging. Biosens Bioelectron 124:143–149

Zhong L, Sun J, Gan Y, Zhou S, Wan Z, Zou Q et al (2019) Portable smartphone-based colorimetric analyzer with enhanced gold nanoparticles for on-site tests of seafood safety. Anal Sci 35(2): 133–140

Zhou Q, Tang D (2020) Recent advances in photoelectrochemical biosensors for analysis of mycotoxins in food. TrAC Trends Anal Chem 124:115814

Zhu Z, Feng M, Zuo L, Zhu Z, Wang F, Chen L, Luo SZ (2015) An aptamer-based surface plasmon resonance biosensor for the detection of ochratoxin a in wine and peanut oil. Biosens Bioelectron 65:320–326

Zhu Y, Wu J, Han L, Wang X, Li W, Guo H, Wei H (2020) Nanozyme sensor arrays based on heteroatom-doped graphene for detecting pesticides. Anal Chem 92(11):7444–7452

Zinoubi K, Majdoub H, Barhoumi H, Boufi S, Jaffrezic-Renault N (2017) Determination of trace heavy metal ions by anodic stripping voltammetry using nanofibrillated cellulose modified electrode. J Electroanal Chem 799:70–77

Chapter 14
Packaging Solutions for Monitoring Food Quality and Safety

Johnsy George, Ranganathan Kumar, Basheer Aaliya, and Kappat Valiyapeediyekkal Sunooj

1 Introduction

Food safety is a critical issue and all stakeholders involved in the food business, such as manufacturers, retailers, consumers, and regulatory bodies, are concerned about the quality and safety of food products. As the standard of living advances, the consumer's demands and expectations also increase due to lifestyle, increased awareness regarding the importance of nutrients and other quality parameters (Fung et al. 2018). Over a period of time, food packaging has evolved from the simple function of just containing and protecting the food to some important factors that can play an active role in monitoring and ensuring the quality and safety of the food it contains (Risch 2009). Packaging plays a vital role in the preservation and maintenance of food quality during transportation, distribution, storage, retailing and final consumption. Packaging is an art, science, and technology of ensuring the wholesomeness, quality, integrity, and safety of a food product (Kalpana et al. 2019). Due to the increased demand for healthy, safe, convenient, and cost-effective processed foods, the food and beverage industry is compelled to look for more advanced packaging solutions that monitor and ensure food quality and safety. This paved the way for emerging opportunities in scientific and industrial sectors for the development of varied novel technologies in food packaging systems. Among different emerging approaches for developing such food packaging systems, active packaging (AP) and intelligent packaging (IP) technologies are the most promising ones. Hence, owing to the growing interest in the need for innovations in packaging technologies, the chapter details notable and emerging IP systems that enhance food product quality and ensure its safety while being delivered to consumers.

J. George (✉) · R. Kumar
Food Engineering and Packaging Division, Defence Food Research Laboratory, Mysore, India

B. Aaliya · K. V. Sunooj (✉)
Department of Food Science and Technology, Pondicherry University, Puducherry, India

H. U. Hebbar et al. (eds.), *Engineering Aspects of Food Quality and Safety*, Food Engineering Series, https://doi.org/10.1007/978-3-031-30683-9_14

2 Active and Intelligent Packaging

The advancement in packaging technology is mainly driven by the demanding needs of consumers, which resulted in the development of sophisticated packaging systems with capabilities not only to protect the food it contains but communicate the changes in food or the environment (Yucel 2016). AP is mainly developed by the deliberate incorporation of active components into the packaging itself to release or absorb these active substances into the food or the environment surrounding it and improve the food product quality and shelf life. On the other hand, the IP system does not directly extend the shelf life of food but senses many properties of food and its environment and provides information about the status of food quality to the manufacturer, retailer, and consumer (Gregor-Svetec 2018). IP aims to constantly move together with the entire supply chain and communicate about food and packaging conditions to all stakeholders. IP systems facilitate easy decision-making by providing real-time information about the food products and hence play an important role in ensuring the quality and safety of food products. Figure 14.1 provides an overview of the application of IP systems to maintain the quality and safety of food and food products.

IP systems have the functionality to sense, detect, and/or record internal and/or environmental changes occurring in a food product. The IP systems help in the detection of physicochemical quality, storage condition, microbial stability, and product expiration and freshness of food. IP provides information about the quality, traceability, safety, or tamper indication with the help of different mechanisms (Kuswandi 2017). IP contributes to the improvement in Quality Analysis and Critical Control Points (QACCP) and Hazard Analysis and Critical Control Points (HACCP) systems that are utilized to detect, prevent, control, or eliminate any possible factors that affect the safety and quality of the final food product (Siracusa and Lotti 2019). The IP system functions like an ON/OFF switch that responds to external or internal stimuli and immediately communicates the product status. An IP system is manufactured by incorporating an external component into the food packaging. Most of the IP systems are manufactured by using three main technologies such as (i) sensors, (ii) indicators, and (iii) data carriers.

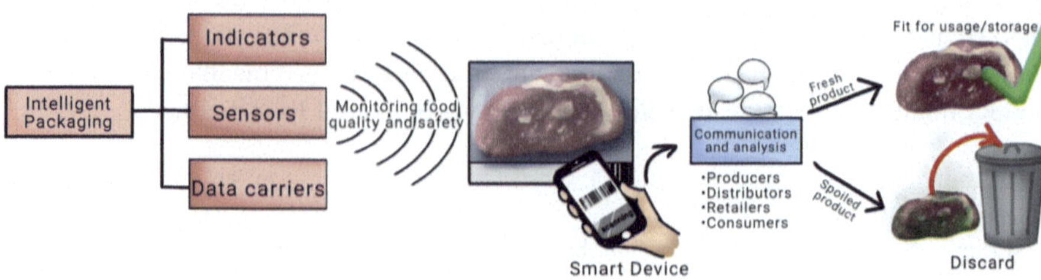

Fig. 14.1 Illustration of the application of IP systems for monitoring food quality and safety

3 Sensors

The sensor is "a device or tool used to detect, locate, or quantify energy or matter, by giving a signal for the detection or measurement of a physical or chemical property to which the device responds" (Kress-Rogers 1998). A sensor helps in detecting a change in the environment surrounding a food product. The sensor mainly comprises receptor and transducer, where the receptor helps in converting chemical or physical information to an energy form. The function of the transducer is to change the energy to an analytical signal (Sohail et al. 2018). On the basis of transducer working type, the sensors are divided into passive and active types. If a transducer operates with an external power supply for measurement, it is called "active", or else it is called "passive" (Vanderroost et al. 2014). Sensors also have a signal processing unit and display unit to process the output of transducer and display the quantifiable results in analog or digital form. An ideal sensor should possess high selectivity and sensitivity, rapid response, good reliability, wider dynamic range, complete reversibility, and long-term stability, within miniature size and low manufacturing cost (Ghaani et al. 2016). Mostly, sensors are classified according to the type of receptor, transducer, and their applications, as chemical sensors, gas sensors, and biosensors. Nanosensors and edible sensors are the other newly established sensors. The edible sensor is a novel concept of non-destructive detection of food spoilage. Dudnyk et al. (2018) fabricated a pectin-based edible sensor by combining anthocyanins (red cabbage extract) for principle colorimetric changes and variation in total volatile basic nitrogen (TVB-N) for determining the spoilage in meat-based products.

3.1 Chemical Sensors

Chemical sensors are analytical tools or devices which utilize chemical reagents as recognition elements or receptors. Receptors can identify the presence, composition, concentration, or activity of specific gases or chemical analytes by surface adsorption resulting in an alteration in surface properties. Later, the transducer converts the change in surface property to a quantifiable signal which is proportionate to the target analyte (Kuswandi et al. 2011). Chemical sensors are categorized into gravimetric, optical, electrical, and electrochemical sensors based on their transduction principle (Azeredo and Correa 2021). During IP, chemical sensors are usually used to identify the pH, gases, and volatile organic compounds (VOC), such as carbon dioxide (CO_2) (Borchert et al. 2013), hydrogen sulfide (H_2S) (Sukhavattanakul and Manuspiya 2021), ammonia (NH_3) (Matindoust et al. 2017), dimethylamine (DMA) and trimethylamine (TMA) (Chang et al. 2017) produced during storage and distribution, which could be generated by deterioration of food or loss in integrity of the package. Thus, chemical sensors help in the direct detection of food quality inside the package. Electrochemical sensors are also found to be helpful in quantifying potential toxic additives that migrate from the packaging materials to the food or their toxic products that can form upon contact with the food. For instance, bisphenol-A, that may cause endocrine disorders produced from epoxy resins and polycarbonate bottles (Karthika et al. 2021) and primary aromatic amines generated from isocyanate monomers of polyurethane adhesives are detected with the help of electrochemical sensors (Ghaani et al. 2018). Kuswandi et al. (2012) fabricated a smart packaging system using polyaniline (PANI) film as a chemical sensor for real-time monitoring of the products of microbial breakdown taking place in the head-space of packaged fish. The on-package indicator of PANI film responded with a visible color change to TVB-N released during fish spoilage. The PANI film-based packaging system was recyclable and can be considered as a low-cost sensor for monitoring fish spoilage.

3.2 Gas Sensors

The composition of gas in a food package changes with time and/or temperature as an outcome of physicochemical or biological reactions or leakages (Kuswandi and Jumina 2020). Therefore, the concentration of gas inside a food package gives indirect information about the quality of food. The gas sensors are considered for detecting and indicating gases or volatile compounds like volatile amines, CO_2, H_2S, and other specific gases. Some common gas sensors are electrochemical sensors, optical sensors, field-effect transistors, piezoelectric crystal sensors, metal oxide semiconductors, and organic conducting polymer sensors (Kress-Rogers 1998; Kerry et al. 2006). There are mainly three types of optical gas sensors; fluorescence-based on pH-sensitive indicator, absorption-based colorimetric sensor, and energy transfer approach based on phase fluorimetric detection (Kuswandi and Jumina 2020).

The optical oxygen (O_2) sensors are based on various principles like lifetime decay of O_2-sensitive dyes, photoluminescence absorbance or quenching, and polymeric host quenching. In a study, the phosphorescent dye incorporated reversible optical sensor was used in modified atmosphere packed and vacuum-packed beef. The sensor monitored alteration in O_2 levels, which is kept at 4 °C for about 15 to 35 days for checking the effect of O_2 content responsible for lipid oxidation. In the first instance, the O_2 content was 1.15% and 0.07% and later increased to 1.26% and 0.55% in modified atmosphere packed and vacuum packed beef samples, respectively. The quality, freshness, and safety of beef and chicken stored for 11 days at 4 ° C were evaluated using O_2 sensors. After tenth and ninth day of storage, an intense colour change to red from green was observed in the sensory array of chicken and beef fillets, respectively. The experiment indicated that O_2 sensor can be potentially used for the spoilage detection in meat by revealing promising evidence from the results of freshness, physicochemical, and microbiological tests (Morsy et al. 2014). A new optical O_2 sensor employing micro-structured platinum(II)-5,10,15,20-tetrakis-(2,3,4,5,6-pentafluorophenyl)-porphyrin / polydimethylsiloxane (PtTFPP/PDMS) pillar arrays sensing layers possessing significant sensitivity was fabricated by Mao et al. (2017). The ultrasensitive sensor film exhibited a meagre O_2 detection limit (0.10 μmol/L), which helps in determining dissolved O_2 under the nanomolar concentration range. Also, the improved light intensity-changing characteristics at lower O_2 partial pressure in the sensor aid in the detection of O_2 levels with naked eyes. Similarly, luminescent O_2 sensors based on porous sensing films exhibited highly enhanced sensitivity with improved O_2 accessibility and photoluminescence (Lee and Park 2017). OxySense, a commercially established fluorescence-quenching sensor, utilizes O_2 sensor (OxyDot) placed inside the transparent or semi-transparent sealed package for measuring dissolved or headspace O_2. It possesses non-destructive rapid action to withstand pasteurization temperature without losing sensitivity (Kerry et al. 2006). OxySense is available in varied forms such as tablet, label, or laminate in a polymer film.

The optical CO_2 sensors are used for monitoring the integrity of MAP products. The optochemical sensing techniques usually involve a fluorescence-based system and absorption based on colorimetric sensing. Borchert et al. (2013) manufactured optochemical CO_2 sensors made up of colorimetric pH indicator α-naphtholphthalein and phosphorescent Pt-porphyrin reporter dye, PtTFPP embedded in a plastic matrix along with a phase transfer agent tetraoctyl- or cetyltrimethylammonium hydroxide. The study indicated that the sensor showed the efficacy in measuring headspace CO_2 in MAP foods by retaining its CO_2 sensitivity for 21 days at 4 °C. A very sensitive squaraine-based system for fluorescently and calorimetrically sensing CO_2 in dimethyl sulfoxide (DMSO) containing fluoride ion was developed by Sun et al. (2016). The colour change in response to CO_2 level was visible to naked eyes, with an immense blue shift in absorption (134 nm) and fluorescence (126 nm) spectra. Similarly, an unsymmetrical squaraine-based chemosensor by UV-visible spectroscopy and proton nuclear magnetic resonance spectroscopy in DMSO was also synthesized for detecting CO_2 gas (Xia et al. 2015). Sun et al. (2017), with the help of a highly sensitive "naked-eye" cationic squaraine-based chemosensor, detected CO_2 in an aqueous medium, which might possess a remarkable role in the meat packaging industry. A fluorimetric assay was developed by Khandare et al. (2015) for detecting dissolved CO_2 using an ion-induced assembly of tetraphenylethylene derivatives. Chitosan, because of its amine functionality, was used for the ion-induced assay, and the degree of the aggregation depends on the charge density, which can be compared to the dissolved CO_2 concentration. Chang et al. (2017) conducted a freshness test by quantitatively measuring DMA, TMA, and NH_3 produced by three different types of fresh fish using an amine gas sensor system in contrast to the pre-detected results obtained from organic solid-state semiconductor having 1 min detection period. Likewise, a highly sensitive, flexible, and less energy-consuming NH_3 gas sensor for protein-rich foods was prepared by oxidative polymerization employing polyaniline (a conducting polymer) (Matindoust et al. 2017). Besides, pH-sensitive dyes are also employed for developing gas sensors to detect basic volatile amines in protein-rich foods. A pH-based gas sensing edible film was prepared using anthocyanin extract from red radish, gellan gum, and gelatin, which shows a change in colour from orange-red to yellow in 2–12 pH range (Zhai et al. 2018). The electrochemically written multi-coloured patterns were helpful in real-time monitoring the spoilage in fish and milk. The film indicated black carp fish freshness by the change of film colour induced by the volatile basic gases such as DMA, TMA, and NH_3 produced due to the protein decomposition by enzymes and bacteria. Similarly, the freshness of milk was indicated using film by sensing the gas generated by anaerobic bacteria present in the milk. In another study, an on-package dual-sensor label was developed using bromocresol purple (BCP) and methyl red (MR) as two pH indicators for monitoring the freshness in beef. The decay in beef was determined when BCP changed from yellow to purple, and MR changed from red to yellow. The label responded precisely to the beef spoilage due to pH change in room and chiller conditions (Kuswandi and Nurfawaidi 2017).

3.3 Biosensors

Biological sensor is defined as "a small analytical device or tool that are capable of detecting and recording specific biochemical reactions and converting their presence or concentration into electrical, thermal, or optical signals that can be easily analyzed". Biosensors are effectively used for the freshness indication in meat and fish products by performing pathogen detection and safety systems during food packaging and storage. A representative biosensor comprises a bioreceptor, a transducer, and an electronic system. The bioreceptor recognizes the targeted analyte (microbe, enzyme, nucleic acid, or antibodies), and the transducer converts the biochemical signals into measurable responses (Lloyd et al. 2019). Based on transducer type, biosensors can be categorized as calorimetric biosensors, optical biosensors (luminescent, colorimetric, fluorescent, and interferometric), electrochemical biosensors (potentiometric, amperometric, and conductometric), and mass-based biosensors (piezoelectric and acoustic wave) (Firouz et al. 2021). Among these sensors, an electrochemical biosensor is a promising tool to continuously monitor food quality in IP, generating electrical signals proportional to the concentration of an analyte. The enzyme-based biosensors are the simplest, inexpensive, and user-friendly approach among varied electrochemical biosensors.

The most favourable characteristic for utilizing biosensors in IP is for the identification of volatile compounds like amines, volatile alcohols, and ethylene. In meat packaging, biosensors are used for detecting biogenic amines. Histamines, diamines, and biogenic amines found in rainbow trout meat, poultry, and fish can be identified by employing a putrescine oxidase reactor besides the amperometric hydrogen peroxide electrodes (Ahmed et al. 2018). Biosensors for detecting xanthine (adenine nucleotide degradation product) in animal tissues were manufactured by immobilizing xanthine oxide on electrodes made of silver, platinum, and pencil graphite (Dolmacı et al. 2012; Devi et al. 2013; Realini and Marcos 2014). Xanthine molecules serve as a meat and fish spoilage indicator. For instance, Dervisevic et al. (2015) developed a novel amperometric xanthine biosensor by immobilizing xanthine oxidase by glutaraldehyde over a pencil graphite electrode. The manufactured electrochemical polymerized electrode calculated the xanthine content in the chicken meat, which checked the potentiality of the developed biosensor having high stability, sensitivity, and selectivity. Electrochemical biosensors are employed as glucose sensors to detect the glucose content in varied beverages (Scampicchio et al. 2010). Pesticide detecting acetylcholinesterase biosensor prepared from nafion modified nanoporous pseudo carbon, gold nanoparticles, and chitosan exhibited quick response, agreeable sensitivity, and stability to methyl parathion and organophosphate pesticides at a lower detection limit (Deng et al. 2016). Biosensors have also shown their efficiency in detecting food contaminants like pathogens and toxins.

Several types of biosensors are accessible on a commercial scale for packaging food products. Toxin Guard™ (Toxin Alert Inc. Ontario, Canada) is a biosensor whose functional system is dependent on the integration of antibodies with plastic packaging made of polyvinyl chloride or polyolefins for the detection of pathogens such as *Escherichia coli*, *Listeria* spp., *Salmonella* spp., and *Campylobacter* spp. (Bodenhamer et al. 2004). Food Sentinel System™ (SIRA Technologies, California, USA) is manufactured for certain pathogen antibodies attached to a barcode which is utilized for membrane development. This biosensor helps in detecting pathogenic contaminants by making the barcode unscannable if pathogenic bacteria are present (Food Sentinel System 2019). A flexible biosensor, Flex Alert (Canada), was commercially developed for the detection of toxins in packed foods along with the supply chain. It was explicitly developed for detecting *Salmonella* spp., *Listeria* spp., *E. coli* O157, and aflatoxins (Flex Alert Company 2022). Bioett (Bioett AB, Sweden) is a system technology which integrates electronics and biochemistry for monitoring the temperature of food products in the course of refrigerated transport. It is comprised of a biosensor affixed to the food container, a detector for reading the information obtained from the biosensor, and a database for storing the data regarding the product. The chief components of a Bioett system are a built-in biosensor and a chip-less radio-frequency (RF) circuit (Sjöholm and Erlandsson 2003). However, commercial application of biosensor for IP are insubstantial and needs further development to merge them into food packaging. Nevertheless, the significant challenges in establishing biosensors in IP are the complexity of food structure and trouble in direct measurement of degradation markers in an enclosed package with no antecedent treatment of the food samples.

3.4 Nanosensors

Nanosensors are the sensors made at nanoscale size, having size indicated in nanometers (1 nm = 10^{-9} m) which have structural and functional devices (Kuswandi 2017). These type of sensors helps in IP by involving the use of varied types of nanomaterials like nanoparticles, nanofibers, nanotubes, nanocylinder, nanosheets, and fullerenes (Fuertes et al. 2016). Like sensors, nanosensors are embedded in a food package for controlling the internal and external conditions of the product. Thus, the tiny chip-like nanosensors which are not visible to the human eye aid in detecting chemical contaminants, pesticides, pathogens, spoilage, product tampering, tracing ingredients, and tracking food along the processing chain (Lloyd et al. 2019). For identifying biogenic amines in beer, a very selective optical sensor dependant on covalent interaction between tryptamine molecule and active vinyl groups was developed by Ramon-Marquez et al. (2016). Luminescence signals were generated by the covalently bounded amines because of their intrinsic phosphorescence nature. The developed optical sensor film based on functional non-woven nanofiber mat was known to be highly sensitive and selective for determining the tryptamine in beer. In another study, the nanofibers (150 μm thickness and 300 nm diameter) containing a high concentration of active vinyl groups (330 μmol/g) detected tryptamine in 10 varied types of beers, within 6 ng/ml detection limit. The solid-state polyamide 66 (PA66)/polyaniline nanofiber sensor based on reversible non-redox acid/base doping process detects L-ascorbic acid. It induces visible colour variations even at a low concentration of up to 50 ppb, which can be read using an iPhone (Wen et al. 2015). Such inexpensive sensing films with superior quality can be potentially employed in IP.

Several nanosensors were developed for freshness detection in protein-based foods. For instance, a protein-based halochromic nanosensor was fabricated to evaluate the quality of rainbow trout fillets. The indicator dye, alizarin containing zein nanofibers were prepared by electrospinning. The colour of the nanosensor changes from light purple to magenta by the 12th day of cold storage indicating spoilage. The colorimetric results from the sensor also showed a correlation with chemical and microbial changes in the fish fillets (Aghaei et al. 2020). Recently, intelligent pH-responsive colour indicator films based on cellulose nanofibers (CNF) and carboxymethyl cellulose (CMC) were prepared from shikonin extracted from *Lithospermum erythrorhizon* roots to monitor the freshness of fish. The film exhibited a reddish-pink colour for fresh fish (pH = 5.7) and bluish-violet colour for spoiled fish (pH = 6.9). The shikonin incorporated films could be likely used for monitoring the quality and freshness of seafood (Ezati et al. 2021). Besides, Ge et al. (2020) prepared a pH-sensitive green nanocomposite film based on black rice bran anthocyanins (BACNs) incorporated oxidized chitin nanocrystals/gelatin. The films exhibited significant colour changes in varied buffer solutions, which could be utilized for monitoring the freshness of hairtail fish and shrimp. Films prepared with low BACNs concentration showed more sensitivity towards basic volatile amines emitted during storage. Hence, the nanocomposite film is a promising sensor for determining the freshness of high-protein foods. Recently, to monitor the H_2S gas concentration, a gas sensor in the form of a hybrid nanocomposite thin film based on bacterial cellulose nanocrystals (BCNCs) was developed. The films were prepared by spraying the hybrid nanocomposite suspension comprising alginate-molybdenum trioxide and silver nanoparticles (AgNPs)-loaded BCNCs over PET substratum. It was observed that the film developed with 1% w/v suspension exhibited the highest sensitivity to H_2S with a 10.94 ppm limit of quantification (LOQ) and 3.27 ppm limit of detection (LOD) when subjected to meat spoilage detection (Sukhavattanakul and Manuspiya 2021).

Carbon nanomaterials are utilized in the manufacture nanosensors because of their light-weight, great mechanical and electrical properties, high flexibility, and high specific surface area (Biji et al. 2015). Carbon nanomaterials are highly sensitive, having a detection limit at ppm levels of gas molecule concentration. Mirica et al. (2013) fabricated an uncomplicated and fast prototyping method by developing selective chemical sensors with graphite and carbon nanotubes (CNTs) on a paper surface. The system depended on the mechanical abrasion of pencils comprising CNTs and small molecules, which may interrelate with certain gases. These nanosensors were able to detect and differentiate vapors and gases at a ppm concentration level. Similarly, the application of printed CNT-based gas sensors was successful with an outstandingly high and quick response to CO_2 and NH_3 (Abdellah et al. 2013). In the foods like pickles and sausages, the nitrate content could be electrochemically sensed with the help of N-doped carbon nanofiber membrane decorated with N-doped graphene quantum dots. The addition of N-doped graphene quantum dots enhances the electron transfer rate, and the N-doped carbon nanofibers imparted a free-standing film structure, greater electroactive area, and electrical conductivity, making the composite suitable for the electrochemical sensors. This quantum dots-based nanofiber sensor with high selectivity, wide linearity range, and outstanding reproducibility exhibited improved efficiency in sensing nitrite with a low detection limit of up to 3 μM (Li et al. 2017). Extensive research in the area of nanosensor has led to significant scientific advancement which paves the way to produce new generation nanosensors that could be promisingly incorporated in IP. However, there are still few concerns regarding the uncertainty in the behaviour and toxicity of nanomaterials with the human body and the further complication it might possess.

4 Indicators

The indicators are semi-quantitative or qualitative devices or tools that give information about the changes taking place in the food or its environment, such as alteration in pH, temperature, time, etc., mainly by a colour change (Yucel 2016). It indicates the presence or absence or the concentration of an analyte or the reaction among two or more analytes (Lloyd et al. 2019). The indicators can be divided into internal and external indicators. The internal indicators are affixed inside the food package (with the lid or in the headspace) like gas indicators or pathogen indicators. In contrast, external indicators are placed on the outer side of the package, such as mechanical shock indicators and time-temperature indicators (TTIs).

4.1 Time-Temperature Indicators (TTIs)

Temperature is a critical environmental factor which determines the kinetics of physical, chemical, and microbial spoilage in foods (Biji et al. 2015). TTIs are the first generation indicators (Bajpai 2019). TTIs are small labels or tags which track the time-temperature of perishable products from production to consumption. TTIs function depending on the temperature changes in the product or the package. It measures and displays the same on the IP system. TTIs determine and exhibit information about when the food has been exposed to a higher temperature than the desired range during storage and transportation, which is very important to know the temperature abuse for frozen or chilled food products (Pavelková 2013). TTIs can be categorized into three main classes on the basis of their function, temperature history, and operating principle. Based on the function, TTIs are further divided into critical temperature indicators, time-temperature indicators, and critical temperature/ time integrators. Critical temperature indicators depict the organoleptic changes and deterioration of food when exposed to temperatures below or above the specified ranges for more than a mentioned time interval via visible but irreversible colour changes of the indicator. Critical temperature/time integrators exhibits a response above a reference critical temperature that could be interpreted with regard to equivalent exposure time at that critical temperature, thus indicating the safety and quality of the product (Taoukis and Labuza 2003). The third type, TTIs shows the overall influence of temperature history on the quality of the product from the manufacture to utilization by the consumers (Janjarasskul and Suppakul 2018). On the basis of temperature history, TTIs are divided into partial history indicators and full history indicators. A partial history indicator gives an indication only when the food is exposed to a higher temperature than its critical temperature. In contrast, a full history indicator furnishes information regarding the food packaging over time (Kerry et al. 2006). Lastly, TTIs are categorized on the different working principles, such as chemical, enzymatic, mechanical, electrochemical, and microbiological principles, usually shown with a change in colour or change in intensity of the colour (Müller and Schmid 2019). In the previous years, the industrial use of different indicators, for instance, diffusion-based TTIs, enzymatic TTIs, microbial TTIs, polymer-based TTIs, and photochromic TTIs have been increased significantly for evaluating the time-temperature of varied commercially developed perishable foods products. Table 14.1 depicts some prominent TTIs that were commercially utilized for varied food products based on different working principles.

Kim et al. (2016) developed isopropyl palmitate (IPP) diffusion-based TTI for determining the microbiological quality in unpasteurized angelica juice. The experiment reported that IPP diffusion up to 7 mm in the TTI showed effective identification of microbial spoilage in the sample when the TTI was used at a temperature 13 °C or higher. Besides, an economical and accurate polymer-based TTI functioning on pH changes was prepared from chitosan, polyvinyl alcohol (PVA), and *Brassica oleracea* (red cabbage) extract. The prepared TTI was able to sense the

Table 14.1 List of some commercial TTIs

Product name	Manufacturer	Operational principle
3M™ MonitorMark®	3 M Company, USA	Molecular diffusion
3M™ Freeze Watch™	3 M Company, USA	Molecular diffusion
CheckPoint®	Vitsab International AB, Sweden	Enzymatic reaction
CoolVu™	Evigence Sensors™, Haifa, Israel	Dissolution process of a fine aluminium layer
Fresh-check®	Temptime Corporation, USA	Solid-state polymerization reaction
Keep-it®	Keep-it technologies, Oslo Norway	Chemical reaction
TempDot®	DeltaTrak, California, USA	Indicates cumulative exposure above temperature threshold
Timestrip® PLUS	Timestrip Plc, Cambridge, UK	Solid-state polymerization and enzymatic reactions
TopCryo™	TRACEO, France	Microbiological reaction
OnVu™	Ciba Speciality Chemicals, Switzerland and Freshpoint, Switzerland	Photochemical reaction
WarmMark®	DeltaTrak, California, USA	Indicates cumulative exposure above temperature threshold

change in pH in the pasteurized milk and showed visual indication by changing the colour (Pereira et al. 2015). Lee et al. (2019b) developed an air-activated TTI for monitoring the shelf life of sandwiches stored for 63 h at 5 °C. The developed TTI worked based on the redox reaction between colorimetric material (leuco methylene blue) and redox reaction repressor (a mixture of L-cysteine and L-ascorbic acid; AC) and in the presence of O_2.

The microbial TTI response shows direct relation with microbial spoilage in food. A correlation can be determined between microbial growth in food and microbial growth and its metabolism in the respective microbial TTI. Microbial TTIs are now increasingly applied to determine the shelf life and quality of perishable foods stored in the cold chain. A microbial TTI was fabricated for vacuum-packed giant grouper (*Epinephelus lanceolatus*) fillets by Hsiao and Chang (2017). The team selected *Lactobacillus sakei* as the main spoilage bacteria for the vacuum-packed fish fillets and observed the biochemical and microbiological changes in the samples. The microbial TTI was reported to be effective for monitoring the freshness of marine products, as it indicated three grades of freshness, such as "very fresh", "fresh", and "spoiled", on the basis of TVB-N levels. Mataragas et al. (2019) developed an efficient microbial TTI using *Janthinobacterium* sp. which has the potential to form a violet pigment, violacein during its early growth by depending on the intrinsic properties and temperature of the growth medium. The bacterium was spot-inoculated in 1% glycerol incorporated tryptic soy agar to fabricate the microbial TTI. Under dynamic and isothermal storage, the TTI device was used to validate spoilage in minced beef employing the data generated from the experimental analysis. They observed a noticeable effect on the initial concentration of

Janthinobacterium sp., spot quantity, and pH of the growth medium related to the different endpoint times and Arrhenius activation energy (Ea) of the TTIs, indicating the flexibility of the intelligent device.

The enzymatic TTIs are more susceptible to changes in environmental temperature and show increased precision than physical diffusion TTIs and microbial TTIs. Giannoglou et al. (2019) conducted a study to select appropriate enzymatic TTIs for monitoring the shelf life and quality of ready-to-eat smoked fish products in refrigerated storage. The enzymatic TTIs response was kinetically modeled and correlated with the quality and shelf life of cold-smoked salmon, and vacuum-packed hot smoked rainbow trout, and European eel in the cold chain. The experiment stated that the developed TTI designs, M-17 U, LP-17 U, and M-5 U determined the shelf life of vacuum-packed smoked salmon slices, smoked trout, and eel fillets at 5 °C for 2, 11, and 7.5 weeks, respectively. A novel solid-state enzymatic TTI after isothermal verification was made by 5 g amylose, 0.02 g glucoamylase microcapsules, 0.1 M iodine solution, and 2 mm thickness agar cover. The formulated enzymatic TTI was used to know the time-temperature history of chilled fresh pork with the help of colour indication. The kinetic properties and spoilage mechanism of pork samples were determined, and the Ea was found to be 64.7 kJ/mol (Meng et al. 2018).

There are different types of TTIs that are commercially developed and patented. They are primarily based on diffusion, chemical, enzymatic, biological, microbiological, and polymerization reactions. Figure 14.2 represents a schematic representation of commercially employed TTIs in the market. Nonetheless, the TTIs based on diffusion systems govern the IP market due to certain limitations in enzymatic TTIs, such as enzymatic instability and high cost. Even though different TTIs have been developed for meat, fish, dairy, and frozen products by evaluating the storage characteristics, the research regarding TTIs for liquid foods is still limited and requires further exploration.

4.2 Gas Indicators

Gas composition inside a food package should be carefully monitored, as any alteration in it from the standard composition may lead to product spoilage. The gas indicators in the form of labels are usually inserted in the package for determining the changes of the inside atmosphere (Ghaani et al. 2016). Alteration in the composition of the gas is caused by various factors such as leakage in the package, respiration of the produce, chemical or enzymatic reaction of the food matrix, microbial permeation through package, or production of gas by microorganisms (Sharma and Ghoshal 2018). Thus, a gas indicator is a small device placed inside or printed outside a package that responds to gas composition changes and indicates the integrity, quality, and safety of packaged food products (Fang et al. 2017). Mostly, the concentration of O_2, CO_2, or ethylene (C_2H_4) is indicated by colour change of the label (Sohail et al. 2018). However, water vapour, H_2S, ethanol (C_2H_5OH), NH_3, DMA, and TMA are accessed for spoilage determination.

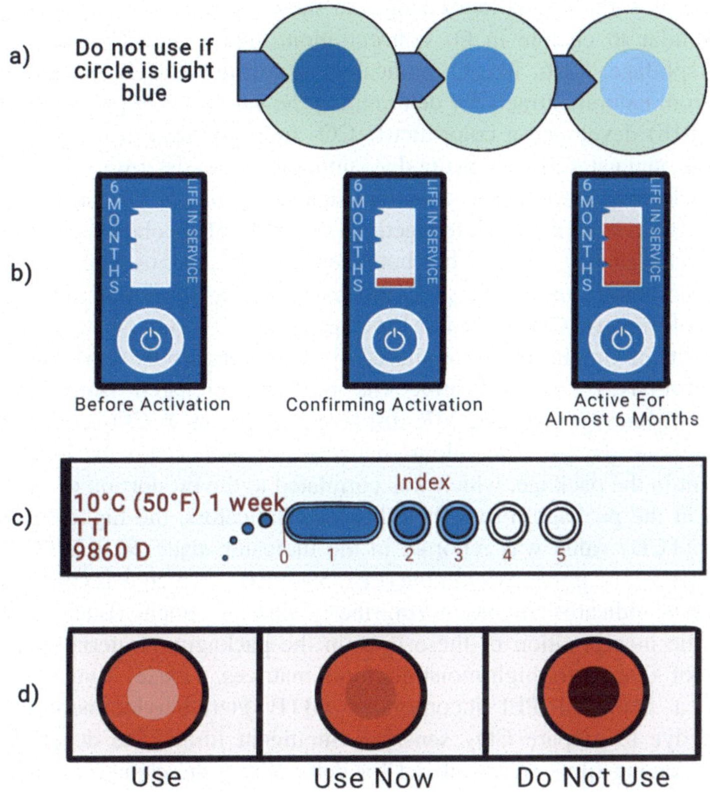

Fig. 14.2 Schematic representation of commercially used TTIs

The most widely employed gas indicator is the O_2 indicator as it indicates the quality of food by colour change, oxidative rancidity, and microbial spoilage (Fang et al. 2017). The O_2 indicators are of two types; colorimetric and luminophoric. The colorimetric system works on the principle of colour change by redox reaction, light-activated redox reaction, or O_2 binding reaction, while the luminescence intensity is an indicator of the presence of O_2 inside the package in a luminophoric system (Sharma and Ghoshal 2018). For example, Yılmaz and Altan (2021) developed a colorimetric oxygen indicator using functionalized electrospun polystyrene (PS) fibers by the process of electrospinning. Meatball samples that were packed with the indicator exhibited quick and remarkable changes in the colour of the indicator in the presence and absence of O_2. The fabricated oxygen indicator which is resistant to dye leakage also showed minimal in-pack activation time. CO_2 indicators are commonly employed for monitoring the CO_2 concentrations in MAP foods. It helps in identifying the incorrectly packaged products for immediate repacking, reducing time-consuming and labour-intensive quality control procedures (Fang et al. 2017). Lately, many works have been conducted for fabricating CO_2-sensitive smart packaging films. As the CO_2 percent inside the headspace of the package elevates, the film colour changes to show a visual index of packaged food quality. Similar to change in O_2 concentration, change in CO_2 also accelerates microbial spoilage. Also, like O_2 indicators, colorimetric CO_2 indicators can be prepared from natural extracts for determining the variations in pH. Saliu and Della Pergola (2018) developed a colorimetric CO_2 indicator from a mixture of lysine, ε-polylysine, and naturally occurring dye, anthocyanin, to determine the pH changes in MAP packages. The indicator functions depending on the formation of carbonic acid derivatives via an irreversible reaction. The researchers observed visible colour change with the change in CO_2 when both the label and aqueous type indicators were tested against poultry meat. The study indicates that the developed indicator can be used as a colorimetric CO_2 indicator for cold preserved foods.

The quality of kimchi can be monitored by developing a poly (ether-block-amide) (PEBA) film-based CO_2 indicator. The indicator prepared from PEBA film, bromothymol blue (BTB), and MR displayed alteration in CO_2 concentration in the package's headspace. The colour change of the indicator is the function of the CO_2 present in the package, which was correlated to the proportion of the materials employed in the package. From the fabricated indicators, the highest total colour difference (TCD) value was reported in the indicator made of PET/PEBA + dye (MR + BTB) (3:7) + polyethylenimine (PEI) 5%/PEBA, which was concluded as the most efficient indicator for monitoring the quality of kimchi (Baek et al. 2018). Howbeit, the incorporation of these dyes in the packaging materials leads to the migration of dyes into high-moisture food matrices. Hence, Lyu et al. (2019) fabricated a branched PEI incorporated (BTB^-)/tetrabutylammonium (TBA^+) ion-paired dye to prepare CO_2-sensitive intelligent films. The developed multi-layered packaging films were utilized for determining the change in CO_2 level in packaged kimchi during its fermentation process. The CO_2-mediated colour change was visible to the human eye, explaining its potentiality in developing commercial IP.

The TVB-N content is a crucial factor when observing the freshness of pork. For example, a colorimetric label was made to evaluate the shelf life of packaged lean pork in cold storage for 8 days. The indicators were made from three varied pH-sensitive dyes; BCP, BTB, and a blend of BTB and MR. The freshness indicators were prepared by combining polyethylene glycol-6000, methylcellulose, and dye solution in distilled water. The biochemical reactions and microbial spoilage are two major factors that affect the freshness of pork. The result of principal component analysis and TCD performed using colorimetric data of various indicators reported that indicator made up of BTB and MR (3:2) discriminated pork as "fresh" (red), "medium fresh" (golden rod), and "spoiled" (green) during cold storage (Chen et al. 2019). Recently, to monitor the freshness of chicken breast stored at 4 °C, Yildiz et al. (2021) developed an IP system from natural halochromic curcumin loaded chitosan/polyethylene oxide (PEO) nanofibers. Depending on the freshness of chicken breast, the pH sensor film changed the colour from bright yellow to red, easily detectable by the naked eyes. After the storage period of 5 days, the TVB-N concentration was 23.45 ± 3.35 mg/100 g indicating the sample was at the edge of the acceptance level. As a result, the curcumin loaded-nanofibers help to perform the real-time monitoring of chicken.

4.3 Freshness Indicators

The chemical and microbial deterioration affects the freshness of food; thus, chemical and microbiological spoilage in food occurring during handling and storage of food are determined by a freshness indicator (Sharma and Ghoshal 2018). Freshness indicators are primarily internal indicators placed inside a package for indicating the freshness and quality of food by changing their colour. Besides determining the food quality and deterioration, freshness indicators are utilized to estimate the shelf life of the product. The quality indicating metabolites are CO_2, C_2H_5OH, glucose, organic acids, volatile nitrogen compounds, biogenic amines, sulphuric compounds, and ATP degradation products (Müller and Schmid 2019). Freshness indicators are mostly used in products like fresh meat, seafood, and fruits. Each freshness indicator is specific to the product. For instance, a freshness indicator developed for seafood will show a response when volatile amines are produced. Similarly, freshness indicators for chicken, beef, and other protein-based foods are prepared using biogenic amines like histamine, tyramine, cadaverine, and putrescine. TMA and TVB-N also aid in indicating the freshness of meat products.

Freshness indicator in the form of a label using BTB and MR (2:3) was used for real-time monitoring of green bell pepper stored at 7 ± 1 °C. As the CO_2 in the IP increases, the indicator exhibited an intense change in colour to orange from yellow-green (Chen et al. 2018). Similarly, a pH-sensitive dye-based colorimetric indicator label was developed to monitor the freshness in skinless chicken breast. During storage, the CO_2 produced was considered a spoilage indicator and showed higher concentration than TVB-N. The indicator correlated with an increase in CO_2 level and microbial growth in the stored sample and exhibited the variation in CO_2 level by colour change in the IP (Rukchon et al. 2014). Lee et al. (2019a) manufactured an inexpensive and simple freshness indicator for monitoring spoilage in chicken breast made of three-layered structure and porous Tyvek® sheet for improved gas and water vapour permeation. The indicator solution composed of BCG was coated over the low-density polyethylene (LDPE) films and later laminated on Tyvek® sheets. The colour change of the freshness indicator containing chicken breast was correlated with bacterial growth and TVB-N and CO_2 contents in the chicken breasts. The changes in the indicator were captured as digital images using a smartphone camera and examined with the help of RGB pixel intensities. The colour change from green to yellow indicated sample spoilage by showing a remarkable difference in RGB values. Thus the developed high-confidence freshness indicator is a reliable, user-friendly approach to monitor the quality changes in fresh chicken breasts.

The aging in cod flesh was examined using an indicator made of beetroot, grape peel, and curcumin extracts. The volatile amine was monitored by the indicators prepared from these natural dyes and compared with the functionality of indicator designed with artificial dye, MR. Out of all the three natural dyes, the indicator made of grape peel extract and curcumin showed comparable results with the MR-based indicator by indicating a colour change with TCD values equal to approximately 30 units correlating the cod flesh spoilage (Tichoniuk et al. 2017). For determining the freshness of silver carp (*Hypophthalmichthys molitrix*) stored at refrigeration temperature, Zhai et al. (2017) fabricated roselle (*Hibiscus sabdariffa* L.) anthocyanins incorporated starch/PVA colorimetric film. The films represented as an IP device by indicating visible colour changes over time showing the occupancy of TVB-N. The rancidity reaction of O_2-sensitive dairy products was determined using a colorimetric indicator. An indicator made up of MR and BTB was used for accelerated shelf life in the milk powder formula stored at 30 °C. The indicator responded via a visible colour change from light green to orange to the volatile compounds, acetic acid, and hexanal developed during hydrolytic and oxidation reaction during storage. The accelerated shelf life in milk powder was observed to be 26 days, and the orange colour indicated a sign of either rejecting or warning (Kulchan et al. 2016).

The pH-based indicators work as freshness indicators by changing the colour for indicating the presence of various metabolites. A direct-contact type freshness indicator was used for real-time monitoring chicken breasts based on the variation in pH. The indicator comprised of BCP was immobilized with PVA on a high absorbance pad (Kim et al. 2017). This type of intelligent device can be utilized as either a freshness indicator or a high sensitivity shelf life determination tool. Liu et al. (2019) developed a *Lycium ruthenicum* Murr. extract incorporated κ-carrageenan colorimetric films as a pH indicator in intelligent packaged milk for assessing its freshness. The indicator showed a reversible colour change from pink to colourless over the pH range, 2–10, specifically pink to colourless with the pH, 2–6, and blue-purple to yellow with the pH, 7–10. The film was also acceptable for indicating the freshness of aquatic food products.

Another intelligent tool for indicating the food quality and freshness is the ripe sensor. With regard to fruits, their freshness is related to the degree of ripeness they possess. One such commercially employed freshness indicator is the RipeSense®, an ethylene gas indicator manufactured by the Jenkins Group and the Plant & Food Research, Auckland, New Zealand (RipeSense® 2004). RipeSense® is the world's first intelligent label sensor which indicates the change in ripeness of a fruit of the package by changing the colour of the package. The fruit generates a characteristic aroma as the process of ripening initiates. The sensor element responds to the released aroma compounds and results in changing the colour of the intelligent device. Firstly, the colour of the label is red if the fruit is unripe, and later on, during ripening, it changes to orange and turns yellow, indicating a fully ripe fruit. The illustration of the function of the RipeSense® freshness indicator in fruits is depicted in Fig. 14.3. Currently, RipeSense® is used for monitoring the freshness of various fruits like pears, apples, kiwi, melon, avocado, mango, and stone fruit. Table 14.2 summarizes commercially manufactured gas and freshness indicators used for monitoring food safety and quality.

Fig. 14.3 Illustration of the function of RipeSense® freshness indicator in fruits

Table 14.2 List of some commercially developed gas and freshness indicator

	Product name	Manufacturer
Gas indicators	Ageless Eye®	Mitsubishi Gas Chemical Inc., Japan
	Best-by™	FreshPoint Lab
	Freshilizer	Toppan Printing Co., Tokyo, Japan
	Novas®	Insignia Technologies Ltd., West Sussex, UK
	O_2 Sense™	FreshPoint Lab
	Shelf Life Guard	UPM-Kymmene Corporation, Finland
	Tell-tab	IMPAK Corporation, Los Angeles, USA
	Tufflex GS	Sealed Air Ltd., USA
	Vitalon	Toagosei Chemical Inc., Tokyo, Japan
Freshness indicators	Food Fresh™	Vanprob Solutions
	Food Sentinel System	SIRA Technologies Inc., California, USA
	Fresh Tag®	COX Technologies, USA
	SensorQ™	Food Quality Sensor International Inc., Massachusetts, USA and DSM NV, Heerlan, Netherland
	Raflatac	UPM Raflatac, Scarborough, UK and VIT Technical Research Centre, Finland
	RipeSense®	Jenkins Group, New Zealand and Plant & Food Research, Auckland, New Zealand
	Toxin Guard	Toxin Alert Inc., Ontario, Canada

5 Data Carriers

Automatic identification devices also known as data carriers, helps in ensuring the automatization, traceability, and counterfeit protection throughout the supply chain by storing and transmitting the product information regarding storage and distribution (Müller and Schmid 2019). Traceability assures food safety by attaining a better market as it provides the complete history of a package with the help of barcodes and radio-frequency identification (RFID) tags (Chen et al. 2008).

5.1 Barcodes

The most popular data carrier is the barcode, which is an inexpensive simplest machine-readable storage database which works on the optical phenomenon of vertical code bars placed at systematic thickness and width (Kalpana et al. 2019). On scanning a barcode, the laser beam moves over the symbol and measure the relative time it uses to scan the dark bars and light spaces (Ghaani et al. 2016). Mostly the barcodes on the products are assigned in numbers and character forms. The barcodes are widely classified into two types; 1D and 2D. 1D barcode is a simple linear arrangement of black and white bars which are simultaneously placed to store data and information. This type of barcode can only hold about 10–13 characters or 2953 bytes of information (Firouz et al. 2021). On the other hand, a 2D barcode uses two-dimensional geometrical patterns. 2D barcodes have the capacity to store about 7089 numeric and 4296 alphanumeric characters. Presently, the most frequently used 2D barcodes are Data Matrix, QR code, and PDF 417 (Bajpai 2019). The first commercialized barcode, Universal Product Code (UPC), having a pattern of lines and spaces, is still successfully utilized in the market. Another type of the barcode is the Reduced Space Symbology or GS1 DataBar™, which was invented to meet the reduced space usage requirement in the product packages (Uniform Code Council 2014). Besides, Content Idea of Asia Company (Kuwana-shi, Mie-ken, Japan) in 2006 launched a barcode named PM code. It is a three-dimensional colourful QR code, where its third dimension is coloured (PM Code 2006). Chen et al. (2017) developed a simple, cost-effective barcode, a colour-based sensor array formed in silica beads using the dyes, zinc tetraphenylporphyrin (TPP), MR, and Nile red in three dissimilar geometric shapes for detecting the spoilage in chicken. The detection of spoilage was found by the separation of green, blue, and red dyes which is read with the help of a built-in app designed on a smartphone. On storage, if the pathogen develops inside the package, it can be detected by the barcode and with the colour changes, which makes the barcode unreadable.

5.2 *Radio-Frequency Identification (RFID) Tags*

RFID tags utilize electromagnetic fields for storing and communicating real-time product information for automatic product identification and traceability (Realini and Marcos 2014). It is widely obtained as a form of chip made of tags used for storing data (Kalpana et al. 2019). RFID has shown its application in almost all food products including, meat, seafood, fruits and vegetables, beverages, bakery, and dairy products. RFID systems provide several advantages such as automatization, product traceability, prevention of product recalls, minimized counterfeit, theft prevention, reducing labour cost, and inventory management. These systems with about 1 MB storage capacity can hold more complex information like relative humidity and temperature, nutritional information and cooking instructions (Ahmed et al. 2018). An RFID system consists of three main elements; tag, reader, and middleware. A tag is made up of a microchip for storing data which is joined to an antenna. The reader sends radio signal and receives the response from the tag. The middleware is the web server or local network responsible for connecting the RFID hardware with their applications (Kuswandi and Jumina 2020). An RFID middleware filter, integrate data, coordinate reader, and manage the scheduled processes (Chen et al. 2008). Figure 14.4 shows a schematic representation of a RFID system. The RFID systems are classified into three types; active RFID tags, which require battery-powered tags for broadcasting signal to the reader; semi-passive RFID tags, which employs battery to power the tag for modulating the waves emitted by the reader; and also for maintaining the memory of the tag; and passive RFID tags, which are simple and short-range, are powered by a reader that does not require any battery (Ilie-Zudor et al. 2006).

The sensor-enabled RFID tags have shown their promising application in storing and transporting perishable commodities like fruits and vegetables, milk, fish, meat, and poultry which need strict environmental conditions. The combination of sensor technology and RFID helps in improving the efficiency of supply chain management and reduces waste production. Sen et al. (2013) developed a RFID system along with gas and temperature sensor for detecting the freshness and quality of pork meat. The

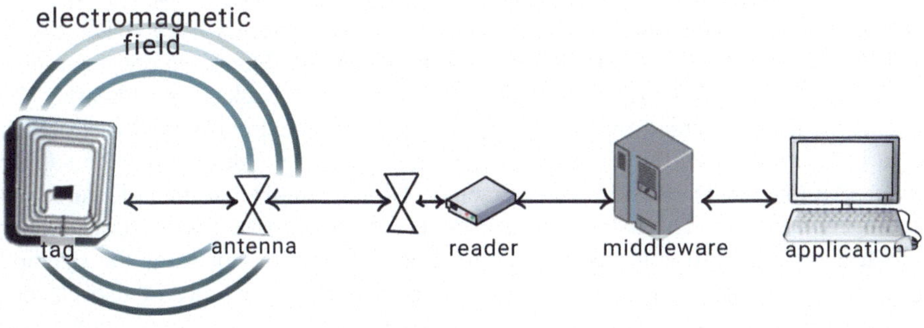

Fig. 14.4 Schematic representation of RFID system

system categorized meat quality as "great", "medium", "low" and "corrupt". In the study, the concentration of H_2S generated from the meat was evaluated at varying temperatures to find its relation with meat spoilage. Correspondingly, the reader received the H_2S concentration and sent the data to the display screen. Thus, the seller or the consumer can read the meat quality. In the case of seafood, a printable smart RFID tag was used for detecting and examining the fish spoilage by knowing the levels of volatile amines and humidity. Besides, anisotropic conductive adhesives (ACA) was employed as the integrated chips. Smits et al. (2012) developed an RFID system of low power-consuming MSP430G microprocessors for studying the freshness of cod fish by evaluating the humidity, temperature, and volatile amine compounds at 5 °C for an interval of 2 min. The study reported that the developed system did not depend on the type and quantity of fish but relied on the freshness indicating factors. Similarly, Abad et al. (2009) demonstrated a RFID smart tag for cold chain monitoring and real-time traceability of intercontinental cold logistic chain of fresh fish. The experiment revealed that the developed system showed supremacy compared to conventional traceability tools by showing no human participation, reusability, more memory, ability to read many tags at a time, and more aversion to humidity and other environmental conditions.

A unique chipless RFID sensor system for wireless sensing was manufactured by Amin et al. (2016). The system was easy to use, requires no maintenance, and works without any electrical source unlike other RFID systems. Recently, Alfian et al. (2020) developed a RFID-based traceability system furnished with internet of thing (IoT) sensor for monitoring and knowing environmental conditions, namely, humidity and temperature for perishable food distribution during storage and transportation. This system communicated counterfeiting and low-quality food products along the supply chain by enhancing the traceability with the developed machine-learning model into RFID gate for automatic identification of the direction of tagged products. RFID systems have also been assisted with indicators for ensuring quality and traceability of foods. For instance, RFID was integrated with critical temperature indicators for fresh-cut fruits (Lorite et al. 2017); RFID was integrated with optical indicator for tracing kiwi fruit (Gautam et al. 2017); and for monitoring O_2 concentration (Martínez-Olmos et al. 2013).

A few reusable TT sensor tags fabricated to show the temperature history of products throughout the cold chain process are the TempTRIP sensor tag (TempTRIP 2012), sensor tag CS8304 (Convergence Systems Ltd.) (CLS 2013), and Easy2log© (CAEN RFID Srl) (CAEN RFID 2014). Manufacturers have also integrated the RFID systems into the food box in the packaging industry. The Intelligent Box presented by Mondi Plc is a RFID-enabled corrugated case which is furnished with a RFID tag at the case level for tracing it throughout the supply chain (Mondi 2011). Besides, the Craemer Group GmbH has manufactured an intelligent fish box. The fish box is composed of an integrated RFID transponder which helps in identifying, tracing, and tracking information regarding the size and quality of fish, and fishing grounds to ensure absolute traceability of fish catches (Craemer 2014). NXP® Semiconductors Company (Eindhoven, Netherlands) fabricated wireless sensors based on RFID systems to monitor packaged food products.

The environmental parameters like C_2H_4, CO_2, and O_2 were evaluated, and the documented information was sent to the central system (NXP® Semiconductors 2022).

6 Other Intelligent Tools

Another type of intelligent devices drawing increasing interest to ensure food safety and quality, and to prevent counterfeiting and tampering are the thermochromic inks, holograms, internet of everything (IoE), carbon photonics, organic photonics and electronics, and printed electronics.

6.1 Thermochromic Inks

Thermochromic ink is an activated ink which changes its colour when exposed to varying temperatures. These inks are usually employed in microwave food products or beverage packaging that allows the consumers to be aware if the food is to be served hot or cold (Sohail et al. 2018). The colour change of the ink is either reversible or irreversible. Irreversible inks are not visible until it reaches a specified temperature, and once the colour appears, it stays unchanged if there is no shift in recorded temperature. If the temperature shoots up, the colour changes and leaves a temperature change indication (Roya and Elham 2016). Whereas reversible thermochromic inks change its colour when heated and returns to the initial state as the temperature decreases. There are also few other variations in the type of thermochromic inks. The cold-activated thermochromic inks are employed on a packaging label for creating a change in colour when the food is cooled. The touch-activated thermochromic inks show an image or another colour printed below once touched or rubbed. High-temperature thermochromic inks switch their colour just beneath the pain threshold, alerting customers regarding a safety hazard. Thermochromic inks are manufactured by different companies like QCR Solutions Corp. (USA), CTI Inks (USA), LCR Hallcrest (USA), B&H Colour Change (UK), and Siltech Ltd. (UK).

6.2 Hologram

A hologram is an emerging and attractive tool applied in food IP, which aims to safeguard the brand name of the product and avoid counterfeiting and tampering with the product (Sohail et al. 2018). The pattern changing of the hologram deprives the counterfeiters to change the product label or the product. However, if the counterfeiters try to remove the hologram, the upper polyester film of the package should be taken off, thus leaving a mark that indicates product tempering (Pareek and Khunteta 2014). Even though holograms are widely used for expensive drugs in the pharmaceutical industry, their utilization in food IP is limited. The existing technological development of IP system in food industries promises more comprehensive future application of holograms to ensure food safety.

6.3 Internet of Everything (IoE)

In the era of IoE, besides the basic functions of packaging, including storage, transportation, and protection, the design of modern transport packaging should be considerably made using eco-friendly and recyclable packaging systems (Zhang and Peng 2019). The IoE is a relatively novel concept that aims at a world-wide network of interconnected objects. Such a system is made possible by the integration of varied state-of-the-art technologies and communication devices like wired and wireless sensors, RFID tags, GPS, and enhanced communication protocols (Vanderroost et al. 2014). Rather than remotely monitoring the food quality and package integrity, it is foreseen that IoE in due course, result in the advancement in food safety management, QACCP, and HACCP systems (Takhistov 2009). This development in IP may gradually help in identifying potential safety hazards and conducting biohazard analysis, recommending controls and critical limits, and monitoring food loss on an international scale. Although IoE is a relatively growing technology in food industries, more comprehensive application of this technology can be expected in the near future.

6.4 Carbon Photonics

Carbon-based nanomaterials are known to show excellent mechanical and electrical properties and exhibit distinctive optical properties that can be utilized to manufacture future optical sensors as a substitute to silicon photonics (Vanderroost et al. 2014). For instance, Kruss et al. (2013) reported the first research work using carbon nanomaterials for the development of optical biosensors. Carbon-based nanomaterials can also be modified chemically so that biologically relevant molecules can be determined with excellent selectivity and sensitivity. The role of external electric fields, electrical and optical mechanisms of their production, non-radiative and radiative decay modes, and their potentiality in technological use as nano-sized light sources, photodetectors or photovoltaic devices (Avouris et al. 2008). Besides, the so-called carbon dots (CDs) are regarded as the class of strongly fluorescent and emission colour tuning carbon nanomaterials with high analytical and bioanalytical potential (Esteves da Silva and Gonçalves 2011).

6.5 Organic Photonics and Electronics

Organic photonics and electronics investigate how electrical and optical circuits could be combined with organic materials (e.g., polymers) instead of silicon with no compromise on the circuit size. Organic photonics and electronics aim to exhibit similar or advanced electrical or optical characteristics with improved mechanical properties (Vanderroost et al. 2014). The photonics system as printed elements ensures the functionality of novel packaging systems, AP and IP by informing the status of a packaged food by altering the internal or external sensor properties. The changes in the packaging system can be instrumentally or visually registered using internal or external devices. Development of photonics system and future simulation, understanding the printing techniques of printed elements, and preparation of functional surfaces containing these systems need to be convincingly carried out, ensuring the functions of IP systems (Sarapulova et al. 2015). Organic photonics and electronics hence enable energy-efficient creation of IP merged with ICT. This area of research also leads to the design and development of novel organic materials for IP.

6.6 Printed Electronics

Printed electronics are the flexible printed sensors holding a receptor on top of a printed transducer (Lloyd et al. 2019). Light-weight, portability, bendability, and foldability are few individual properties of printed electronic sensors (Biji et al. 2015). This packaging solution allows printing over varied substrates (such as steel, paper, polyethylene terephthalate (PET), polyimide, polyether ether ketone (PEEK), transparent conductive polyester, etc.) by electrically functional inks and shows them as a unique and tailor-made sensor for packaging food products (Lloyd et al. 2019). The flexible printed chemical sensor, comprising a receptor printed over a transducer, shows promising possibilities to revolutionize the development and utilization of IP systems. The printing techniques for producing printed electronics are screen printing, ink-jet printing, and gravure, where each method shows different advantages and limitations, depending on the product and the purpose.

7 Global Market and Legislative Aspects of Intelligent Packaging

The primary objective of innovative packaging solutions is to adopt varied packaging technologies to reduce food spoilage and food waste. Such technologies are adopted to cope with the augmenting demands in food safety, brand differentiation, and stock management (Realini and Marcos 2014). From the studies so far, it is evident that IP helps enhance the quality of food, reduce food wastage, and improve overall food production efficiency. The market for intelligent, active, and advanced packaging has increased at a compound annual growth rate (CAGR) of approximately 5.8%, in which IP showed around $ 5.3 billion in sales in 2017 (BCC Research 2013). However, the application of IP in the food sector is still limited because of its high cost and restricted integration with other packaging systems. The legislative aspects and perceptions of consumers are considered before implementing a new IP in the market.

One major issue which deprives the acceptance of intelligent devices in the market is the reluctance of consumers to encourage the non-edible items separate from the package. The prejudiced approach of consumers to innovative packaging applications might misguide them about the actual product quality. The beneficial part of IP systems is still unclear as the consumers find the inserts, sachets, dots, and spots to be unnecessary in a food package. The contaminants that could be observed from IP materials include inks, adhesives, resins, pigments, oils, solvents, stabilizers, surfactants, plasticizers, antimicrobials, antioxidants and other additives (Mirza Alizadeh et al. 2021). The risk of accidental consumption or leaching of active components (e.g., inks) to food from an intelligent device when affixed with the primary packaging material creates a concern of food safety and health among the consumers. Consequently, the identification of substances that can possibly be released from the packaging system is important due to their impact on product quality and consumer safety. The release of intelligent substances from IP should be considered in terms of their toxicity, and their migration should comply with food legislative aspects. The migration of materials can be determined by migration tests or by adapting mass transfer modelling tools.

Japan, Australia, and the United States have shown the widespread application of IP systems in food industries, while Europe has implemented certain legislative regulations which hinder the integration of novel packaging technologies into the market. The first law regarding the materials considered to come in contact with food came in 2004 (European Commission 2004). Specifically, the law aimed to focus the human health versus the safety of the packaging materials, stating that any foreign component in the package should not transfer to the food in impermissible quantity, making unsatisfactory changes in the composition and organoleptic properties of food. Besides, IP systems employed in food packaging should not provide misleading information to the customers, and their suitability to food contact and appropriate use should be clearly mentioned. Nevertheless, in 2009 a list was published containing the authorized substances which can be allowed to have food contact and can be used for the production of active and IP systems (European Commission 2009). In addition, this law stated that when either active or intelligent devices are employed in a food package, it is compulsory to mention the phrase "DO NOT EAT", and if manageable, insert a specific symbol to indicate the intelligent device on the package. Points that need to be considered in determining the risk associated with the consumption of IP food products are the migration of intelligent substances or their reaction products, toxicological properties of intelligent substances, and interaction of intelligent substances with food matrices (European Food Safety Authority 2009). Thus, to alleviate consumer acceptance towards IP, it is necessary to reduce or deprive any potential risk of food contamination. The advancement in the fields of chemistry, biotechnology, material science, and microelectronics aid in the significant development of novel economic IP solutions.

8 Limitations and Future Perspectives

The integration of IP in food industries has led to significant improvement in food quality, food safety, shelf life and gradually reduces food loss or wastage. However, few major drawbacks of IP systems cannot be ignored and need further investigation to overcome the same. The higher development cost or production cost is one of the main limiting factors for extensive commercialization of IP systems. Secondly, the larger size of the existing intelligent devices or tools to integrate into food packaging is problematic and should be bought into consideration. Apart from these factors, the lack of public awareness or the limited demand in the market regarding the utility of IP systems is a notable factor affecting the growth of such systems in the commercial market. An ideal intelligent device for food packaging should be user-friendly, simple and reproducible, accurate, reliable, non-toxic, eco-friendly, cost-effective relative to the value of the food item, and compatible with printing technology for mass production.

Even though significant innovations have evolved in IP over the past two decades, there is an opportunity for further research and development. Some points to be considered for future works concerning IP systems are:

- The size and operation steps in most intelligent devices are complex. More research is required to minimize the size of the intelligent device and simplify the intelligent system operations.
- Different kinds of active compounds are infused with the packaging systems to attain new functionalities such as antioxidant and antimicrobial activities. Howbeit, the interaction of these active compounds with the package and product still needs thorough investigation. The effect of these extraneous compounds on the nutritional value, cooking properties, and sensory attributes need to be explored in detail.
- Another vital point to be considered is the nontoxicity of materials and packages used for IP systems. The integration of materials for IP should be performed by amending the necessary safety regulations due to the migration ability of few compounds to the food.
- The IP systems are found to be commonly used for a large class of fresh foods like meat, fish, fruits, and vegetables. However, extensive research on the contribution of IP in liquid foods like milk, fruit juices, and other beverages is still minimal.

- Synergistic technology is to be encouraged for ensuring improved quality and safety of foods. Combination of AP and IP help in the creation of smart packaging. The changes which occur inevitably in the packaged food can be managed by AP, and the condition of the product and its environment can be continuously monitored with the help of IP.
- The introduction of multifunctional sensing devices into smartphones and other gadgets is needed to make the IP systems user-friendly and reliable. The monitoring of food products with these devices should not be limited to colour change, but also with origin, authentication, nutritional information, etc.
- Eco-friendly IP comprising of biodegradable materials which can be reused, renewed, recycled, or repurposed should be encouraged for future food packaging production. Extensive work should be conducted to manufacture IP systems that can carry more complex information and reduce packaging waste.

9 Conclusion

Intelligent packages can be considered as communicative packages that help in monitoring, evaluating and informing suppliers, retailers, and consumers about the quality and safety of different food products such as meat, fish, dairy, fruits, and vegetables. The food producers, food processors, logistic operators, retailers, and consumers are the subjects of innovations. Howbeit, the innovative packaging solutions must amend the regulatory requirements and subdue major expenses of such newly developed packaging systems. Technological problems are the primary hurdle in commercializing the IP systems. A multidisciplinary approach involving researchers and industrialists from varied disciplines, including food science, food engineering, microbiology, material science, electronics, and electrochemical engineering, is necessary to successfully design, develop, and implement potential IP systems in the market. Besides, IP and its materials have to be sustainable with regard to design, production, and application. The challenges of environmentally sustainable technology; interest versus reusability, reversibility, and multi-functionality should be overcome. In conclusion, continuous research in the field of food packaging help in reducing the cost and complexity of IP systems and development of novel packaging solutions to enhance the quality, safety, and shelf life of food products.

Acknowledgments The authors are thankful to Dr. Shijin A, veterinary surgeon, and Ms. K V Jiya Fathima.

References

Abad E, Palacio F, Nuin M et al (2009) RFID smart tag for traceability and cold chain monitoring of foods: demonstration in an intercontinental fresh fish logistic chain. J Food Eng 93:394–399. https://doi.org/10.1016/j.jfoodeng.2009.02.004

Abdellah A, Abdelhalim A, Loghin F et al (2013) Flexible carbon nanotube based gas sensors fabricated by large-scale spray deposition. IEEE Sensors J 13:4014–4021. https://doi.org/10.1109/JSEN.2013.2265775

Aghaei Z, Ghorani B, Emadzadeh B et al (2020) Protein-based halochromic electrospun nanosensor for monitoring trout fish freshness. Food Control 111:107065. https://doi.org/10.1016/j.foodcont.2019.107065

Ahmed I, Lin H, Zou L et al (2018) An overview of smart packaging technologies for monitoring safety and quality of meat and meat products. Packag Technol Sci 31:449–471. https://doi.org/10.1002/pts.2380

Alfian G, Syafrudin M, Farooq U et al (2020) Improving efficiency of RFID-based traceability system for perishable food by utilizing IoT sensors and machine learning model. Food Control 110:107016. https://doi.org/10.1016/j.foodcont.2019.107016

Amin EM, Karmakar NC, Jensen BW (2016) Fully printable chipless RFID multi-parameter sensor. Sensors Actuators A Phys 248:223–232. https://doi.org/10.1016/j.sna.2016.06.014

Avouris P, Freitag M, Perebeinos V (2008) Carbon-nanotube photonics and optoelectronics. Nat Photonics 2:341–350. https://doi.org/10.1038/nphoton.2008.94

Azeredo HMC, Correa DS (2021) Smart choices: mechanisms of intelligent food packaging. Curr Res Food Sci 4:932–936. https://doi.org/10.1016/j.crfs.2021.11.016

Baek S, Maruthupandy M, Lee K et al (2018) Preparation and characterization of a poly(ether-block-amide) film–based CO2 indicator for monitoring kimchi quality. React Funct Polym 131:75–83. https://doi.org/10.1016/j.reactfunctpolym.2018.07.007

Bajpai P (2019) Recent trends in packaging of food products. In: Biobased polymers. Elsevier Inc., pp 139–169

BCC Research (2013) The advanced packaging solutions market value for 2017 is projected to be nearly $44.3 billion. http://www.bccresearch.com/pressroom/fod/advanced-packaging-solutions-market-value-projected-nearly-$44.43-billion-2017. Accessed 10 Jan 2022

Biji KB, Ravishankar CN, Mohan CO, Srinivasa Gopal TK (2015) Smart packaging systems for food applications: a review. J Food Sci Technol 52:6125–6135. https://doi.org/10.1007/s13197-015-1766-7

Bodenhamer WT, Jackowski G, Davies E (2004) Surface binding of an immunoglobulin to a flexible polymer using a water soluble varnish matrix. U.S. Patent No. 6,692,973. Washington, DC: U.S. Patent and Trademark Office.

Borchert NB, Kerry JP, Papkovsky DB (2013) A CO2 sensor based on Pt-porphyrin dye and FRET scheme for food packaging applications. Sensors Actuators B Chem 176:157–165. https://doi.org/10.1016/j.snb.2012.09.043

CAEN RFID (2014) CAEN RFID easy2log© RT0005. In: R® CAEN RFID. https://www.caenrfid.com/en/?mypage=3&parent=65&idmod=780. Accessed 10 Jan 2022

Chang LY, Chuang MY, Zan HW et al (2017) One-minute fish freshness evaluation by testing the volatile amine gas with an ultrasensitive porous-electrode-capped organic gas sensor system. ACS Sensors 2:531–539. https://doi.org/10.1021/acssensors.6b00829

Chen RS, Chen CC, Yeh KC et al (2008) Using RFID technology in food produce traceability. WSEAS Trans Inf Sci Appl 5:1551–1560. https://doi.org/10.1016/b978-0-08-100596-5.22853-9

Chen Y, Fu G, Zilberman Y et al (2017) Low cost smart phone diagnostics for food using paper-based colorimetric sensor arrays. Food Control 82:227–232. https://doi.org/10.1016/j.foodcont.2017.07.003

Chen H-z, Zhang M, Bhandari B, Guo Z (2018) Applicability of a colorimetric indicator label for monitoring freshness of fresh-cut green bell pepper. Postharvest Biol Technol 140:85–92. https://doi.org/10.1016/j.postharvbio.2018.02.011

Chen H-z, Zhang M, Bhandari B, Yang C-h (2019) Development of a novel colorimetric food package label for monitoring lean pork freshness. Lwt 99:43–49. https://doi.org/10.1016/j.lwt.2018.09.048

CLS (2013) CS8304 cold chain temperature logging tag. In: CLS Cover. Syst. Ltd. https://www.convergence.com.hk/. Accessed 10 Jan 2022

Craemer (2014) New RFID supported fish box enables seamless tracking & tracing. http://www.craemer.com/en/service-portal/news/detail/article/craemer-group-presents-product-innovation-for-the-fishing-industry/. Accessed 14 May 2021

Deng Y, Liu K, Liu Y et al (2016) An novel acetylcholinesterase biosensor based on nano-porous pseudo carbon paste electrode modified with gold nanoparticles for detection of methyl parathion. J Nanosci Nanotechnol 16:9460–9467. https://doi.org/10.1166/jnn.2016.13059

Dervisevic M, Custiuc E, Çevik E et al (2015) Electrochemical biosensor based on REGO/Fe3O4 bionanocomposite interface for xanthine detection in fish sample. Food Control 57:402–410. https://doi.org/10.1016/j.foodcont.2015.05.001

Devi R, Yadav S, Nehra R et al (2013) Electrochemical biosensor based on gold coated iron nanoparticles/chitosan composite bound xanthine oxidase for detection of xanthine in fish meat. J Food Eng 115:207–214. https://doi.org/10.1016/j.jfoodeng.2012.10.014

Dolmacı N, Çete S, Arslan F, Yaşar A (2012) An amperometric biosensor for fish freshness detection from xanthine oxidase immobilized in polypyrrole-polyvinylsulphonate film. Artif Cells, Blood Substitutes, Biotechnol 4:275–279. https://doi.org/10.3109/10731199.2011.646410

Dudnyk I, Janeček ER, Vaucher-Joset J, Stellacci F (2018) Edible sensors for meat and seafood freshness. Sensors Actuators B Chem 259:1108–1112. https://doi.org/10.1016/j.snb.2017.12.057

Esteves da Silva JCG, Gonçalves HMR (2011) Analytical and bioanalytical applications of carbon dots. TrAC - Trends Anal Chem 30:1327–1336. https://doi.org/10.1016/j.trac.2011.04.009

European Commission (2004) Commission regulation (EC) no. 1935/2004 of 27 October 2004 on materials and articles intended to come into contact with food and repealing directive 80/590/EEC. Off J Eur Union L 338(4):4–17

European Commission (2009) Commission regulation (EC) no 450/2009 of 29 may 2009 on active and intelligent materials and articles intended to come into contact with food. Off J Eur Union 3–11

European Food Safety Authority (2009) Guidelines on submission of a dossier for safety evaluation by the EFSA of active or intelligent sub- stances present in active and intelligent materials and articles intended to come into contact with food. EFSA J 7:1208

Ezati P, Priyadarshi R, Bang YJ, Rhim JW (2021) CMC and CNF-based intelligent pH-responsive color indicator films integrated with shikonin to monitor fish freshness. Food Control 126:108046. https://doi.org/10.1016/j.foodcont.2021.108046

Fang Z, Zhao Y, Warner RD, Johnson SK (2017) Active and intelligent packaging in meat industry. Trends Food Sci Technol 61:60–71. https://doi.org/10.1016/j.tifs.2017.01.002

Firouz MS, Mohi-Alden K, Omid M (2021) A critical review on intelligent and active packaging in the food industry: research and development. Food Res Int 141:110113. https://doi.org/10.1016/j.foodres.2021.110113

Flex Alert Company (2022) FLEX-ALERT biosensor technology solutions. In: Flex Alert Co. Ltd. https://flex-alert.com/applications/. Accessed 10 Jan 2022

Food Sentinel System (2019) SIRA technologies food sentinel system thermal barcode for packaging. In: SIRA Technol. http://www.sustainableisgood.com/sira-technologies-food-sentinel-system-thermal-barcode-for-packaging/. Accessed 10 Jan 2022

Fuertes G, Soto I, Carrasco R et al (2016) Intelligent packaging systems: sensors and nanosensors to monitor food quality and safety. J Sensors 2016. https://doi.org/10.1155/2016/4046061

Fung F, Wang HS, Menon S (2018) Food safety in the 21st century. Biom J 41:88–95. https://doi.org/10.1016/j.bj.2018.03.003

Gautam R, Singh A, Karthik K et al (2017) Traceability using RFID and its formulation for a kiwifruit supply chain. Comput Ind Eng 103:46–58. https://doi.org/10.1016/j.cie.2016.09.007

Ge Y, Li Y, Bai Y et al (2020) Intelligent gelatin/oxidized chitin nanocrystals nanocomposite films containing black rice bran anthocyanins for fish freshness monitorings. Int J Biol Macromol 155:1296–1306. https://doi.org/10.1016/j.ijbiomac.2019.11.101

Ghaani M, Cozzolino CA, Castelli G, Farris S (2016) An overview of the intelligent packaging technologies in the food sector. Trends Food Sci Technol 51:1–11. https://doi.org/10.1016/j.tifs.2016.02.008

Ghaani M, Pucillo F, Olsson RT et al (2018) A bionanocomposite-modified glassy carbon electrode for the determination of 4,4′-methylene diphenyl diamine. Anal Methods 10:4122–4128. https://doi.org/10.1039/c8ay01376d

Giannoglou M, Evangelopoulou AM, Perikleous N et al (2019) Time temperature integrators for monitoring the shelf life of ready-to-eat chilled smoked fish products. Food Packag Shelf Life 22:100403. https://doi.org/10.1016/j.fpsl.2019.100403

Gregor-Svetec D (2018) Intelligent packaging. In: Nanomaterials for food packaging. Elsevier, pp 203–247

Hsiao HI, Chang JN (2017) Developing a microbial time–temperature indicator to monitor total volatile basic nitrogen change in chilled vacuum-packed grouper fillets. J Food Process Preserv 41:e13158. https://doi.org/10.1111/jfpp.13158

Ilie-Zudor E, Kemeny Z, Egri P, Monostori L (2006) The modern information technology. In: The innovation processes of the industrial enterprises-MITIP. Wiley, New York, pp 29–36

Janjarasskul T, Suppakul P (2018) Active and intelligent packaging: the indication of quality and safety. Crit Rev Food Sci Nutr 58:808–831. https://doi.org/10.1080/10408398.2016.1225278

Kalpana S, Priyadarshini SR, Maria Leena M et al (2019) Intelligent packaging: trends and applications in food systems. Trends Food Sci Technol 93:145–157. https://doi.org/10.1016/j.tifs.2019.09.008

Karthika P, Shanmuganathan S, Viswanathan S, Delerue-Matos C (2021) Molecularly imprinted polymer-based electrochemical sensor for the determination of endocrine disruptor bisphenol-a in bovine milk. Food Chem 363:130287. https://doi.org/10.1016/j.foodchem.2021.130287

Kerry JP, O'Grady MN, Hogan SA (2006) Past, current and potential utilisation of active and intelligent packaging systems for meat and muscle-based products: a review. Meat Sci 74:113–130. https://doi.org/10.1016/j.meatsci.2006.04.024

Khandare DG, Joshi H, Banerjee M et al (2015) Fluorescence turn-on Chemosensor for the detection of dissolved CO2 based on ion-induced aggregation of Tetraphenylethylene derivative. Anal Chem 87:10871–10877. https://doi.org/10.1021/acs.analchem.5b02339

Kim JU, Ghafoor K, Ahn J et al (2016) Kinetic modeling and characterization of a diffusion-based time-temperature indicator (TTI) for monitoring microbial quality of non-pasteurized angelica juice. LWT – Food Sci Technol 67:143–150. https://doi.org/10.1016/j.lwt.2015.11.034

Kim D, Lee S, Lee K et al (2017) Development of a pH indicator composed of high moisture-absorbing materials for real-time monitoring of chicken breast freshness. Food Sci Biotechnol 26:37–42. https://doi.org/10.1007/s10068-017-0005-6

Kress-Rogers E (1998) Terms in instrumentation and sensors technology. In: Instrumentation and sensors for the food industry, pp 673–691

Kruss S, Hilmer AJ, Zhang J et al (2013) Carbon nanotubes as optical biomedical sensors. Adv Drug Deliv Rev 65:1933–1950. https://doi.org/10.1016/j.addr.2013.07.015

Kulchan R, Boonsupthip W, Jinkarn T, Suppakul P (2016) Developing a novel colorimetric indicator for monitoring rancidity reaction and estimating the accelerated shelf life of oxygen-sensitive dairy products. Int Food Res J 23:1092–1099. https://doi.org/10.1016/B978-0-323-51271-8.00008-5

Kuswandi B (2017) Freshness sensors for food packaging. Ref Modul Food Sci 1–11

Kuswandi B, Jumina (2020) Active and intelligent packaging, safety, and quality controls. In: Fresh-cut fruits and vegetables: technologies and mechanisms for safety control. Elsevier Inc., pp 243–294

Kuswandi B, Nurfawaidi A (2017) On-package dual sensors label based on pH indicators for real-time monitoring of beef freshness. Food Control 82:91–100. https://doi.org/10.1016/j.foodcont.2017.06.028

Kuswandi B, Wicaksono Y, Jayus J et al (2011) Smart packaging: sensors for monitoring of food quality and safety. Sens & Instrumen Food Qual 5:137–146. https://doi.org/10.1007/s11694-011-9120-x

Kuswandi B, Jayus RA et al (2012) A novel colorimetric food package label for fish spoilage based on polyaniline film. Food Control 25:184–189. https://doi.org/10.1016/j.foodcont.2011.10.008

Lee S, Park JW (2017) Luminescent oxygen sensors with highly improved sensitivity based on a porous sensing film with increased oxygen accessibility and photoluminescence. Sensors Actuators B Chem 249:364–377. https://doi.org/10.1016/j.snb.2017.04.112

Lee K, Baek S, Kim D, Seo J (2019a) A freshness indicator for monitoring chicken-breast spoilage using a Tyvek® sheet and RGB color analysis. Food Packag Shelf Life 19:40–46. https://doi.org/10.1016/j.fpsl.2018.11.016

Lee SB, Kim DH, Jung SW, Lee SJ (2019b) Air-activation of printed time–temperature integrator: a sandwich package case study. Food Control 101:89–96. https://doi.org/10.1016/j.foodcont.2019.02.021

Li L, Liu D, Wang K et al (2017) Quantitative detection of nitrite with N-doped graphene quantum dots decorated N-doped carbon nanofibers composite-based electrochemical sensor. Sensors Actuators B Chem 252:17–23. https://doi.org/10.1016/j.snb.2017.05.155

Liu J, Wang H, Guo M et al (2019) Extract from Lycium ruthenicum Murr. Incorporating κ-carrageenan colorimetric film with a wide pH–sensing range for food freshness monitoring. Food Hydrocoll 94:1–10. https://doi.org/10.1016/j.foodhyd.2019.03.008

Lloyd K, Mirosa M, Birch J (2019) Active and intelligent packaging. In: Encyclopedia of Food Chemistry, pp 177–182

Lorite GS, Selkälä T, Sipola T et al (2017) Novel, smart and RFID assisted critical temperature indicator for supply chain monitoring. J Food Eng 193:20–28. https://doi.org/10.1016/j.jfoodeng.2016.06.016

Lyu JS, Choi I, Hwang KS et al (2019) Development of a BTB−/TBA+ ion-paired dye-based CO2 indicator and its application in a multilayered intelligent packaging system. Sensors Actuators B Chem 282:359–365. https://doi.org/10.1016/j.snb.2018.11.073

Mao Y, Gao Y, Wu S et al (2017) Highly enhanced sensitivity of optical oxygen sensors using microstructured PtTFPP/PDMS-pillar arrays sensing layer. Sensors Actuators B Chem 251: 495–502. https://doi.org/10.1016/j.snb.2017.05.081

Martínez-Olmos A, Fernández-Salmerón J, Lopez-Ruiz N et al (2013) Screen printed flexible radiofrequency identification tag for oxygen monitoring. Anal Chem 85:11098–11105. https://doi.org/10.1021/ac4028802

Mataragas M, Bikouli VC, Korre M et al (2019) Development of a microbial time temperature indicator for monitoring the shelf life of meat. Innov Food Sci Emerg Technol 52:89–99. https://doi.org/10.1016/j.ifset.2018.11.003

Matindoust S, Farzi A, Baghaei Nejad M et al (2017) Ammonia gas sensor based on flexible polyaniline films for rapid detection of spoilage in protein-rich foods. J Mater Sci Mater Electron 28:7760–7768. https://doi.org/10.1007/s10854-017-6471-z

Meng J, Qian J, Tang Y (2018) A solid-state time-temperature indicator used in chilled fresh pork monitoring. Packag Technol Sci 31:353–360. https://doi.org/10.1002/pts.2328

Mirica KA, Azzarelli JM, Weis JG et al (2013) Rapid prototyping of carbon-based chemiresistive gas sensors on paper. Proc Natl Acad Sci U S A 110:E3265–E3270. https://doi.org/10.1073/pnas.1307251110

Mirza Alizadeh A, Masoomian M, Shakooie M et al (2021) Trends and applications of intelligent packaging in dairy products: a review. Crit Rev Food Sci Nutr 62:383–397. https://doi.org/10.1080/10408398.2020.1817847
Mondi (2011) Nor®Absorbit makes your food nice and crispy. http://www.mondigroup.com/desktopdefault.aspx/tabid-2179/2472_read-22564/. Accessed 14 May 2021
Morsy MK, Khalaf HH, Sharoba AM, El-Tanahi HH (2014) Applicability of biosensor and oxygen sensor for monitoring spoilage and bacterial contaminants of packed minced beef and poultry. 2nd Int Conf Biotechnol Appl Agric:8–12. https://doi.org/10.1016/j.fm.2017.06.007
Müller P, Schmid M (2019) Intelligent packaging in the food sector: a brief overview. Foods 8:16. https://doi.org/10.3390/foods8010016
NXP® Semiconductors (2022) RFID Ecosystem. In: NXP® Semicond. Co. https://www.nxp.com/products/rfid-nfc/rfid-ecosystem:RFID_ECOSYSTEMS#:~:text=NXP delivers leading-edge RFID, system integrators around the world. Accessed 10 Jan 2022
Pareek V, Khunteta A (2014) Pharmaceutical packaging: current trends and future. Int J Pharm Pharm Sci 6:480–485. https://doi.org/10.1016/B978-0-12-816184-5.00012-4
Pavelková A (2013) Time temperature indicators as devices intelligent packaging. Acta Univ Agric Silvic Mendelianae Brun 61:245–251. https://doi.org/10.11118/actaun201361010245
Pereira VA, de Arruda INQ, Stefani R (2015) Active chitosan/PVA films with anthocyanins from brassica oleraceae (red cabbage) as time-temperature indicators for application in intelligent food packaging. Food Hydrocoll 43:180–188. https://doi.org/10.1016/j.foodhyd.2014.05.014
PM Code (2006) PM-Code's world. In: Content Idea Asia Co. http://pmcode.co-site.jp/. Accessed 10 Jan 2022
Ramon-Marquez T, Medina-Castillo AL, Fernandez-Gutierrez A, Fernandez-Sanchez JF (2016) Novel optical sensing film based on a functional nonwoven nanofibre mat for an easy, fast and highly selective and sensitive detection of tryptamine in beer. Biosens Bioelectron 79:600–607. https://doi.org/10.1016/j.bios.2015.12.091
Realini CE, Marcos B (2014) Active and intelligent packaging systems for a modern society. Meat Sci 98:404–419. https://doi.org/10.1016/j.meatsci.2014.06.031
RipeSense® (2004) How ripe do you like it? In: Ripesense Ltd. http://www.ripesense.co.nz/ripesense_why.html. Accessed 10 Jan 2022
Risch SJ (2009) Food packaging history and innovations. J Agric Food Chem 57:8089–8092. https://doi.org/10.1021/jf900040r
Roya AQ, Elham M (2016) Intelligent food packaging: concepts and innovations. Int J ChemTech Res 9:669–676. https://doi.org/10.1039/d0ra07898k
Rukchon C, Nopwinyuwong A, Trevanich S et al (2014) Development of a food spoilage indicator for monitoring freshness of skinless chicken breast. Talanta 130:547–554. https://doi.org/10.1016/j.talanta.2014.07.048
Saliu F, Della Pergola R (2018) Carbon dioxide colorimetric indicators for food packaging application: applicability of anthocyanin and poly-lysine mixtures. Sensors Actuators B Chem 258:1117–1124. https://doi.org/10.1016/j.snb.2017.12.007
Sarapulova O, Sherstiuk V, Shvalagin V, Kukhta A (2015) Photonics and Nanophotonics and information and communication Technologies in Modern Food Packaging. Nanoscale Res Lett 10. https://doi.org/10.1186/s11671-015-0939-7
Scampicchio M, Arecchi A, Lawrence NS, Mannino S (2010) Nylon nanofibrous membrane for mediated glucose biosensing. Sensors Actuators B Chem 145:394–397. https://doi.org/10.1016/j.snb.2009.12.042
Sen L, Hyun KH, Kim JW et al (2013) The design of smart RFID system with gas sensor for meat freshness monitoring. Adv Sci Technol Lett 41:17–20. 110.14257/astl.2013.41.05
Sharma R, Ghoshal G (2018) Emerging trends in food packaging. Nutr Food Sci 48:764–779. https://doi.org/10.1108/NFS-02-2018-0051
Siracusa V, Lotti N (2019) Intelligent packaging to improve shelf life. In: Food quality and shelf life. Academic, pp 261–279

Sjöholm J, Erlandsson P O (2003) Biosensor and its use to indicate the status of a product. U.-S. Patent No. 6,642,016. Washington, DC: U.S. Patent and Trademark Office.

Smits E, Schram J, Nagelkerke M et al (2012) Development of printed RFID sensor tags for smart food packaging. 14th Int Meet Chem Sensors:403–406. https://doi.org/10.5162/IMCS2012/4.5.2

Sohail M, Sun DW, Zhu Z (2018) Recent developments in intelligent packaging for enhancing food quality and safety. Crit Rev Food Sci Nutr 58:2650–2662. https://doi.org/10.1080/10408398.2018.1449731

Sukhavattanakul P, Manuspiya H (2021) Influence of hydrogen sulfide gas concentrations on LOD and LOQ of thermal spray coated hybrid-bacterial cellulose film for intelligent meat label. Carbohydr Polym 254:117442. https://doi.org/10.1016/j.carbpol.2020.117442

Sun J, Ye B, Xia G et al (2016) A colorimetric and fluorescent chemosensor for the highly sensitive detection of CO2 gas: experiment and DFT calculation. Sensors Actuators B Chem 233:76–82. https://doi.org/10.1016/j.snb.2016.04.052

Sun J, Zheng X, Wu X et al (2017) A squaraine-based sensor for colorimetric detection of CO2 gas in an aqueous medium through an unexpected recognition mechanism: experiment and DFT calculation. Anal Methods 9:6830–6838. https://doi.org/10.1039/c7ay02218b

Takhistov P (2009) Biosensor technology for food packaging applications, vol 121. Wiley Encycl Packag Technol John Wiley Sons Inc. https://doi.org/10.1016/j.ultsonch.2017.04.019

Taoukis PS, Labuza TP (2003) Time-temperature indicators (TTIs). In: Novel food packaging techniques, pp 103–126

TempTRIP (2012) Temperature monitoring system. https://www.temptrip.com/. Accessed 10 Jan 2022

Tichoniuk M, Radomska N, Cierpiszewski R (2017) The application of natural dyes in food freshness indicators designed for intelligent packaging. Stud Oeconomica Posnaniensia 5:19–34. https://doi.org/10.18559/soep.2017.7.2

Uniform Code Council (2014) GS1 databar family. In: GS1 US. https://www.gs1us.org/what-we-do/standards/barcodes/gs1-databar-family. Accessed 10 Jan 2022

Vanderroost M, Ragaert P, Devlieghere F, De Meulenaer B (2014) Intelligent food packaging: the next generation. Trends Food Sci Technol 39:47–62. https://doi.org/10.1016/j.tifs.2014.06.009

Wen Y, Li Y, Si Y et al (2015) Ready-to-use strip for L-ascorbic acid visual detection based on polyaniline/polyamide 66 nano-fibers/nets membranes. Talanta 144:1146–1154. https://doi.org/10.1016/j.talanta.2015.07.086

Xia G, Liu Y, Ye B et al (2015) A squaraine-based colorimetric and F- dependent chemosensor for recyclable CO2 gas detection: highly sensitive off-on-off response. Chem Commun 51:13802–13805. https://doi.org/10.1039/c5cc04755b

Yildiz E, Sumnu G, Kahyaoglu LN (2021) Monitoring freshness of chicken breast by using natural halochromic curcumin loaded chitosan/PEO nanofibers as an intelligent package. Int J Biol Macromol 170:437–446. https://doi.org/10.1016/j.ijbiomac.2020.12.160

Yılmaz M, Altan A (2021) Optimization of functionalized electrospun fibers for the development of colorimetric oxygen indicator as an intelligent food packaging system. Food Packag Shelf Life 28:100651. https://doi.org/10.1016/j.fpsl.2021.100651

Yucel U (2016) Intelligent packaging. Ref Modul Food Sci. https://doi.org/10.1016/B978-0-08-100596-5.03374-6

Zhai X, Shi J, Zou X et al (2017) Novel colorimetric films based on starch/polyvinyl alcohol incorporated with roselle anthocyanins for fish freshness monitoring. Food Hydrocoll 69:308–317. https://doi.org/10.1016/j.foodhyd.2017.02.014

Zhai X, Li Z, Zhang J et al (2018) Natural biomaterial-based edible and pH-sensitive films combined with electrochemical writing for intelligent food packaging. J Agric Food Chem 66:12836–12846. https://doi.org/10.1021/acs.jafc.8b04932

Zhang D, Peng H (2019) New paradigm of packaging design in internet scenario. IOP Conf Ser Mater Sci Eng 573. https://doi.org/10.1088/1757-899X/573/1/012111

Index

H. U. Hebbar et al. (eds.), *Engineering Aspects of Food Quality and Safety*, Food Engineering Series, https://doi.org/10.1007/978-3-031-30683-9

The manufacturer's authorised representative in the EU is Springer Nature Customer Service Centre GmbH, Europaplatz 3, 69115 Heidelberg, Germany. If you have any concerns regarding our products, please contact ProductSafety@springernature.com

Printed and bound by CPI Group (UK) Ltd, Croydon, CR0 4YY
01/10/2026
02225851-0018